Astrophysics and Space Science Library

Volume 422

More information about this series at http://www.springer.com/series/5664

Wayne Orchiston

Exploring the History of New Zealand Astronomy

Trials, Tribulations, Telescopes and Transits

Springer

Wayne Orchiston
National Astronomical Research Institute
of Thailand
Chiang Mai
Thailand

ISSN 0067-0057 ISSN 2214-7985 (electronic)
Astrophysics and Space Science Library
ISBN 978-3-319-22565-4 ISBN 978-3-319-22566-1 (eBook)
DOI 10.1007/978-3-319-22566-1

Library of Congress Control Number: 2015947953

Springer Cham Heidelberg New York Dordrecht London

Cover illustration: Engraving of the 1874 German transit of Venus camp at Ross Cove, Auckland Islands (after *The Illustrated Australian News for Home Readers*, 19 April 1875).

Printed on acid-free paper

Springer International Publishing AG Switzerland is part of Springer Science+Business Media
(www.springer.com)

Contents

List of Figures

List of Tables

Chapter 1
Introduction

1.1 An Emerging Interest in Observational Astronomy

When my father scored a lectureship at Canterbury Agricultural College (as it was in those days)[1] and the family moved from Christchurch out to Lincoln in New Zealand my life entered a new direction and one which has stayed with me through to the present day. Lincoln (Fig. 1.1) was 14 miles (22.5 km) from the light pollution of Christchurch, and on a dark moonless night the skies over the College were simply stunning. Is it little wonder that I instantly fell in love with astronomy? How could anyone with an enquiring mind do otherwise?

With the aid of a small telescope that my father temporarily released from one of the instruments in his soil science laboratory I was able to observe planets and the Moon telescopically for the first time, and see some of the double stars, gaseous nebulae and star clusters that had enthralled me when I read about them in books. Lincoln Primary School had a number of inspiring teachers, and when I had to give a half-hour talk to my classmates and chose "Astronomy" my fate was sealed.

Later, at 12, I joined the Canterbury Astronomical Society and two years later I gave my first society lecture, on "Solar flares and their terrestrial effects". About this time (1957) Comets Arend-Roland and Mrkos (Figs. 1.2 and 1.3) were grand spectacles in our southern skies (see Hendrie 1996) and there also were two very favourable oppositions of Mars, so I became a regular Friday night visitor at the University's observatory located atop an 80-ft (24.4-m) tower (these were pre-metric days) at the old campus diagonally across the road from the Canterbury Museum in Christchurch (Fig. 1.4). My schedule was always the same: if a clear night looked likely, after school I would go into Whitcombe and Tombs, Christchurch's leading

[1]Canterbury Agricultural College, which we preferred to call 'Lincoln College' at the time, began as a School of Agriculture in 1878. As such, it was the oldest tertiary agricultural institution in the Southern Hemisphere. In 1961 it became an affiliated college of the University of Canterbury, then in 1990 it became autonomous, as Lincoln University (see http://en.wikipedia.org/wiki/Lincoln-University-(New-Zealand).

W. Orchiston, *Exploring the History of New Zealand Astronomy*,
Astrophysics and Space Science Library 422,
DOI 10.1007/978-3-319-22566-1_1

Fig. 1.1 Stately Ivy Hall was the most prominent historic building at 'Lincoln College' during the years our family lived on campus (Orchiston collection)

bookshop, and peruse the shelves of new astronomy books, having already decided which would be my next purchase when I had saved sufficient funds. Then it was off to the Canterbury Public Library for some light astronomy reading before buying fish and chips and finding a quiet spot beside the Avon River (Fig. 1.5) where I could eat my dinner in the company of a few inquisitive sea gulls that liked to share my chips. Finally, a short walk took me to the University, and up the long spiral staircase to the dome with its historic 6-in (15.2-cm) equatorially-mounted Cooke refractor (see Tobin 1996),[2] then under the care of the ever-informative Walter K. Roth, a lecturer in Physics at the University and a very familiar figure thanks to meetings of the Canterbury Astronomical Society.

Through this routine, society meetings and reading a succession of astronomy books I hankered to become a regular observational astronomer, but this only came to pass in 1959 when my father accepted a new position, this time in Australia, and we all moved to Sydney. Once there, I soon met the astronomers at Sydney Observatory (Fig. 1.6)—which in those days was still a research observatory—and joined Sydney's leading and oldest astronomical society, the New South Wales Branch of the British Astronomical Society, which was founded in 1895 (see Orchiston 1988). Through the Branch I was able to borrow a 3.65-in (9.3-cm) equatorially-mounted refractor, and start making regular observations of sunspots

[2]On 24 September 2010 and 22 February 2011 Christchurch suffered two major earthquakes which were followed by numerous aftershocks. The historic observatory building was damaged by the first earthquake and collapsed following the second earthquake. Remarkably, the tube and objective of the Cooke refractor survived intact, and there are now plans afoot to re-erect the telescope at an alternative site and continue public astronomy sessions (personal communication, John Hearnshaw 2015).

Fig. 1.2 Comet Arend-Roland, with its distinctive anti-tail, photographed by the American astronomer Alan McClure on 24 April 1957 (http://stony-ridge.org/AlanMcClure.html)

Fig. 1.3 Comet Mrkos, photographed by the American astronomer Alan McClure on 13 August 1957 (http://stony-ridge.org/AlanMcClure.html)

Fig. 1.4 The tower observatory at the old University of Canterbury campus in Christchurch, New Zealand. Following the major earthquakes of 2010 and 2011 the tower collapsed (www.phys.canterbury.ac.nz/Townsend/townsend-1896.png)

and faculae (Fig. 1.7), the planets, lunar craters and other features (Fig. 1.8), and particularly variable stars (e.g. see Fig. 1.9). I also enjoyed making naked eye observations of meteors and bolides (Orchiston 1963, 1964a). Later I worked one evening a week at Sydney Observatory, taking visitors on a tour of part of the main building and—if it was clear and they were in the sky—showing them the Moon and one or more planets, along with double stars, globular and galactic clusters and perhaps a gaseous nebula through the historic 11.5-in (29.2-cm) Schroeder refractor (Fig. 1.10). After the visitors left I was free to use the telescope for my own observations, the details of Jupiter's belts being a favourite study (e.g. see Fig. 1.11). The Observatory's Director, Dr Harley Weston Wood (1911–1984; Fig. 1.12), also allowed me access to a 6-in (15.2-cm) Grubb refractor housed in one of the other domes, in order to visually monitor selected flare stars.

These latter observations were carried out for Dr Bruce Slee (b. 1924) from the CSIRO's Division of Radiophysics who was conducting a search for radio flares from these enigmatic objects. He was successful, and along with Professor Bernard (later Sir Bernard) Lovell from Jodrell Bank, is credited with being the first to prove

Fig. 1.5 An autumn view of the picturesque Avon River, which flows through Christchurch (www.trekearth.com/gallery/)

Fig. 1.6 A nineteenth century sketch of Sydney Observatory, which was founded in 1858 (Orchiston collection)

that stars other than the Sun are capable of generating radio emission. I got to know Bruce well, and after leaving school worked with him at the Division of Radiophysics, then much later we teamed up again at the Division's successor, the Australia Telescope National Facility (see Fig. 1.13). Bruce was one of those who used radar during WWII to independently discover solar radio emission (see Orchiston and Slee 2002a; Orchiston et al. 2006b). Later he worked on an amazing variety of research projects and built an international reputation in radio astronomy (Orchiston 2004b, 2005b). While he continues to carry out astrophysical research,

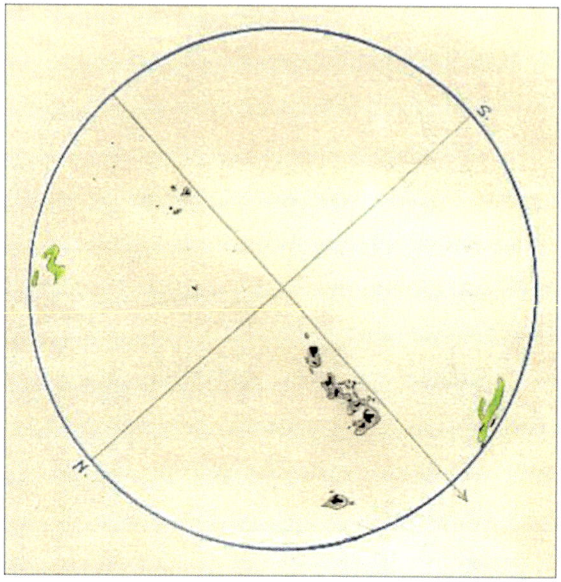

Fig. 1.7 One of my full-disk drawings of the Sun made using the BAA New South Wales Branch's 3.65-in refractor. The green areas are faculae (Orchiston collection)

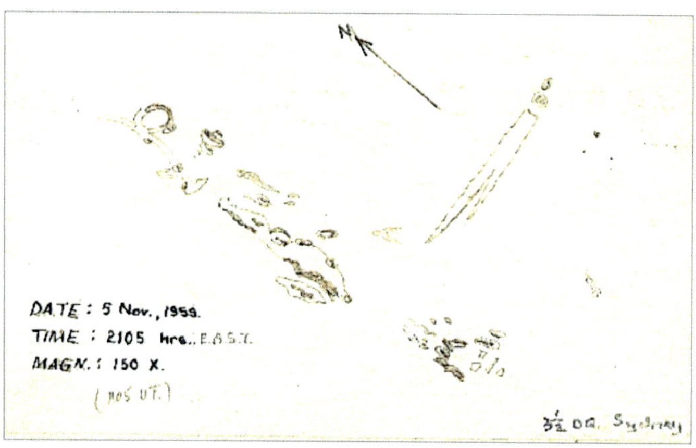

Fig. 1.8 A drawing of the lunar craters Messier and Pickering and surroundings made on 5 November 1959 with the 3.65-in refractor at a magnification of 150× (after Orchiston 1964b)

over the past decade or so he also has published extensively on the early history of Australian radio astronomy, generally in association with me and a number of our former doctoral students (see Slee 2005; George et al. 2015a, b; Kellermann et al. 2005; Orchiston and Slee 2002b, c, 2005a, b, 2006; Orchiston et al. 2015d, e, 2016a; Robertson et al. 2010, 2014; Stewart et al. 2010a, b, 2011a, b, c; Wendt et al. 2008a, b, c, 2011a, b, c).

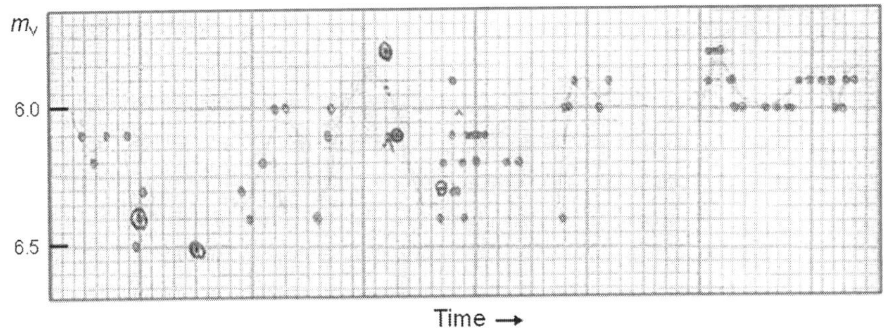

Fig. 1.9 Magnitude estimates of the irregular variable star Theta Apodis from 22 November 1959 to 29 April 1960. The five circled dots in each case represent the average of two successive observations made on the same evening (Orchiston collection)

Fig. 1.10 The 11.5-in Schroeder refractor at Sydney Observatory, which was acquired in 1874 (after Russell 1892)

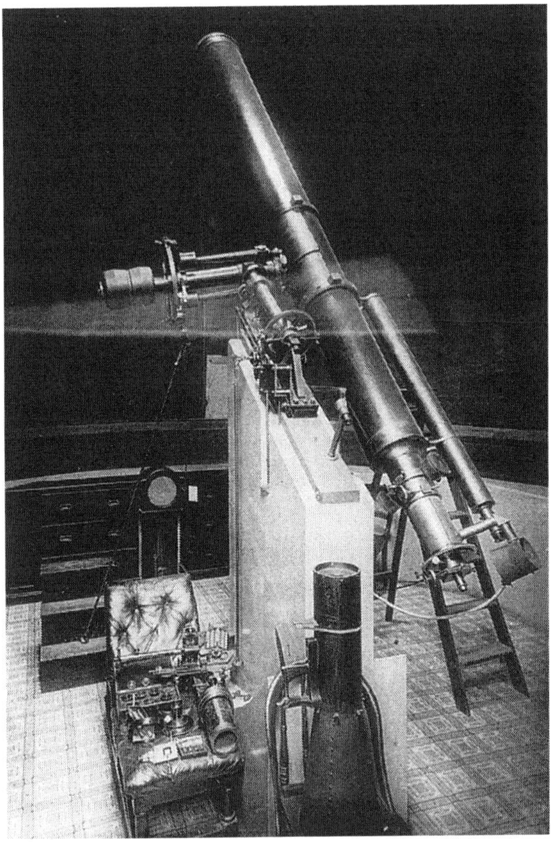

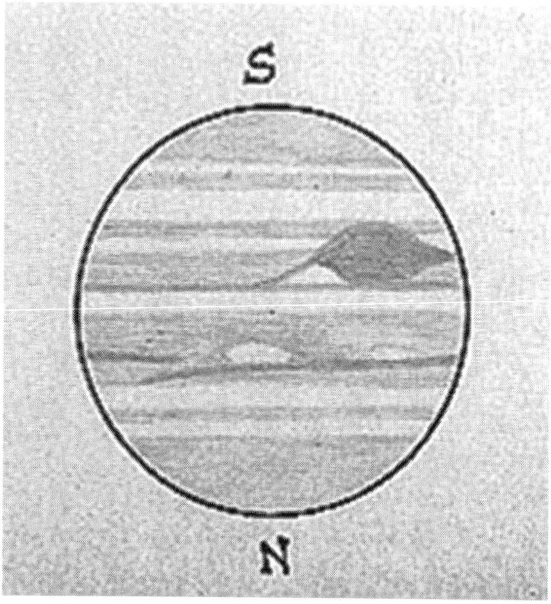

Fig. 1.11 A drawing of Jupiter made on 28 February 1968 using the Sydney Observatory 11.5-in refractor at a magnification of 240×. The Great Red Spot and two white spots in the North Equatorial Belt are conspicuous (Orchiston collection)

Fig. 1.12 Dr Harley Wood, who was Director of Sydney Observatory from 1943 to 1974 (Orchiston collection)

Fig. 1.13 Dr Bruce Slee at the control desk of the Australia Telescope Compact Array, Narrabri, during our observing run to image chromospherically-active stars in September 2002 (*Photograph* Wayne Orchiston)

1.2 John Tebbutt and Astronomical History

As the aforementioned comments indicate, non-observational astronomy, and particularly the history of astronomy, also beckoned. While still a school boy in Lincoln I was captivated by the remarkable achievements of the amateur, John Tebbutt (Fig. 1.14), who was Australia's foremost astronomer at the end of the nineteenth century (Orchiston 2002) and featured in several of the books that were in my library. From 1863 up until his death in 1916 Tebbutt maintained an impressive private observatory at Windsor (Fig. 1.15), near Sydney (Orchiston 2001), where he carried out a truly amazing range of astronomical observations (e.g. see Orchiston 2000, 2004d). He also discovered two of the 'Great Comets' of the nineteenth century (Orchiston 1998b, 1999). Soon after settling in Sydney I discovered that two of the original Windsor Observatory buildings had survived, and that Tebbutt's son had donated a veritable treasure-trove of records to the Mitchell Library in Sydney (see Orchiston 2004c). Thus began an interest in researching the history of astronomy, which grew as the years passed, so by 1990 most of my research efforts were directed towards astronomical history rather than radio or optical observational astronomy.

Fig. 1.14 In 1984, John Tebbutt featured on the new Australian $100 bank note (Orchiston collection)

Fig. 1.15 An 1880 woodcut of the Windsor Observatory (Orchiston collection)

Thanks largely to the Tebbutt Collection in the Mitchell Library my historical studies expanded from Tebbutt to other Australia amateur and professional astronomers who featured in his letter books, and then to New Zealand and South African-based astronomers who maintained contact with Tebbutt (e.g. see Orchiston 1985). My research on New Zealand astronomical history received an added impetus when I spent 11 years back in New Zealand, from 1988, first as Director of the Museum and Arts Centre in Gisborne (Fig. 1.16) and then at the Carter Observatory in Wellington (Fig. 1.17). At that time the Carter Observatory was the designated National Astronomical Observatory of New Zealand, and I started there as the Manager of the New Zealand Centre, which was established

Fig. 1.16 Gisborne, on the east coast of the North Island of New Zealand, fringes Poverty Bay and is the first city in the world to greet the new day. This is where Lieutenant James Cook (who inspired the statue) first anchored when the *Endeavour* reached Aotearoa/New Zealand in October 1769 (http://www.flightsaustralia.com.au/Cheap-Flights-Gisborne)

Fig. 1.17 The Carter Observatory in June 2012. The large dome on the left, which is mostly hidden behind the tree, houses the planetarium; the central dome has a 41-cm reflector and the right-hand dome an historic 24.8-cm refractor (*Photograph* Wayne Orchiston)

when the Wellington Planetarium Society relocated their Zeiss planetarium to the Carter Observatory and gifted it to the Observatory Board (van Dijk 1992). Then, after one year, I was offered and accepted the position of Executive Director of the Observatory (see New Director for Carter Observatory 1995).

While associated with the Carter Observatory I was able to research different aspects of New Zealand astronomy, with emphasis on Maori astronomy, the development of professional astronomy in Wellington, leading amateur astronomers, historically-significant telescopes, the 1874 and 1882 transits of Venus and early achievements in radio astronomy. Many of the chapters of this book draw on these studies.

I also used my position at the Carter Observatory to reactivate a much earlier interest (see Orchiston 1971) and began a detailed investigation of the nautical astronomy that characterized Cook's three voyages to the South Seas, and his association with the 1769 transit of Venus (e.g. see Fig. 1.18). This long-term multi-faceted project is on-going, and some of the resulting publications relate directly to New Zealand and are included in this book, but others that have a specific Pacific focus (i.e. Orchiston 1998a, 2004a, 2005a, 2015a) are not.

After returning to Australia at the end of 1999 I expanded the geographical scope of my research beyond New Zealand and Australia to include the early development of radio astronomy in France (see Débarbat et al. 2007; Encrenaz et al. 2011; Lequeux et al. 2010; Orchiston and Steinberg 2007; Orchiston et al. 2007, 2009; Pick et al. 2011—see Fig. 1.19) and later Japan (see Ishiguro and Orchiston 2016;

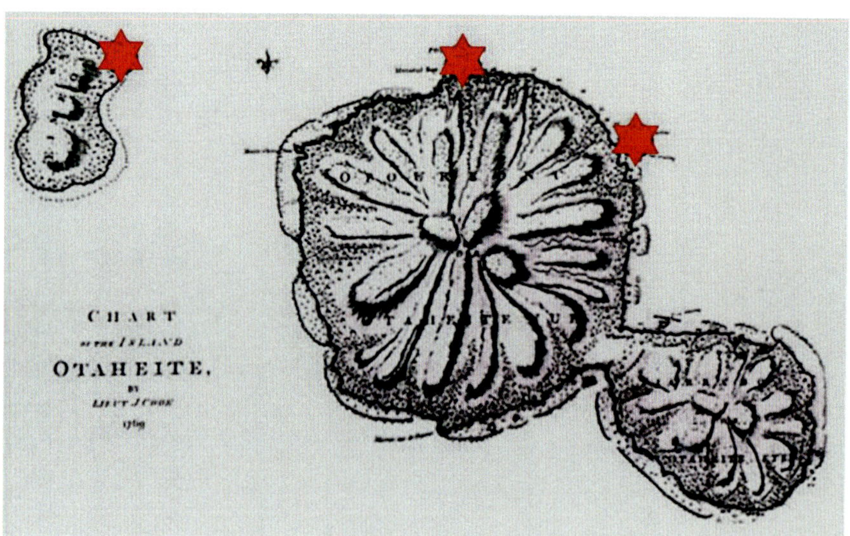

Fig. 1.18 A modified version of Cook's original map of Tahiti, showing the locations of the three different sites where the 1769 transit of Venus was successfully observed. From left to right they were Irioa Island, just off the northeastern tip of Morea; Fort Venus at Matavai Bay, Tahiti; and Taaupiri Island, just off the coast of Tahiti (after Orchiston 2015a)

Fig. 1.19 A recent photograph of the sole surviving Würzburg antenna at the Nançay Radio Astronomy Field Station in France. Originally there were two of these dishes, mounted on rails, and they operated as an interferometer. They were used, mainly, to measure the angular sizes and structure of discrete radio sources (see Orchiston et al. 2007)

Ishiguro et al. 2012; Nakajima et al. 2014; Shimoda et al. 2013—see Fig. 1.20). These studies were conducted under the umbrella of the Historic Radio Astronomy Working Group of the International Astronomical Union (IAU), which I formed in 2003. And through James Cook University (Townsville, Australia), and later the University of Southern Queensland (Toowoomba, Australia), I have supervised five Ph.D. students and one Master of Astronomy student who researched aspects of early Australian (see Fig. 1.21), British and U.S. radio astronomy.

In 2000 I established a Transits of Venus Working Group of the IAU, with a view to encouraging worldwide research on historic transits of Venus in the lead-up to the 2004 and 2012 events. As well as looking at Cook's involvement in the 1769 transit I

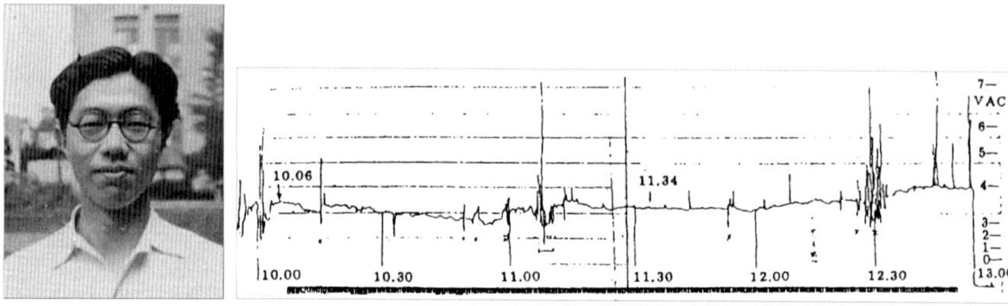

Fig. 1.20 As a young graduate student in physics at the University of Tokyo, Koichi Shimoda carried out the first radio astronomical observations in Japan when he recorded a partial solar eclipse on 9 May 1948. Unfortunately, he never published these pioneering observations at this time, and they only entered the public domain in 2013 when they were the subject of the second of our papers in the series on early Japanese radio astronomy. At the time we were delighted to find that Professor Shimoda was still alive, and to commemorate his historic observations we offered him first authorship on our paper. The photograph above shows Koichi Shimoda one year before he carried out the observations, and the plot on the right shows the oscilloscope display during the eclipse (after Shimoda et al. 2013)

have also reviewed the 1761, 1769, 1874 and 1882 transits (see Clark and Orchiston 2004; Darlington and Orchiston 2015; Dick et al. 1998; Orchiston 2004e, f, 2015a; Orchiston and Darlington 2015; Orchiston and Howse 2000) and written on the surviving relics in Tasmania of the U.S. 1874 transit campaign (Orchiston and Buchanan 1993, 2004; Orchiston et al. 2015a—e.g. see Fig. 1.22). Meanwhile, through James Cook University I supervised a U.S. Ph.D. student who looked at the 1874 and 1882 transits and three different nineteenth century total solar eclipses and the ways in which they were used by the media to help popularize astronomy in the U.S.A. (see Cottam and Orchiston 2014; Cottam et al. 2011a, b, 2012).

Through yet another U.S. James Cook University Ph.D. student I was able to pursue my research interest in nineteenth century solar eclipses and the development of solar physics (e.g. see Orchiston et al. 2006a) when we investigated the various Lick Observatory solar eclipse expeditions and the scientific instruments associated with them (see Orchiston and Pearson 2011; Pearson and Orchiston 2008, 2011; Pearson et al. 2011—see Fig. 1.23). This research also is on-going and further publications are anticipated.

James Cook University closed down Astronomy at the end of 2012, and I then moved to Thailand and a full-time position as a Senior Researcher at the National Astronomical Research Institute of Thailand in Chiang Mai. This not only allowed me to continue my previous research projects—including the preparation of this and other books—but also to initiate new projects pertaining to Thai astronomy. Thus far, my main focus has been on the lunar and solar observations conducted by French Jesuit missionary-astronomers in the 1680s (Orchiston et al. 2015c, 2016b—e.g. see Fig. 1.24) and on French observations of the 1868 total solar eclipse (Orchiston and Soonthornthum 2016). In addition, currently I am in the process of conducting

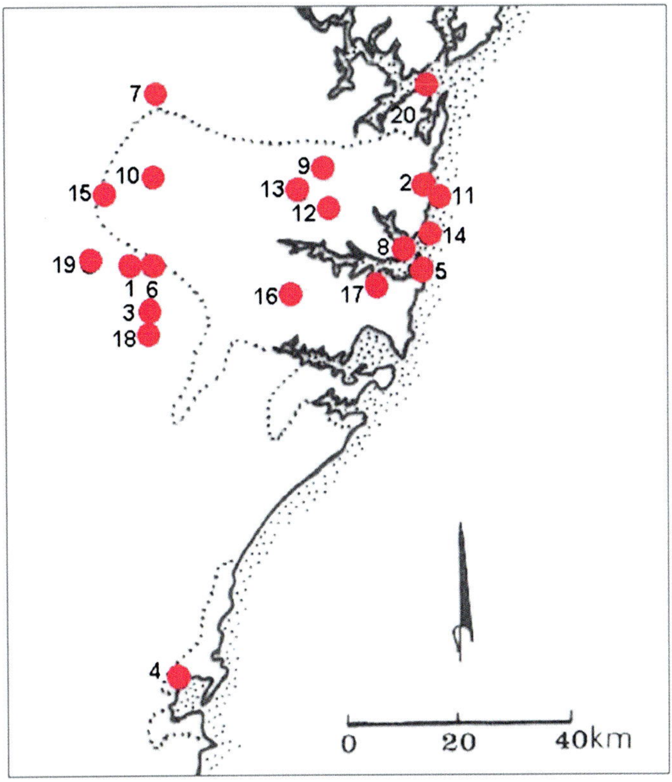

Fig. 1.21 This map shows the locations of the Radiophysics Laboratory (site #17) and the scattered field stations and remote sites in the Sydney and Wollongong areas used by the CSIRO's Division of Radiophysics at one time or another between 1945 and 1965. Over the years, my close friend and role model, Dr Bruce Slee, was associated with all of these sites except numbers 2, 4, 9, 13, 14, 15 and 16, but he spent most of his time at the Dover Heights (#5) and Fleurs (#6) field stations. From November 1961 I also was based at the Fleurs field station, and initially worked with Bruce on the Shain Cross and on the long baseline interferometry project he and Dr Peter Sheuer were conducting at Fleurs and the six remote sites immediately to the north, south and west of Fleurs to investigate the sizes of discrete sources (thus, Peter and I used to visit Llandilo, site #10), but eventually I was transferred to the Solar Group and operated the Chris Cross radio telescope and produced the daily isophote maps of solar emission at 1420 MHz (see Orchiston and Mathewson 2009). Three of my graduate students have researched selected Radiophysics field stations for their Ph.D. theses: for Harry Wendt it was Potts Hill (#16) and Murraybank (#13); for Ron Stewart is was Penrith (#15) and Dapto (#4); and for Peter Robertson it was Dover Heights (#5). The upper dotted boundary shows the approximate present-day extent of suburban Sydney and the lower dotted boundary of 'greater Wollongong' (*Map* Wayne Orchiston)

historical analyses of Thai and Japanese meteorites, by combining my interest in astronomical history with an academic and vocational background in geology (see Orchiston 1991, 1997; Theodossiou et al. 2002; Warnes et al. 1998). As part of this agenda, recently yet another part-time off-campus U.S. graduate student of mine

Fig. 1.22 The distinctive gateposts at the entrance to The Grange in Campbell Town, Tasmania, are the two metal piers that supported the horizontal photographic telescope used by the Americans in 1874. Other transit of Venus relics at The Grange are the brick pier that supported the transit telescope, and the octagonal observatory that housed the 5-in (12.7-cm) Clarke refractor (see Orchiston et al. 2015a)

Fig. 1.23 A scene showing the Lick Observatory party ready to observe the 22 January 1898 total solar eclipse from Jeur, India. All of the instruments are in place and the volunteers are at their practice stations. The 40-ft Schaeberle Camera with its twin wooden towers dominates the scene (after Orchiston and Pearson 2011: 199)

Fig. 1.24 A drawing showing Siam's King Narai and French Jesuit missionary-astronomers observing the 11 December 1685 total lunar eclipse from the King's country retreat which was on an island in the water reservoir that was located to the northeast of his palace in Lop Buri. In our analysis of this drawing we have demonstrated that it contains considerable artistic licence (see Orchiston et al. 2015c)

completed a University of Southern Queensland Ph.D. on the history of impact cratering in Tennessee (Ford et al. 2012, 2013, 2014, 2015—see Fig. 1.25).

Finally, I should mention that other current or former Ph.D. students have taken me out of my original 'comfort zone' and introduced me to elements of Armenian, Georgian, German, Iranian and Italian astronomy with which I originally was not familiar (e.g. see Cunningham et al. 2011; Cunningham and Orchiston 2013, 2015; Hafez et al. 2011, 2015a, b; Sauter et al. 2015a, b), thereby broadening my astronomical research horizons.

The foregoing review foreshadows the fact that over the past two decades or so the geographical focus of my research has grown from a myopic view of Australia and New Zealand to accommodate Armenia, Australia, Britain, France, Georgia, Germany, India, Indonesia, Iran, Italy, Japan, Thailand, Turkey, South Africa and the USA—yet New Zealand astronomical history remains my sentimental favourite, hence this book.

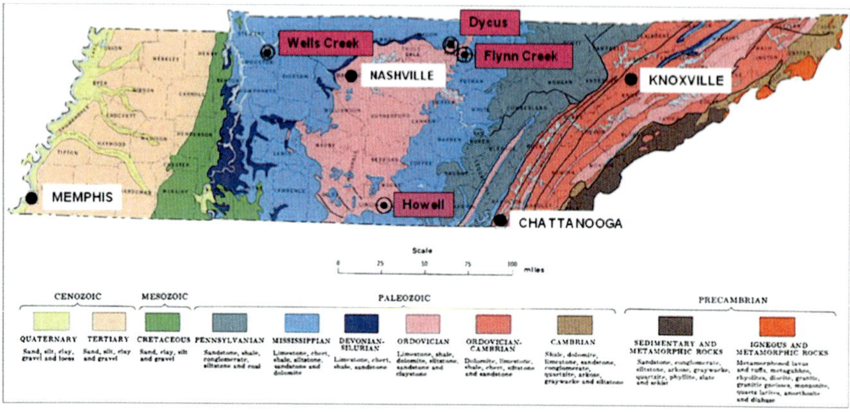

Fig. 1.25 A generalized geological map of Tennessee, U.S.A., showing the locations of the four largest cities (*black dots*) and the two confirmed and two suspected meteorite impact sites (*small black dots with circles*). Note that these sites are located on the Highland Rim (Wells Creek), a Highland Rim outlier remnant (Howell), or on the Highland Rim escarpment (Dycus and Flynn Creek). The Highland Rim is the sky blue region on the map (after Ford et al. 2012: 160)

1.3 About This Book: What to Include and What to Exclude?

Since 1983 I have published many research papers on aspects of New Zealand astronomical history, plus the monograph *Nautical Astronomy in New Zealand: The Voyages of James Cook* (Orchiston 1998c), but considerable culling was necessary as this book could not include everything. In deciding on what to include and what to omit I had to consider what was already available. There were autobiographical and biographical books published by Bateson (1989) and Hayes (1987) respectively; Mackrell's *Halley's Comet Over New Zealand* (1985; but see Orchiston 1986); and invaluable review papers written by Eiby (1971, 1978), Hearnshaw (2006), McIntosh (1970) and Rumsey (1974). There also were many papers published by other astronomers that dealt with specific aspects of New Zealand astronomical history, and there was no point in replicating their contents in my book. There also were time and space constraints to consider.

After careful consideration I ended up with this introductory chapter, followed by 23 thematic chapters that chronologically span from the initial Polynesian settlement of New Zealand/Aotearoa through to the 1960s, and explore Maori astronomy, the New Zealand legs of Cook's three voyages to the South Seas, historic telescopes and observatories, the 1874 and 1882 transits of Venus, the 1885 total solar eclipse, other aspects of observational astronomy and, finally, the birth of radio astronomy in New Zealand. Post-1960 developments in the history of New Zealand astronomy, particularly at the University of Canterbury and Mt John Observatory, are well documented on web sites maintained by the Royal

Astronomical Society of New Zealand and various universities, and through the attractive new book, *Mt John – The First Fifty Years. A Celebration of Half a Century of Optical Astronomy at the University of Canterbury* by Hearnshaw and Gilmore (2015).

My book is quite unlike the Hearnshaw/Gilmore tome in that many of the chapters deal with amateur astronomers and astronomy, for until quite recently New Zealand made only a token commitment to professional astronomy (e.g. see Bateson 1984). However, this amateur bias should not deter us, for dedicated amateur astronomers, if properly equipped (mentally and instrumentally), were—and still are—capable of making valuable contributions to research astronomy, not to mention the promotion of astronomy in schools and among the general public. Historically, there is abundant evidence of this internationally, especially before astrophysics began to eclipse positional astronomy (e.g. see Ashbrook 1984; Chapman 1998; Clerke 1893; Dunlop and Gerbaldi 1988; Lankford 1981; Macpherson 1906; Meadows and Henbest 1981; Orchiston 1989; Williams 1987). In 1956, the Director of Carter Observatory was moved to write: "… a great deal of the astronomical history of New Zealand is the history of the amateur astronomer." (Thomsen 1956: 11, cf. 1954: 80). For the purposes of this book, I define an amateur astronomer as "Someone who was involved in astronomy *for the love of it*, and normally did not derive a *primary income* from astronomy."

Back in the late 1970s, the University of Calgary sociologist, Robert A. Stebbins (b. 1938), pioneered a whole new field of research, the sociology of 'amateurs' (e.g. Stebbins 1977, 1978, 1979, 1980a, 1992, 2004, 2007). In his analysis, amateurs straddle the nebulous boundary between work and leisure, and are 'amators' in the strict sense in that they love their hobby and are happy to invest time, money and effort in it for the sake of the expected 'rewards'. Five of Stebbins' later papers (1980b, 1981, 1982a, b, 1987) deal specifically with the avocation of amateur astronomy, and these form an important contribution to our discipline.

Using 'dedication' as a criterion, Stebbins distinguishes 'devotees' from 'dabblers'. New Zealand's leading amateur astronomers during the nineteenth century and first half of the twentieth century were all devotees, individuals who were happy to make a substantial commitment to the science in terms of both time and money. Using another dimension, 'knowledge and involvement', Stebbins differentiates between 'active' and 'armchair' amateur astronomers. In the nineteenth and twentieth centuries, leading active amateur astronomers were engaged in observational or mathematical astronomy, or in instrument-making. Stebbins also was able to categorize individual active astronomers within an 'apprentice – journeyman – master' continuum. Apprentices were beginning their astronomical 'careers', while masters were the acknowledged experts who were making a meaningful contribution to science whatever their area(s) of astronomical involvement. Most of the New Zealand amateur astronomers you will meet in this book were devotees who were active masters. You will also encounter a special group of New Zealand amateur astronomers, those who were able to graduate from amateur to professional ranks and make their marks as professional astronomers. The amateur-turned-professional (ATP) syndrome was a distinctive feature of nineteenth century world astronomy (see Orchiston 2015b), and as we shall see, New Zealand was well endowed with ATPs.

Back in 1954, Ivan Thomsen, the Director of the Carter Observatory, offered the following perceptive comments about New Zealand astronomy and New Zealanders:

> For the size of its population, New Zealand must be one of the most interested countries in the world on the subject of astronomy. It is not easy to obtain exact figures, but from rough comparisons, there must be more members of astronomical societies per 1,000 of the population than anywhere else. Furthermore, from the numbers of people visiting observatories that are opened to them, there seems to be a fairly high general appreciation of the subject among the population as a whole. This has always impressed me as being surprising, in view of the fact that we are completely lacking in any formal instruction on the subject at our schools and colleges … (Thomsen 1954: 79–80).

This now brings us to the subtitle of this book "Trials, Tribulations, Telescopes and Transits", which was carefully chosen. As we all know, most astronomical observing is performed with 'telescopes' and for a small nation New Zealand has more than its far share of historically-significant telescopes. Most of these are discussed in this book. During the nineteenth century 'transits' of Venus arguably inspired the most ambitious international astronomical expeditions ever undertaken, and because of its favourable geographical position New Zealand attracted observing teams from England, France, Germany and the U.S.A., and its own home-grown observers, collectively producing results out of all proportion to the nation's meagre population. These transits are discussed in two of the chapters. What of the other components of the subtitle, "Trials" and "Tribulations"? Particularly when researching the chapters on Maori astronomy, Cook voyage astronomy, the astronomy carried out by the first European settlers, and Grigg's cometary work, time and again I was struck by the testing conditions under which these pioneering astronomers worked, and the challenges they often faced in successfully carrying out their observations. Thus, 'trials' and 'tribulations' permeated early New Zealand astronomy, and this reminds us that astronomy is not just about telescopes, observatories, observations and publications. It is also about people, individuals who succeeded to varying degrees as astronomers, notwithstanding the cultural, economic and political obstacles they encountered. Sometimes these obstacles—especially the political ones—could prove almost insurmountable. Meanwhile, observational astronomers also had to contend with "Murphy's Law"! Challenges indeed ….

We have spoken about what is in this book, but what has been omitted and why? Not included are chapters on the history of New Zealand's numerous astronomical societies, and particularly the Royal Astronomical Society of New Zealand and its famous Variable Stars Section, as there are others far more knowledgeable and better qualified then I am who should write these. Also missing in the Section on Historic Telescopes and Observatories is a chapter on Henry Skey and his pioneering construction of liquid mirror telescopes in England and New Zealand, as this has already been adequately dealt with by Dr Duncan Steel (see Steel 1984; Olsson-Steel 1986). Another notable personality missing from this book is

New Zealand's first 'astrophysicist', Alexander William Bickerton, who emigrated from England to New Zealand in 1874 in order to take up the inaugural Chair of Chemistry at Canterbury University College in Christchurch (Burdon 1956). Despite leading an unconventional lifestyle, his public lectures on astronomy—and especially his 'Partial Impact Theory'—enthralled audiences throughout New Zealand. While still an astronomy graduate student at the University of Canterbury, a youthful Gerry Gilmore—now a Fellow of the Royal Society and the Professor of Experimental Philosophy in the Institute of Astronomy at the University of Cambridge—published a fascinating account of Bickerton's astronomical exploits in *Southern Stars*, the journal of the Royal Astronomical Society of Society of New Zealand (see Gilmore 1985), and readers are referred to this very entertaining paper.[3]

Meanwhile, mainly because of space limitations, this book merely mentions—and sometimes only in passing—the achievements of many other notable New Zealand amateur and professional astronomers (e.g. C.E. Adams, Frank Bateson, Murray Geddes, Charlie Gifford, Albert Jones, Charles Michie, Arthur Stock, Ivan Thomsen, etc.) who deserve more detailed treatment, and hopefully will feature in a second volume. This planned volume also will allow the inclusion of chapters on notable New Zealand astronomers who mainly made their names whilst working overseas and are now no longer with us (e.g. Sir Ian Axford, Leslie Comrie, Ben Gascoigne, etc.), and the pioneering efforts of the late Cliff Ellyett and Colin Keay in radar meteor astronomy while at the University of Canterbury. There also should be room to display the achievements of the nation's main astronomical societies; relate early site-testing activities in New Zealand that preceded Bateson's successful efforts; and recount the sorry saga of the much-anticipated Cawthron Solar Observatory.

The present book, therefore, is merely a start in the right direction—it is not designed as a definitive tome on the history of New Zealand astronomy during the pre-Mount John/University of Canterbury era. Rather, it is a sampling of aspects of New Zealand astronomical history, drawn primarily from my own published papers and a monograph, and previously-unpublished conference presentations (e.g. Chaps. 16 and 19). My initial plan was simply to repackage all of these previously-published works in a consistent format and submit them to Springer, but many early New Zealand newspapers have been digitized since I wrote the original versions of these chapters and by accessing PastPapers (see http://paperspast.natlib.govt.nz/cgi-bin/paperspast) I quickly discovered a wealth of new information that previously was not readily or easily available to me. In addition, the development of the web allowed me to source a large number of images that were not included in my original papers. Consequently, all of my original papers and chapters of my Cook Voyage Astronomy monograph used in this book were revised, many of them very substantially, in the preparation of this book.

---

[3]We also plan to reprint this paper in the New Zealand section of a forthcoming book on the early development of astrophysics in Asia (see Nakamura and Orchiston 2016).

Finally, we should note that this is not the first time that an attempt has been made to publish a book like this. In 1940 the New Zealand Astronomical Society (forerunner to the Royal Astronomical Society of New Zealand) decided to produce a 'Centennial Bulletin' on the history of New Zealand astronomy, and Carter Observatory's Ivan Thomsen was appointed editor. At the 18th AGM of the Society, in 1941, he

> … outlined the progress of this work and gave an account of the material that has been collected. The scope of the Bulletin is considerable and much detailed material is on hand. Valuable research on Maori astronomy by Mr. Geddes and the occurrence of an ancient eclipse of the sun, may throw light on the date and place of the Maori migration to New Zealand and provide a confirmation of the present chronology of the Maori people. The early transits of Venus and the work of Captain Cook are to be investigated. The history of longitude determination is receiving thorough consideration. Reports from numerous helpers on many subjects, including the Yale offer and the Cawthron telescope and many of the older personalities in N.Z. astronomy have provided interlocking and confirmatory accounts of great reliability. (Report of Council for the year ended 30 September 1940: 37).

With the benefit of hindsight we know that Thomsen was quite the wrong choice as editor of this volume, for he had difficulty writing, and over his long 'reign' as head of professional astronomy in New Zealand he produced comparatively few publications. Consequently, despite an anonymous donor contributing the considerable sum of £100 towards of the cost of producing the "… long-awaited publication …" (Annual address by the President 1942: 91), the President of the Society, Auckland's R.A. McIntosh, reported that the project

> … for several excellent reasons is now slightly overdue. It is a most important project, and one which, I trust, will be completed during my term of office. We are all of us far too ignorant of the pioneers in this young country who, working under insuperable difficulties, paved the way for the present-day widespread interest in the grandest of the sciences evinced in this country. By all means in our power, therefore, let us press forward with the editing and publishing of this record of the first 100 years of astronomy in this land. (ibid.).

Despite good intentions, the 'Centennial Bulletin' never was published, and we can but hope that the present book is, in the eyes of some, a worthy substitute.

Acknowledgements
Although separate acknowledgements are listed at the ends of the individual chapters, this book could not have been published were it not for the generosity of the Alexander Turnbull Library (Wellington, NZ), The National Maritime Museum (Greenwich, England), the Royal Astronomical Society (London, England), the Ward Wanganui Observatory (Wanganui, NZ), and Dr Mary Harris (London, England) in making so many images available.

I also particularly wish to thank Dr Richard Dodd (Martinborough, NZ), John Drummond (Gisborne, NZ) and Gordon Hudson (Wellington, NZ) for reading and commenting on many of the chapters in this book. I also wish to thank Emeritus Professor John Hearnshaw for his encouragement. The original plan was to launch of this book at the Mt John 50th Birthday Conference in Tekapo in May 2015 in conjunction with John's new book on the development of Astronomy at the University of Canterbury and Mt John Observatory. Unfortunately on-going illness prevented me from completing the manuscript in time for this to happen.

Finally, I would like to thank Maury Solomon and Nora Rawn at Springer for agreeing to publish this book and for seeing it successfully through the press, and Darunee Lingling Orchiston for looking after me while I was working day and night on this book, for encouraging and supporting me when illness dramatically reduced my productivity, and for putting up with this book when its preparation severely impacted on the planning of our wedding ceremonies in November 2014. Lingling, it is true to say that without your love and support this book would never have been completed. Thank you darling.

References

Annual address by the President. *Southern Stars, 8*, 90–92 (1942).

Ashbrook, J. (1984). *The Astronomical Scrapbook: Skywatchers, Pioneers, and Seekers in Astronomy*. Cambridge, MA: Sky Publishing Corporation.

Bateson, F. M. (1984). Astronomy in New Zealand: amateurs take the lead. *Mercury*, January/February, 14–20.

Bateson, F. M. (1989). *Paradise Beckons*. Waikanae: Heritage Press.

Burdon, R. M. (1956). *Scholar Errant. A Biography of Professor A. W. Bickerton*. Christchurch: Pegasus Press.

Chapman, A. (1998). *The Victorian Amateur Astronomer. Independent Astronomical Research in Britain 1820–1920*. Chichester: Wiley.

Clark, B. A. J., & Orchiston, W. (2004). The Melbourne observatory Dallmeyer photoheliograph and the 1874 transit of Venus. *Journal of Astronomical History and Heritage, 7*, 44–49.

Clerke, A. (1893). *A Popular History of Astronomy during the Nineteenth Century*. London: Adam and Charles Black.

Cottam, S., & Orchiston, W. (2014). *Eclipses, Transits, and Comets in the Nineteenth Century: How America's Perception of the Skies Changed*. New York: Springer.

Cottam, S., Orchiston, W., & Stephenson, R. (2011a). The 1874 transit of Venus, and the popularisation of astronomy in the U.S.A. as reflected in *The New York Times*. In W. Orchiston, T. Nakamura, & R. Strom (pp. 225–241).

Cottam, S., Orchiston, W., & Stephenson, F. R. (2012). The 1882 transit of Venus and the popularization of astronomy in the USA as reflected in *The New York Times*. *Journal of Astronomical History and Heritage, 15*, 183–199.

Cottam, S., Pearson, J., Orchiston, W., & Stephenson, R. (2011b). The total solar eclipses of 7 August 1869 and 29 July 1878 and the popularisation of astronomy in the U.S.A. as reflected in *The New York Times*. In W. Orchiston, T. Nakamura, & R. Strom (pp. 339–375).

Cunningham, C., Marsden, B. G., & Orchiston, W. (2011). Giuseppe Piazzi: the controversial discovery and loss of Ceres in 1801. *Journal for the History of Astronomy, 42*, 283–306.

Cunningham, C., & Orchiston, W. (2013). Olbers' planetary explosion hypothesis: genesis and early nineteenth-century interpretation. *Journal for the History of Astronomy, 44*, 187–206.

Cunningham, C. J., & Orchiston, W. (2015). The clash between William Herschel and the great German 'amateur' astronomer Johann Schroeter. In W. Orchiston, D. A. Green, & R. Strom (pp. 205–222).

Darlington, V., & Orchiston, W. (2015). North Queensland and the 1882 transit of Venus: our final role in the quest for the Astronomical Unit. In W. Orchiston & A. McGregor (in press).

Débarbat, S., Lequeux, J., & Orchiston, W. (2007). Highlighting the history of French radio astronomy. 1: Nordmann's attempt to observe solar radio emission in 1901. *Journal of Astronomical History and Heritage, 10*, 3–10.

Dick, S. J., Orchiston, W., & Love, T. (1998). Simon Newcomb, William Harkness and the nineteenth century American transit of Venus expeditions. *Journal for the History of Astronomy, 29*, 221–255.

Dunlop, S., & Gerbaldi, M. (1988). *Stargazers: The Contribution of Amateurs to Astronomy*. Berlin: Springer.

Eiby, G. A. (1971). Astronomy in New Zealand. *Sky and Telescope, 42*(1), 18–20.

Eiby, G. (1978). History, astronomy and New Zealand. *Southern Stars, 27*, 118–124.

Encrenaz, P., Gómez-González, J., Lequeux, J., & Orchiston, W. (2011). Highlighting the history of French radio astronomy. 7: The genesis of the Institute of Radioastronomy at Millimeter Wavelengths (IRAM). *Journal of Astronomical History and Heritage, 14*, 83–92.

Ford, J. R. H., Orchiston, W., & Clendening, R. (2012). The Wells Creek meteorite impact site and changing views on impact cratering. *Journal of Astronomical History and Heritage, 15*, 159–178.

Ford, J. R. H., Orchiston, W., & Clendening, R. (2013). The Flynn Creek meteorite impact site and changing views on impact cratering. *Journal of Astronomical History and Heritage, 16*, 127–183.

Ford, J. R. H., Orchiston, W., & Clendening, R. (2014). An historical perspective on the suspected meteorite impacts sites of Tennessee. 1: The Dycus Structure. *Journal of Astronomical History and Heritage, 17*, 353–364.

Ford, J. R. H., Orchiston, W., & Clendening, R. (2015). An historical perspective on the suspected meteorite impacts sites of Tennessee. 2: The Howell Structure. *Journal of Astronomical History and Heritage, 18*, 65–97.

George, M., Orchiston, W., Slee, B., & Wielebinski, R. (2015a). The history of early low frequency radio astronomy in Australia. 2: Tasmania. *Journal of Astronomical History and Heritage, 18*, 14–22.

George, M., Orchiston, W., Slee, B., & Wielebinski, R. (2015b). The history of early low frequency radio astronomy in Australia. 3: Ellis, Reber and the Cambridge field station near Hobart. *Journal of Astronomical History and Heritage, 18*, 177–189.

Gilmore, G. (1985). Alexander William Bickerton: New Zealand's colourful astronomer. *Southern Stars, 29*, 87–108.

Hafez, I., Stephenson, R., & Orchiston, W. (2011). Abdul-Rahman al-Sufi and his *Book of the Fixed Stars*: a journey of re-discovery. In W. Orchiston, T. Nakamura, & R. Strom (pp. 121–138).

Hafez, I., Stephenson, F. R., & Orchiston, W. (2015a). Abdul-Rahman al-Sufi and his 3-step magnitude system. In W. Orchiston, D. A. Green, & R. Strom (pp. 169–177).

Hafez, I., Stephenson, F. R., & Orchiston, W. (2015b). The investigation of stars, star clusters and nebulae in ʻAbd al-Raḥmān al-Ṣūfī's *Book of the Fixed Stars*. In W. Orchiston, D. A. Green, & R. Strom (pp. 143–168).

Hayes, M. (1987). *In Spite of His Time. A Biography of R. C. Hayes. Pioneer Seismologist Astronomer Musician*. Wellington: The New Zealand Geophysical Society.

Hearnshaw, J. (2006). Astronomy in New Zealand. In A. Heck (Ed.), *Organizations and Strategies in Astronomy* (pp. 63–86). Dordrecht: Springer.

Hearnshaw, J., & Gilmore, A. (2015). *Mt John – The First Fifty Years. A Celebration of Half a Century of Optical Astronomy at the University of Canterbury*. Christchurch: Canterbury University Press.

Hendrie, M. J. (1996). Two bright comets of 1957. *Journal of the British Astronomical Association, 106*, 315–330.

Ishiguro, M., & Orchiston, W. (2016). The early development of radio astronomy in Japan. In T. Nakamura & W. Orchiston.

Ishiguro, M., Orchiston, W., Akabane, K., Kaifu, N., Hayashi, M., Nakamura, T., et al. (2012). Highlighting the history of Japanese radio astronomy. 1: An introduction. *Journal of Astronomical History and Heritage, 15*, 213–231.

Kellermann, K. I., Orchiston, W., & Slee, B. (2005). Gordon James Stanley and the early development of radio astronomy in Australia and the United States. *Publications of the Astronomical Society of Australia, 22*, 13–23.

Lankford, J. (1981). Amateurs and astrophysics: a neglected aspect in the development of a scientific speciality. *Social Studies in Science, 11*, 275–303.

Lequeux, J., Steinberg, J.-L., & Orchiston, W. (2010). Highlighting the history of French radio Astronomy. 5: The Nançay large radio telescope. *Journal of Astronomical History and Heritage, 13*, 29–42.

Mackrell, B. (1985). *Halley's Comet over New Zealand*. Auckland: Reed-Methuen.
Macpherson, H. (1906). *A Century's Progress in Astronomy*. Edinburgh: Blackwood.
McIntosh, R. A. (1970). Early New Zealand astronomy. *Southern Stars, 23*, 101–108.
Meadows, J., & Henbest, N. (1981). Astronomy: how amateurs got there first. *New Scientist, 90*, 170–171.
Nakajima, H., Ishiguro, M., Orchiston, W., Akabane, K., Enome, S., Hayashi, K., et al. (2014). Highlighting the history of Japanese radio astronomy. 3: Early solar radio research at the Tokyo Astronomical Observatory. *Journal of Astronomical History and Heritage, 17*, 2–28.
Nakamura, T., & Orchiston, W. (Eds.), (2016). *The Emergence of Astrophysics in Asia: Opening a New Window on the Universe*. New York: Springer.
New Director for Carter Observatory. *Southern Sky, 9*, 12–14 (1995).
Olsson-Steel, D. (1986). A note on the discovery of the liquid mirror telescope. *Journal of the Royal Astronomical Society of Canada, 80*, 128–133.
Orchiston, W. (1963). The return of the 1956 Phoenicid meteor shower. *Southern Stars, 20*, 23–25.
Orchiston, D. W. (1964a). A minor meteor radiant in Libra. *Southern Stars, 20*, 162–163.
Orchiston, W. (1964b). Topographical changes on the lunar surface. *Southern Stars, 20*, 142–148.
Orchiston, W. (1971). Astronomy in early New Zealand. *Bulletin of the New South Wales Branch of the British Astronomical Association, 509*(February), 1–2.
Orchiston, W. (1985). A survey of manuscript material in the Mitchell Library (Sydney) relating to the history of astronomy in New Zealand. *Southern Stars, 31*, 225–238.
Orchiston, W. (1986). Towards an accurate history of early New Zealand astronomy. *Southern Stars, 31*, 257–266.
Orchiston, W. (1988). Illuminating incidents in Antipodean astronomy: formation of the New South Wales Branch of the British Astronomical Association. *Journal of the British Astronomical Association, 98*, 75–84.
Orchiston, W. (1989). The role of the amateur in early Australian astronomy. *Search, 20*, 16–23.
Orchiston, W. (1991). *The Waingaromia Meteorite*. Gisborne: Gisborne Museum & Arts Centre (Information Sheet No. 17) (p. 4).
Orchiston, W. (1997). The Waingaromia Meteorite: a IIIAB iron from New Zealand. *Meteorite! 3* (4), 20–21 (plus Front Cover).
Orchiston, W. (1998a). From the South Seas to the Sun: the astronomy of Cook's voyages. In M. Lincoln (Ed.), *Science and Exploration in the Pacific. European Voyages to the Southern Oceans in the Eighteenth Century* (pp. 55–72). Woodbridge: Boydell Press (in association with the National Maritime Museum).
Orchiston, W. (1998b). Illuminating incidents in Antipodean astronomy: John Tebbutt and the Great Comet of 1861. *Irish Astronomical Journal, 25*, 167–178.
Orchiston, W. (1998c). *Nautical Astronomy in New Zealand: The Voyages of James Cook*. Wellington: Carter Observatory.
Orchiston, W. (1999). C/1881 K1: a forgotten "Great Comet" of the nineteenth century. *Irish Astronomical Journal, 26*, 33–44.
Orchiston, W. (2000). John Tebbutt of Windsor, New South Wales: a pioneer southern hemisphere variable star observer. *Irish Astronomical Journal, 27*, 47–54.
Orchiston, W. (2001). 'Sentinel of the Southern Heavens': the Windsor observatory of John Tebbutt. *Journal of the Antique Telescope Society, 21*, 11–23.
Orchiston, W. (2002). Tebbutt vs Russell: passion, power and politics in nineteenth century Australian astronomy. In S. M. R. Ansari (Ed.), *History of Oriental Astronomy* (pp. 169–201). Dordrecht: Kluwer (IAU Joint Discussion 17).
Orchiston, W. (2004a). Astronomical results of Cook's voyages. In J. Robson (Ed.), *Captain Cook Encyclopaedia* (pp. 31–35). London: Chatham Publishing.
Orchiston, W. (2004b). From the solar corona to clusters of galaxies: the radio astronomy of Bruce Slee. *Publications of the Astronomical Society of Australia, 21*, 23–71.
Orchiston, W. (2004c). Highlighting the history of nineteenth century Australian astronomy: the Tebbutt Collection in the Mitchell Library, Sydney. *Journal of Astronomical Data, 10*, 77–103.

Orchiston, W. (2004d). John Tebbutt and observational astronomy at Windsor Observatory. *Journal of the British Astronomical Association, 114*, 141–154.

Orchiston, W. (2004e). New South Wales observations of the 1874 transit of Venus. *AAO Newsletter, 104*, 12–14.

Orchiston, W. (2004f). The nineteenth century transits of Venus: an Australian and New Zealand overview. *Journal of Astronomical Data, 10*, 219–308.

Orchiston, W. (2005a). James Cook's 1769 transit of Venus expedition to Tahiti. In D. W. Kurtz (Ed.), *Transits of Venus: New Views of the Solar System and Galaxy* (pp. 52–66). Cambridge: Cambridge University Press.

Orchiston, W. (2005b). Sixty years in radio astronomy: a tribute to Owen Bruce Slee. *Journal of Astronomical History and Heritage, 8*, 3–10.

Orchiston, W. (2015a). Captain Cook and the 1769 transit of Venus: the Australian connection. In W. Orchiston & A. McGregor (in press).

Orchiston, W. (2015b). The amateur-turned-professional syndrome: two Australian case studies. In W. Orchiston, D. A. Green, & R. Strom (pp. 259–350).

Orchiston, W., Ahn, Y.-S., Lee, E.-H., & Chen, K.-Y. (2006a). British observations of the 1868 total solar eclipse from Guntoor, India. In K.-Y. Chen, W. Orchiston, B. Soonthornthum, & R. Strom (Eds.), *Proceedings of the Fifth International Conference on Oriental Astronomy* (pp. 23–34). Chiang Mai: University of Chiang Mai.

Orchiston, W., & Buchanan, A. (1993). Illuminating incidents in Antipodean astronomy: Campbell Town, and the 1874 transit of Venus. *Australian Journal of Astronomy, 5*, 11–31.

Orchiston, W., & Buchanan, A. (2004). 'The Grange', Tasmania: survival of a unique suite of 1874 transit of Venus relics. *Journal of Astronomical History and Heritage, 7*, 34–43.

Orchiston, W., Buchanan, A., Price, G., & Sprent, T. (2015a). Highlighting our international astronomical heritage: Tasmanian relics of the 1874 transit of Venus. *Journal of Astronomical History and Heritage, 18*, 149–160.

Orchiston, W., & Darlington, V. (2015). The most noble problem in nature: encounters with the historic transits of Venus. In W. Orchiston & A. McGregor (in press).

Orchiston, W., George, M., Slee, B., & Wielebinski, R. (2016a). Early low frequency radio astronomy in Australia. In Y.-L. Shi (Ed.), *Astronomical Heritages in Asia-Pacific Areas: Proceedings of the Eighth International Conference on Oriental Astronomy*. Hefei: University of Science and Technology of China (in press).

Orchiston, W., Green, D., & Strom, R. (Eds.). (2015b). *New Insights from Recent Studies in Historical Astronomy: Following in the Footsteps of F. Richard Stephenson*. New York: Springer.

Orchiston, W., & Howse, D. (2000). From transit of Venus to teaching navigation: the work of William Wales. *Journal of Navigation, 53*, 156–166.

Orchiston, W., Lequeux, J., Steinberg, J.-L., & Delannoy, J. (2007). Highlighting the history of French radio astronomy. 3: The Würzburg antennas at Marcoussis, Meudon and Nançay. *Journal of Astronomical History and Heritage, 10*, 221–245.

Orchiston, W., & Mathewson, D. (2009). Chris Christiansen and the Chris Cross. *Journal of Astronomical History and Heritage, 12*, 11–32.

Orchiston, W., & McGregor, R. (Eds.). (2015). *Scientific Alignments: James Cook, the 1769 Transit of Venus and the Colonisation of North Queensland*. Brisbane: Queensland Museum (in press).

Orchiston, W., Nakamura, T., & Strom, R. (Eds.). (2011). *Highlighting the History of Astronomy in the Asia-Pacific Region*. Springer: New York.

Orchiston, W., Orchiston, D. L., Soonthornthum, B., & George, M. (2015c). Exploring the first scientific observations of lunar eclipses made in Siam. *Journal of Astronomical History and Heritage, 18* (in press).

Orchiston, W., & Pearson, J. (2011). The Lick Observatory solar eclipse expedition to Jeur (India) in 1898. In: T. Nakamura, W. Orchiston, M. Soma, & R. Strom (Eds.), *Mapping the Oriental Sky. Proceedings of the Seventh International Conference on Oriental Astronomy* (pp. 197–206). Tokyo: National Astronomical Observatory of Japan.

Orchiston, W., & Slee, B. (2002a). The Australasian discovery of solar radio emission. *AAO Newsletter, 101*, 25–27.

Orchiston, W., & Slee, B. (2002b). The flowering of Fleurs: an interesting interlude in Australian radio astronomy. *ATNF News, 47*, 12–15.

Orchiston, W., & Slee, B. (2002c). Ingenuity and initiative in Australian radio astronomy: the Dover Heights 'hole-in-the-ground' antenna. *Journal of Astronomical History and Heritage, 5*, 21–34.

Orchiston, W., & Slee, B. (2005a). The Radiophysics field stations and the early development of radio astronomy. In W. Orchiston (Ed.), *The New Astronomy: Opening the Electromagnetic Window and Expanding our View of Planet Earth* (pp. 119–168). New York: Springer.

Orchiston, W., & Slee, B. (2005b). Shame about Shain! Early Australian radio astronomy at Hornsby Valley. *ATNF News, 55*, 14–16.

Orchiston, W., & Slee, B. (2006). Early Australian observations of historic supernova remnants at radio wavelengths. In K.-Y. Chen, W. Orchiston, B. Soonthornthum, & R. Strom (Eds.), *Proceedings of the Fifth International Conference on Oriental Astronomy* (pp. 43–56). Chiang Mai: University of Chiang Mai.

Orchiston, W., Slee, B., & Burman, R. (2006b). The genesis of solar radio astronomy in Australia. *Journal of Astronomical History and History, 9*, 35–56.

Orchiston, W., Slee, B., George, M., & Wielebinski, R. (2015d). The history of early low frequency radio astronomy in Australia. 1: The CSIRO Division of Radiophysics. *Journal of Astronomical History and Heritage, 18*, 3–13.

Orchiston, W., Slee, B., George, M., & Wielebinski, R. (2015e). The history of early low frequency radio astronomy in Australia. 4: Kerr, Shain and Higgins and the Hornsby Valley field station near Sydney. *Journal of Astronomical History and Heritage, 18* (in press).

Orchiston, W., & Soonthornthum, B. (2016). French observations of the total solar eclipse of 1868 from Thailand. In T. Nakamura & W. Orchiston.

Orchiston, W., Soonthornthum, B., & Komonjinda, S. (2016b). Seventeenth century Jesuit astronomical activities in Siam. In Y.-L. Shi (Ed.), *Astronomical Heritages in Asia-Pacific Areas: Proceedings of the Eighth International Conference on Oriental Astronomy*. Hefei: University of Science and Technology of China (in press).

Orchiston, W., & Steinberg, J.-L. (2007). Highlighting the history of French radio astronomy. 2: The solar eclipse observations of 1949–1954. *Journal of Astronomical History and Heritage, 10*, 11–19.

Orchiston, W., Steinberg, J.-L., Kundu, M., Arsac, J., Blum, E.-J., & Boischot, A. (2009). Highlighting the history of French radio astronomy. 4: Early solar research at the École Normale Supérieure, Marcoussis and Nançay. *Journal of Astronomical History and Heritage, 12*, 175–188.

Pearson, J., & Orchiston, W. (2008). The 40-foot solar eclipse camera of the Lick Observatory. *Journal of Astronomical History and Heritage, 11*, 25–37.

Pearson, J., & Orchiston, W. (2011). The Lick Observatory solar eclipse expedition to Padang (Indonesia) in 1901. In Nakamura, T., Orchiston, W., Soma, M., & Strom, R. (Eds.), *Mapping the Oriental Sky. Proceedings of the Seventh International Conference on Oriental Astronomy* (pp. 207–216). Tokyo: National Astronomical Observatory of Japan.

Pearson, J., Orchiston, W., & Malville, J. M. (2011). The Lick Observatory solar eclipse expeditions. In W. Orchiston, T. Nakamura, & R. Strom (pp. 243–337).

Pick, M., Steinberg, J.-L., Orchiston, W., & Boischot, A. (2011). Highlighting the history of French radio astronomy. 6: The multi-element grating arrays at Nançay. *Journal of Astronomical History and Heritage, 14*, 57–77.

Report of Council for the year ended 30 September 1940. *Southern Stars, 7*, 37 (1941).

Robertson, P., Cozens, G., Orchiston, W., & Slee, B. (2010). Early Australian optical and radio observations of Centaurus A. *Publications of the Astronomical Society of Australia, 27*, 402–430.

Robertson, P., Orchiston, W., & Slee, B. (2014). John Bolton and the discovery of discrete radio sources. *Journal of Astronomical History and Heritage, 17*, 283–306.

Rumsey, N. J. (1974). Astronomy in New Zealand. *Southern Stars, 25*, 171–207.
Russell, H. C. (1892). *Observations of the Transit of Venus, 9 December, 1874: Made at Stations in New South Wales*. Sydney: Government Printer.
Sauter, J., Simonia, I., Stephenson, F. R., & Orchiston, W. (2015a). Historical astronomy of the Caucasus: sources from Georgia and Armenia. In W. Orchiston, D. A. Green, & R. Strom (pp. 103–117).
Sauter, J., Simonia, I., Stephenson, F. R., & Orchiston, W. (2015b). The legendary fourth-century total solar eclipse in Georgina: fact or fantasy? In W. Orchiston, D. A. Green, & R. Strom (pp. 25–45).
Shimoda, K., Orchiston, W., Akabane, K., & Ishiguro, M. (2013). Highlighting the history of Japanese radio astronomy. 2: Koichi Shimoda and the 1948 solar eclipse. *Journal of Astronomical History and Heritage, 16*, 98–106.
Slee, B. (2005). Early Australian measurements of angular structure in discrete radio sources. *Journal of Astronomical History and Heritage, 8*, 97–116.
Stebbins, R. A. (1977). The amateur: two sociological definitions. *Pacific Sociological Review, 20*, 582–606.
Stebbins, R. A. (1978). Towards amateur sociology: a proposal for the profession. *The American Sociologist, 13*, 239–252.
Stebbins, R. A. (1979). *Amateurs: On the Margin Between Work and Leisure*. Beverley Hills: Sage Publications.
Stebbins, R. A. (1980a). "Amateur" and "hobbyist" as concepts for the study of leisure problems. *Social Problems, 27*, 413–417.
Stebbins, R. A. (1980b). Avocational science: the amateur routine in archaeology and astronomy. *International Journal of Comparative Sociology, 21*, 34–48.
Stebbins, R. A. (1981). Science amators? Rewards and cost in amateur astronomy and archaeology. *Journal of Leisure Research, 13*, 289–304.
Stebbins, R. A. (1982a). Amateur and professional astronomers: a study of their interrelationships. *Urban Life, 10*, 433–454.
Stebbins, R. A. (1982b). Looking downwards: sociological images of the vocation and avocation of astronomy. *Southern Stars, 29*, 177–188.
Stebbins, R. A. (1987). Amateurs and their place in professional science. In D. S. Hayes, R. M. Genet, & D. R. Genet (Eds.), *New Generation Small Telescopes* (pp. 217–225). Mesa: Fairborn Press.
Stebbins, R. A. (1992). *Amateurs, Professionals, and Serious Leisure*. Montreal: McGill-Queens University Press.
Stebbins, R. A. (2004). *Between Work and Leisure: The Common Ground of Two Separate Worlds*. New Brunswick: Transaction Publications.
Stebbins, R. A. (2007). *Serious Leisure: A Perspective for Our Time*. Piscataway: Transaction Publishers.
Steel, D. (1984). The liquid mirror telescope—an early example of Kiwi ingenuity? *Southern Stars, 31*, 32–40.
Stewart, R., Orchiston, W., & Slee, B. (2011a). The contribution of the Division of Radiophysics Dapto field station to solar radio astronomy, 1952–1964. In W. Orchiston, T. Nakamura, & R. Strom (pp. 481–526).
Stewart, R., Orchiston, W., & Slee, B. (2011b). The Sun has set on a brilliant mind: Paul Wild (1923–2008). In W. Orchiston, T. Nakamura, & R. Strom (pp. 527–542).
Stewart, R., Wendt, H., Orchiston, W., & Slee, B. (2010a). Highlighting our history: the world's first solar radiospectrograph—Penrith 1948–1949. *ATNF News, 68*, 8–11.
Stewart, R., Wendt, H., Orchiston, W., & Slee, B. (2010b). The Radiophysics field station at Penrith, New South Wales, and the world's first solar radiospectrograph. *Journal of Astronomical History and Heritage, 13*, 2–15.
Stewart, R., Wendt, H., Orchiston, W., & Slee, B. (2011c). A retrospective view of Australian solar radio astronomy 1945 to 1960. In W. Orchiston, T. Nakamura, & R. Strom (pp. 589–629).

Theodossiou, E. T., Niarchos, P. G., Manimanis, V. N., & Orchiston, W. (2002). The fall of a meteorite at Aegos Potami in 467/466 BC. *Journal of Astronomical History and Heritage, 5*, 135–140.

Thomsen, I. L. (1954). The amateur in astronomy. *Southern Stars, 16*, 78–82.

Thomsen, I. L. (1956). The future of astronomy in New Zealand. *Southern Stars, 17*, 7–14.

Tobin, W. (1996). The Townsend Telescope—a century of university stargazing. In W. Tobin & G. M. Evans (Eds.), *Stars in a Cluster* (pp. 78–83). Christchurch: Department of Physics & Astronomy, University of Canterbury.

Van Dijk, G. (1992). The New Zealand Astronomy Centre. *Southern Stars, 34*, 403–406.

Warnes, P., Orchiston, W., & Englert, P. (1998). A reported tektite transported from Australasia and found at Gabriel's Gully mining camp, Central Otago, New Zealand. *Journal of the Royal Society of New Zealand, 28*, 329–331.

Wendt, H., Orchiston, W., & Slee, B. (2008a). The Australian solar eclipse expeditions of 1947 and 1949. *Journal of Astronomical History and Heritage, 11*, 71–78.

Wendt, H., Orchiston, W., & Slee, B. (2008b). W.N. Christiansen and the development of the solar grating array. *Journal of Astronomical History and Heritage, 11*, 173–184.

Wendt, H., Orchiston, W., & Slee, B. (2008c). W.N. Christiansen and the initial Australian investigation of the 21cm hydrogen line. *Journal of Astronomical History and Heritage, 11*, 185–193.

Wendt, H., Orchiston, W., & Slee, B. (2011a). An overview of W.N. Christiansen's contribution to Australian radio astronomy, 1948–1960. In W. Orchiston, T. Nakamura, & R. Strom (pp. 547–587).

Wendt, H., Orchiston, W., & Slee, B. (2011b). The contribution of the Division of Radiophysics Murraybank field station to international radio astronomy. In W. Orchiston, T. Nakamura, & R. Strom (pp. 433–479).

Wendt, H., Orchiston, W., & Slee, B. (2011c). The contribution of the Division of Radiophysics Potts Hill field station to international radio astronomy. In W. Orchiston, T. Nakamura, & R. Strom (pp. 379–431).

Williams, T. R. (1987). Criteria for classifying an astronomer as an amateur. Paper presented at IAU Colloquium 98, Paris (Note: the paper published in the *Proceedings* is merely a precis of the full paper).

Part I
Pre-European Astronomy in The Pacific

Chapter 2
The Skies Over Aotearoa/New Zealand: Astronomy from a Maori Perspective

Abstract In this chapter, which is largely based on Orchiston (2000), we critically examine the sources of information on Maori astronomy available to us, briefly describe the archaeological evidence for the settlement of Aotearoa/New Zealand and hence the origins of Maori astronomical knowledge, and examine the life of the Maori astronomer. We then detail the range of astronomical objects and events that graced the Maori sky between AD 1200 and 1768, and see what evidence of these is preserved in the oral histories and early literature recorded by Elsdon Best. We then query whether astronomical motifs are represented in Maori rock art, summarise Maori concepts of time, and conclude with some comments about what the future may hold for those wishing to carry out further studies of Maori astronomical lore.

2.1 Introduction

To the Maori of Aotearoa/New Zealand, astronomy used to be an integral part of everyday life. Before the arrival of the Europeans, it foretold the seasons and the weather, dictated a variety of food-quest activities, and played a vital role in oceanic voyaging. Thus, the Maori had names for the Sun, the Moon, all of the naked eye planets, some of the brighter stars, the Milky Way, the Coal Sack, both Magellanic Clouds, and even the Zodiacal Light. There also were names for comets and meteors. But Kingsley-Smith (1967: 5) reminds us that

> The Maori, like other races, endowed the stars and constellations with human attributes and wove fantastic stories about their conflicts, loves, hatreds, and achievements … They possessed great potency for good or ill to the human family.

Yet our knowledge of Maori astronomy is superficial and plagued by severe source limitations.

W. Orchiston, *Exploring the History of New Zealand Astronomy*,
Astrophysics and Space Science Library 422,
DOI 10.1007/978-3-319-22566-1_2

2.2 Source Limitations: The Nature of the Evidence

This study is concerned with ethnoastronomy,

> … that peculiar intermingling of astronomy, history, anthropology and folklore … Ethnoastronomy is a fascinating discipline on two counts, its novelty and its potential … [and] remains largely underexplored and underdeveloped. Consequently, it is full of promise. (Snedegar 1992: 41).

The special place of ethnoastronomy within the discipline of astronomy is discussed by Erny (1992).

The study of Maori astronomy is endowed with its own special problems and challenges. Like many other indigenous peoples of the world, the Maori were non-literate, and so we had to await the arrival of the first European explorers for written accounts of their knowledge systems. Although Cook, Surville and Du Fresne were all in Aotearoa/New Zealand waters between 1769 and 1777 and recorded a great deal of detail about Maori culture in different coastal localities (see Salmond 1991), their various logs, journals and published accounts say almost nothing about Maori astronomy.

Personally, I find this especially sad in the case of the Cook voyages, for astronomers were on all three voyages, and spent considerable periods ashore, particularly at Queen Charlotte Sound, carrying out astronomical observations (see Orchiston 1998, and Chaps. 4, 5 and 6 in this book). For the most part there were very amiable Maori-European relations, and some of the Maoris who visited the Europeans would have learnt what the astronomers were doing. Yet none of the Maoris thought to discuss their own astronomical activities and beliefs and compare them with those of the visiting Europeans. This may have been because such knowledge was jealously guarded and was not to be shared with Europeans, or because such knowledge was the domain of specialists and Maori astronomers were not among those interacting with the Europeans at this time.

Although there are snippets of information on Maori astronomy included in a number of books and research papers published during the nineteenth century, much of this material is superficial, contradictory or downright unreliable, and it was only towards the end of that century that a detailed study was carried out. The scholar responsible for this was Elsdon Best (1856–1931; Fig. 2.1), who was later to work as Ethnologist at the Dominion Museum in Wellington (Craig 1964). During the last decade of the nineteenth century and first decade of the twentieth, he carried out fieldwork, particularly in the Ureweras in the East Coast region of the North Island (see Fig. 2.2 for Aotearoa/New Zealand localities mentioned in the text). Best subsequently published a succession of Monographs and Bulletins of the Dominion Museum (as it was then known) on aspects of Maori culture, and in 1922 his 80-page Monograph No. 3, *The Astronomical Knowledge of the Maori*, appeared in print (Fig. 2.3). This contained a synthesis of information drawn from his own fieldwork and that culled from earlier published accounts. Subsequently, Best included astronomical material in two other Monographs, *The Maori Division of Time* (No. 4) and *Polynesian Voyagers. The Maori as a Deep-sea Navigator,*

Fig. 2.1 A photograph of Elsdon Best by Stanley Polkinghorne Andrew (*Courtesy* Alexander Turnbull Library, Ref. 1/1-018778-F)

Explorer, & Colonizer (No. 5). To this day, these remain the only substantive works written on the 'traditional' astronomical systems of the Maori, and they have been through numerous reprintings; I made use of copies printed in the 1950s for this study (i.e. Best 1954, 1955, 1959).

Invaluable as these books may be, we cannot ignore their serious limitations (Orchiston 1986). All of the published sources he drew on date to the nineteenth century, and Best carried out his fieldwork at the very end of that century, more than one hundred years after initial European contact. In the interim, *major* changes had taken place in 'traditional' Maori culture. This was particularly so of Maori religious systems, which invited the wrath of the early evangelical missionaries, and we must not forget that Maori astronomy was intricately inter-woven with mythology and religion. In other words, the 'Maori astronomy' that Best recorded may have differed significantly from the practises and beliefs in vogue in 1768, before the arrival of Cook and other Europeans.

Another important consideration is that Best's astronomical field data were geographically biased in that they came—in the main—from just the one area of the North Island of Aotearoa/New Zealand. Studies have shown that at any one point in time there were major regional variations in Maori culture and material culture

Fig. 2.2 Aotearoa/New Zealand topography, with localities mentioned in the text shown in *red*

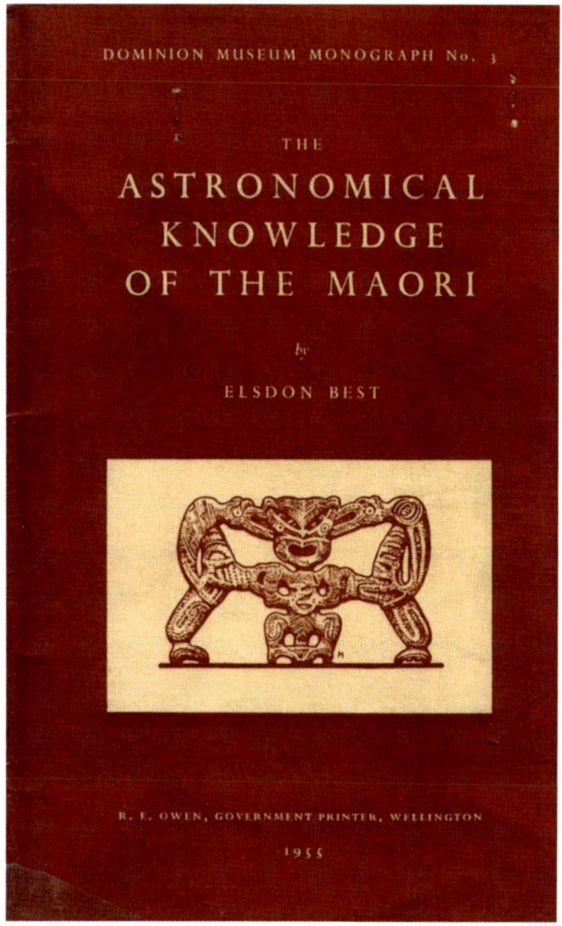

Fig. 2.3 Front cover of the 1955 edition of *The Astronomical Knowledge of the Maori* (*Photograph* Wayne Orchiston)

(e.g. see Skinner 1974: 18–26), and that significant changes occurred through time. Given the data at his disposal, all that Best could do was attempt to provide an overview of Maori astronomy based on his own fieldwork and supplementary data provided by earlier writers. The result is a synthesis, which indicates the general trends and patterns prevalent in the nineteenth century. This may be fine for basic astronomical knowledge, but when it comes to the Maori perception of the cosmos, its creation and its evolution, how much is truly 'traditional' and how much is an amalgamation of pre-European concepts and elements of Christian-based religion and European scientific precepts?

2.3 The Initial Settlement of Aotearoa/New Zealand and the Origin of Maori Astronomy

Cultural, linguistic, floral and faunal evidence reveals that most Polynesian peoples are genetically related, and this is also reflected in the similarities of the various astronomical knowledge systems found in the different island groups at the time of initial European settlement (see Best 1955).

The present-day Polynesians can be traced back to the first settlers of Samoa and Tonga, who made their initial appearance on the western boundary of the 'Polynesian Triangle' by 1000 BC. They were associated with the distinctive Lapita pottery tradition, also found further to the west, in Fiji, New Caledonia, the New Hebrides, the Solomon Islands and islands in the Bismarck Archipelago off New Guinea (see Green 1992: Fig. 1.3).

The evolution of this 'Lapita cultural complex' into a distinctly 'Polynesian' culture occurred in the Samoa-Tonga region. As the late Professor Roger Curtis Green (1932–2009) pointed out long ago, Polynesian

> … language, biological make-up, and culture was not brought from elsewhere already fully formed, but rather evolved in that region from a proto-language (Central Pacific) ancestral to the present-day daughter languages of Fiji, Rotuma and Polynesia (Pawley and Green 1973, pp. 42–43, 53), from a pre-Polynesian parental stock … and from the Eastern Lapita cultural complex which can be associated with this language and population. (Green 1992: 41).

By the first few centuries BC an ancestral Polynesian culture was recognisable, and it was people from this region, armed with what we may call an Ancestral Polynesian astronomical knowledge system, who then colonised the Cook, Society and Marquesas Islands. These, in their turn, became the primary dispersal points from which the remaining island groups of Polynesia were settled.

Aotearoa/New Zealand is situated in the far southwestern corner of the Pacific Ocean at one apex of the 'Polynesian triangle' (see Fig. 2.4). Current evidence favours an initial settlement during the thirteenth century AD (Anderson 1991; Davidson 1987; Higham et al. 1999; Holdaway and Jacomb 2000), while archaeological and linguistic evidence points clearly to the general region of the Cook, Society and Marquesas Islands as the source of the initial settlers (see various papers in Sutton 1992).

When they arrived in Aotearoa/New Zealand the ancestral Maori discovered a pristine alien environment that was very different indeed from the small high islands of central or eastern Polynesia whence they came. They found two very large islands and many smaller ones, extending from sub-tropical north to sub-antarctic south. Both islands were bisected longitudinally by high mountains, and in the central North Island some of these were volcanically active. Most of the land was forested, and the principal terrestrial inhabitants were a range of giant birds, most of them flightless (e.g. see Fig. 2.5) and nearly all of them now extinct (Cumberland 1962; McGlone et al. 1992).

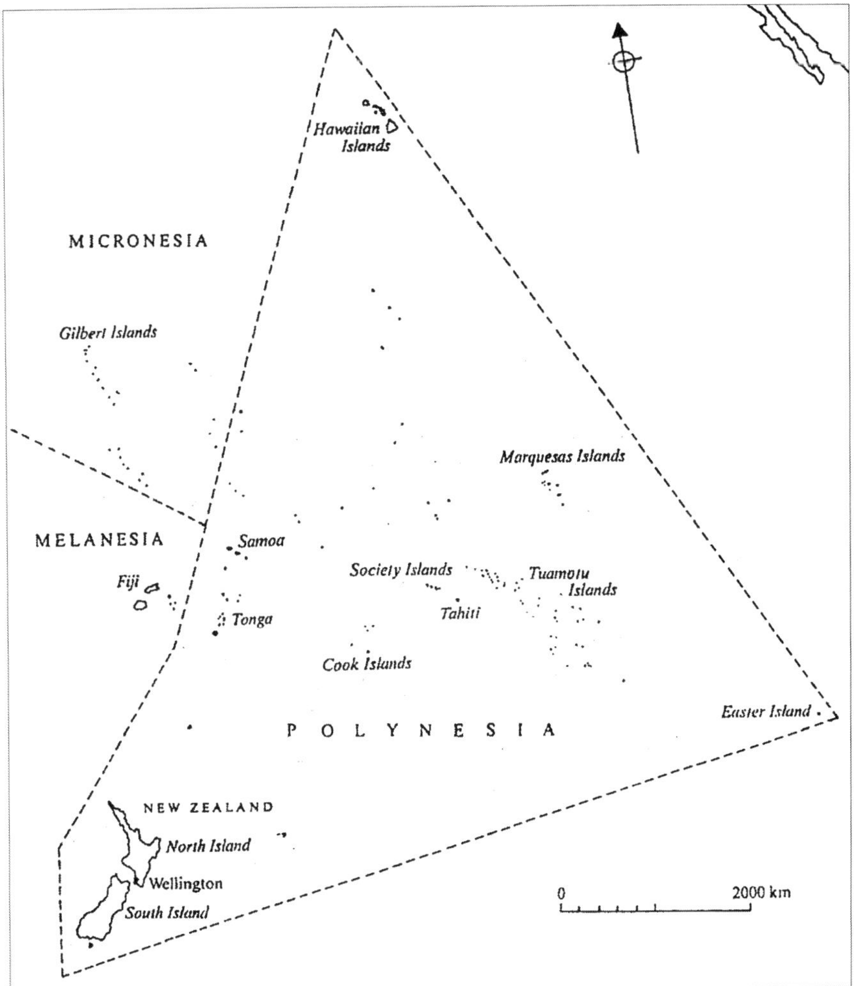

Fig. 2.4 The main islands of Polynesia and eastern Micronesia and Melanesia

Yet the earlier settlers were not totally unprepared for this new environment, for they

> … brought plants and other material from their tropical home that were needed for the successful colonization (Davidson 1987; Anderson 1991; Higham et al. 1999; Holdaway and Jacomb 2000). These included kumara (*Ipomoea batatus*), yam (*Dioscorea* sp.), taro (*Colocasia esculanta*), gourd (Family Cucurbitaceae), paper mulberry (*Broussonetia terminalis*), banana (*Musa* sp.), breadfruit (*Artocarpus communis*), tropical cabbage tree (*Cordyline terminalis*), and coconuts (*Cocos nucifera*) (Best 1925). The cooler, seasonal temperate climate of New Zealand meant that many of these crops failed to thrive, or, as in the case of coconuts, bananas, and breadfruit, did not grow at all. Taro, paper mulberry, and yam growing were restricted to the warmer North Island. (Bassett et al. 2004).

Fig. 2.5 The range of flightless birds that occupied Aotearoa/New Zealand at the time of the initial Maori settlement included several different species of moas (https://en.wikipedia.org)

The ancestral Maori responded to this horticultural challenge by focusing some of their attention on the moa and other large flightless birds, and soon drove most of these to extinction (e.g. see Anderson 1989; Holdaway and Jacomb 2000). They also used fire to rapidly change the landscape in the course of their exploration (McGlone 1983), and to clear areas for cultivation once they developed the ability to over-winter the *kumara* and discovered that it could then grow successfully as far south as the sheltered microclimates found on Banks Peninsula (Bassett et al. 2004). The *kumara* then became a dietary staple over large areas of Aotearoa/New Zealand (Yen 1961), supplemented on a seasonal basis by the ubiquitous fernroot tubers; various edible fruits and berries; a myriad of forest and shore birds; endless supplies of shellfish, fish and crayfish; and an abundance of seals and dolphins.

The astronomical knowledge base that reached Aotearoa/New Zealand with the first settlers was the one then current in central Polynesia, and was part of their prevailing 'world view'. However, it had its roots further to the west, and was derived from an earlier Lapita prototype. In this regard, it is important to realise that the major stars and constellations visible from Aotearoa/New Zealand, and other prominent features such as the Magellanic Clouds, had been equally visible from island Polynesia *and* the islands of Melanesia occupied by the Lapita people. So although they moved to a more southerly land, the ancestral Maori settled under a familiar sky—even if the Magellanic Clouds and Milky Way rose higher in the sky at culmination. In other words, there was no need to instantly develop a totally new or even substantially revised astronomical system specifically for Aotearoa/New Zealand.

Yet no knowledge base is unchanging, for cultures are dynamic and automatically evolve (or devolve) with time. The archaeological record clearly documents that the initial Maori culture underwent substantial modification in the course of the next 800 years (Davidson 1987; Golson 1959; Green 1975; Trotter and McCulloch 1997). This is best seen in elements of material culture, but it is reasonable to suppose that the astronomical knowledge base also would have shared in this process, especially as the economic basis of Maori society evolved with time.

2.4 The World of the Maori Astronomer

In pre-European Maori society, membership of various social groups—as elsewhere in Polynesia—was based on kinship and descent. Three different groupings are conventionally recognised: the *whanau* (extended family), the *hapu* (clan) and the *iwi* (tribe). However, Davidson (1987) provides a timely reminder:

> It is important to realise, however, that this is a generalised (and idealised) summary based on nineteenth century studies. The system encompassed considerable individual and regional variations.

Nonetheless, the local group usually comprised the *whanau*, typically 20–50 individuals who lived together in a hamlet or village. Only in times of threat would members of several *whanau* flee to a common *pa* (or fortified settlement), but at other times the *whanau* was the group involved in day-to-day activities (Orchiston 1979).

If astronomy underpinned not only horticulture, but also other food-quest activities—as documented by Best (1955)—then every local group would have needed one or more individuals with specialist celestial knowledge. These Maori astronomers were always men of high rank, and were termed *tohunga kokorangi*. Their duty was to study the heavens, and utilize their astronomical knowledge for the welfare of the community. Best (1955: 6) recounts how

> One famed old wise man of the Wairarapa district, of last century, devoted much of his time to studying the stars and planets. His contemporaries have told me that they have often known him to pass the greater part of the night on the summit of a hillock near his hut, gazing continuously at the heavens.

Like other Maori astronomers, he had to rely solely on the naked eye for his studies. The only reference I could find to any sort of observing aid—if it may be termed that—comes from Beattie (1918: 145), who was told by an informant that "When he was a lad at Temuka [during the nineteenth century] he had seen his father put sticks in the ground and observe the stars." This would appear to have been an isolated incident, or perhaps an idiosyncrasy of that particular astronomer. It certainly was not the norm.

Training to be a *tohunga kokorangi* was no trivial task. It involved years of study to acquire the full gamut of celestial lore, which was passed down from generation to generation by word of mouth. Consequently, most Maori astronomers were respected senior members of their local groups, but to think of them as 'elderly' in the context of present-day populations is misleading, for Houghton (1980: 95–97) has shown that in pre-European Maori society the average lifespan was only 31–32 years, and that very few individuals lived beyond 50 years of age. It would seem that most Maori astronomers were 'venerable old men' in their late twenties or in their thirties!

Although astronomy was the domain of experts, in 1922 Best believed that "It is assuredly a fact that in former times the average Maori knew much more about the stars than does the average man among us [today]." (Best 1955: 8).

2.5 The Creation of the Maori Universe

Best (1955: 5–6) stresses that

> Maori beliefs concerning the heavenly bodies were very different from our own … much of the star-lore of the Maori was empirical - astronomy and astrology were intermingled in his beliefs and teachings … there was much of sentiment in the Maori mind in connection with the stars …

Mythological accounts were used to explain various astronomical objects and events, and these myths were passed down by word of mouth, and in songs, charms, chants, and as sayings.

To the East Coast Maoris there was a system of twelve separate and distinct heavens, which were termed *nga rangi tuhaha* (the bespaced heavens). Some other tribes reported just ten heavens, while two earlier writers, White and Davis, mention twenty heavens. Despite these variations, Best (1955: 8) gives greatest credence to the East Coast accounts.

Of these twelve heavens, closest to the Earth's surface was the mythological *Ranginui*, and it was upon his body that the stars and other celestial objects moved. East Coast accounts reveal that it was *Turangi* and *Moe-ahura*, two of the offspring of *Ranginui* (the Sky Father) and *Papatuanuku* (the Earth Mother), who gave birth to the Sun, the Moon and the stars, as depicted in the following genealogy (see Fig. 2.6). The stars were produced after the Sun and Moon, and were referred to as the 'younger members' of the family (Best 1955: 9–10).

Variations on this genealogy occur throughout Aotearoa/New Zealand, generally in the names assigned to the parents of the Sun, Moon and stars [e.g. see (Orchiston 1996: 326) for another version of this genealogy]. To the Te Awa people of the Bay of Plenty they were *Tangotango* and *Wainui*, respectively, while other tribes claim the father's name was *Tongatonga*. The mother's name has also been given as

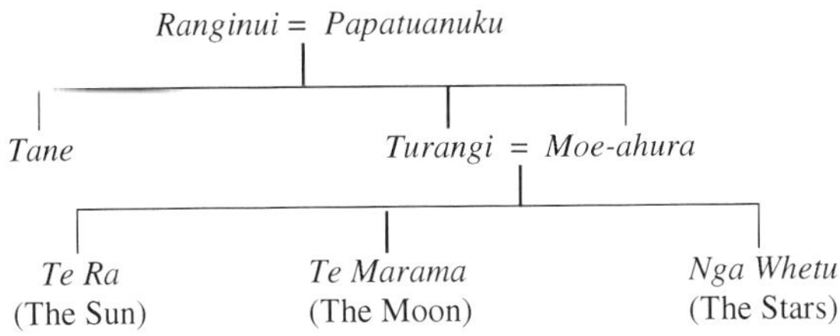

Fig. 2.6 Genealogy documenting the Maori origins of the Sun, Moon and stars (after Best 1955)

Moe-te-ahura and *Hine-te-ahura*. Regardless of the names used, the offspring were generally the same, except that the Te Awa included *Hinetore* (phosphorescence) as a fourth sibling (ibid.).

One interesting variant, supplied to Best by Nepia Pohutu (d. 1882) from the Wairarapa, substitutes *Uru-te-ngangana* for *Turangi* and has him marrying two different women. From the first, *Hine-te-ahura*, he begot *Te Ra-kura* (the red Sun) and *Te Marama-i-whanake* (the waxing Moon), and from the other, *Hine-turama*, came the stars (Best 1955: 10–11).

Various other options are mentioned by Best (1955: 11–13), providing just grounds for claiming that regional variations in Maori 'star lore' did indeed occur in pre-European times.

Originally, *Tane*, *Turangi* and *Moe-ahura* lived in darkness, but through the birth of the *whanau marama* (children of the light) he brought night and day to the world. It was *Tane* who was responsible for placing his nieces and nephews up in the sky, across the body of their grandfather, *Ranginui*. In a Bay of Plenty version of the myth, *Tane* began by positioning *Hinatore* on *Ranginui's* chest, but

> Feeble indeed was the light emitted … and darkness held fast. Tane procured the stars, and now dim light was seen. He next brought the moon, and light became stronger. Then Tane placed the sun on high, and bright light entered the world. (Best 1955: 13).

Other myths assign different ancestors with the important task of placing the stars in the sky. In one of these, *Tama-rereti* took them aboard his canoe for transportation. The basket for the stars was named *Te Ikaroa* (the Milky Way), and instead of going into a basket with all the other stars, *Atutahi* (Canopus) clung to the outside. This explains why it now resides beyond the bounds of the Milky Way (Best 1955: 14).

Because of the conflicting accounts, the precise identity of the Milky Way is unclear. In some accounts *Te Ikaroa* is equated with *Turangi*, and in others is identified as his younger brother (Best 1955: 10–15). But in all versions, *Te Ikaroa* was charged with their well-being and "… was placed in the middle of the little suns (stars) in order that he might protect and cherish them." (Best 1955: 14).

The Milky Way and two guardians of the seasons were then appointed to lay down the courses of the different heavenly bodies in order to regulate the seasons, and so that they would not end up interfering with one another. *Te Ikaroa* was responsible for guarding the *ara matua* (the main road), which is the path followed by the Sun, Moon and the planets (Best 1955: 15).

2.6 Celestial Objects and Events, and Documentation of Maori Astronomical Knowledge

2.6.1 *Introduction*

In this section, we examine the range of objects and events that were known to have been visible in the skies over Aotearoa/New Zealand between AD 1200 and 1768, and then examine Best's account for any evidence of these.

We also look for evidence of regional variations in Maori astronomical practices and beliefs, to determine whether these would have been a feature of the geographically-discrete 'culture areas' that the late Dr Henry Devenish Skinner (1886–1978), long-time Director of the Otago Museum, found were typical of Maori culture at the time of early European settlement (see Skinner 1974: 18–26).

2.6.2 *The Sun*

The Sun is the most conspicuous star visible from the Earth, and is seen as vital for human survival. This was also the view of the Maori astronomer, who referred to the Sun as *Te Ra*, an ancient term found throughout Polynesia (and even further afield, in Egypt, though this should not be seen to imply cultural links between these two regions). Best (1955: 9) also mentions the terms *Komaru* and *Mamaru* for the Sun, but notes that they are seldom heard. *Ra kura*, the 'red Sun', is a term that was sometimes used to describe the Sun, and would have been most applicable at sunset. When the Sun was on the meridian it was known as *Poutumaro*, and two different personified names for the Sun were *Tama-nui-te-ra* and *Tama-uawhiti* (Best 1955: 16–17). Various songs included reference to the Sun. For example:

> *E whiti, e te ra, e maene ki te kiri*. (Shine, oh Sun, in pleasing manner at the skin [of man]). (Best 1955: 7).

Alternatively:

> *E to, e te ra, rehurehu ki te rua*. (Decline, oh Sun, and set in the abyss.) (ibid.).

Sunspots are a well-known photospheric phenomenon, and vary in abundance with a mean period of 11.1 years between successive maxima (Fig. 2.7). By systematically observing the Sun through thin cloud or when it was near the horizon, Mossman (1989) has empirically established that near sunspot maximum, experienced observers should be able to see naked eye sunspots almost every day, and

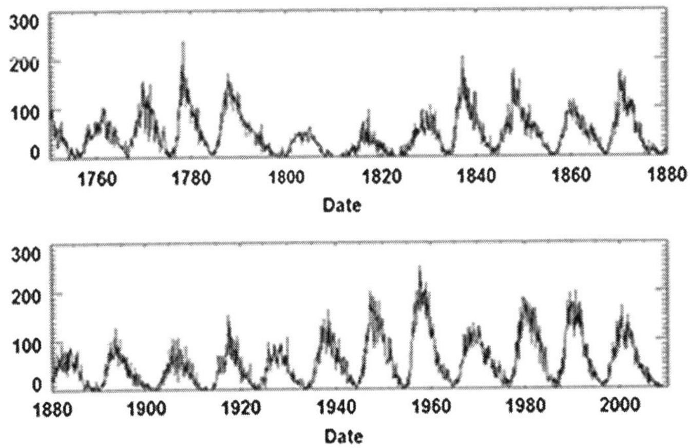

Fig. 2.7 A plot of monthly average sunspot numbers over the past 250 years illustrating the 11.1-year solar cycle (www.solarsystemcentral.com/sunspot_cycles_page.html)

even an inexperienced person would easily notice large spots (cf. Keller and Friedli 1992). Moreover, the general shapes of large individual sunspots or groups of smaller spots were discernible, which provides confirmation for some of the descriptions contained in Oriental records (see Clark and Stevenson 1978; Willis et al. 1996; Yau and Stephenson 1988), even if the actual numbers of spots reported greatly understated their actual abundance (Eddy 1987). Mossman (1989: 62) also found that spots were most conspicuous at sunset, "… when viewed against a naturally reddened disk …" Upon applying Mossman's parameters to Oriental records, Eddy et al. (1989) believe that sunspots were visible within ±3 years of each solar maximum. Assuming an average cycle period of 11.1 years, this would imply that there were about 52 solar maxima between AD 1200 and 1768, and that naked eye spots would have been obvious to the Maori astronomer.

This, of course, assumes that the sunspot cycle has not changed appreciably during the last 800 years, but we now know that such was not the case and that during the Maunder Minimum (from about AD 1645 to 1715) sunspots were absent or very rare (Eddy 1983; Hoyt and Schatten 1996; Ribes and Nesme-Ribes 1993). On the basis of Chinese and Korean sunspot records, Clark and Stephenson (1978) suggest that reduced sunspot numbers may also characterise the Spörer Minimum and the Medieval Minor Minimum, both of which also occurred during Maori times (see Fig. 2.8). Even so, there would still have been numerous occasions during the following intervals when Maori astronomers could have observed naked eye sunspots: AD 1200–1280, 1350–1420, 1550–1645 and 1710–1768. Yet no evidence of spotty imperfections or blemishes—of sunspots—exists in Best's account of Maori astronomy. *Te Ra* was perfection personified!

Grant (1992) has combined data drawn from the sunspot record, speleothems, tree-rings and erosional-alluviational sequences to reconstruct the palaeoclimate of Aotearoa/New Zealand for the last 2000 years, and he identifies two different

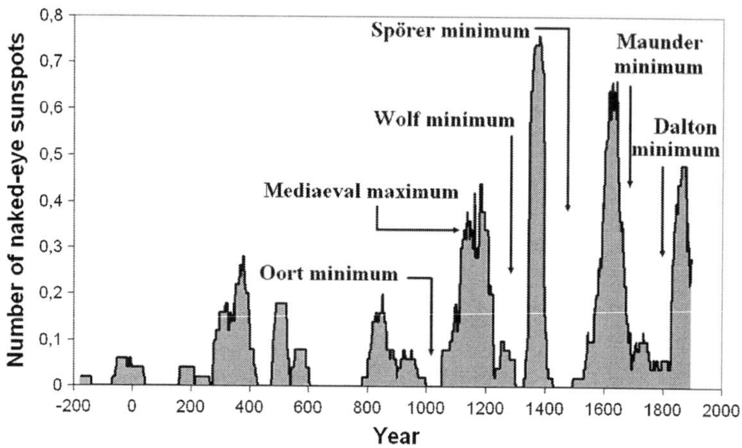

Fig. 2.8 A plot showing the incidence of naked eye sunspots and intervals with low to zero sunspot activity (after Vaquero et al. 1997)

periods of warmer weather during the pre-European Maori occupation of Aotearoa/New Zealand. They are the Warm Waihirere Period (immediately prior to the Medieval Minor Minimum) and the Warm Matawhero Period (between the Spörer Minimum and the Maunder Minimum). Each of these Periods was characterised by higher temperatures than at present, stormy weather that caused widespread gale damage to forests and coastlines, heavy rainfall and associated flooding. Maori mythology contains an account that could conceivably have been inspired by one, or both, of these warm Periods. Soon after the Sun was placed in the sky

> It was now found that the heat of the sun was unbearable. The body of the Earth Mother dried up and became dust; the eye of man could see naught. For at that period the body of Papa, the Earth Mother, was without covering. So now Tane said to Te Ikaroa (Milky Way), "Space out the courses of the little suns and the moon that we may obtain sleep. Move the sun forward, there to traverse his course, while you and the younger ones follow behind …" This was done … But the heat of the red sun was still intolerable, and all the offspring of the Earth Mother wailed aloud. Rangi was afflicted sorely by the great heat, and moaned in anguish; his head was scorched by the fierce rays of the sun … So the sun was removed to the back of Rangi, and all things were content. (Best 1955: 15).

Judging from Chinese analogies (see Xu 1987) and more recent statistics (Neidig and Cliver 1983), another naked eye phenomenon which a dedicated Maori astronomer was bound to see sooner or later was a white light flare. Data from the last century and a half suggest that during those intervals with sunspots, about 50 white light flares were potentially visible each century, but because no mention of them appears in Best's account there is no way of knowing whether any were witnessed by Maori astronomers.

Solar eclipses are rare and spectacular events (Fig. 2.9), and no fewer than 10 total eclipses and 16 annular eclipses were potentially visible from

Fig. 2.9 A drawing of a solar eclipse showing the pearly-white corona and four red/orange-coloured prominences (after Sands 1871)

Table 2.1 Total and annular solar eclipses visible from Aotearoa/New Zealand, AD 1200–1768

Total eclipse		Annular eclipse	
1301	August 5	1270	September 17
1409	October 9	1275	December
1430	February 23	1293	December 29
1463	November 11	1330	January 20
1491	November 2	1359	June 26
1545	December 4	1368	June 16
1610	June 21	1384	February 22
1735	April 23	1424	January 2
1739	February 8	1433	December 12
1748	January 30	1444	May 18
		1485	September 9
		1488	January 14
		1507	July 10
		1516	June 30
		1581	December 26
		1666	December 26

Aotearoa/New Zealand between AD 1200 and 1768. These are listed in Table 2.1 (after J. Meeus, personal communication, 1999), and each event would only have been visible for a few minutes, and in every case from a very localized region of the country. In addition to these 35 eclipses, a large number of partial solar eclipses were visible from Aotearoa/New Zealand in pre-European times.

In perusing Table 2.1, the century and a half from 1400 to 1550 stands out as being particularly rich in such events (there were 5 total eclipses and 7 annular eclipses), with less than one generation in each case separating the 4 different eclipses that occurred between 1424 and 1444, and between 1485 and 1507. Inclement weather probably prevented observation of some of these eclipses, but it is reasonable to expect that information about those that were observed, and others that occurred during the fifteenth century, would have spread by word of mouth.

Collectively, the eclipses listed in Table 2.1 would have provided Maori astronomers with ample examples of each type of phenomenon, yet Best (1955: 20) only records a single term for a solar eclipse, which is *Ra Kutia*. There is no indication whether this applies solely to total eclipses, or whether it also was intended to include annular, and even partial, solar eclipses.

Nor does the standard Maori explanation of solar eclipses provide any clarification: "An eclipse of the sun was caused by its being attacked and devoured by demons, from which attacks, however, it invariably recovers." (ibid.). Obviously these demons were not always successful in their attacks, given the comparative prevalence of partial solar eclipses. In this regard, it is interesting to note that there is considerable debate as to just how obvious partial solar eclipses would have been to people without optical aid (e.g. see Mostert 1989).

Currently, the most prominent naked eye feature of a total solar eclipse is undoubtedly the corona (see Fig. 2.9), and its form and extent varies in the course of a solar cycle. However, Eddy (1983) has shown that the corona was absent during periods without sunspots. Therefore, coronae would only have accompanied 5 or at most 6 total solar eclipses visible from Aotearoa/New Zealand between AD 1200 and 1769, and if weather conditions prevented observation of some of these events the corona would have been viewed but rarely. It is therefore no surprise that Best fails to mention the existence of the corona in his account.

Prominences (Fig. 2.9) were another conspicuous feature of total solar eclipses, but most would have been invisible or at very best inconspicuous to the naked eye, despite their distinctive red colour. Nor are they mentioned by Best.

A comparatively rare terrestrial atmospheric phenomenon was a solar halo (Fig. 2.10), and one form of this (Best suggests a sun-dog of several colours) was know to the Maori as *kura hau awatea*. Such haloes were believed to portend approaching bad weather. If the different colours were bright and distinct then the storm was nearby; a dim halo showed that the storm was still some way off. It was also believed that certain gifted men could generate solar haloes at will and use these as a means of communication with other Maori (Best 1955: 19).

The Sun was originally important in everyday Maori life. To quote Best (1955: 20–21):

> A considerable amount of respect was paid to the sun in Maori ritual performances … All higher classes of knowledge are connected with the sun; they emanated from Tane. The cultus of Tane represents the Maori form of sun-worship. It is marked by deference to Tane as representing the fertilizing-qualities of the sun, and by placatory gifts made to him. Thus all ritual formulae and offerings were made to the personified form of the sun.

Fig. 2.10 Example of a solar halo seen from Brocken (Harz), Germany on December 28, 2012 (https://en.wikipedia.org)

2.6.3 *The Moon*

The Moon is our nearest natural celestial neighbour, and its highlands and plains (*maria*) are obvious to the naked eye, as are its changing phases.

In everyday speech the Moon was known to the Maori as *Te Marama*, although the terms *Ahoroa*, *Atarau* and *Mahina* were occasionally used (Best 1955: 23). From time to time, the Moon appeared in the songs of the Maori, as in the opening lines of these two songs:

> *Tera te marama e ata haere ana* (Yonder the Moon drifts slowly along) (Best 1955: 7).

and

> *Tera te marama ka mahuta ake i te pae* (Yonder the Moon rises over the horizon). (ibid.).

Different terms were used to distinguish the phases of the Moon: *kua toriwha te marama* described the crescent Moon; *marama taiahoaho* the full moon; and *kua tohi te marama* the waning Moon. Best also lists other terms for unspecified phases of the Moon. Maori mythology provides a logical explanation for the phases. Unlike men, the Moon never dies but at certain times it

> … approaches its elder brother, the sun, and the two move together for a period. The moon belittles itself in the presence of its more important elder; its importance (brightness) is lost in the superior magnificence of the sun. After a time the moon leaves the sun behind; then it is said by men, 'The moon is again seen'. Best (1955: 25–26).

When the Moon reappears in crescent form, she has just re-enervated herself by bathing in the life-giving waters of *Tane*, and returns to view young and beautiful!

The personified form of the Moon was *Hina* or *Hine-te-iwaiwa*, and other variants were *Hina-uri*, *Hina-te-iwaiwa*, *Hine-nui-te-po* and *Hine-to-otaota* (Best 1955: 22). *Hina*

> … is said to have flourished in the days of the gods, and to have been a kind of patroness of the female sex and of all labours peculiar to women, such a weaving. Female children were dedicated to her, and, most significant of all, she presided over childbirth. (Best 1955: 22).

Best (1955: 22–23) specifically refers to "… the cult of Hine-te-iwaiwa …", and mentions that the Moon was believed to exert considerable influence on the birth of a child (Best 1955: 27). Meanwhile, Taylor relates that when the new moon appeared, women assembled and used the following lament to bewail those relatives who had died during the previous month:

> Alas! O moon! Thou has returned to life, but our departed beloved ones have not. Thou has bathed in the *waiora a Tane*, and had thy life renewed, but there is no such fount to restore life to our departed ones. Alas! (ibid.).

Our 'Man in the Moon' to the Maori was a 'Woman in the Moon' called *Rona*. One myth recounts that *Rona*

> … was originally a woman of this world … She was going to a spring for water one night with her gourd water-vessels when the moon became obscured, which caused her to apply a most offensive epithet to it. She was at once taken away by the moon, and she is still seen in it with *rururu taha*, or bundle of gourd vessels. (Best 1955: 25).

Sometimes *Rona* also was known by her longer name, *Rona-whakamau-tai* (*Rona* the tide-controller), which would indicate that the Maori knew of the connection between the Moon and the tides. Whether or not this reflects pre-European knowledge is a moot point.

Compared to eclipses of the Sun, total lunar eclipses were common, and hundreds of them were observed by Maori astronomers prior to the arrival of Europeans. Moreover, they were easily explained:

> The common view of an eclipse of the moon is that Rona, a malignant being, is attacking and destroying it. When the moon does not appear the twain are battling with each other, and so cannot be seen. After the combat, the moon bathes in the *waiora a Tane*, and so returns to us young and beautiful. (Best 1955: 24–25).

There are some similarities here to the disappearance of the Moon near the new moon phase.

One of the remarkable features of a totally eclipsed Moon is its colour, and it is interesting that Best does not allude to this. Those of us who have witnessed many lunar eclipses never know which particular hue the Moon will take on at totality. Red, pink, orange, golden and coppery-coloured Moons all occur (e.g. see Fig. 2.11), each of which may have had a different significance to the Maori astronomer.

Fig. 2.11 Examples of orange and pink moons during the lunar eclipses of August 28, 2007 and December 21, 2010 respectively (*Courtesy* John Drummond, Patutahi, New Zealand and https://en.wikipedia.org respectively)

Partial lunar eclipses were far less prominent than their solar counterparts, but Stephenson and Fatoohi (1994) found that they were regularly observed by Chinese and Arab astronomers nonetheless. There is no reason to suppose that they would not have elicited the same attention from the pre-European Maori astronomer. In their study, Stephenson and Fatoohi (1994: 93) noted an interesting feature, that "The magnitudes of small eclipses are consistently overestimated, while for large eclipses the reverse is generally true." (and they define "magnitude" as the maximum degree of obscuration of the Moon's disk). Since this seems to have been a universal phenomenon, it should also characterise the observations of the Maori astronomer.

One other Moon-related phenomenon which deserves to be mentioned is the lunar occultation. Typically between ten and twenty of these would be visible to the naked eye in any one year. Lunar occultations were observed by the Maori, and were reputed to have portended the outcome of military action. Best (1898: 233) was told by one of his informants that

> The moon represents a pa and the star a war party attacking that fort. The star knows all about the coming trouble. Should it pass to the other side of the moon [i.e. be occulted] the pa will fall. Just before the battle of Orakau we saw this sign, and we saw the star reappear at the other side of the moon. As we were a war party of course our warriors made much of this omen.

Much rarer were lunar occultations of planets. Although these would have been even more notable phenomena, Best is not forthcoming on how these were explained by the Maori.

2.6.4 *The Planets*

Apart from the Earth, five different planets were known to ancient peoples (Mercury, Venus, Mars, Jupiter and Saturn), and they stood out from the background stars by their motion along the ecliptic. Mars was also conspicuous because

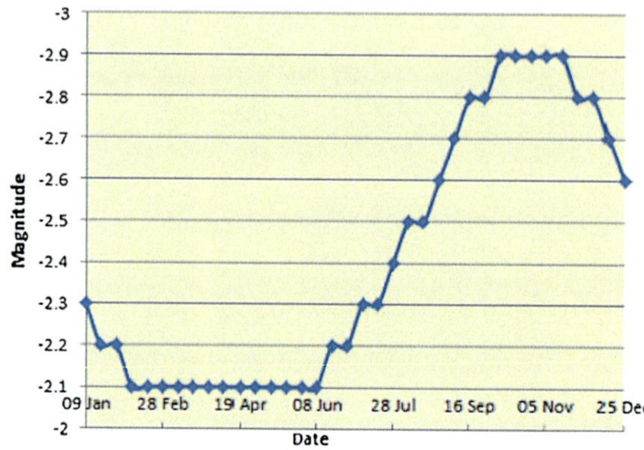

Fig. 2.12 This plot shows the way in which the magnitude of Jupiter varied in the course of just one year (in this case 2011), between −2.1 and −2.9 (www.sydneyobservatory.com.au/wp-content/…)

of its red colour. In addition, all five planets underwent noticeable changes in brightness in the course of a year (Fig. 2.12). Jupiter and Venus were always brighter than the brightest stars, and Mars sometimes was. Venus spent part of a year as the 'Evening Star' and the remainder as the 'Morning Star', while Mercury never seemed to be far from the Sun, low in the western sky at sunset or similarly positioned in the east before sunrise.

Among the myriads of stars in the night sky, the Maori astronomers also recognised these same five 'deviant' stars, which were referred to as *Whetu Ao* (Best 1955: 9). Mercury was called *Whiro*, while Mars was generally known as *Matawhero* although the name *Maru* has also been proposed (Best 1955: 49). Given the significance of the colour red in Maori society, it is remarkable that Mars' distinctive red hue had no particular meaning to the Maori.

Jupiter was generally known as *Kopu-nui* and perhaps *Parearau* (Best 1955: 40–42), but there is some controversy surrounding the latter name. Four Bay of Plenty Maori assigned it to Jupiter, but

> … Stowell says that it is Saturn; that Parearau is a descriptive name for that planet, and describes its appearance, surrounded by a ring. The word *pare* denotes a fillet or head-band; *arau* means "entangled" - perhaps "surrounded" in this case, if natives really can see the *pare* of Saturn with the naked eye. If so, then the name seems a suitable one. (Best 1955: 43).

In fact, Saturn's ring is not visible to the naked eye, even to those with acute vision, so if this name does indeed refer to Saturn and its rings then it must be of nineteenth century derivation, when telescopic studies of the planet and its rings were very much in vogue (e.g. see Proctor 1882).

Parearau, whatever her true identity, was also sometimes known as *Hine-i-tiweka*, and was often spoken of as a companion of *Kopu* (Venus) and in one version as the wife of *Kopu* (Best 1955: 43–44). In another variant, far from being a wife, she is a widow! Best (1955: 44) recounts that "Seafarers consulted *Parearau* when a storm was threatening, for if she appeared to be of a light misty aspect the storm would pass by."

When the Italian astronomer Galileo Galilei (1564–1642) first viewed Jupiter through a telescope on 7 January 1610 he was amazed to observe four small adjacent stars which he quickly came to realise were satellites of the giant planet (Fig. 2.13). They are now termed the 'Galilean satellites'. During the nineteenth century, William Colenso (1811–1899) was told that some Maori astronomers had such phenomenal eyesight that they could see these four moons of Jupiter (Best 1955: 35). Peak (1958: 256) gives the mean visual magnitudes of Io (I), Europa (II), Ganymede (III) and Callisto (IV) at opposition as 5.5, 6.1, 5.1 and 6.2 respectively, and notes that "… I and III would be comparatively easy naked eye objects, were it not for the proximity of the brilliant planet; II and IV would be near the limit but might be seen on occasions …" (ibid.). These comments by an authority on visual observations of Jupiter must throw doubt on Colenso's account—which, after all, was obtained long after the existence of the Galilean moons had become common knowledge.

Best known and brightest of all the planets was Venus, which unlike the other *whetu ao* was blessed with a number of names. A Tuhoe informant from the Ureweras told Best (1955: 50):

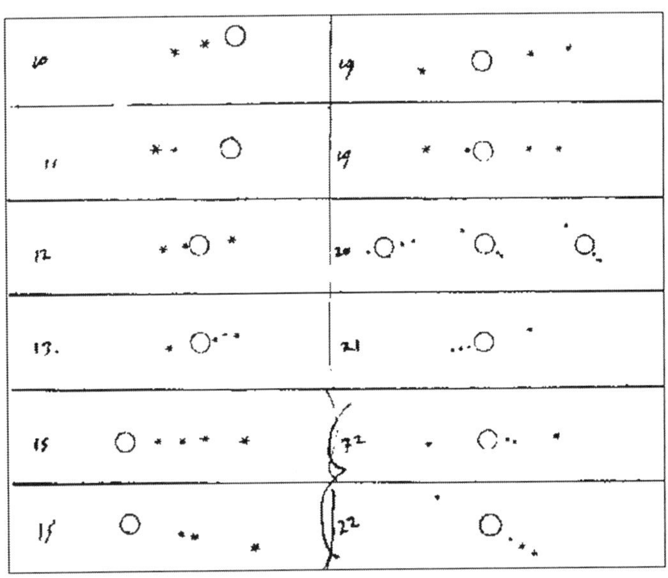

Fig. 2.13 Sketches that Galileo made of Jupiter and the changing positions of its four (Galilean) moons from night to night (www.astro.umontreal.ca/~paulchar/grps/site/images/galileo.4.html)

> … Venus has three names - Kopu, Tawera, and Meremere. As an evening star in summer it is called Meremere-tu-ahiahi; in the winter, as a morning star, it is Kopu. In other districts Venus as a morning star is called Tawera; as an evening star, Meremere and Meremere-tu-ahiahi.

Other terms listed by Best (1955: 40) are *Rangi-tu-ahiahi*, *Rere-ahiahi*, *Tu-ahiahi*, and possibly *Puaroa*—although Best (1955: 50–51) has some doubts regarding this last name. The first three names all relate to Venus as an evening star.

Of all the planets, Venus was most revered for her beauty, and what better way of complimenting an attractive woman than to describe her by way of the following popular saying:

> *Me te mea ko Kopu e rere i te pae.* (Like Venus as she appears above the horizon.) (Best 1955: 50).

Yet in Maori mythology, Venus as a morning star is a male (Best 1955: 51).

Conjunctions of two or three of the five naked eye planets are beautiful events (especially if the Moon also is in the region), and many of these must have been observed by Maori astronomers.

Much rarer were quintuple planetary groupings, when all five known planets were visible together in the sky within a circle of 25° diameter or less. These events were seen as portends of disaster in Europe, and the spectacle would also have drawn Maori concern and comment (though Best is silent on this topic). Quintuple planetary groupings that occurred during pre-European Maori times are listed in Table 2.2 (after de Meis and Meeus 1994).

Although these dates relate to the closest approaches of the planets, all five planets would have been in comparatively close proximity for several successive days—certainly long enough to guarantee their observation regardless of prevailing weather conditions. It is a safe assumption that all seven planetary groupings would have been seen, marvelled at, and remarked upon by Maori astronomers.

However, given the spacing of these apparitions, and assuming an average Maori lifespan of 31–32 years, most quintuple planetary groupings would have been once-in-a-lifetime events, except for those lucky enough to witness the 1564–1584 pair of events.

Table 2.2 Quintuple planetary groupings, AD 1200–1768

1284	December 23
1483	October 23
1524	February 19
1564	June 21
1584	May 1
1624	August 26
1662	December 6

2.6.5 Comets

Comets, those dreaded 'bearded stars' to some ancients, are commonly referred to in the historical records (e.g. see Hasegawa 1980), but precisely which ones were observed by Maori astronomers is not known. Burnham (2000), Hughes (1987), Seargent (2009) have written about 'great' or 'remarkable' comets, which passed close to the Sun and the Earth, and consequently were conspicuous to the naked eye for several successive weeks or months.

By drawing on Hughes' paper, Guillemin's (1877) book and data in Hasegawa (1980), Hughes (1985), Jansen (1991), Kronk (1984), Marsden and Williams (1996), Vsekhsvyatskii (1964) and Yeomans (2007), I have been able to assemble a list of all naked eye comets with maximum apparent visual magnitudes brighter than +2 that were definitely visible from Aotearoa/New Zealand between AD 1200 and 1769 (Table 2.3). A minor shortcoming of this list is that not all of these comets may have still been in Aotearoa/New Zealand skies when at their brightest or when their tails were of maximal length, although those of 1577, 1652, 1668, 1686, 1689, 1695 and 1758 certainly were. However, Masse (1995: 465) reminds us that

> The incredible variable nature of comets cannot be overstated in terms of impact on early culture. A single comet can change from being short-tailed (or even non-tailed) to long-tailed … from single straight tails to branching, twisting, chain-like or even multiple tails … and from dull to brilliant over the course of a few days or even hours of time.

As well as the comets listed above, there would have been others of comparable brightness visible from Aotearoa/New Zealand which had faded beyond Hughes' magnitude threshold by the time they were detected by northern observers and so were omitted from his list—and hence from Table 2.3.

In addition, other fainter comets of less spectacular appearance would also have been visible to Maori astronomers. It is significant that Marsden and Williams (1996) list no fewer than 46 different non-periodic comets that were visible from the northern hemisphere between AD 1200 and the advent of the telescope, which gives some idea of the occurrence frequency of naked eye comets (as opposed only to 'remarkable' comets, of which there are just 11 in Table 2.3). Of the 46 comets, 15 (33 %) were observed for more than one month.

Naked eye comets certainly were known to the Maori, who generally referred to them as *Auahi-roa* or *Auahi-turoa*. The former translates literally as 'long smoke', and according to one legend, *Auahi-roa* was the son of *Te Ra* (the Sun) and was sent down to the Earth to supply the people there with fire. After reaching the Earth *Auahi-roa* married, and his wife subsequently gave birth to five mythical 'Fire Children' who produced fire for mankind. Best (1955: 66) succinctly summarises the connection:

> So when you see the comet in the heavens, know that it is Auahi-turoa, he who brought fire to mankind. And fire is often called Te Tama a Auahi-roa, or Te Tama a Upoko-roa (the son of Auahi-roa, or of Upoko-roa), because it is the offspring of a comet.

Table 2.3 'Great Comets' visible from Aotearoa/New Zealand, AD 1200–1768

Name	Designation	Maximum magnitude	Naked eye visibility (days)	Maximum tail length (°)
1222	1P/Halley	1–2	Visible in daylight	30
1264	C/1264 N1	0–1	~110	100
1301	1P/Halley	1–2		15
1378	1P/Halley	+1.2	44	>20
1402	C/1402 D1	−5	~82; visible in daylight	>15
1456	1P/Halley	−0.2	43	57
1468	C/1468 S1	1–2	82	30
1471	C/1471 Y1	−3	59; visible in daylight	>30
1500	C/1500 H1		>72; visible in daylight	10
1531	1P/Halley	+1.4	36	>15
1532	C/1532 R1	+1	120; visible in daylight	15
1533	C/1533 M1	0	78	15
1556	C/1556 D1	−2	73	4
1577	C/1577 V1	−7	87; visible in daylight	80
1585	C/1585 T1	−4	46	2
1607	1P/Halley	+1.6	23	Long
1618	C/1618 W1	+1 to +3	69; visible in daylight	104
1652	C/1652 Y1		26	8
1664	C/1664 W1	+1	~90	37
1665	C/1665 F1	−2 to −3	25	30
1668	C/1668 E1	1–2	28	37
1680	C/1680 V1	+1.5	~90	90
1682	1P/Halley	+1.6	26	30
1686	C/1686 R1	+1	>37	18
1689	C/1689 X1		~31	68
1695	C/1695 U1		22	40
1744	C/1743 X1	−5	~112; visible in daylight	90
1758	C/1758 K1	−3	~67	Short
1759	1P/Halley	−0.9	69	47

Best also gives a number of other names for comets, namely *Auroa*, *Manu-i-te-ra*, *Meto*, *Puaroa*, *Puihiihi-rere*, *Purereahu*, *Tiramaroa*, *Tunui-a-te-ika*, *Taketake-hikuoa*, *Upoko-roa*, *Wahieroa*, *Whetu* and *Whetu-kaupo*. Some of these probably were generic names of regional rather than national relevance; others were descriptive, for particular kinds of comets (e.g. those with prominent tails); and others again probably referred to specific comets. Hongi (1920: 27) is most likely referring specifically to its spectacular apparition in 1910 when he says that Comet 1P/Halley (Fig. 2.14) was known as *Awa-nui-a-rangi*, or the 'celestial river of light'.

Fig. 2.14 Halley's Comet photographed at the Yerkes Observatory on May 29, 1910 (https://en.wikipedia.org)

Two features of all bright comets which excite present-day observers are the nature and behaviour of the head (that is, the nucleus and the coma), and the form and evolution of the dust and ion tails. While the head does not seem to have been of much interest to the Maori astronomer, four of the different 'comet' names listed above specifically refer to their tails. *Meto* denoted a *whetu puhihi*, a star which emitted 'rays' (a tail), and Best (1955: 67) was told that "The rays or tail of Meto extend upwards … if its body be below the horizon, as a range of hills, its *puhihi* extend up above the horizon (*Ka hihi ake nga puhihi*)." The second term is *Puaroa*, which Best (ibid.) tells us, "… is said to possess or emit mist-like emanations, referred to by the name of *hiku makohurangi*, or misty tail." *Tiramaroa* referred to a comet with long *puhihi* (rays), which sometimes pointed upwards and sometimes downwards (Best 1955: 65), while *Tunui-a-te-ika* reputedly possessed a long tail (Best 1955: 68). The Great Comet of 1843 (C/1843 D1) certainly possessed a long tail (see Fig. 2.15) and was readily visible from Aotearoa/New Zealand. Maoris in the Wellington region "… hailed it as an evil omen, and commenced howling very pathetically." (*New Zealand Gazette and Wellington Spectator* 1843), but whether they—or other Maoris—specifically referred to it as *Tunui-a-te-ika* is not known.

Finally, for centuries comets were seen by the lay public in many areas of the world as ill omens—as 'harbingers of doom' (e.g. see Chambers 1910;

Fig. 2.15 A painting of the Great Comet of 1843 (C/1843 D1) by Charles Piazzi Smyth (https://en.wikipedia.org)

Proctor 1896). Generally, the Maori also saw comets as signs of impending doom. Thus, when *Taketake-hikuroa* was seen in the sky "… it was viewed as an evil portent for the tribe." (Best 1955: 68), and the appearance of *Tunui-a-te-ika* announced the death or impending death of some person so "… priestly adepts performed the *matapuru* rite, in order to avert the threatened evil, whatever it may be." (ibid.). But apparently not all comets portended disaster, for the appearance of *Meto* gave forewarning of a hot summer (Best 1955: 67).

2.6.6 Meteors and Meteorites

Meteors, those sand-sized pieces of cometary debris that light up when they penetrate the Earth's atmosphere, are visible on every clear night, more so if the Moon is absent. To the Maori, these transient celestial visitors were commonly referred to as *Kotiri*, *Kotiri-tiri*, *Matakokiri* or *Tumatakokiri*, and Best (1955: 69) states that the Bay of Plenty Maori used the name *Tamarau*.

Meteors were explained as wayward stars which wandered from their places in the sky, and were struck by their elder siblings, the Sun and Moon. Some Maori regarded the appearance of a meteor as portending the death of a chief, but others thought that if a meteor appeared to fly towards them this was a good sign. One old informant told Best (1955: 70) that

> Another ancestor is Tumatakokiri, who is seen darting at night. His appearance is that of a star flying through space. His task, as he so flies, is to foretell the aspect and conditions of the heavenly bodies, of winds, and of seasons. If he swoops downwards, the following season will be a windy one. If he just flies through space, a fruitful season follows; a season of plenty lies before the people.

Sporadic meteors (Fig. 2.16) were present throughout the year, but at certain times they paled into insignificance in the presence of shower meteors. A meteor shower occurs when the Earth passes through debris that has been ejected by a comet and spread around the orbital path of that comet (Jenniskens 2006). There are a relatively small number of very active meteor showers that occur every year, and a great many minor showers. When a meteor shower occurs, the associated meteors are seen to radiate from a common point, known as the 'radiant'. Major meteor showers are usually named after the constellations in which these radiants are located. In pre-European times, many showers would have been visible from Aotearoa/New Zealand in the course of any one year (e.g. see Rada and Stevenson 1992) and the major ones are listed below in Table 2.4. Most meteor showers typically had zenith hourly rates of between 5 and 50 meteors, but judging from twentieth century experiences, certain showers (such as the Pi Puppids and the Phoenicids) occasionally exhibited greatly enhanced activity (e.g. see Jenniskens and Lyytinen 2005).

A totally different scale of activity, however, was sometimes associated with the Leonid meteor shower. This shower currently is active each November, and is associated with Comet 55P/Tempel-Tuttle. Every 33 years there is a chance of

Fig. 2.16 Photograph of an atypical meteor—most meteors are white (*Courtesy* John Drummond, Patutahi, New Zealand)

Table 2.4 Major meteor showers visible from Aotearoa/New Zealand in pre-European times (*Courtesy* American Meteor Society)

Name	Maximum activity	Radiant		Maximum zenith hourly rate	Parent body
		RA (h, min)	Dec (°)		
Pi Puppids	Late April	07:20	−45	Variable	Comet 26P/Grigg-Skjellerup
Eta Aquariids	Early May	22:36	−00.6	60	Comet 1P/Halley
Southern Delta Aquarids	Late July–early August	22:42	−16.4	20	Comet 96P/Machholz, Marsden and Kracht Comet group complex
Alpha Capricornids	Late July	20:20	−10.2	5	Comet 169P/NEAT
Orionids	Late October	06:24	+15.5	20	Comet 1P/Halley
Southern Taurids	Early November	02:06	+08.7	5	Comet 2P/Encke
Northern Taurids	Mid-November	03:54	+22.5	5	Minor planet 2004 TG10 and others
Alpha Monocerotids	Mid-November	07:48	+01	Variable	Unknown
Leonids	Mid-November	10:16	+21.6	Variable	Comet 55P/Tempel-Tuttle
Phoenicids	Early December	01:12	−53	Variable	Comet 289P/Blanpain

greatly enhanced meteor activity for just the one year or two successive years, with zenith hourly rates of 1000 s or even tens of 1000 s of meteors (Littman 1998). When this occurs the sky literally 'rains meteors', providing a truly unforgettable experience (e.g. see Fig. 2.17). These short-lived events are termed *meteor storms*, and are only visible for a few days in any one of these 'storm years'. By searching through historical sources Dick (1998) was able to identify the occurrence dates of such storms, and these are listed in Table 2.5.

All of these Leonid meteor storms would have been visible from Aotearoa/New Zealand, weather-permitting, and for many Maori would have afforded an unforgettable once-in-a-lifetime spectacle. Even at its best, the point from which the meteors were seen to emanate would have been very low in the northern sky, with most meteors shooting up into the sky from this radiant point, and others disappearing below the horizon. The presence of many bright and colourful meteors would have added to the drama of the occasion.

Very bright meteors—whether sporadics or shower-derived—are known as fireballs, while exploding fireballs are termed 'bolides'. Such meteors never fail to excite present-day confirmed meteor observers and would have provided the same appeal for Maori astronomers. To the Maori, fireballs were personified as *Rongomai*,

Fig. 2.17 A nineteenth century woodcut of the 1833 Leonid meteor storm (https://en.wikipedia.org)

Table 2.5 Leonid meteor storms, AD 1200–1768 (after Dick 1998)

Date range		Where documented
1202	October 18–19	Middle East and China
1237	October 18–19	Japan
1238	October 18–19	Japan
1366	October 21–22	Europe and China
1532	October 24–26	China and Korea
1533	October 24–26	Europe, China, Korea, Japan
1566	October 25–27	China and Korea
1601	November 5–6	China
1602	November 6–7	China
1666	November 6–7	China
1698	November 8–9	Europe and Japan
1766	November 11–12	South America

and occasionally one of these would escape from the sky and crash to the Earth as a meteorite. Evidence of these may be indicated by Maori place names, for there is a locality near Wellington named Te Hapua o Rongomai, where *Rongomai* is said to have descended to Earth at some time in the past (Best 1955: 67).

Known Aotearoa/New Zealand meteorites are listed in Table 2.6 (updated from Orchiston 1997). Those identified as 'falls' in the second column are recent phenomena, but most if not all of the remainder must date to Pleistocene or Holocene times given their proximity in each case to the land surface when recovered and the

Table 2.6 Known Aotearoa/New Zealand meteorites

Name[a]	Discovery year	Discovery mode	Type	Weight (kg)
Manaia (Masterton)	1863	Farming	Stony	5.8
Makarewa	1879	Railway construction	Stony	2.3
Mokoia	1908 (fall)	Search	Stony	~5
Waingaromia	1915	Farming	Iron	9.2
Berhampore (Wellington)	Early 1920s (fall)	Search	Stony	Cricket ball size
Morven	1925	Farming	Stony	7
View Hill	1952	Farming	Iron	33.6
Dunganville	1976	Prospecting	Iron	50.2
Kimbolton	1976	Farming	Stony	7.5
'Otago'	19??	Laboratory analysis	Stony	?
Opotiki	1989 (fall)	Search	Stony	Fusion crust only
Ellerslie (Auckland)	2004 June 12 (fall)	Search	Stony	?
Whakamarama	2013 May 13	Search	Iron	?

[a]For localities see Fig. 2.2

types of sediments in which they were found. It is very likely that some of these fell during Maori times and that their impacts were witnessed.

This table only lists recovered meteorites, yet we know of other meteorites that have definitely fallen in recent times but have not been recovered (despite systematic searches, including those carried out by the author of this book and his research associates). Moreover, the frequency of meteorite finds in agricultural areas of the USA, Canada and Europe and the areal extent of comparable agricultural land in Aotearoa/New Zealand indicates that Table 2.6 in no way reflects the true meteorite population. Clearly, many more meteorite impacts occurred between AD 1200 and 1768 than the mere handful reflected in this table.

Meanwhile, in more recent times, Maori oral histories record one example of what appears to be a meteorite impact:

> … when the Pakakutu *pa* at Otaki was being besieged [during the 1830s] Rongomai was seen in broad daylight, a fiery form rushing through space. It struck the ground and caused dust to rise. (Best, 1955: 67).

From Smith's (1910: 517) account of the hostilities between the Ati Awa tribe and the Ngati Raukawa and their allies from the Waikato and Taupo regions in 1834 we can pinpoint the precise location of Pakakutu Pa: "… on the north side of the Otaki river, not very far from the sea-coast, and between there and the Rangi-uru (or Whakarangirangi) stream." This location lies on a narrow coastal flat between the Tasman Sea and the escarpment marking the rise of the Tararua Range of mountains, and if a meteorite impact in fact took place during the siege of the *pa* and was visible from the vicinity of the *pa* then this most likely occurred somewhere on the triangular-shaped coastal flat which extends from the sea near Paraparaumu and Waikanae and becomes progressively broader as one proceeds to the northeast. At Otaki this coastal flat is only 5 km wide but when one reaches Levin, ~40 km from Waikanae, it has broadened to about 15 km. It is notable that there has been very extensive disturbance of the terrain throughout this whole area since 1834, in the form of farming, house and building construction, the forming of roads and a railway, etc., but as yet no discovery of a meteorite has been reported.

2.6.7 Stars and Asterisms

On a clear moonless night away from city lights, the average astronomer can nowadays see about 2500 stars (Moore 1985: 9), down to a limiting apparent visual magnitude of between 6 and 6.5 depending upon age, experience and visual acuity. But arriving at a proper understanding of Maori 'star-lore' is a considerable challenge. In the first place, Best (1955: 29–30) found accounts written by others of little use:

> There exists no monograph on the subject of Maori star-lore - no paper of any importance. Such matter as has been placed on record is in the form of brief or incomplete notes in a number of publications. Taylor's star-notes in *Te Ika a Maui* are sadly jumbled. Few men

> have been field-workers in Maori lore; thus many of the works dealing with such material simply contain rewritten data from previous publications.

And then when he tried to investigate Maori astronomy in the course of his own fieldwork, Best encountered serious difficulties:

> The fixing of Maori star-names is by no means always easy, for the average person among us needs a planisphere [or a star atlas] to refer to when making inquiries, and such is not always to hand. Nor is it often convenient to have one's native authority at one's side at night-time. (Best 1955: 29).

Compounding this was the fact that "Star-names differ, in some cases, among different tribes." (ibid.), reflecting the regional variations that occurred in Maori astronomical knowledge.

Despite these shortcomings, Best was able to assemble an interesting body of data. To the Maori, stars in general, were known as *Nga Whetu*, and occasionally as *Whanau Riki* and *Whanau Punga* (Best 1955: 9). All of the stars were collectively referred to as *kahui-o-te-rangi*, or "… the flock or assembly of the heavens …" (Best 1955: 42), and like the Sun, Moon and planets, the principal stars were assigned mythological attributes. Best (1898: 239) refers to these as 'high-born' stars, while "All the small stars were common people." High-born stars included Aldebaran, Altair, Antares, Arcturus, Canopus, Capella, Castor, Procyon, Rigel, Sirius, Spica and Vega, and all were known by one or more names (Best 1955: 38–42). Note that all of these are bright (first magnitude) stars.

In addition to naming individual stars, the Maori recognised various groupings of stars and assigned names to these. A few of these asterisms corresponded to our present-day constellations (e.g. the Southern Cross, Leo) or widely-recognised open clusters (such as the Hyades and the Pleiades—see Fig. 2.18), but others did not. Examples of the latter include part of Orion likened to a bird snare, a triangle of stars in Canis Major, and the tail of the scorpion in Scorpius (ibid.). According to

Fig. 2.18 The Maori grouping of stars known as *Matariki* (the Pleiades). The blue nebulosity round the brighter stars is common on photographs, but was not obvious to pre-European naked eye observers (*Courtesy* John Drummond, Patutahi, New Zealand)

Best, the Southern Cross had no fewer than six Maori names; Orion's Belt five; the Pleiades four; and the Hyades three. Some of these names were regional variants. The Maori recognised that stars in some of these groups were circumpolar and never set (Best 1955: 46).

Perhaps the most remarkable of all Maori constellations was the fabled 'Canoe of Tama-rereti' which *in toto* extended all the way from Taurus and Orion to Crux! More specifically:

> The Pleiades form the bow of this starry vessel, and the three bright stars in Orion's Belt represent the stern. The sail, the Ra o Tainui, is perhaps the Hyades. The cable is seen in the Pointers [Alpha and Beta Centauri], and the anchor is the Punga a Tama-rereti, the Southern Cross … The position of the cable in relation to the far-flung anchor is somewhat unusual. (Best 1955: 60).

This was the mythical ancestral canoe of the Tainui people.

Of all the stars known to the Maori, Canopus and Rigel had a special place, because of their intimate links with *kumara* cultivation, but they also were used to foretell the weather (Best 1955: 42, 48). Canopus was known variously as *Aotahi*, *Atutahi*, *Atutahi-ma-Rehua*, *Autahi*, *Kauanga*, *Paepae-poto*, and possibly *Makahea* (Best 1955: 38), and Best (1955: 42) was told that "… its appearance is a sign for the task of planting the crop to be commenced." Rigel was generally referred to as *Puanga*, but another name for it was *Pua-tawhiwhi o Tautoro*, and in the South Island it was known as *Poaka* or *Puaka* (Best 1955: 39). The *kumara* was a dietary staple throughout Aotearoa/New Zealand to the north of Banks Peninsula, and this is reflected in the combined geographical distributions of modified soils, borrow pits (which supplied the soil), stone structures and slope trenches. Meanwhile, covered pits like those shown in Fig. 2.19 where designed for the successful storage of *kumara*.

Two prominent asterisms also were associated with *kumara* cultivation, and they were the Pleiades and the three stars comprising Orion's Belt (see Fig. 2.20). The latter were known variously as *Tautoro*, *Tata o Tautoro*, *Te Tira o Puanga*, *Te Tuke o Maui* and *Te Tuke o Tautoro* (Best 1955: 39). But of all the asterisms recognised by the Maori, the Pleiades held pride of place, just as they did in Polynesia and in many other parts of the world (e.g. see Ammarell 1987; Ceci 1978; Hugh-Jones 1982; Snedegar 1995).

To the Maori, the Pleiades were generally referred to as *Matariki*, but they also were known by three other names, *Ao-kai*, *Hoho-kumara*, *Te Huihui o Matariki* (Best, 1955: 40)—and the first two of these alternatives clearly reflect their horticultural affinities. Whatever the name used, their role was to stay together and paddle their canoe across the sky (across the body of *Ranginui*), whilst ensuring that mankind was supplied with food. The primary 'food' was of course *kumara*, and Best was told that

> Our forebears consulted those sign-giving stars in connection with the planting of the *kumara* crop. The principal stars so relied on were Rigel, the Pleiades, Orion's Belt (Tautoro, and Whakaahu [Castor]. According to the manner of their rising, the crops would be planted early or late. (Best 1955: 48).

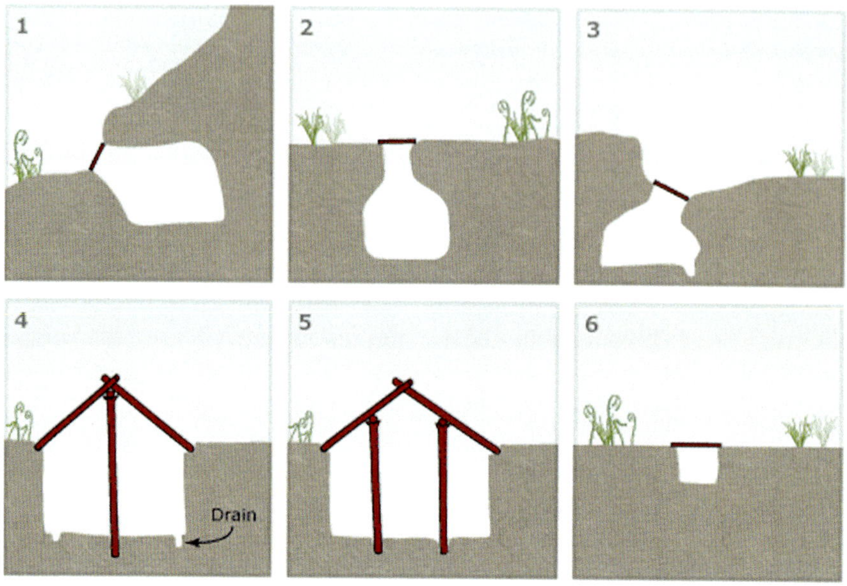

Fig. 2.19 Different types of *kumara* storage pits (after Davidson 1987)

As well as its horticultural connotations, the appearance of *Matariki* heralded the start of the birding season: "An old saying is, "When Matariki is seen, then game is preserved"; for it marked the season when such food-supplies have been procured and preserved in fat in certain vessels." (Best 1955: 53). In addition, the arrival of *Matariki* indicated that it was time to go in search of the lamprey, *Geotria Australia Gray* 1851. At first sight lampreys can be mistaken for eels (Fig. 2.21) except that they have a circular sucker *in lieu* of a regular mouth, lack a backbone and have seven gills behind the head. Lampreys spend most of their lives at sea, but when it is time to breed they return to rivers and especially areas near river mouths. If their current distribution (Fig. 2.22) reflects their availability in pre-European times, they would have been a popular seasonal food source for Maori people.

There is one further application of Maori astronomical knowledge which we must discuss, and this is oceanic voyaging. I have suggested elsewhere that "After the Vikings, they [the Polynesians] were the world's most accomplished mariners. Two-way voyaging was endemic throughout the central Polynesian region, and celestial objects were habitually used for navigation." (Orchiston 1996: 324). But just how closely the Polynesian norm applied to pre-contact Aotearoa/New Zealand is debatable. It is now apparent that the first settlement of Aotearoa/New Zealand occurred as a result of a planned voyage involving systematic exploration in a southwesterly direction, but there is considerable debate as to the extent of two-way voyaging between Aotearoa/New Zealand and island Polynesia following this initial settlement (e.g. see Finney 1992: 68–73). If two-way voyaging was rare—as some authors suggest—then much of the 'traditional navigational knowledge' of

Fig. 2.20 A photograph showing the two asterisms associated with *kumara* cultivation, *Matariki* (1 the Pleiades) and *Tautoro* (2 the three stars in Orion's Belt). Also included in this photograph are three 'high-born' stars, *Puanga* (3 Rigel), *Putara* (4 Betelgeuse) and *Taumata Kuku* (5 Aldebaran) (*Courtesy* John Drummond, Patutahi, New Zealand)

the Maori (e.g. see Best, 1954) could have been obtained from fellow-Polynesians by Maoris who served on whaling, sealing and other European vessels during the nineteenth century.

Fig. 2.21 The lamprey (www.niwa.conz/our-science/freshwater/tools/fishatlas/species/fish-species/lamprey)

This brings our account of Maori 'star-lore' to a close. What is remarkable is that so little was recorded, for of the 2500 naked eye stars that were visible to the Maori, Best was only able to supply names for 12! We have already alluded to the problems that non-astronomers faced in trying to identify different stars and asterisms, and Best (1955: 65) succinctly sums up the situation when he writes: "Such are the Maori star-names collected, and a poor showing it is, compared with what might have been obtained, for so few have been identified."

2.6.8 Variable Stars

To the Maori astronomer most stars were unchanging in relative position, brightness and colour from week to week or month to month, but they must have noticed that just a few altered appreciably in brightness. These are 'variable stars', and we now know that there are very large numbers of them in our Galaxy. There are various classes of variable stars (see Kholopov 1985), and most exhibit cyclic variations in brightness with well-defined periods and magnitude variations.

However, few variable stars remain within naked eye range throughout their full cycle, or if they do vary sufficiently in brightness for this to be apparent to a naked eye observer. Nowadays, variable stars are a favourite quarry of the amateur astronomer, and experienced observers are able to reliably discern a magnitude variation of 0.1.

Table 2.7 provides a list of naked eye variable stars whose magnitude variations would have been conspicuous to any observer (see Figs. 2.23 and 2.24), but on the basis of the limited information supplied by Best there is no evidence that Maori astronomers actually systematically monitored any of these stars. Indeed, Best is silent on the very existence of variable stars!

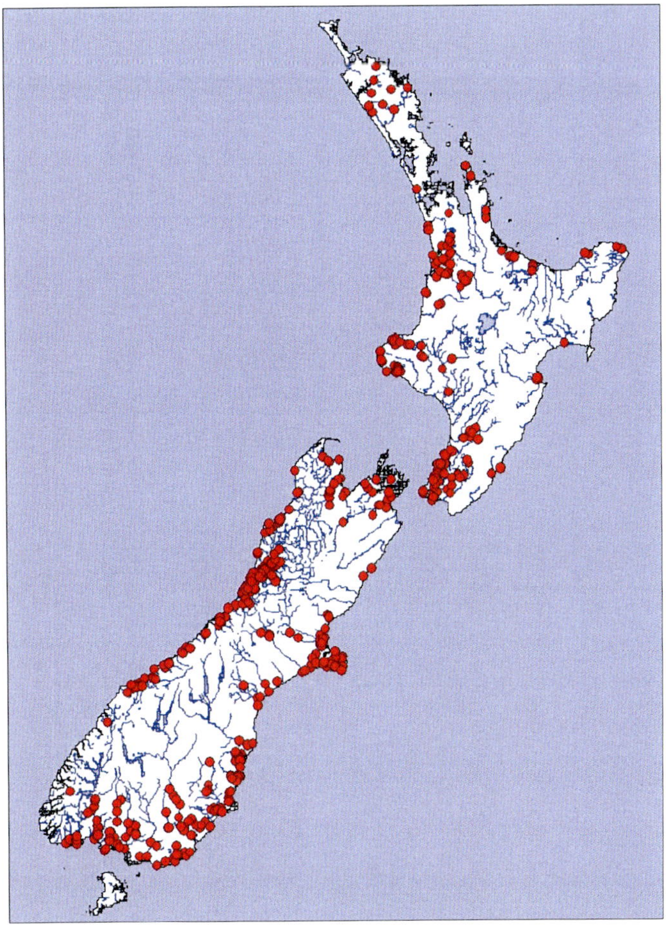

Fig. 2.22 Current distribution of lampreys (www.niwa.co.nz/our-science/freshwater/tools/fishatlas/species/fish-species/lamprey)

Apart from variable stars, which would have been constant companions of the Maori astronomer, from time to time he would have encountered two different types of temporary stars, or 'guest stars' as they were termed by the Chinese: supernovae and novae.

Liller (1992: 28) explains the cause of supernovae:

> The supernova explosion, catastrophic, or nearly so, to the star, results from the collapse of a very massive star that follows when it runs out of nuclear fuel ... when there is no more energy being generated at the star's centre, the outer layers fall in. Some supernovae (Type II) appear to be isolated supergiant stars that evolve naturally to this brilliant flash finish; others (Type I) are theorised to follow the pattern of novae involving a white dwarf, but ones that grow disastrously from an immense over-accumulation of material from a close-by companion star. Whereas novae recur over and over again, a supernova is reduced, once and for all to a super-compact neutron star and a rapidly expanding cloud of gas that once made up the star. Sometimes no visible star is left at all.

Table 2.7 Examples of conspicuous naked eye variable stars visible from Aotearoa/New Zealand (after Norton 1957)

Star name	Magnitude variation	Period (days)	Popular name
η Aquilae	3.7–4.5	7.2	
l Carinae	3.6–5.0	35.5	
o Ceti	1.7–9.6	331	Mira
α Orionis	0.5–1.1	?	Betelgeuse
κ Pavonis	4.0–5.5	9.1	
L2 Puppis	4.6–6.2	140	
λ Tauri	3.3–4.2	3.9	

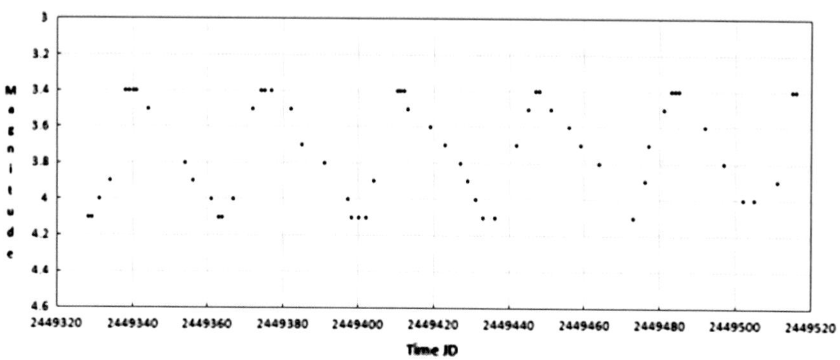

Fig. 2.23 Rod Stubbings' light curve of l Carinae (http://rodstubbbingsobservatory.wordpress.com/lightcurves/classical-cepheid/l-car)

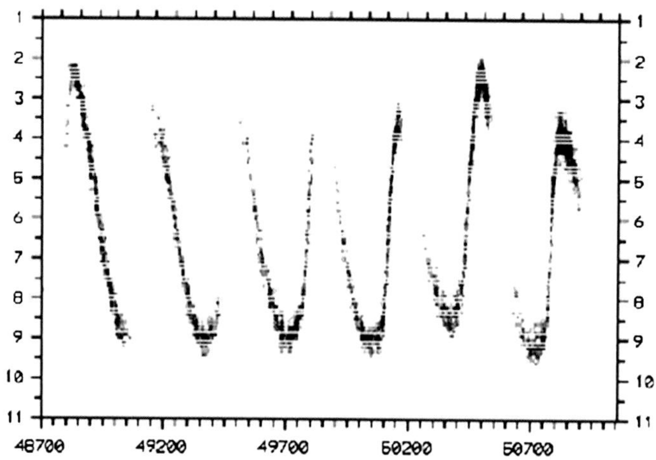

Fig. 2.24 The AAVSO light curve of o Ceti (www.hyperphysics.phy-astr.gsu.edu/hbase/starlog/ceti.html)

In all supernova explosions the total energy output is prodigious. In ancient times, not only were supernovae conspicuous by their brightness, but also by their longevity, for most remained visible to the naked eye for more than a year.

Because of their prominence, historic supernovae were widely observed by Chinese, Japanese, Korean and European astronomers, but only one of these is definitely known to have been visible from Aotearoa/New Zealand during the past 800 years. This was SN 1604 (also referred to as 'Kepler's Supernova') in Ophiuchus, which was first observed on 8 October 1604, and remained visible for 12 months. It was recorded by Chinese, Korean and European observers (Stephenson and Green 2002). However, in all probability SN 1604 was joined by a small number of far southern supernovae visible to the Maori, but for which no Arabic, Asian or European records exist.

Unlike a supernova, a nova is a star which rapidly increases in brightness over just a few hours and then slowly fades, usually over several months (see Fig. 2.25). Not all novae reach naked eye visibility, but the light curve follows the same pattern as for a supernova, although the magnitude increase is much less. It is believed that novae are associated with close binary stars, where one component is a white dwarf and the other a red giant. The two stars are so close together that matter can flow from the giant to the dwarf star, and after this has occurred for a time the accumulated material is ignited explosively and some of it is blasted into space. When this happens we see a nova (e.g. see Zeilik 1985).

Stephenson and Green (2002) surveyed the historical records of the past 2000 years, and found evidence of numerous 'new stars'. A few of these were supernovae, but the majority undoubtedly were novae. Those that were visible from Aotearoa/New Zealand are listed in Table 2.8, and all would have been obvious to the Maori astronomer even though the one at +20° declination would have been low in the northern sky.

This has to be regarded as a minimal list for Ho (1962) includes many more 'guest stars' and 'new stars' drawn from Korean and Japanese sources, but unfortunately he does not pinpoint their positions so it is not known how many of these would have been accessible to Aotearoa/New Zealand observers (although undoubtedly some were).

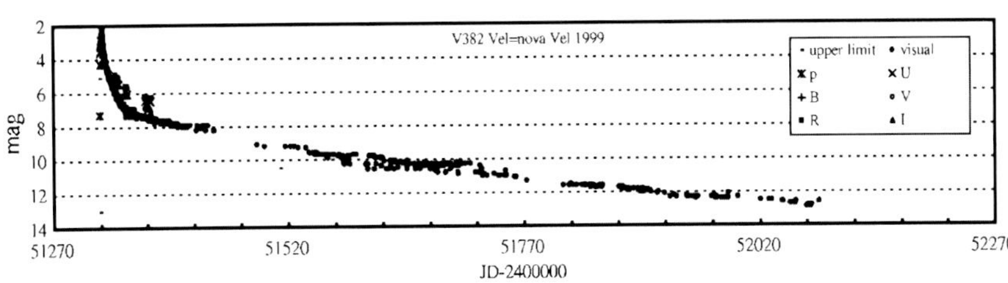

Fig. 2.25 The light curve of Nova Vel 1999, an example of a classical nova that was a naked eye object at maximum (after Kiyota et al. 2004: Fig. 1)

Table 2.8 Historic novae visible from Aotearoa/New Zealand, AD 1200–1700 (after Ho 1962; Hsi 1958)

Date recorded	Constellation	R.A.	Dec.
1224 July 17	Scorpius	17	−40
1240 August 17	Scorpius	17	−40
1388 March 29	Pegasus-Andromeda	00	+20
1430 September 9	Canis Minor	07 20	+07
1431 January 4–April 3	Eridanus	05	−10
1461 July 30–August 2	Ophiuchus	17 40	00
1584 July 11	Scorpius	15 40	−25
1676 February 18	Eridanus	04	−10
1690 October 18	Sagittarius	18	−30

In addition to these, and the novae listed in Table 2.8, it is certain that there were other novae that were visible from Aotearoa/New Zealand but were beyond the view of Chinese, Japanese and Korean astronomers. This is indicated by the fact that novae tend to be most abundant in the general region of the Milky Way (see Fig. 2.26), which is particularly conspicuous in the region of Carina, Crux and Centaurus.

Even without allowing for all these additional novae, the fifteenth century stand out for the relative abundance of these 'new stars', but more than this, the occurrence of the novae of 1430 and 1431 within just four months of one another would have been of special note to the Maori astronomer. Further, there are several other

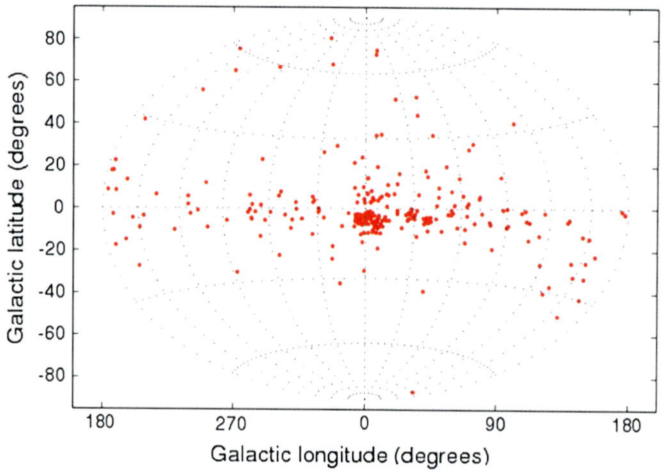

Fig. 2.26 A map showing the distribution of classical novae, and their concentration along or near the plane of our Galaxy (after Richmond 2007)

instances represented in the table where two or more novae would have been observed in the course of a typical human lifespan of 31–32 years.

Clearly novae, and to a much lesser extent supernovae, would have been notable intruders in an otherwise largely unchanging night sky, and would have featured in the star lore of the Maori. Yet the only possible references to 'temporary stars' or 'new stars' in Best's account of Maori astronomy relate to *Mahutonga*, which is described as "… a star of the south that remains invisible…" (Best 1955: 46). Other names which include the term *Mahu* refer to the Southern Cross and the Coal Sack, suggesting that this 'invisible star' was located in or near Crux (and therefore on the very southern visibility limit for Chinese, Japanese and Korean astronomers). Given the relative prevalence of novae, the Maori reference to *a single star* could suggest that a supernova was involved (see Green and Orchiston 2005, and Chap. 3).

2.6.9 Nebulae

The only gaseous nebulae *clearly* visible to the naked eye in the southern skies on a moonless night are the dark nebulae scattered throughout the Milky Way (Fisher 1963). The most conspicuous of these is undoubtedly the Coal Sack (Fig. 2.27), flanking Crux (the Southern Cross).

Best (1955: 39–41) lists six different Maori names for the Coal Sack, which are *Manako-uri*, *Naha*, *Te Patiki*, *Te Rua-patiki*, *Te Rua o Mahu* and *Te Whai-a-titipa*, but nowhere does he expound on how this feature was interpreted in Maori

Fig. 2.27 A photograph showing the Coal Sack, and the adjacent constellation of Crux, a recognised Maori asterism with multiple names (*Courtesy* John Drummond, Patutahi, New Zealand)

Fig. 2.28 *Mangaroa*, the Milky Way, also was known by a number of other Maori names (*Courtesy* John Drummond, Patutahi, New Zealand)

mythology. However, Sirius is said to have made his way into the sky "… *down through* the Coal Sack." (Hongi 1920: 26), which suggests that the Coal Sack was regarded as a hole in the heavens, similar to a viewpoint that at one time also was prevalent in Europe (e.g. see Pannekoek 1961: 476).

Other 'holes' or 'dark patches' in the sky were scattered along the length of the southern Milky Way (see Fig. 2.28), and had great importance to some indigenous cultures. For example, the Aboriginal Australians linked a succession of them mentally to identify an 'emu in the sky' (e.g. see Fuller et al. 2014). Unfortunately, there is no evidence that the Maoris interpreted dark nebulae in this way and saw 'a moa in the sky'!

2.6.10 The Milky Way and the Magellanic Clouds

The Milky Way, that amazing concentration of stars, dust and gas, is seen at its best in the southern sky, as beautifully depicted in John Drummond's photo mosaic presented in Fig. 2.28. Telescopic examination promptly reveals a plethora of star clusters and double stars, but pre-European Maori astronomers were not afforded this privilege.

Nonetheless, the Milky Way was a conspicuous feature in the Maori night sky, especially on clear moonless nights, and was known by numerous names (see Table 2.9). Of all of these options, Best (1955: 45) informs us that *Mangaroa* was the most commonly-used. Other names, like *Te Kupenga a Taramainuku* and *Whitikaupeka* were regional terms, and it is likely that a number of the other names listed also were regionally-specific. In addition, some similar terms may reflect regional dialect variations, that are found in Maori language (see Harlow 1979).

According to Maori legend, the Milky Way was one of the offspring of *Ranginui* and *Papatuanuku*, and its role—as we have seen—was to care for the stars.

The Large and Small Magellanic Clouds (see Fig. 2.29) were seen as separate entities to the Milky Way, despite their similar appearance, and a number of different names were applied to them. As Table 2.10 indicates, some of these terms

Table 2.9 Different Maori names for the Milky Way (after Best 1955)

Name	Region/source
Te Ika a Maui	
Te Ika-matua a Tangaroa	
Te Ika-o-te-rangi	
Te Ikaroa	
Te Ikaroa-o-te-rangi	
Te Ika-whenua-o-te-rangi	
Te Kupenga a Taramainuku	Ureweras
Te Mangoroa	
Mangoroiata	
Mokoroa-i-ata	
Te Paeroa o Whanui	
Te Tuahiwi-nui-o-rangi	
Tuahiwi o Rangi-nui	
Whiti-kaupeka	South Island

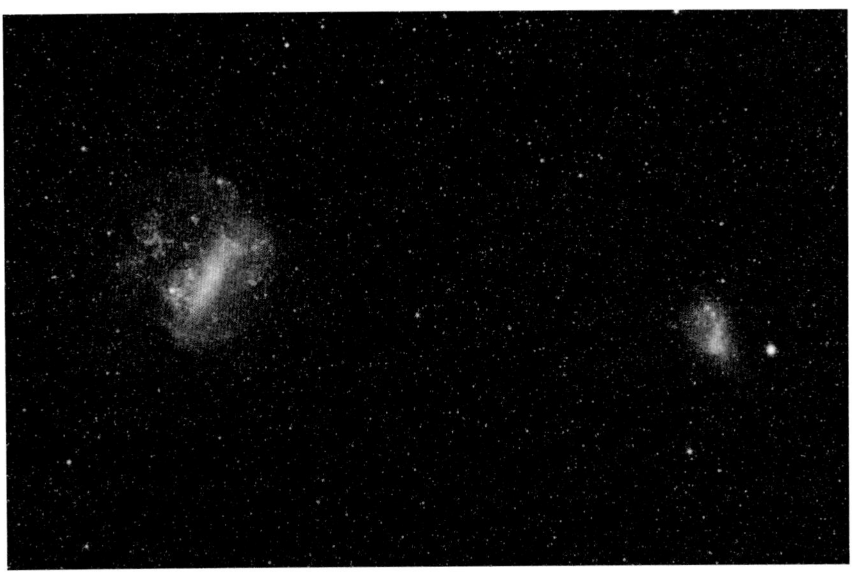

Fig. 2.29 Maori astronomers recognized the Large and Small Magellanic Clouds and many different names were assigned to them, as Table 2.10 indicates (*Courtesy* John Drummond, Patutahi, New Zealand)

applied collectively to both Magellanic Clouds and others to either the Large or the Small Magellanic Clouds. Again, there is evidence here of regional variations in terminology.

Table 2.10 Different Maori names for the Magellanic Clouds (after Best 1955)

Name	Association[a]	Source/region
Nga Patari	The MCs	
Nga Pataritari-hau	The MCs	
Nga Patari-kai-hau	The MCs	
Nga Patari-hau	The MCs	
Te Purangi	The MCs	Williams
Whakaruru-hau	The MCs	
Kokirikiri	LMC	White
Patari-rangi	LMC	Williams
Rangi-matanuku	LMC	White
Tioreore	LMC	
Patari-kaihau	SMC	Williams
Tikatakata	SMC	
Ao-tea	One of the MCs	Stowell
Ao-uri	One of the MCs	Stowell
Kokouri	One of the MCs	
Kokotea	One of the MCs	
Manako-tea	One of the MCs	White; Williams
Manako-uri	One of the MCs	White
Nonoko-tea	One of the MCs	Taranaki
Nonoko-uri	One of the MCs	Taranaki
Tiripua	One of the MCs	Williams
Tiritiripua	One of the MCs	
Tuputuputu	One of the MCs	
Pioriori	Upper MC	White

[a]Key MC = Magellanic Cloud; LMC = Large Magellanic Cloud; SMC = Small Magellanic Cloud

The large number of names employed for these two systems probably demonstrates the importance of the Magellanic Clouds in Maori society, but unfortunately, Best does not divulge precisely what their importance was.

2.7 The Astronomy-Meeting House Connection, and Astronomical Motifs in Maori Rock Art

Thus far, our account of Maori astronomy has been based solely upon the oral record, but in other pre-literate societies around the world astronomical information also was preserved in rock art, house design and orientation, portable artifacts, clothing, and even on personal tattoos or temporary body decoration. With the exceptions of meeting house design and possibly rock art, there is little if any evidence that the Maori made regular use of any of these options.

2.7.1 Meeting Houses and Astronomy

Harris et al. (2013: 331) explain that according to tradition

> Māori meeting houses [see Fig. 2.30] always face the rising Sun. Houses needed to be positioned so that the Sun encroached upon the porch, and if this did not occur it was considered an *aitua*, or a sign of miscalculation that could lead to death … Where the Sun rises and sets are two primary spatial designations from which all subsequent orientations and calculations regarding building construction are made.

The actual internal design and decoration of meeting houses (see Fig. 2.31) also had astronomical dimensions:

> … central to the construction of the meeting house is the Milky Way, which in traditions is often referenced by its shape as a fish … During the Māori New Year in mid-winter and at the time of the winter solstice the *tāhuhu* [ridge beam of the meeting house] lies in the same position as the 'great fish' … in the sky. This significant representation is adorned as patterns on the ridge-beam of the meeting house. Extending from the ridge-beam … are the rafters which connect to wall panels, which are sometimes carved to represent ancestors. In pre-colonial times the rafters were often painted in patterns that represented shapes of star groupings that were observed in the sky. These patterns also related to foods and resources that were accessed seasonally by the tribe. (ibid.).

Fig. 2.30 Example of a Maori meeting house, in this case Tamatekapua at Ohinemutu, Rotorua (https://en.wikipedia.org)

Fig. 2.31 This photograph of the interior of the Waitangi Meeting House at the Waitangi National Reserve at the Bay of Islands includes the following architectural features: the *tāhuhu* (ridge beam along the *top* of the meeting house), the decorated rafters and the carved wall panels. In pre-European times the *tāhuhu* and rafters sometimes were decorated with astronomical motifs (https://en.wikipedia.org)

2.7.2 Are There Astronomical Motifs in Maori Rock Art?

Rock art is an important feature of many cultures, and was executed for a number of reasons. In some instances, the actual painting or engraving of the rock walls was imbued with symbolic significance, and once accomplished the final product was of no significance. Conversely, rock paintings were sometimes made at the time of a significant event or activity to commemorate its occurrence or observation. Thirdly, some rock art was prepared in order for the site to be used in future for ceremonies associated with the motifs depicted there.

Solar eclipses, comets, meteor storms and supernovae were all spectacular apparitions, and there is good reason for believing that these sometimes were depicted in the rock paintings and engravings of pre-literate peoples (e.g. see Ronan 1996: 250–251, and 'Physical representations of celestial objects' in Warner 1996).

During the first five centuries of South Island settlement, rock art was a feature of Maori culture, particularly in the limestone regions of north and south Canterbury and northern Otago (see Trotter and McCulloch 1971: 28). Many of the motifs have been interpreted as humans, dogs and moas, and appear to depict moa-hunting expeditions, but there are a small number of paintings which do not seem to relate to the natural landscape and some of these may represent celestial objects.

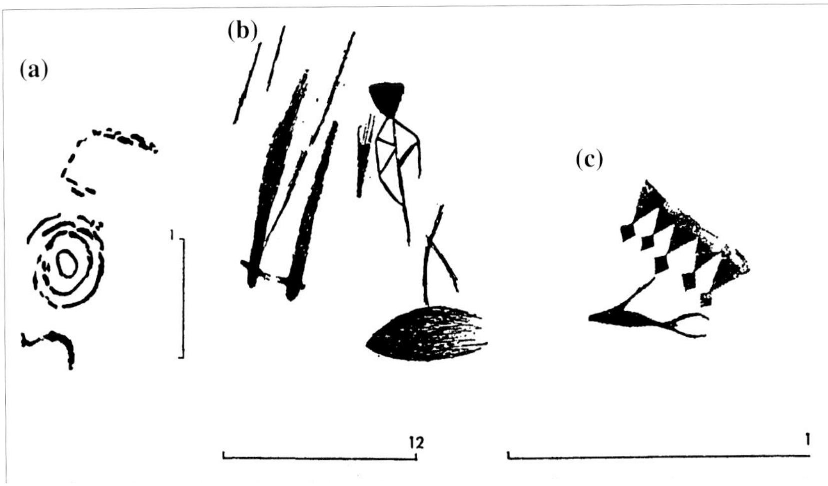

Fig. 2.32 Possible rock art depictions of celestial objects and events; scale in inches (after Ambrose 1970)

For example, Ambrose (1970) includes the concentric circles shown in Fig. 2.32a, which may possibly depict the Sun or a supernova. Similar concentric circles are found in other sites, but are rare (Trotter and McCulloch, 1971: 40). Of similar antiquity to the Fig. 2.32a painting are the six linear motifs on the left hand half of Fig. 2.32b, which may represent a meteor shower, the two most prominent objects being bolides. The sixth object in this group, on the right hand side also is reminiscent of a comet, and may portray a comet that was visible at the time of the meteor shower. Finally, Fig. 2.32c shows another composition from the same rock shelter, and the five aligned triangles with 'tails' may document the celestial movement of a conspicuous comet over a finite time interval. Obviously, all of these interpretations must remain speculative.

2.8 Maori Concepts of Time

The Maori developed their own calendar system, based in the main on the motion of the Earth round the Sun and on the movement of the Moon. In this way, they could identify seasons, months of the year, and days of the month. What they could not do, apparently, was count individual years: "The Maori had no tale of years as Europeans have …" (Best 1959: 11).

To the Maori, there were two main seasons of the year, summer (termed *Raumati*) and winter (*Takurua* or *Hotoke*), but autumn (*Ngahuru*) and spring (*Koanga* or *Mahuru*) also were recognised (Best 1959: 46). The first of the two words for winter is also a name for the star Sirius, while *Koanga* (spring) announces

Fig. 2.33 This painting by Gottfried Lindauer shows the preparation of the ground for the planting of the *kumara*, which occurred in *Koanga* or spring (https://en.wikipedia.org)

the 'digging season'—a time to plant the *kumara* and other crops (see Fig. 2.33). In contrast, *Ngahuru* (autumn) is the season for harvesting the crops; it was a time when food was plentiful, hence the terms *Ngahuru-kai-paenga* and *Ngahuru-kai-paeke*, where *kai* refers to 'food'.

Best (1959: 49) relates that

> Summer and winter were personified in two beings named Hine-raumati, the Summer Maid, and Hine-takurua, the Winter Maid. These damsels … were taken to wife by Te Ra, the sun. The Winter Maid dwells out on the ocean and controls the food supplies of that region … The Summer Maid dwells on land, her task being to foster the food products of the earth. Ra, the sun, spends half a year with each of his two wives.

The Maori year (with its four seasons) was made up of lunar months, but there were considerable variations throughout Aotearoa/New Zealand in the precise number of months recognised and in their names. Most tribes recognised twelve months in a Maori year, but Best (1959: 21–22) also cites 10 and 13 as rare alternatives. In regard to the naming of the months, Best (1959: 18–19) observes:

> … no common system of naming months existed. Several series of names were in use, even in the North Island. Each tribe recognised proper names for the months, but also, and apparently more commonly, employed a series of names consisting partially or entirely of ordinal numbers, as Te Tahi (The First), Te Rua (The Second), and so on. The remarkable point is that the proper names of the month did not agree. Two distinct series of such names were in use on the east coast of the North Island.

Table 2.11 Months of the Tuhoe Maori year (after Best 1959: 19)

Month	Name	Features
1	*Piripi*	All things on Earth come together owing to the cold; likewise man
2	*Hongonui*	Man is now extremely cold, and so kindles fires before which he basks
3	*Hereturi-koka*	The scorching effect of fire on the knees of man is seen
4	*Mahuru*	The Earth has now acquired warmth, as also have herbage and trees
5	Whiringa-nuku	The Earth has now become quite warm
6	Whiringa-rangi	It has now become summer, and the Sun has acquired strength
7	Hakihea	Birds are now sitting in their nests
8	Kohi-tatea	Fruits have now set, and man eats of the new food products of the season
9	Hui-tanguru	The root of Ruhi now rests upon the Earth
10	Poutu-te-rangi	The crops are now taken up
11	Paenga-whawha	All stakes are now stacked at the borders of the plantations
12	Haratua	Crops have now been stored in the store pits. The tasks of man are finished

This is yet another example of the way in which regional variations permeated Maori astronomical lore.

Regardless of the number of months used and their names, the Maori year was intimately associated with a succession of ecological situations and associated food-quest activities. This is illustrated below in reference to the Tuhoe tribe from the Ureweras (see Table 2.11). Because they lived in a mountainous somewhat inhospitable inland region of the North Island, there is greater emphasis than elsewhere on forest products (e.g. see Fig. 2.34) and of course there is no mention of maritime resources.

Identifying months by number rather than name was all very well so long as a clearly-defined start to the Maori year was universally adopted, but this was not the case:

> Now in some districts, as the east coast of the North Island, the Pleiades year was a permanent institution, but in others the heliacal rising of Puanga (Rigel in Orion) marked the commencement of the year. This was the case in the far North, in the South Island, and at the Chatham Isles. It is possible that the two systems were introduced by different bands of migrants, and possibly from different regions of the Pacific. (Best 1959: 6).

For some the year commenced in May and for others in June, depending upon whether Orion or the Pleiades controlled the Maori year. The actual starting date varied from year to year, and depended on the occurrence of the first new moon after the reappearance of the Pleiades or Rigel in the morning sky.

The month, in its turn, was composed of different named 'nights of the Moon' rather than 'days of the month'. Most sources recognised 30 days per month, but others give 29 or 31, and one even lists 33.

Fig. 2.34 Pai Kanohi of the Ngāi Tūhoe tribe seated on the porch of his *whare* (house) at Ruatahuna in the early 1900s. Around him are *tahā huahua*, calabashes for pigeons preserved in their own fat. These were a vital part of the Maori diet, especially in inland forested areas, and sometimes were traded between tribes (*Courtesy* Alexander Turnbull Library, 1/2-019482-F)

Best (1959: 30) observes that the "Names of nights the moon differ to some extent in different districts, as also does the order in which the names occur." This is clearly illustrated by reference to Table 2.12, where the successive columns list the names supplied by Otaki, Tuhoe and Kahungunu informants (after Best 1959: 28–35).

In each instance, "The moon disappears on the Mutuwhenua night; it acquires form on the Whiro night and its radiance is seen." (Best 1959: 31). In other words, the first night of the Moon correlates with the initial appearance of the crescent Moon low in the western sky immediately after sunset.

One major problem with this calendar system was that the true lunar month was in fact less than 30 days (or nights), which sooner or later would have led to discrepancies between the stated night of the Moon and the required phase of the Moon. Best's Otaki informant explained how this dilemma was resolved:

> The fifteenth night is an Ohua, but in certain months it is the sixteenth night, and sometimes it is the seventeenth night – that is, ere the condition of full moon is obtained. If the moon does not become full until the seventeenth night, then the fifteenth, sixteenth and seventeenth nights are all termed Ohua, and then the last three nights of the moon, Orongonui, Maurea, and Mutu, are omitted, because a new moon has appeared. (cited in Best 1959: 29).

Table 2.12 Maori names for nights of the Moon from different regions of the North Island

Night	Region		
	Otaki	Ureweras	Wairarapa
1	*Whiro*	*Whiro*	*Whiro*
2	*Tirea*	*Tirea*	*Tirea*
3	*Ohoata*	*Hoata*	*Hoata*
4	*Oue*	*Oue*	*Ouenuku*
5	*Okoro*	*Okoro*	*Okoro*
6	*Tamatea*	*Tamatea-tutahi*	*Tamatea-ngana*
7	*Tamatea-ngana*	*Tamatea-ngana*	*Tamatea-kai-ariki*
8	*Tamatea-aio*	*Tamatea-aio*	*Huna*
9	*Tamatea-whakapau*	*Tamatea-kai-ariki-whakapa*	*Ari-roa*
10	*Huna*	*Ari-matanui*	*Maure*
11	*Ari*	*Huna*	*Mawharu*
12	*Hotu*	*Mawharu*	*Ohua*
13	*Mawharu*	*Maure*	*Hotu*
14	*Atau*	*Ohua*	*Atua*
15	*Ohau*	*Atua*	*Turu*
16	*Turu*	*Hotu*	*Rakau-nui*
17	*Rakau-nui*	*Turu*	*Rakau-matohi*
18	*Rakau-matohi*	*Rakau-nui*	*Takirau*
19	*Takirau*	*Rakau-matohi*	*Oike*
20	*Oike*	*Takirau*	*Korekore-te-whiwhia*
21	*Korekore*	*Oika*	*Korekore-te-rawea*
22	*Korekore-turua*	*Korekore-whakatehe*	*Korekore-hahani*
23	*Korekore whakapiri ki nga Tangaro*	*Korekore-piri-ki-te-Tangaroa*	*Tangaroa-amua*
24	*Tangaroa-a-mua*	*Tangaroa-a-mua*	*Tangaroa-aroto*
25	*Tangaroa-o-roto*	*Tangaroa-a-roto*	*Tangaroa-kiokio*
26	*Tangaroa-kiokio*	*Tangaroa-kiokio*	*Otane*
27	*Otane*	*Otane*	*Orongonui*
28	*Orongonui*	*Orongonui*	*Mauri*
29	*Maurea*	*Mauri*	*Omutu*
30	*Mutu*	*Mutuwhenua*	*Mutuwhenua*

This flexibility not only allowed corrections to be made to accommodate the changing phase of the Moon, but it also goes some way towards explaining why identical names in Table 2.12 sometimes denote very different nights of the Moon.

This flexibility aside, the Maori were able to use their calendar system to pinpoint particular dates during the year with great precision, by simply specifying the month in question and the relevant night of the Moon.

2.9 Concluding Remarks

The foregoing account indicates that Maori astronomy was intricately inter-woven with mythology, ritual and religion, and that various celestial bodies were attributed human qualities—good and bad. The Sun, the Moon and certain stars (and groupings of stars) played an on-going role in everyday life on Earth, and their surveillance and study was therefore a matter of considerable importance.

It is equally apparent when comparing the various objects and events that we know were visible in the Aotearoa/New Zealand night sky between AD 1200 and 1768 with what is actually recorded in Best's accounts that either

1. the Maori astronomer was basically disinterested in much of what was happening in his night sky, or
2. the account provided by Best is superficial and in no way reflects the actual interests, skills and knowledge-base of the pre-European Maori astronomer.

I believe the second alternative best represents the true situation and that Best's work simply presents the tip of the Maori 'astronomical iceberg'.

Given the passage of time and the degree of acculturation that has occurred in Aotearoa/New Zealand, I initially wondered (Orchiston 2000) whether there was still a chance of capturing details of pre-contact Maori astronomy and reconstructing the morphology of that all-elusive 'astronomical iceberg', with its regional variations and ramifications. Back in 1922 Best noted that: "The knowledge … of this subject is meagre and unsatisfactory, but it is now too late to remedy the deficiency." But maybe he was overly pessimistic. Currently there is a resurgence of interest in all aspects of Maori astronomy, and the initiatives of groups like SMART (see Harris et al. 2013)[1] are already leading to a greater understanding of the various regional astronomical systems that existed in Aotearoa/New Zealand at the time of initial European contact. Harris et al. (2013: 334) note that

> As our communities engage in dialogue, the celestial knowledge of the past continues to emerge, enabling researchers to record details that were hidden from the early ethnographers, thus countering Best's claim that all sources of information on Māori astronomy have been exhausted.

The challenge now surely is to delineate the ways in these pre-contact regional astronomical systems developed from the initial proto-Polynesian astronomical beliefs and practices introduced by the first Maori settlers, and the ways in which the Maori astronomical knowledge base has evolved and adapted since European settlement. We therefore must document contemporary Maori astronomical knowledge, note any evidence of regional variations, and see if we can establish which components are drawn directly from Best's accounts and which definitely

[1]SMART is the acronym for the Society of Maori Astronomy Research and Traditions, which was formed in the late 2000s. This group has "… dedicated itself to the preservation and revitalization of Māori astronomical knowledge … [and] consists of Maori knowledge experts, educators, navigators and scientists." (Harris et al. 2013: 326).

derive from the post-contact period and reflect European religious and scientific views. We must also quantify the various ways in which this contemporary astronomical knowledge-base is utilised in everyday life. This is our challenge.

Acknowledgements I am grateful to the following for their assistance: Professor Peter Bellwood (Canberra, Australia), Jenni Chrisstoffels (Alexander Turnbull Library, Wellington, NZ), John Drummond (Patutahi, NZ) and Jan Meeus (Belgium). I also wish to thank John Drummond and Okkie Symmonds (Wellington, NZ) for reading and commenting on the first draft of this chapter, and the Alexander Turnbull Library and John Drummond for kindly supplying Figs. 2.1, 2.11, 2.16, 2.18, 2.20, 2.27, 2.28, 2.29 and 2.34.

References

Ambrose, W. (1970). Archaeology and rock drawings from the Waitaki Gorge, central South Island. *Records of the Canterbury Museum, 8*, 383–437.

Ammarell, G. (1987). Sky calendars of the Indo-Malay archipelago. In Swarup, Bag, and Shukla, pp. 241–247.

Anderson, A. (1989). *Prodigious Birds: Moas and Moa-hunting in Prehistoric New Zealand*. Cambridge: Cambridge University Press.

Anderson, A. (1991). The chronology of colonization in New Zealand. *Antiquity, 65*, 767–795.

Bassett, K. N., Gordon, H. W., Nobes, D. C., & Jacomb, C. (2004). Gardening at the edge: documenting the limits of tropical Polynesian kumara horticulture in southern New Zealand. *Geoarchaeology: An International Journal, 19*, 185–218.

Beattie, H. (1918). Traditions and legends. Collected from the natives of Murihiku (Southland, New Zealand). *Journal of the Polynesian Society, 27*, 137–161.

Best, E. (1898). Omens and superstitious beliefs of the Maori. *Journal of the Polynesian Society, 7*, 233–243.

Best, E. (1925). *Maori Agriculture. The Cultivated Food Plants of the Natives of New Zealand, With Some Account of Native Methods of Agriculture, its Ritual and Origin Myths*. Wellington: Whitcombe and Tombs (Dominion Museum Bulletin 9).

Best, E. (1954). *Polynesian Voyagers. The Maori as a Deep-sea Navigator, Explorer, & Colonizer*. Wellington: Government Printer (Dominion Museum Monograph No. 5) [Originally published in 1923].

Best, E. (1955). *The Astronomical Knowledge of the Maori*. Wellington: Government Printer (Dominion Museum Monograph No. 3) [Originally published in 1922].

Best, E. (1959). *The Maori Division of Time*. Wellington: Government Printer (Dominion Museum Monograph No. 4) [Originally published in 1922].

Burnham, R. (2000). *Great Comets*. Cambridge: Cambridge University Press.

Ceci, L. (1978). Watchers of the Pleiades: ethnoastronomy among native cultivators in northeastern North America. *Ethnohistory, 25*, 301–317.

Chambers, G. F. (1910). *The Story of Comets Simply Told for General Readers*. Oxford: Clarendon.

Clark, D. H., & Stevenson, F. R. (1978). An interpretation of the pre-telescopic sunspot records from the Orient. *Quarterly Journal of the Royal Astronomical Society, 19*, 387–410.

Craig, E. W. G. (1964). *Man of the Mist. A Biography of Elsdon Best*. Wellington: Reed.

Cumberland, K. B. (1962). Moas and men, New Zealand about A.D. 1250. *The Geographical Review, 52*, 151–173.

Davidson, J. M. (1987). *The Prehistory of New Zealand* (2nd ed.). Auckland: Longman Paul.

de Meis, S., & Meeus, J. (1994). Quintuple planetary groupings—rarity, historical events and popular beliefs. *Journal of the British Astronomical Association, 104*, 293–297.

Dick, S. (1998). Observation and interpretation of the Leonid meteors over the last millennium. *Journal of Astronomical History and Heritage, 1*, 1–20.

Eddy, J. A. (1983). The Maunder Minimum—a reappraisal. *Solar Physics, 89*, 159–207.

Eddy, J. A. (1987). Uses of ancient data in modern astronomy. In Swarup, Bag, and Shukla, pp. 253–260.

Eddy, J. A., Stephenson, F. R., & Yau, K. K. C. (1989). On pre-telescopic sunspot records. *Quarterly Journal of the Royal Astronomical Society, 30*, 65–73.

Erny, P. (1992). Comment situer l'ethnoastronomy? In Jaschek, pp. 1–15.

Finney, B. (1992). Experimental voyaging and Maori settlement. In Sutton, pp. 52–76.

Fisher, W. J. H. (1963). Dark nebulae in the region of the Southern Cross. *Southern Stars, 20*, 61–76.

Fuller, R. S., Anderson, M. G., Norris, R. P., & Trudgett, M. (2014). The emu sky knowledge of the Kamilaroi and Euahlayi Peoples. *Journal of Astronomical History and Heritage, 17*, 171–179.

Golson, J. (1959). Culture change in prehistoric New Zealand. In J.D. Freeman & W.R. Geddes (Eds.), *Anthropology in the South Seas* (pp. 29–74). New Plymouth: Avery.

Grant, P. J. (1992). Late Holocene histories of climate, geomorphology and vegetation, and their effects on the first New Zealanders. In Sutton, pp. 164–194.

Green, D. A., & Orchiston, W. (2005). In search of *Mahutonga*: a possible supernova recorded in Maori astronomical traditions? *Archaeoastronomy, 18*, 110–113.

Green, R. C. (1975). Adaptation and change in Maori culture. In G. Kuschel (Ed.), *Biogeography and Ecology in New Zealand* (pp. 591–641). The Hague: W.J Junk.

Green, R. C. (1992). Changes over time: recent advances in dating human colonisation of the Pacific Basin area. In Sutton, pp. 19–51.

Guillemin, A. (1877). *The World of Comets*. London: Sampson Low, Marston, Searle, & Rivington.

Harlow, R. B. (1979). Regional variation in Maori. *New Zealand Journal of Archaeology, 1*, 123–138.

Harris, P., Matamua, R., Smith, T., Kerr, H., & Waaka, T. (2013). A review of Māori astronomy in Aotearoa-New Zealand. *Journal of Astronomical History and Heritage, 16*, 325–336.

Hasegawa, I. (1980). Catalogue of ancient and naked eye comets. *Vistas in Astronomy, 24*, 59–102.

Higham, T., Anderson, A., & Jacomb, C. (1999). Dating the first New Zealanders: the chronology of Wairau Bar. *Antiquity, 73*, 420–427.

Ho, P. Y. (1962). Ancient and medieval observations of comets and novae in Chinese sources. *Vistas in Astronomy, 5*, 127–225.

Holdaway, R., & Jacomb, C. (2000). Rapid extinction of moas (Aves: Dinornithiformes) in New Zealand. *Science, 287*, 2250–2255.

Hongi, H. (1920). The gods of Maori worship. Sons of light. *Journal of the Polynesian Society, 29*, 24–28.

Houghton, P. (1980). *The First New Zealanders*. Auckland: Hodder & Stoughton.

Hoyt, D. V., & Schatten K. H. (1996). How well was the Sun observed during the Maunder Minimum? *Solar Physics, 165*, 181–192.

Hsi, T. (1958). A new catalog of ancient novae. *Smithsonian Contributions to Astrophysics, 2*(6), 109–130.

Hugh-Jones, S. (1982). The Pleiades and Scorpius in Barasane cosmology. *Annals of the New York Academy of Sciences, 385*, 183–201.

Hughes, D. W. (1985). The position of Earth at previous apparitions of Halley's Comet. *Quarterly Journal of the Royal Astronomical Society, 26*, 513–520.

Hughes, D. W. (1987). The criteria for cometary remarkability. *Vistas in Astronomy, 30*, 145–152.

Jansen, J. (1991). Additions and corrections to the I. Hasegawa catalogue. *Vistas in Astronomy, 34*, 179–186.

Jaschek, C. (Ed.), (1992). *Reunion Europeenne d'Astronomie & Sciences Humaines. European Meeting on Archaeoastronomy & Ethnoastronomy*. Strasbourg, L'Observatoire Astronomique de Strasbourg.

Jenniskens, P., & Lyytinen, E. (2005). Meteor showers from the debris of broken comets: D/1819 W_1 (Blanpain), 2003 WY_{25}, and the Phoenicids. *Astronomical Journal, 130*, 1286–1290.

Jenniskens, P. (2006). *Meteor Showers and their Parent Comets*. Cambridge: Cambridge University Press.

Keller, H. U., & Freidli, T. K. (1992). Visibility limit of naked-eye sunspots. *Quarterly Journal of the Royal Astronomical Society, 33*, 83–89.

Kholopov, P. N. (Ed.), (1985). *General Catalogue of Variable Stars*. (Four Volumes). Moscow: Nauka Publishing House.

Kingsley-Smith, C. (1967). Astronomers in puipuis. Maori star lore. *Southern Stars, 22*, 5–10.

Kiyota, S., Kato, T., & Yamaoka, H. (2004). Light curves of novae on VSNET. *Publications of the Astronomical Society of Japan, 56*, S193–S211.

Kronk, G. W. (1984). *Comets: A Descriptive Catalog*. Enslow: Hillside.

Liller, W. (1992). *The Cambridge Guide to Astronomical Discovery*. Cambridge: Cambridge University Press.

Littmann, M. (1998). *The Heavens on Fire: The Great Leonid Meteor Storms*. Cambridge: Cambridge University Press.

McGlone, M. S. (1983). Polynesian deforestation of New Zealand: a preliminary synthesis. *Archaeology in Oceania, 18*, 11–25.

McGlone, M. S., Anderson, A. J., & Holdaway R. N. (1992). An ecological approach to the Polynesian settlement of New Zealand. In Sutton, pp. 136–163.

Marsden, B., & Williams, G.V. (1996). *Catalogue of Cometary Orbits*. Cambridge (Mass.): Minor Planet Center.

Masse, W. B. (1995). The celestial basis of civilization. *Vistas in Astronomy, 39*, 463–477.

Moore, P. (1985). *Stargazing Astronomy Without a Telescope*. London: Arum Press.

Mossman, J. E. (1989). A comprehensive search for sunspots without the aid of a telescope, 1981–1982. *Quarterly Journal of the Royal Astronomical Society, 30*, 59–64.

Mostert, R. (1989). On the naked-eye visibility of a partial solar eclipse. *Quarterly Journal of the Royal Astronomical Society, 30*, 113–114.

Neidig, D. F., & Cliver, E. W. (1983). A catalog of solar white-light flares, including their statistical properties and associated emissions, 1859–1982. (US) Air Force Geophysics Laboratory Technical Report.

New Zealand Gazette and Wellington Spectator, March 8 1843 (p. 2).

Norton, A. P. (1957). *A Star Atlas and Reference Handbook*. London: Gall and Inglis.

Orchiston, D. W. (1979). Settlement or citadel? The basic function of the Maori *pa* in East Coast South Island New Zealand prehistory and protohistory. *Archaeology & Physical Anthropology in Oceania, 14*, 168–183.

Orchiston, W. (1986). Towards an accurate history of early New Zealand astronomy. *Southern Stars, 31*, 257–266.

Orchiston, W. (1996). Australian Aboriginal, Polynesian and Maori astronomy. In Walker, pp. 318–328.

Orchiston, W. (1997). The Waingaromia Meteorite: a IIIAB iron from New Zealand. *Meteorite! 3* (4), 20–21.

Orchiston, W. (1998). *Nautical Astronomy in New Zealand. The Voyages of James Cook*. Wellington: Carter Observatory.

Orchiston, W. (2000). A Polynesian astronomical perspective: the Maori of New Zealand. In H. Selin & X. Sun (Eds.), *Astronomy Across Cultures. The History of Non-Western Astronomy* (pp. 161–196). Dordrecht: Kluwer.

Pannekoek, A. (1961). *A History of Astronomy*. London: Allen & Unwin.

Peak, B. M. (1958). *The Planet Jupiter*. London: Faber and Faber.

Proctor, R. A. (1882). *Saturn and its System*. London: Chatto & Windus.

Proctor, R. A. (1896). *Myths and Marvels of Astronomy*. London: Longmans, Green.

Rada, W. S., & Stephenson, F. R. (1992). A catalogue of meteor showers in Medieval Arab chronicles. *Quarterly Journal of the Royal Astronomical Society, 33*, 5–16.
Ribes, J. C., & Nesme-Ribes, E. (1993). The solar sunspot cycle in the Maunder Minimum AD1645 to AD1715. *Astronomy & Astrophysics, 276*, 549–563.
Richmond, M. (2007). Contributing photometry of novae and supernovae. (http://spiff.rit.edu/richmond/aavso/nova_sn.html).
Ronan, C. (1996). Astronomy in China, Korea and Japan. In Walker, pp. 245–268. Salmond, A. (1991). *Two Worlds. First Meeting Between Maori and Europeans 1642–1772*. Auckland: Viking.
Salmond, A. (1991). *Two Worlds. First Meeting Between Maori and Europeans 1642–1772*. Auckland: Viking.
Sands, B. F. (1871). *Reports on Observations of Total Solar Eclipse of December 22 1870 Conducted Under the Direction of Rear Admiral B. F. Sands, U.S.N.* Washington: U.S. Naval Observatory.
Seargent, D. A. J. (2009). *The Greatest Comets in History: Broom Stars and Celestial Scimitars*. New York: Springer.
Skinner, H. D. (1974). *Comparatively Speaking. Studies in Pacific Material Culture 1921–1972*. Dunedin: University of Otago Press.
Smith, S. P. (1910). *History and Traditions of the Maoris of the West Coast North Island New Zealand Prior to 1840*. New Plymouth: Polynesian Society.
Snedegar, K. V. (1992). Prospects for studying the ethnoastronomy of Southern Africa. In Jaschek, pp. 39–48.
Snedegar, K. V. (1995). Stars and seasons in southern Africa. *Vistas in Astronomy, 39*, 529–538.
Stephenson, F. R., & Fatoohi, L. J. (1994). Accuracy of early estimates of lunar eclipse magnitudes. *Quarterly Journal of the Royal Astronomical Society, 35*, 81–94.
Stephenson, F. R., & Green, D. A. (2002). *Historical Supernovae and their Remnants*. Oxford: Oxford University Press.
Sutton, D. G. (Ed.), (1992). *The Origins of the First New Zealanders*. Auckland: Auckland University Press.
Swarup, G., Bag, A. K., & Shukla, K. S. (Eds.), (1987). *History of Oriental Astronomy*. Cambridge: Cambridge University Press.
Trotter, M., & McCulloch, B. (1971). *Prehistoric Rock Art of New Zealand*. Wellington: A.H. & A.W. Reed.
Trotter, M., & McCulloch, B. (1997). *Digging up the Past: New Zealand's Archaeological History*. Auckland: Penguin Books.
Vaquero, J. M., Gallego, M. C., & García, J. A. (1997). A 250-year cycle in naked-eye observations of sunspot. *Geophysical Research Letters, 29*(20) (doi:10.1029/2002GL014782).
Vsekhsvyatskii, S. K. (1964). *Physical Characteristics of Comets*. Jerusalem: Program for Scientific Translations.
Walker, C. (Ed.), (1996). *Astronomy Before the Telescope*. London: British Museum.
Warner, B. (1996). Traditional astronomical knowledge in Africa. In Walker, pp. 304–317.
Willis, D. M., Davda, V. N., & Stephenson, F. R. (1996). Comparison between Oriental and Occidental sunspot observations. *Quarterly Journal of the Royal Astronomical Society, 37*, 189–229.
Xu, Z. (1987). Work of art of the Han Dynasty unearthed in China and observations of solar phenomena. In Swarup, Bag, and Shukla, pp. 51–56.
Yau, K. K. C., & Stephenson, F. R. (1988). A revised catalogue of Far Eastern observations of sunspots 165 BC to AD 1918. *Quarterly Journal of the Royal Astronomical Society, 29*, 175–197.
Yen, D. E. (1961). The adaptation of kumara by the New Zealand Maori. *Journal of the Polynesian Society, 70*, 338–348.
Yeomans, D. (2007). *Great Comets in History*. Pasadena: Jet Propulsion Laboratory.
Zeilik, M. (1985). *The Evolving Universe*. New York: Harper & Row.

Chapter 3
Mahutonga: Did Proto-Polynesian Astronomers Record the Supernova of AD 185?

Abstract Maori astronomical traditions refer to *Mahutonga,* which can be interpreted as a possible record of a supernova seen in or near Crux. A search for known 'young' supernova remnants in this region does not reveal any obvious candidate to associate with this possible supernova. Relaxing the positional constraint slightly, the supernova of AD 185 near α Centauri is close by. If this is indeed associated with *Mahutonga,* then the Maori term must be a relic from an earlier Proto-Polynesian record.

3.1 Introduction

The Polynesians were great maritime explorers and possessed an intricate astronomical knowledge system, which not only facilitated marine navigation, food-quest activities and ceremonies but also provided mechanisms for measuring time (see the preceding chapter in this book). A basic Proto-Polynesian astronomical knowledge base came to the various Pacific islands with the original settlers, but then began to evolve in slightly different directions as the colonists adapted to the specific environmental demands encountered in their new homelands. Consequently, by the nineteenth century different astronomical knowledge systems were found on the main island groups of Polynesia (e.g. see Chauvin 2000; Liller 2000; Orchiston 2000), but when they were examined there were enough commonly-recognisable words and phrases to indicate the original existence of a single Proto-Polynesian astronomical tradition.

It was mainly during the nineteenth century that Europeans made their first attempts to gather information on the various indigenous astronomical systems of Polynesia. In Aotearoa/New Zealand the leading advocate was the ethnologist, Elson Best, and as we saw in the previous chapter he only carried out field work late in the nineteenth century, mainly in the Urewera region of the North Island and published his little monograph on the *Astronomical Knowledge of the Maori* in 1922.

W. Orchiston, *Exploring the History of New Zealand Astronomy*,
Astrophysics and Space Science Library 422,
DOI 10.1007/978-3-319-22566-1_3

As originally noted by Orchiston (2000, 2002), Best (1922: 46) mentions an object called *Mahutonga*, which is "… a star of the south that remains invisible." Best actually obtained this information from Stowell (1911: 202–203) who described *Maahu* (*Mahu*) as a star of the south, which "… has left its place in pursuit of a female. When it secures the female, it will come back again to its true home." Stowell (1911: 209) also mentions that currently "… *Maahu-Tonga* is invisible."

In this chapter we meet H.M. Stowell and look at possible sources of the term *Mahutonga*, before examining potential astronomical correlates for this object.

3.2 H.M. Stowell and *Mahutonga*

Henry Matthew Stowell (1859–1944; Fig. 3.1; Gibbons 1996) was born in Waimate North, Bay of Islands (for Aotearoa/New Zealand localities mentioned in this chapter see Fig. 3.2) to a sawyer from the USA who had settled in Aotearoa/New Zealand and his wife Huhana (Susan) Rimaumau, a high-ranking Nga Puhi woman. Henry Stowell was educated in Auckland where he excelled as a scholar and as an athlete. Subsequently, he

Fig. 3.1 H.M. Stowell ('Hare Hongi') (*Courtesy* Alexander Turnbull Library, Wellington, Ref. PAColl-7081-25)

Fig. 3.2 Aotearoa/New Zealand localities mentioned in this chapter are shown in *red*

> … worked with a survey party in Northland, but the most important experience of his teens, and in some sense of his whole life, was a period of more than a year at Waitaha village, near Ahipara, absorbing Maori lore from the tohunga Nga Kuku Mumu, whom he later described as 'my first and my grandest old mentor'. Stowell's abiding interest thereafter was the acquisition of traditional Maori knowledge, especially mythology, genealogy and cosmology. (ibid.).

Stowell then worked as a sawyer in the Mangakahia district, before moving to Wellington in 1888, becoming an interpreter in the Native Land Court. He then visited Taranaki, and in July 1891 married Mary Rachel Robson, daughter of a sawyer and a high-ranking Te Ati Awa woman. In the 1890s and early 1900s Stowell and his wife lived at Hawera, but he

> … often travelled widely in the course of both business and collecting information from learned Maori. Assuming the name Hare Hongi as a tribute to his Nga Puhi ancestors, he published articles and some highly rhetorical verse, so impressing members of the government with his erudition that he was given official encouragement to gather Maori lore and legends. In 1908 Stowell was appointed a temporary clerk and interpreter in the Native Department in Wellington, a position which was made permanent in 1913. (ibid.).

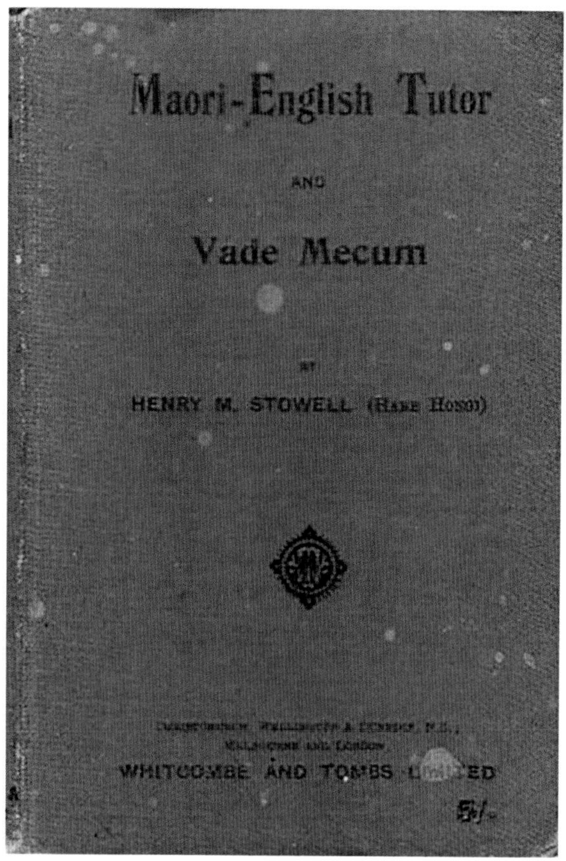

Fig. 3.3 The cover of Stowell's *Maori-English Tutor …* (www.renaissancebooks.co.nz/?page=shop.flypage&product_id=19935&keyword=Stowell,+Henry+M.+(Hare+Hongi)&searchby=author&offset=0&)

He resigned from the Department in 1921.

The word *Mahutonga* entered the public domain when Stowell (1911) published his *Maori-English Tutor and Vade Mecum* in 1911 (Fig. 3.3). Without accessing relevant archival records—if indeed they exist—it is impossible to determine precisely where Stowell learnt about *Mahutonga*. It was probably from Nga Kuku Mumu or from *tohunga* in the Taranaki area where he had close Maori contacts, but there is a possibility that he was told this account while in Northland, in the Mangakahia district; during his travels (presumably in the North Island) while living in Hawera; or while involved with the Native Land Court or Native Department in Wellington. All we can say is that it is likely that the *Mahutonga* story derived from Northland or Taranaki, and it is almost certainly of North Island origin.

3.3 Identifying *Mahutonga*

3.3.1 Introduction

What sort of astronomical object could *Mahutonga* refer to, and where was it located in the sky? The name *'Mahutonga'* immediately provides a clue as to the 'where', for this word, or components of it, specifically refer to the well-known Coal Sack Nebula and to the associated constellation of Crux. What we cannot say—given the lack of additional lexical information—is how far outside the boundaries of Crux *Mahutonga* may have been located. Nonetheless, it is safe to conclude that we should be looking for an optical candidate in or near Crux.[1]

What type of object are we looking for? Clearly *Mahutonga* refers to a star, not to a comet, or a meteor or a bolide, but more than this, it relates to *a specific star*, and one that was deemed to be so important or unusual that it warranted the allocation of a unique name. Thus, we are looking for an unusual variable star in or near Crux, and one that attracted widespread attention when at maximum and then faded below naked eye visibility. There would seem to be three possibilities: (1) Eta Carinae; (2) a nova; or (3) a supernova. Let us examine these three options in turn.

3.3.2 Eta Carinae

As Humphreys et al. (2008: 1249) emphasize,

> Eta Carinae is famous for its many superlatives as the most massive, most luminous star in our region of the Milky Way and for its "great eruption" (1837–1858) more than 160 yr

[1]McIntosh (1965) believes that the Maoris used the name *Mahutonga* to refer to an invisible star that they assumed was at the South Celestial Pole. However, this spot is a considerable distance away from the Crux-Coal Sack region.

> ago, when it ejected 6 $M_\odot$ or more (Davidson and Humphreys 1997; Morris et al. 1999; Smith et al. 2003) forming its bipolar Homunculus Nebula. η Car is also our closest and best-studied example of a "supernova impostor" …

It is 2350 ± 50 pc from the Sun (Smith 2006) at R.A. 10h 45m 04.0s and Dec. −59° 41′ 04″ (J2000), in the constellation of Carina. It is a binary system with a combined mass of >100 $M_\odot$. The dominant member of the binary system is an eruptive Luminous Blue Variable (LBV) with a luminosity of $\sim 4 \times 10^6$ that of the Sun (Davidson and Humphreys 1997).

In 1800 η Carinae was a conspicuous first or second magnitude star, and subsequently it was observed by the British astronomer Sir John Herschel (1792–1871) in the 1830s when he was at the Cape of Good Hope and began surveying the Southern sky (Herschel 1847). Herschel

> … noted that its brightness was relatively constant during this period (V = +1.2 on the modern Pogson scale), before he observed it to rapidly brighten at the close of 1837 to be as bright as Alpha Centauri, before quickly fading again—this brightening is generally considered to be the start of the period of enhanced brightness known as the 'Great Eruption' (Frew 2004; Smith and Frew 2011) … [It] brightened markedly again during 1843; at its peak brightness in March of that year, and again in January 1845, it was the second brightest star in the sky after Sirius. The well-known Homunculus Nebula is the ejected debris from this explosive event … The star returned to its pre-outburst brightness in 1858, and continued to rapidly fade; by 1869, it was invisible to the naked eye. (Hamacher and Frew 2010: 228).

Frew (2004) has made a special study of the optical variability of η Carinae during the nineteenth century, and his light-curve is shown in Fig. 3.4.

On the basis that Stowell's description of *Mahutonga* could describe a recurrent variable star—a star that appears, disappears and is then expected to appear again—Hamacher and Frew (2010: 231) suggest η Carinae as "… a possible, albeit speculative, candidate for *Mahutonga*." However, as the name suggests, η Carinae is in the constellation of Carina, some distance from Crux (see Fig. 3.5). Furthermore,

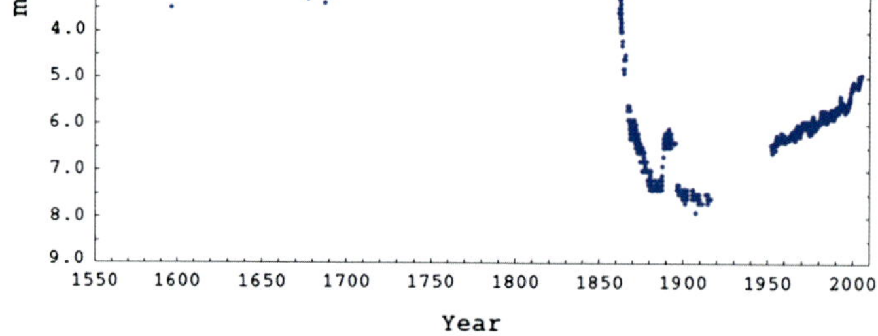

Fig. 3.4 The visual light curve of η Carinae between 1596 and 2000 (after Frew 2004)

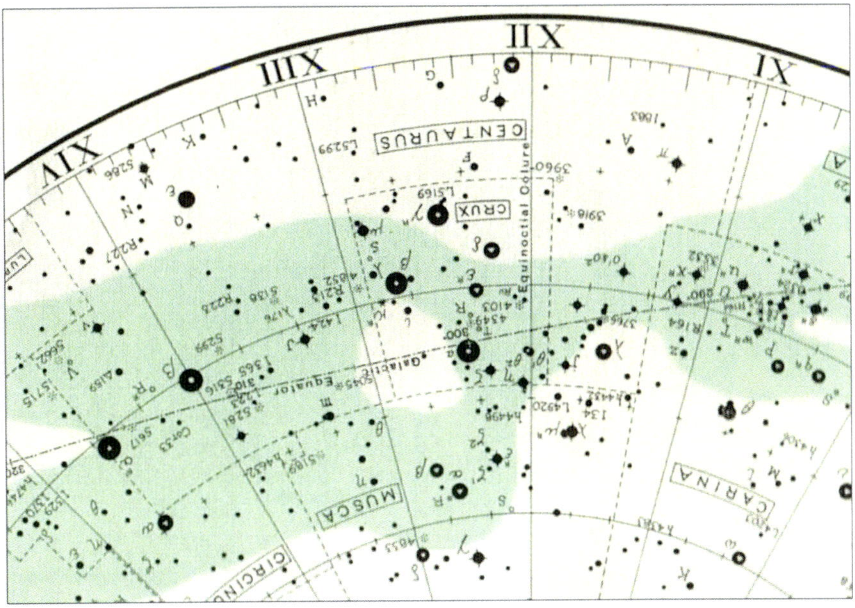

Fig. 3.5 A map showing the relative locations of Centaurus, Crux and Carina. The Milky Way is shown in green. The Coal Sack is the white area within the Milky Way extending from α to β Centauri and to RA 13 h. Eta Carinae is located in a very 'busy' area of the Milky Way at about RA 10h 45m and Dec −59° (after Norton 1957)

although we know that η Carinae exhibited variability prior to the 'Great Eruption' (Frew 2004; Smith and Frew 2011), there is no documentation to indicate that it ever experienced a similar earlier 'Great Eruption' during the interval that the Maori occupied Aotearoa/New Zealand and at that time was one of the brightest stars in the sky. In this context, it is perhaps significant that in his recent review of ethnographic evidence from around the work relating to visible supernovae, Hamacher (2014) does not repeat the suggestion that η Carinae could be associated with *Mahutonga* (even though he specifically mentions *Mahutonga* in his paper).

3.3.3 A Nova

We know from Chinese, Korean and Japanese records that many novae appeared in southern skies prior to the arrival of Europeans in Aotearoa/New Zealand (see Table 2.8). As indicated in the previous chapter, this must be regarded as a minimal list since Ho (1962) includes many more 'guest stars' and 'new stars' drawn from Korean and Japanese sources, but he does not indicate their celestial positions so we do not know how many of these would have been visible from Aotearoa/

New Zealand. In addition, there certainly were other novae that would have been visible from Aotearoa/New Zealand but not from China, Japan or Korea.

It is significant that Best does not report any Maori word for a nova, even though Table 2.8 indicates that there were several instances during pre-European times where two or more novae were visible within the average life-span of the typical Maori astronomer. Since *Mahutonga* refers to a specific star, not a class of stars, it is unlikely that this term refers to a nova.

3.3.4 A Supernova

By a process of logic, and elimination, *Mahutonga* could refer to a supernova (SN), a rare and spectacular event that would have attracted widespread attention. A supernova explosion marks the violent end-point of stellar evolution in stars whose masses exceed 8 $M_\odot$, and is visible optically for months or even years. The explosion releases large amounts of energy and material, which interact with the surrounding interstellar medium and produce a supernova remnant (SNR).

Over the past millennium there have been five historical SNe in our Galaxy that were recorded by Asian, European and Arab astronomers (see Stephenson and Green 2002; Green and Stephenson 2003), and these are listed in Table 3.1. When David Green and I researched our *Mahutonga* paper that was published in 2004 all of these SNe lay within the date-range when Aotearoa/New Zealand was thought to have first been settled by Polynesians, but with the subsequent revision of the accepted settlement date to AD 1200 (Higham et al. 1999; Holdaway and Jacomb 2000) only the first two SNe are relevant. However, neither of these SNe was located in or near Crux.

Nonetheless, *Mahutonga* may refer to a SN that was visible from Aotearoa/New Zealand and was so impressive that it was assigned its own name which was then incorporated into the Maori astronomical knowledge base, but was not recorded by Asian or Arab astronomers. This being the case, is there a SNR in the Crux region (say between RA 11h 30m and 14h and Dec. −50° to −70°), whose formation would appear to post-date AD 1200?

Currently 294 SNRs have been identified in our Galaxy (see Green 2014), and most belong to one of two distinct types: (i) 'shell remnants', which show a more or

Table 3.1 Records of historical supernovae during the last millennium

Date (AD)	Name	Remnant	Chinese	Japanese	Korean	Arabic	European
1604	Kepler's		Few		Many		Many
1572	Tycho's		Few		Two		Many
1181		3C58	Few	Few			
1054		Crab Nebula	Many	Few		One	
1006			Many	Many		Few	Two

After Stephenson and Green (2002)

less complete limb-brightened ring of radio emission; or (ii) 'filled-center remnants', which are centrally brightened, and are powered by a central source—a pulsar—which is the compact remains of the SN explosion. Galactic SNRs are detectable for thousands or up to tens of thousands of years after their formation. For us, the challenge is to identify 'young' remnants, but as Green and Orchiston (2004: 111) have stressed "... it is not easy to definitively determine the age of an SNR from available observations (which reinforces why knowledge of the age of the remnants of historical supernovae is astrophysically so useful." The following criteria have been developed, although none is straightforward:

1. Physical size: A small size is clearly a direct indication of being young (a diameter of less that 10 or 15 pc), but this requires an accurate knowledge of the distance of the remnant, which in practice is usually difficult to determine.
2. Surface brightness: Almost all known historical SNRs are 'bright' in a surface brightness sense, i.e. $\Sigma > 10^{-19}$ W m^{-2} Hz^{-1} sr^{-1} at 1 GHz. However, the remnant of SN 1006 is not very bright, with a Σ of only a few times 10^{-21} W m^{-2} Hz^{-1} sr^{-1}, so this is not a completely clear-cut criterion due to the range of properties shown by SNRs (see Duncan and Green 2000; Green 1991).
3. Morphology: For shell remnants, a symmetrical morphology is a sign of being young (e.g. Cas A—which is thought to be $\sim$300 years old—and the remnants of Kepler's SN, Tycho's SN and SN 1006 all have circular shells), as is a well-defined filled-center remnant morphology (cf. the Crab Nebula, or 3C58, the remnant of SN 1181).

In the region of interest around Crux there are 18 known SNRs (see Fig. 3.6). Only one of these (G296.8-0.3) has a reasonable distance determination in the literature, but it has a diameter of $\sim$50 pc, so is physically too large to be 'young'. None of the 18 is comparable in surface brightness with the majority of the known, bright historical SNRs (except for the remnant of SN 1006). Almost all of the 18 have been imaged by the MOST Galactic Plane Survey at 843 MHz (Whiteoak and Green 1996), but none of these SNRs has a highly-symmetrical circular shell, nor is any of them a filled-center remnant. Two SNRs that were not observed by MOST, because they are a few degrees from the Galactic Plane, are G296.5+10.0 and G299.2−2.9. G296.5+10.0 is not nicely circular, although its high latitude suggests it is relatively nearby, which would be consistent with its parent SN being visible. G299.2−2.9 also lacks a highly-symmetrical shell, although this remnant has not been well studied.[2] Consequently there is no obvious, younger than 800-year old SNR in the prime search region that can be associated with the proposed supernova interpretation of *Mahutonga*.

However, if we relax the chronological and positional constraints a little, there was a somewhat older historical SN nearby. There is a single record from China of a

[2]G299.2-2.9 was reported as a possible young SNR by Busser, Egger and Aschenbach (1996, but their interpretation is not supported by the more detailed discussions of Slane, Vancura and Hughes (1996).

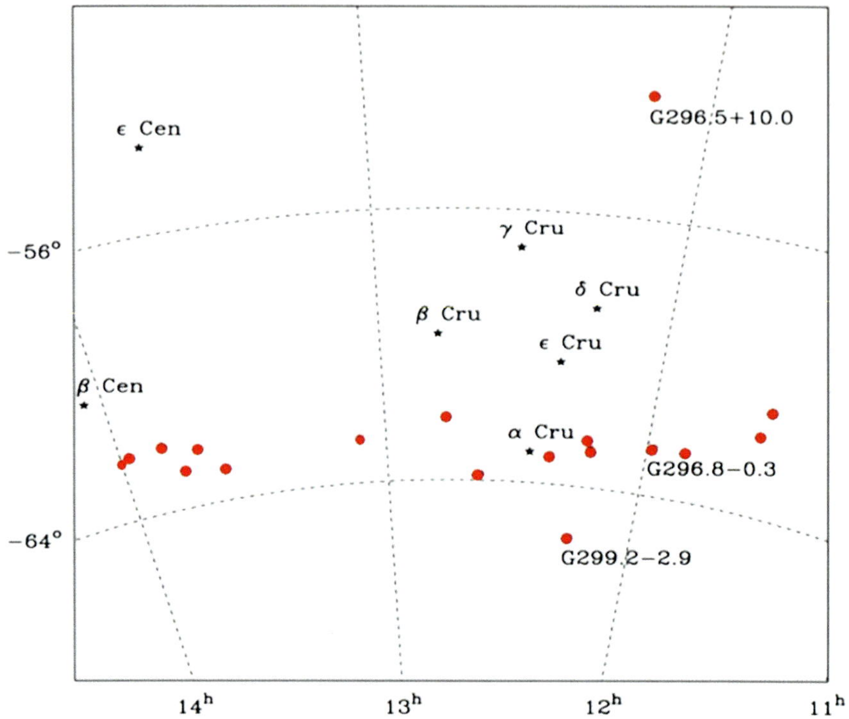

Fig. 3.6 SNRs in the region around Crux are shown in *red* (*Courtesy* David Green)

'guest star' which was seen in AD 185 and was visible for at least eight months (Fig. 3.7).[3] Stephenson and Green (2002), Zhao et al. (2006) are convinced that this was a SN, but this interpretation has not been universally accepted (e.g. see Schaefer 1995). This object was seen near the Chinese asterism *Nanmen* (α and β Centauri; see Fig. 3.8), at about RA 14h 43m and Dec. −62° 27.7′ (J2000), close to α Centauri and near the border of Centaurus and Circinus.

Several SNRs have been proposed as remnants of this SN, of which G315.4−2.3 is the prime candidate (Clark and Stephenson 1975; Martocchia and Polcaro 2009; Stephenson and Green 2002). At radio wavelengths this is a reasonably bright circular shell remnant (Fig. 3.9)[4] and it is also a conspicuous X-ray source, RCW 86 (Fig. 3.10). Its location relative to α Centauri and Crux is shown in Fig. 3.11. However, the association of *Mahutonga* with this known historical SN would require the Maori oral record to be a continuation of an earlier Proto-Polynesian record of this supernova.

---

[3]But see Stothers (1977) for a possible European record of this SN.

[4]Australian radio observations of this SNR are summarised in Orchiston and Slee (2006).

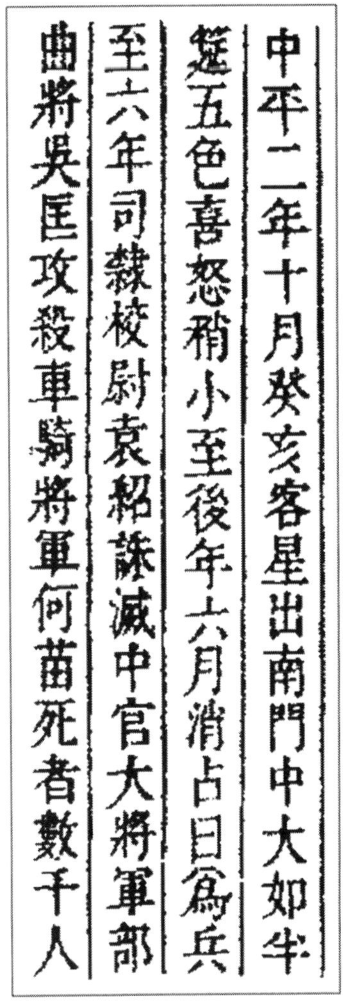
中平二年十月癸亥客星出南門中大如半
筵五色喜怒稍小至後年六月消占曰為兵
至六年司隸校尉袁紹誅滅中官大將軍部
曲將吳匡攻殺車騎將軍何苗死者數千人

Fig. 3.7 The Chinese record of the AD 185 supernova (after Stephenson and Green 2002)

3.4 Concluding Remarks

Upon examining various alternative interpretations we concluded that the Maori term *Mahutonga* most likely referred to a supernova that was seen in or near Crux, but our search for a 'young' supernova remnant in the preferred region did not reveal any suitable candidates. This may be because further young supernova remnants in this region have yet to be discovered, or because *Mahutonga* does not refer to a supernova.

However, by relaxing the positional constraint on the identification a little, we noted that the supernova of AD 185 was visible nearby (close to α Centauri). In this context, the only part of Polynesia that was already settled by AD 185 was the

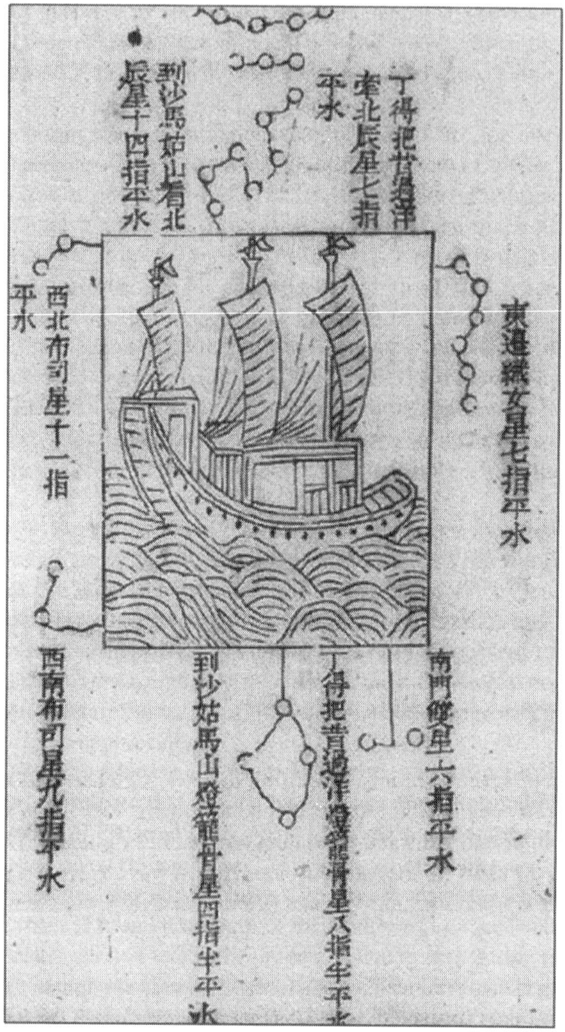

Fig. 3.8 A Chinese star map, which below the maritime scene includes the asterisms of Crux and *Nanmen* (after Stephenson and Green 2002)

Samoa-Tonga region (from which α and β Centauri were readily visible). Moreover, it is thought that the evolution of a specifically Proto-Polynesian culture evolved in this region at this time. Several hundred years later, these ancestral Polynesians settled the Cook Islands, Society Islands and Marquesas Islands, and this central Polynesian area then served as the primary dispersal centre for the settlement of the remaining major island groups of Polynesia—including the Hawaiian Islands,

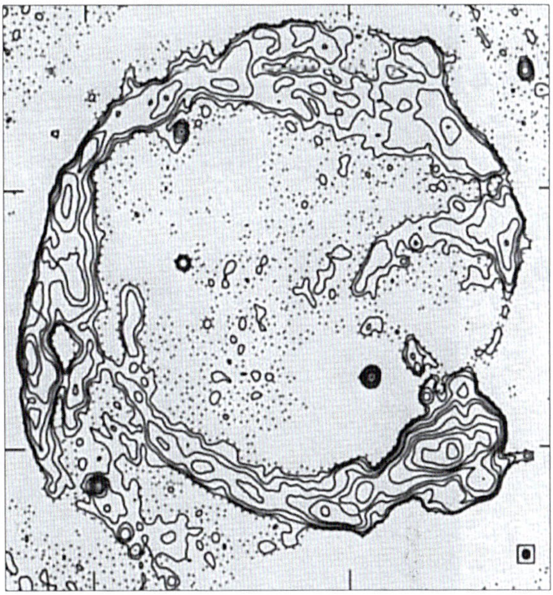

Fig. 3.9 843 MHz isophote map of the G315.4−2.3 SNR (after Whiteoak and Green 1996: 353)

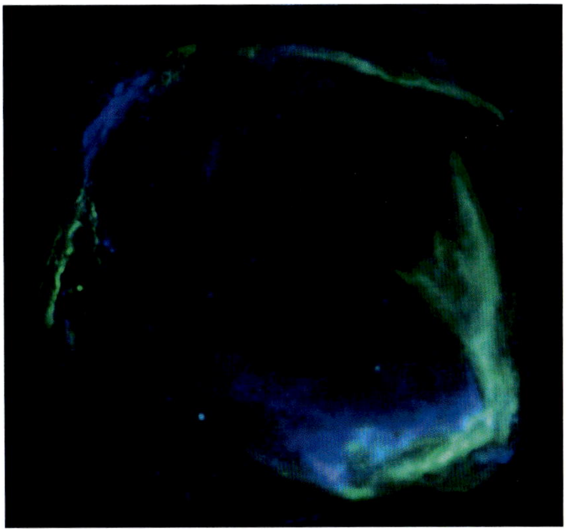

Fig. 3.10 Chandra X-ray image of RCW 85 (*Courtesy* NASA/CXC/SAO & ESA)

Easter Island and Aotearoa/New Zealand. So if *Mahutonga* does indeed refer to the SN of AD 185, then this is a relic of a Proto-Polynesian record that was passed down to become part of Maori astronomical lore and was eventually documented

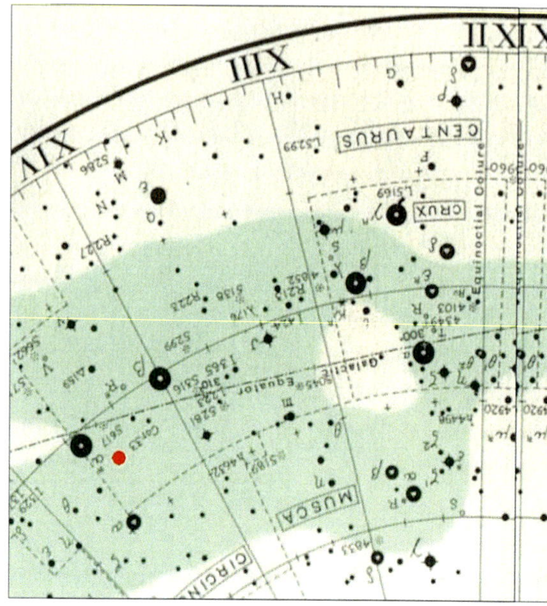

Fig. 3.11 The location of the G315.4−2.3 SNR, shown in *red*, relative to *Nanmen* and Crux (base map after Norton 1957)

by Henry Stowell and Elsdon Best. We therefore might expect to find other Polynesian records relating to this supernova, although none has yet been documented (see Liller 2000: 133). However, the fact that the Easter Island term, *maáhu*, means 'to vanish' (William Liller, personal communication, 2003) may be more than a coincidence.

Acknowledgements I am grateful to the following for their assistance: Jenni Chrisstoffels (Alexander Turnbull Library, Wellington, NZ), Dr David Green (Cambridge University, England), Dr Duane Hamacher (University of New South Wales, Sydney) and Professor Richard Stephenson (Durham University, England). I also wish to thank Duane Hamacher for reading and commenting on the first draft of this chapter, and the Alexander Turnbull Library and David Green for kindly supplying Figs. 3.1 and 3.6.

References

Best, E. (1922). *Astronomical Knowledge of the Maori.* Wellington: Government Printer (Dominion Museum Monograph No. 3).

Busser, J.-U., Egger, R., & Aschenbach, B. (1996). G299.2−2.9: A new galactic supernova remnant. *Astronomy and Astrophysics, 310*, L1–L4.

Chauvin, M. (2000). Useful and conceptual astronomy in ancient Hawaii. In H. Selin & X. Sun (pp. 91–125).

Clark, D. H., & Stephenson, F. R. (1975). The remnants of the supernovae of AD 185 and AD 393. *The Observatory, 95*, 190–195.

Davidson, K., & Humphreys, R. (1997). Eta Carinae and its environment. *Annual Review of Astronomy and Astrophysics, 35*, 1–32.

Duncan, A. R., & Green, D. A. (2000). The supernova remnant RX J0852.0–4622: radio characteristics and implications for SNR statistics. *Astronomy and Astrophysics, 364*, 732–740.

Frew, D. J. (2004). The historical record of Eta Carinae. I: the visual light curve, 1595–2000. *Journal of Astronomical Data, 10*(6), 1–76.

Gibbons, P. J. (1996). Stowell, Henry Matthew 1859–1944. In *Dictionary of New Zealand Biography* (Vol. 3). www.teara.govt.nz/en/biographies/3s38/stowell-henry-matthew.

Green, D. A. (1991). Limitations imposed on statistical studies of Galactic supernova remnants by observational selection effects. *Publications of the Astronomical Society of the Pacific, 103*, 209–220.

Green, D. A., & Stephenson, F. R. (2003). Historical supernovae. In K. W. Weiler (Ed.), *Supernovae and Gamma-Ray Bursters* (pp. 7–19). Berlin: Springer.

Green, D. A., & Orchiston, W. (2004). In search of *Mahutonga*: a possible supernova recorded in Maori astronomical traditions? *Archaeoastronomy, 18*, 110–113.

Green, D. A. (2014). *A Catalogue of Galactic Supernova Remnants (2014 May version)*. Electronic document http://www.mrao.cam.ac.uk/surveys/snrs/.

Hamacher, D. W. (2014). Are supernovae recorded in Indigenous astronomical traditions? *Journal of Astronomical History and Heritage, 17*, 161–170.

Hamacher, D. W., & Frew, D. J. (2010). An Aboriginal Australian record of the Great Eruption of Eta Carinae. *Journal of Astronomical History and Heritage, 13*, 220–234.

Herschel, J. F. W. (1847) *Results of Astronomical Observations Made at the Cape of Good Hope*. London: Smith, Elder & Co.

Higham, T., Anderson, A., & Jacomb, C. (1999). Dating the first New Zealanders: the chronology of Wairau Bar. *Antiquity, 73*, 420–427.

Ho, P. Y. (1962). Ancient and medieval observations of comets and novae in Chinese sources. *Vistas in Astronomy, 5*, 127–225.

Holdaway, R., & Jacomb, C. (2000). Rapid extinction of moas (Aves: Dinornithiformes) in New Zealand. *Science, 287*, 2250–2255.

Humphreys, R. M., Davidson, K., & Koppelman, M. (2008). The early spectra of Eta Carinae 1892 to 1941 and the onset of its high excitation emission spectrum. *Astronomical Journal, 135*, 1249–1263.

Hsi, T. (1958). A new catalog of ancient novae. *Smithsonian Contributions to Astrophysics, 2*(6), 109–130.

Liller, W. (2000). Ancient astronomical monuments in Polynesia. In H. Selin & X. Sun (pp. 127–160).

McIntosh, R. A. (1965). Maoris had name for Pole star. *New Zealand Herald*, August 16.

Martocchia, A., & Polcaro, V. F. (2009). SN 185 and its proposed remnants. *Memorie della Società Astronomica Italiana, 80*, 52–53.

Morris, P. W., Waters, L. B. F. M., Barlow, M. J., Lim, T., de Koter, A., Voors, R. H. M., et al. (1999). Discovery of a massive equatorial torus in the η Carinae stellar system. *Nature, 402*, 502–504.

Norton, A. P. (1957). *A Star Atlas and Reference Handbook*. London: Gall and Inglis.

Orchiston, W. (2000). A Polynesian astronomical perspective: rhe Maori of New Zealand. In H. Selin & X. Sun (pp. 161–196).

Orchiston, W. (2002). Southern Hemisphere observations. In H. Rickman (Ed.), *Highlights of Astronomy* (Vol. 12, pp. 322–325). San Francisco: Astronomical Society of the Pacific.

Orchiston, W., & Slee, B. (2006). Early Australian observations of historical supernova remnants at radio wavelengths. In K.-Y. Chen, W. Orchiston, B. Soonthornthum, & R. Strom (Eds.), *Proceedings of the Fifth International Conference on Oriental Astronomy* (pp. 43–56). Chiang Mai: Chiang Mai University.

Schaefer, B. E. (1995). 'Supernova' 185 is really a nova plus Comet P/Swift-Tuttle. *Astronomical Journal, 110*, 1793–1795.

Selin, H., & Sun, X. (Eds.), (2000). *Astronomy Across Cultures: A History of Non-Western Astronomy*. Dordrecht: Kluwer.

Slane, P., Vancura, O., & Hughes, J. P. (1996). A study of the evolutionary state of the supernova remnant G299.2−2.9. *Astrophysical Journal, 465*, 840–844.

Smith, N. (2006). The structure of the Homunculus. I. Shape and latitude dependence from H_2 and [Fe II] velocity maps of η Carinae. *Astrophysical Journal, 644*, 1151–1163.

Smith, N., & Frew, D. J. (2011). A revised historical light curve of Eta Carinae and the timing of close periastron encounters. *Monthly Notices of the Royal Astronomical Society, 415*, 2009–2019.

Smith, N., Gehrz, R. D., Linz, P. M., Hoffmann, W. F., Hora, J. L., Mamajek, E. E., & Meyer, M. R. (2003). Mass and kinetic energy of the Homunculus Nebula around η Carinae. *Astronomical Journal, 125*, 1458–1466.

Stephenson, F. R., & Green, D. A. (2002). *Historical Supernovae and their Remnants*. Oxford: Oxford University Press.

Stothers, R. (1977). Is the supernova of AD 185 recorded in ancient Roman literature? *Isis, 68*, 443–447.

Stowell, H. M. (Hare Hongi), (1911). *Maori-English Tutor and Vade Mecum*. Christchurch: Whitcombe & Tombs.

Whiteoak, J. B. Z., & Green, A. J. (1996). The MOST supernova remnant catalogue (MSC). *Astronomy and Astrophysics Supplement, 118*, 329–380.

Zhao, F. Y., Strom, R. G., & Jiang, S. Y. (2006). The guest star of AD 185 must have been a supernova. *Chinese Journal of Astronomy and Astrophysics, 6*, 635–640.

Part II
Cook Voyage Astronomy and New Zealand

Chapter 4
Astronomy on Cook's First Voyage: Mercury Bay and Queen Charlotte Sound, 1769–1770

Abstract Cook's First Voyage to the South Seas primarily was undertaken in order to observe the 3 June 1769 transit of Venus from Tahiti. For this purpose, two official astronomers were assigned to the *Endeavour*, an ex-Whitby collier. One was Charles Green and the other was none other than the commander of the vessel, Lieutenant James Cook, who therefore wore two 'hats' during the voyage. Following a successful transit campaign Cook sailed southwest in search of the Great Southern Continent, and eventually came upon New Zealand. As they sailed along the coast Cook and Green used astronomical observations to determine the latitude and longitude of the *Endeavour*, and they carefully mapped the coastline. They also anchored in what is now known as Mercury Bay in order to observe the 9 November 1769 transit of Mercury. This was the first time that scientific astronomical observations were made from New Zealand soil. By the time the *Endeavour* completed its circumnavigation of New Zealand Cook and Green had established that although this landmass consisted of two large mountainous islands it was not a viable candidate for the missing continent.

4.1 Introduction

The eighteenth century was an important era in world history as the nations of Europe and the Americas acquired an enlightened 'world view'. In the Pacific region, voyages of exploration and discovery undertaken for the most part by the British and French revealed a myriad of 'island universes' inhabited by populations embodying the best elements of Rouseau's 'noble savage'. Of course, some groups were deemed to be more 'noble' than others (Campbell 1980)! One direct result of these voyages was that they triggered wholesale changes—often with devastating consequences—in the indigenous cultures that were encountered (see Howe 1984).

W. Orchiston, *Exploring the History of New Zealand Astronomy*,
Astrophysics and Space Science Library 422,
DOI 10.1007/978-3-319-22566-1_4

In addition to written accounts of Pacific peoples, some of which were to appear in the form of popular books, these European voyages produced an abundance of maps, charts and other pictorial records (e.g. Joppien and Smith 1985–1997; Skelton 1969). They also provided a veritable treasure trove of exotic flora, fauna, geological specimens and indigenous 'curiosities' for the public museums and private collections of Europe and Britain (for the Cook voyage ethnographic collections alone see Kaeppler 1978 and references therein). Only now are we starting to realise the full research potential of these widely-scattered collections.

Of all the European voyages to the South Seas it is those associated with James Cook which have left us the richest overall scientific legacy. Cook made three different voyages, visiting New Zealand on each occasion, and Queen Charlotte Sound was his favourite revictualling centre (see Table 4.1). It featured an excellent protected anchorage; readily-available supplies of fresh water, timber, edible plants, fish, shellfish and birds; and a local population of Maoris which, for the most part, was very friendly and supportive. Fish were often acquired by trade from these people, who were seen to be "… far more expert in catching [them] than we." (Beaglehole 1969: 288). The Sound also offered 'spruce beer', native celery and 'scurvy grass', which Cook soon discovered were useful in combatting the dreaded scurvy. Meanwhile, Ship Cove proved an ideal setting for a makeshift 'hospital', and in this bay and nearby ones the wooden casks could be repaired and decaying bread and mouldy ship's biscuit dried out while those on board ship were busy repairing the sails, ironwork and rigging, and cleaning and caulking the hull. Revictualling stop-overs were rather hectic affairs! The location of Queen Charlotte Sound, other Table 4.1 anchorages, and New Zealand localities mentioned elsewhere in this chapter are shown in Fig. 4.1.

Table 4.1 New Zealand anchorages during Cook's three voyages to the South Seas

Voyage	Date	Location
First	1769 October	Poverty Bay
	1769 October	Anaura Bay
	1769 October	Tolaga Bay
	1769 November	Mercury Bay
	1769 November	Thames
	1769 November–December	Bay of Islands
	1770 January–February	Queen Charlotte Sound
Second	1773 March–May	Dusky Sound
	1773 April–June	Queen Charlotte Sound
	1773 November	Tolaga Bay
	1773 November–December	Queen Charlotte Sound
	1774 October–November	Queen Charlotte Sound
Third	1777 February	Queen Charlotte Sound

Fig. 4.1 New Zealand localities mentioned in this chapter are shown in *red*

Astronomy had a crucial role to play on all three Cook voyages to the South Seas and the astronomers therefore were important members of the ships' crews. In addition to their invaluable observations in such fields as meteorology, oceanography and seismology (not to mention botany, ethnology, geography, geology and zoology), they used nautical astronomy to pin-point the positions of the scattered islands encountered. Astronomy and navigation were mutually inseparable (Cotter 1968; May 1973; Taylor 1968), and without the astronomers these voyages could have ended in tragedy.

The major pre-occupation of the various Cook voyage astronomers was therefore the calculation of latitude and longitude, and each relied on astronomical observations. Latitude was most easily determined by recording the altitude of the Sun as it crossed the meridian. Knowing the Sun's declination, which was recorded in the *Nautical Almanac*, it was then a comparatively easy matter to calculate the latitude of the point of observation.

Finding longitude was a more taxing problem, as it depended upon accurate time-keeping (Atkinson 1963; Howse 1983). One commonly-used procedure involved recording the occurrence times of particular astronomical events in local time and comparing these, in each instance, with the known times of the events at Greenwich. Lunar occultations, planetary transits, and phenomena of Jupiter's satellites were most favoured, but the actual longitude computations involved many corrections and minor calculations, and chances of error were high. Nevertheless, the principle involved was simple, and is clearly illustrated by the following example, recorded by Cook on 29 June 1770:

> This night Mr Green and I observed an Emersion of Jupiter first Satellite which happen'd at 2h58′53″ in the AM, the same Emersion happend at Greenwich according to calculation on the 30th at 5h17′43″ in the PM; the difference is 14h18′50″ equal to 214°42′30″ of Longitude which this place is west of Greenwich … (Beaglehole 1968: 353).

An alternative method was to measure the angular separation of the Moon from the Sun or selected stars, and to obtain the longitude by reading off the listed figures in the *Nautical Almanac*. This was the so-called 'lunars' method.

A basic trouble with these methods was that they depended upon a very accurate knowledge of local time, but the clocks available to voyagers at the start of the eighteenth century were far from reliable. Back in 1714 the Board of Longitude in London had offered a substantial prize to whoever was able to devise a means of accurately determining longitude at sea, "… and during the 1760s and 1770s both the lunar distance method and the chronometer were perfected." (Gloria Clifton, personal communication 2014). The 'lunars' method was used on all three Cook voyages, but chronometers only went on the Second and Third Voyages.

In addition to its place in latitude and longitude determinations, astronomy had another interesting role to play in the Pacific during the eighteenth century. A basic 'yardstick' of astronomical measurement was the distance from the Earth to the Sun, and at that time the best-known method of calculating this was to make accurate telescopic measurements of a transit of Venus (see Woolley 1970).

Transits of Venus are rare events, and take place when the Sun, Venus and the Earth (in that order) are in alignment and Venus is seen to cross the disk of the Sun. Such transits occur in pairs eight years apart, with more than one hundred years separating successive pairs of transits. In 'recent' times, there have been transits of Venus in 1761 and 1769, and in 1874 and 1882. The most recent pair of transits occurred in 2004 and 2012.

Attempts to determine the Earth-Sun distance on the basis of observations made during the 1761 transit of Venus proved unsuccessful (see Orchiston 2005; Woolf 1959), and the Royal Society of London formed a Transit of Venus Committee which discussed the possibility of sending an expedition to the South Seas to observe the 1769 transit (Beaglehole 1969: cv). In 1767, the Committee decided to dispatch two observers to Tahiti, and the King agreed to allocate the necessary funds.[1] Thus, Cook's First Voyage to the Pacific came into being, and was largely motivated by astronomical concerns (Orchiston 1998a, 2004, 2005, 2015). Moreover, it was so successful that two further voyages were sanctioned, but this time under the direction of the Admiralty.

As a result of the First Voyage, New Zealand gained its first—albeit transient—scientists, and local astronomy entered a new era. The aim of this chapter is to describe the various astronomers involved in Cook's First Voyage to the South Seas, their scientific instruments, and their New Zealand-based astronomical observations. But first, Sect. 4.2 provides further information about the planning of the voyage and the selection of a suitable vessel. The following section then introduces us to Cook and Green. In Sect. 4.4 we discuss their telescopes, quadrants and sextants; and their astronomical clocks. This leads us into Sect. 4.5 where we examine the ways in which the two astronomers put their instruments to use while visiting New Zealand.

Although much has been written about New Zealand in 1769–1772 based on the reports of the first European explorers and artifacts collected at the time (e.g. see Groube 1965; Jones 1983, 1984, 1989; Kennedy 1969; Mead 1969; Orchiston 1972, 1974a, b, 1975a, b; Salmond 1991; Simmons 1981), until the Carter Observatory in Wellington, New Zealand, published *Nautical Astronomy in New Zealand: The Voyages of James Cook* (Orchiston 1998b) astronomy had been all but ignored, except for superficial references by McIntosh (1957), Mackrell (1985) and Orchiston (1971, 1994, 1997, 1998a, 2004). Much of this chapter is based on material included in my Carter Observatory monograph.

---

[1] It is not surprising that King George III supported the Royal Society's transit of Venus proposal, as he was

> … the first British monarch to have studied science as part of his formal education. He was known 'for his love of the sciences' and had been taught physics and chemistry as a boy, holding a particular interest in *scientific instruments*, *astronomy*, the quest for longitude, botany and *the work of the Royal Society*. (Wulf 2012: 101; my italics).

4.2 Planning for the Expedition

4.2.1 Introduction

As we have seen, Cook's First Voyage to the Pacific ostensibly was a scientific expedition inspired by the 3 June 1769 transit of Venus. By 1769 the Seven Years' War was fast becoming a distant memory (e.g. see Baugh 2011). Britain and Prussia may have won the war (which was against Austria, France, Russia, Saxony and Sweden), but what better way for Britain to demonstrate her scientific supremacy than through astronomy, and particularly the up-coming transit of Venus? Elsewhere (Orchiston 2005: 52) I have used the term "… fighting the peace …" to characterise this philosophy. Thomas Hornsby (1733–1810), the highly-respected Savilian Professor of Astronomy at Oxford University (Wallis 2008), more or less suggested this when he wrote:

> It behoves us therefore to profit as much as possible by the favourable situation of Venus in 1769, when we may be assured the several Powers of Europe will again contend which of them shall be most instrumental in contributing to the solution of this grand problem. (Hornsby 1765: 343).

Clearly, British pride was at stake.

In a very useful 19-page paper, "On the transit of Venus in 1769" that Hornsby (1765) had published in the *Philosophical Transactions of the Royal Society* he identified three ideal sites for the British transit stations: North Cape in Norway, Hudson Bay in Canada and a suitable site in the Pacific Ocean. The trouble was that no known land existed at this Pacific location at the time Hornsby wrote his paper, but fortune favoured the British because in May 1768—just over a year before the all-important transit—*HMS Dolphin* arrived back in England after a lengthy voyage of exploration in the Pacific (see Robertson 1948), and its commander, Samuel Wallis (1728–1795), reported his discovery on 17 June 1767 of 'King George III Island' (now called Tahiti) at precisely the desired Pacific location! He also reported that the climate was pleasant, the local Tahitian people (eventually) were friendly and co-operative, and that Port Royal (now Matavai Bay) offered an excellent harbour. The Transit of Venus Committee then adopted all three observing sites (see Beaglehole 1963(I): 20–21). However, Tahiti (Fig. 4.2) would be the principal station, for two important reasons: the beginning *and* end of the transit could be observed there, and the duration of the transit would be significantly shorter than at the North Cape or Hudson Bay (Hornsby 1765). Now Cook's First Voyage to the Pacific could proceed.

4.2.2 Finding a Suitable Vessel

With funding approved by King George III the Admiralty then spent £2,307.5s.6d purchasing a suitable vessel, the 370 ton *Earl of Pembroke* shown in Fig. 4.3

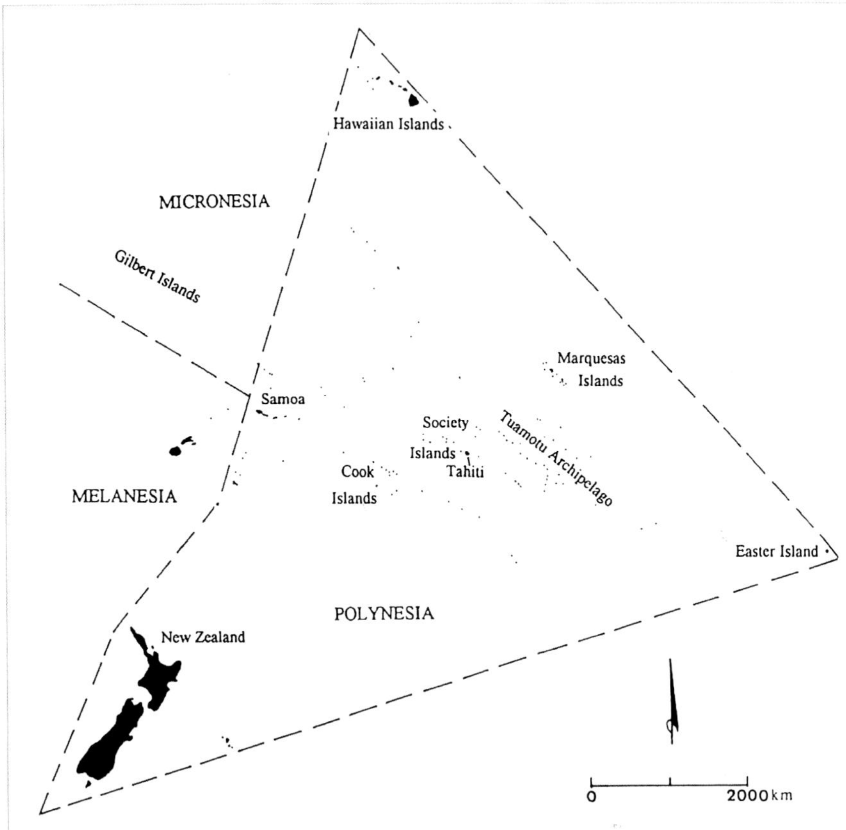

Fig. 4.2 Map showing the location of Tahiti, the largest island in the Society Islands, near the centre of the 'Polynesian Triangle', which is bordered by the Hawaiian Islands, Easter Island and New Zealand (*Map* Wayne Orchiston)

(Deptford Yard Officers 1768a), and a further £2,293.17s.7d modifying it for the voyage (Deptford Yard Officers 1768b). This ex-Whitby collier was renamed *Endeavour*, and was a type of vessel very familiar to Cook from his pre-naval days. However, even after refitting, space was at a premium, and it was to prove cramped quarters for close on 100 men. These comprised commissioned officers, warrant and petty officers, able seamen, servants, marines and supernumeraries. Only these last-mentioned were non-naval personnel, and they consisted of astronomers, other scientists, and their retinues (including servants).

Ship-board life at this time was far from idyllic. Dr Johnson went so far as to suggest that

> No man will be a sailor who has contrivance enough to get himself into jail; for being in a ship is being in a jail, with the chance of being drowned. A man in a jail has more room, better food and commonly better company. (cited in Lloyd 1968: 209).

Fig. 4.3 A painting by Thomas Luny of the *Earl of Pembroke* leaving Whitby Harbour in 1768 (https://en.wikipedia.org)

Apart from the appalling food and frequent danger which typified a maritime lifestyle, there was sub-standard hygiene and ever-present disease and illness to contend with. Pay—when it eventually arrived—was poor, and promotion was, in the main, elusive (see Lloyd 1968, Chap. 11).

Marginally better off were the officers and supernumeraries, who had superior pay and quarters, their own food and drink, and more free time during shore-based stop-overs. The astronomers therefore were in a privileged position.

Now that a suitable vessel had been found it was a matter of finding suitable astronomers for the voyage.

4.3 The Astronomers

4.3.1 Introduction

Two different astronomers were allocated the the *Endeavour* for Cook's First Voyage. Surprisingly, one of these was Lieutenant James Cook who was selected by the Council of the Royal Society to command the expedition, but also would serve as an astronomer alongside Charles Green. Their main objective was to successfully observe the 1769 transit of Venus from Tahiti (see Beaglehole 1968: 619–620; Orchiston 1998a, b, 2005). Cook already had an onerous workload as leader of the expedition, and only agreed to serve also as an astronomer when he was offered a salary supplement of 100 guineas, paid by the Royal Society (Beaglehole 1968: cxxvi). The second astronomer was Charles Green, who was appointed solely as an astronomer, and agreed to a salary of 200 guineas for the voyage itself, plus 100 guineas per year if the voyage extended beyond two years (Beaglehole 1974: 131).

Let us now look in turn at each of the astronomers.

4.3.2 *Lieutenat James Cook (1728–1779)*

James Cook (Fig. 4.4) was born on 27 October 1728 in Marton, Yorkshire (see Fig. 4.5), to a Scottish labourer father and English mother (Beaglehole 1974). Soon after, his father accepted a post of foreman on a farm at Great Ayton, and it was here that young Cook grew up and received a school education (Beaglehole 1968: cvi). To his fellow pupils, Cook

> … had such an obstinate and sturdy way of his own, as made him sometimes appear in an unpleasant light; notwithstanding which, *there was a something* in his manners and deportment, which attracted the reverence and respect of his companions. (Beaglehole 1974: 5).

At age seventeen he was apprenticed to William Sanderson, a grocer and draper at the small nearby fishing port of Staithes. Eighteen months later, in July 1746, he went to sea, serving a three year apprenticeship with John Walker, a respected

Fig. 4.4 Oil painting of Captain James Cook by Nathaniel Dance-Holland, ca. 1775, and now in the National Maritime Museum, Greenwich (https://en.wikipedia.org)

Fig. 4.5 English and Scottish localities mentioned in this chapter are shown in *red*

Quaker ship-owner and coal-shipper in the bustling port of Whitby. During his time serving on colliers engaged in the North Sea coastal trade, Cook received a solid grounding in mathematics and navigation (Beaglehole 1968: cvi).

By 1752, at age 24, he was promoted to Mate, and in 1755 Walker offered him a command. Cook turned down this attractive proposition for he believed that with a war in the offing even better opportunities would emerge if he joined the Royal Navy—and this proved to be so (Beaglehole 1968: cvi–cvii). However, this would surely be a brave move, for in comparison to the merchant service, the Royal Navy was generally not seen as the preferred option:

> Its physical conditions were worse; its pay was worse; its food was worse, its discipline was harsh, its record of sickness was appalling. To the chance of being drowned could be added the chance of being flogged, hanged or being shot, though it was true that deaths in battle were infinitely fewer than deaths from disease. The enemy might kill in tens, scurvy and typhus killed in tens of hundreds. (Beaglehole 1974: 15).

In spite of this far from appealing scenario, on 17 June 1755 Cook joined the Navy as an able seaman. At first he was assigned to the *Eagle* which was soon to be commanded by Captain (later Sir) Hugh Palliser (1723–1796), yet another Yorkshireman destined for high honours. Within a month, Cook was promoted to Master's Mate, and during his two years on the *Eagle* rose to Boatswain and then Mate (Beaglehole 1968: cvii).

On his 29th birthday, in 1757, Cook began a new enterprise that was to profoundly influence his future. In joining the *Pembroke* as the equivalent of a

Navigating-Lieutenant, he was able to chart the St Lawrence River during the siege of Louisbourg. He soon transferred to the *Northumberland* as Master, and began

> … to emerge from that valuable body of persons, the masters of His Majesty's ships, as an unusually valuable person; and that the senior officers with whom he has come in contact are aware of the fact. In the context of naval journals, under their standard headings, he can virtually be classed, along with court martials and Publick Demonstrations of Joy, as a Remarkable Occurrence. (Beaglehole 1974: 55).

After carrying out further cartographic work on the St. Lawrence River, he married in London on 21 December 1762, and four months later embarked for North America once more to chart the coasts of Newfoundland and Nova Scotia. Through Palliser's influence, he was then assigned his own command, in 1764. His vessel was the sixty-ton naval schooner, *Grenville*, with an eventual crew of twenty (Beaglehole 1974: 126), and he was ordered to return to Newfoundland and continue the coastal survey. Each winter for the next three years there were interludes in London, while the survey continued.

Meanwhile, Cook used the survey to extend his education, carefully and successfully absorbing the contents of Leadbetter's two books, *A Compleat System of Astronomy* and *The Young Mathematician's Companion* (Villiers 1971: 39).

Through his North American exploits and the succession of charts that he published, Cook quickly established a reputation as a navigator and hydrographic surveyor (Beaglehole 1971; Ritchie 1978; Skelton 1954; White 1970). This reputation also extended to astronomy (Dyson 1929), and in 1766 he observed a solar eclipse from Newfoundland and published an account of this event in the *Philosophical Transactions of the Royal Society* (Fig. 4.6; Cook 1767). By this stage, Cook had obviously "… acquired the taste for astronomical determination of longitude." (Beaglehole 1974: 90), and so he wrote to the Admiralty suggesting that they should supply him with an astronomical telescope. They were pleased to oblige.

In May 1768 Cook was offered command of the *Endeavour*, with a rank of First Lieutenant, and agreed also to serve as one of the two astronomers. He was the ideal candidate: his name was well known to the Royal Society and the Admiralty, and Palliser—in his capacity as Comptroller of the Navy Board—was also able to provide his support (Beaglehole 1968: cviii). Thus began Cook's affair with the Pacific.

By the end of the First Voyage the Pacific had left its mark on Cook. To quote Beaglehole (1968: cxii–cxiii): "We may say that he departed from England a good sailor, a first-rate marine surveyor, an able mathematician; and returned a great commander, a great discoverer, and a man with a greatly heightened sense of the scope of human thought."

In September 1771 the Admiralty decided on a second South Seas voyage, with Cook in command, and in July 1772 the *Resolution* and *Adventure* sailed from England. Each vessel had its own astronomer, for Cook was not to enjoy the privilege of this dual responsibility once more until his Third and final Voyage.

XXIV. *An Obſervation of an Eclipſe of the Sun at the Iſland of* New-found-land, *Auguſt* 5, 1766, *by Mr.* James Cook, *with the Longitude of the Place of Obſervation deduced from it: Communicated by* J. Bevis, *M. D. F. R. S.*

Read April 30, 1767.

MR. Cook, a good mathematician, and very expert in his buſineſs, having been appointed by the Lords Commiſſioners of the Admiralty, to ſurvey the ſea coaſts of New-found-land, Labradore, &c. took with him a very good apparatus of inſtruments, and among them, a braſs teleſcopic quadrant made by Mr. John Bird.

Being, Auguſt 5, 1766, at one of the Burgeo Iſlands near Cape Ray, latitude 47° 36′ 19″, the ſouth-weſt extremity of New-found-land, and having carefully rectified his quadrant, he waited for the eclipſe of the ſun; juſt a minute after the beginning of which, he obſerved the zenith diſtance of the ſun's upper limb 31° 57′ 00″; and, allowing for refraction and his ſemidiameter, the true zenith diſtance of the ſun's centre 32° 13′ 30″, from whence he concluded the eclipſe to have begun at 0^h 4′ 48″ apparent time, and by a like proceſs to have ended at 3^h 45′ 26″ apparent time.

Fig. 4.6 The first page of Cook's short paper in the *Philosophical Transactions of the Royal Society* reporting his observation of the 5 August 1766 partial solar eclipse

This Third Voyage also involved two vessels, with Cook in charge of the entire expedition. It was on this fateful voyage that James Cook died and England lost one of their favourite sons (Beaglehole 1974).

4.3.3 Charles Green (1734–1771)[2]

The second astronomer on the *Endeavour*, Charles Green, was a Yorkshireman like Cook, but was born in Swinton in 1734 (making him six years Cook's junior). Green's father was a prosperous farmer (Howse 1993). It seems that Green's education fell to his older brother, the Reverend John Green, who was the master of a school, "… the Academy in Denmark Street, London." (Wales 2008). Charles so excelled in mathematics that he was appointed an assistant teacher at the school. While at the school he taught himself astronomy, so successfully as it turned out that towards the end of 1760 he joined the staff of the Royal Observatory at Greenwich as assistant to James Bradley (1692–1762; Fig. 4.7), the Astronomer

[2]This section draws heavily on Orchiston 2015.

Fig. 4.7 A painting of James Bradley by Thomas Hudson, now in the National Portrait Gallery, London (https://en.wikipedia.org)

Royal (Kippis 1788). For many, the idea of working at a professional observatory may have seemed exciting, but for an 'assistant' at the Royal Observatory at this time nothing could be further from the truth:

> Nothing can exceed the tediousness and ennui of the life the assistant leads in this place, excluded from all society, except, perhaps, that of a poor mouse which may occasionally sally forth from a hole in the wall, to seek after crumbs of bread dropt by his lonely companion at his last meal … Here forlorn, he spends days, weeks, and months, in the same long wearisome computations, without a friend to shorten the tedious hours, or a soul with whom he can converse. (cited in Morris 1980).

Bradley planned to observe the 1761 transit from the Royal Observatory but became ill and could not do so. Instead it was his close friend, the Reverend Nathaniel Bliss (1700–1764; Fig. 4.8), Savilian Professor of Geometry at Oxford, who observed it, assisted by Green (see Bliss 1761).

When Bradley died in 1762 Bliss became Astronomer Royal, and Green remained at the Observatory. In 1763 he and the Reverend Dr Nevil Maskelyne (1732–1811) sailed to Barbados in the Caribbean aboard the *Princess Louisa*, by order of the Board of Longitude, so that they could determine the longitude of the capital, Georgetown, by observing Jovian satellite phenomena and using Maskelyne's system of lunar distances, and then compare both results with the longitude offered by Harrison's H4 marine chronometer (see Higgitt 2010).

Fig. 4.8 A painting of the Reverend Nathaniel Bliss by an unknown artist, now in the National Maritime Museum, London (https://en.wikipedia.org)

John Harrison (1693–1776) was keen to claim the Board of Longitude's £20,000 'longitude prize' (see Betts 2006; Sobel 1995), but Maskelyne—with an obvious conflict of interest (see Kippis 1788)—was equally determined the prize should be awarded for his system of 'lunars' that also could be used to determine longitude on land or while at sea (Howse 1989). Green was not involved in this 'competition', and Harrison's chronometer did not go on the *Endeavour's* voyage to the South Seas.

Although Maskelyne apparently wrote a very good report on Green (Paulding 2000), during the Barbados expedition Maskelyne and Green did not see eye-to-eye on all matters because when Bliss died suddenly in 1764 and Maskelyne became the new Astronomer Royal, after a brief interval he and Green argued and Green promptly resigned (Kippis 1788). But, as Higgitt (2010) perceptively points out,

> … because Bradley and Bliss … were both quite unwell while Green was there, he was effectively in charge of the Observatory himself for some of the time. So he was a sort of Astronomer Royal stand-in for some of the period … and he [also] tided over the period between Bliss's death and the arrival of Maskelyne as Astronomer Royal.

After investigating London's water supply, in 1768 Green joined the Royal Navy and served as a Purser on HMS *Auror*a (Kippis 1788). A Purser was a warrant officer whose primary responsibility was meant to be the ship's accounts and supplies, but in reality astronomers were sometimes listed as Pursers (Beaglehole 1968: 119), and they assisted in navigation, where astronomical knowledge was at a premium. So if Green could not work as an official astronomer he did the next best thing.

However, his return to the ranks of 'astronomer' was not long in coming, for later that same year (1768) the Royal Society appointed him as one of the two astronomers on the *Endeavour*. Despite their prior differences, Maskelyne was one of his supporters (Kippis 1788). On 30 July 1769 Cook was sent the following order by the Admiralty:

> You are hereby requir'd and directed to receive the said Mr Charles Green with his Servant [John Reynolds] Instruments and Baggage, on board the said Bark, and proceed in her according to the following Instructions … (cited in Morris 1981).

Following the transit, Green was one of many who were destined to die before the *Endeavour* arrived safely back in England (Fig. 4.9). After safely circumnavigating New Zealand, Green fell ill with scurvy during the exploration and mapping of the east coast of Australia. Cook then headed for Batavia (present-day Jakarta) in the Dutch East Indies (Indonesia) as the *Endeavour* needed further repairs if it was to successfully sail to England. However, Batavia would turn out to be a hell-hole for the *Endeavour* and its crew. Looks can be deceiving, for at the time Batavia was:

> … a rather picturesque city showing evidence in its architecture of the Dutch colonists. However, it was built on swampy ground. The climate was most unappealing - high temperatures and high humidity proved debilitating. There were frequent thunderstorms. Additionally, in 1699 Batavia had suffered a severe earthquake. The rivers about it were choked with mud and flooded the surrounding country. Batavia became notorious for being unhealthy and was in danger of being abandoned. In the 22 years from 1730 to 1752,

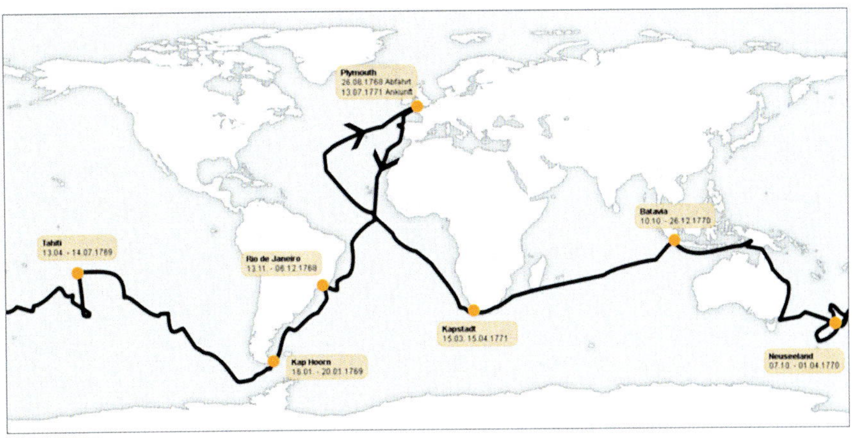

Fig. 4.9 A map showing the route of the *Endeavour* on Cook's First Voyage to the South Seas. Green first was sick as the *Endeavour* explored the east coast of Australia. He then contracted dysentery in Batavia (the pink spot to the west of and a little above Australia), and died soon afterwards while the *Endeavour* was crossing the Indian Ocean. He was one of many on the *Endeavour* who fell sick in Batavia

> 1,100,000 deaths are said to have been recorded. *Endeavour* had been a healthy ship but at Batavia the ship's company were exposed to dysentry, malaria and a variety of other tropical diseases. Green's servant, Reynolds, died of dysentry here on 18 December 1770 ... (Morris 1981).

At the time, Batavia had a thriving astronomical society with an observatory, and Pastor Johan Maurits Mohr (1716–1775) also had an impressive private observatory atop his mansion on the outskirts of the city (Fig. 4.10). People with an interest in astronomy, were delighted by the *Endeavour*'s visit (ibid.), and since Mohr had observed the 1761 *and* 1769 transits (see van Gent 2005; Zuidervaart and van Gent 2004), it is a safe assumption that he met with Green (and maybe also Cook). Subsequently, the *London Evening Post* printed a letter from 'a gentleman' on board the *Endeavour*, which described how "... great respect was paid here to Mr. Green by the principal people of Batavia, but no particular notice was taken of the rest of us by the Dutch." (cited in Morris 1981).

Yet all was not well for Green, as towards the end of his stay in Batavia he also contracted dysentery, and his condition continued to deteriorate. Nearly two weeks after the *Endeavour* left Batavia he was gravely ill and on 29 January 1771 Cook recorded in his journal:

> In the night Died M[r] Charls Green who was sent out by the Royal Society to Observe the Transit of Venus; he had long been in a bad state of hilth, which he took no care to repair but on the contrary lived in such a manner as greatly promoted the disorders he had had long upon him, this brought on the Flux which put a period to his life. (Beaglehole 1968: 448).

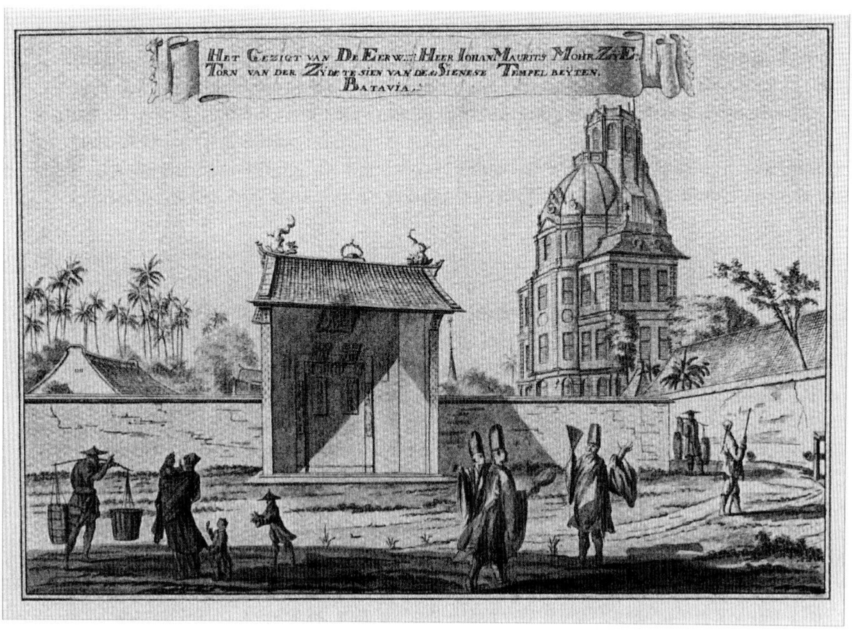

Fig. 4.10 A view of Pastor Mohr's mansion and observatory (after van Gent 2005: 69)

He was buried at sea, in the middle of the Indian Ocean. Between them, scurvy and dysentery almost destroyed the *Endeavour* at this time: Green was but one of twenty-three different crew members (about a quarter of the entire complement) who died within the space of six short weeks (see Watt 1979). Given this mortality rate, elsewhere I have suggested that "… fatalists could be excused for thinking that the Tahitian transit carried a curse of Tutankhamen-like proportions!" (Orchiston 2005: 58).

During his sojourn on the *Endeavour*, Green had proved himself a competent astronomer, and faithfully trained the officers and some of the seamen in the specifics of nautical astronomy, including the calculation of Maskelyne's 'lunars' (Beaglehole 1968: 599). Among his effects was a log-journal (Green 1768–1770) which Beaglehole thought somewhat pedestrian in nature, but it did show that Green thought himself a wit and that he was "… a highly conscientious as well as sprightly person." (Beaglehole 1968: ccxlii). Meanwhile, his brother-in-law, William Wales (who was destined to go on Cook's Second Voyage as an Astronomer), said that Green "… was a most excellent observer … and tolerably well versed in most branches of mathematics." (cited in Morris 1981).

Notwithstanding extensive searches, no portraits of Charles Green are known to exist (ibid.).

4.4 The Instruments

4.4.1 Introduction

The astronomical instruments assigned to the *Endeavour* by the Royal Society clearly reflect the specific needs of the astronomers:

> 2 Reflecting telescopes of two feet focus, with a Dolland's micrometer to one of them and moveable wires for the other … 2 Wooden Stands for the telescopes with polar axes suited to the Equator … An astronomical quadrant of one foot radius, made by Mr Bird … An Astronomical Clock and Alarm Clock … A Brass Hadley's sextant bespoke by Mr Maskelyne of Mr Ramsden … A Journeyman Clock bespoke of Mr Shelton. (Beaglehole 1968: cxliii).

The astronomers also were supplied with instruments for meteorological and geomagnetic measurements, and with such books, maps and charts as were deemed necessary for them to carry out their assigned duties (e.g. see Beaglehole 1967(II): 1499–1500; 1969: 724).

Most of the astronomical instruments taken on Cooks' voyages were state-of-the-art technology, manufactured by the leading practitioners of the day, Bird, Dollond, Ramsden and Short.

John Bird (1709–1776; Fig. 4.11; Hellman 1932) was born at Bishop Auckland. He was famous for his astronomical quadrants (Bird 1768), which Chapman (1995: 75) has described as "… mechanical, not optical, instruments, and aimed to eliminate

Fig. 4.11 A mezzotint of John Bird in the Royal Museums, Greenwich (https://en.wikipedia.org)

every possible source of tension or imbalance in the fabric, for which they employed the simplest optical system." Bird began his adult life as a weaver, but in 1740 moved to London and joined the well-known instrument-maker, Jonathan Sisson (1690–1747), where he gained his training (see Gould 1976). He then decided to open his own business, and this soon became the centre of the astronomical instrument-trade, "… especially after Bradley [the Astronomer Royal], with £1000 available for new instruments, gave Bird his first large order." (King 1979: 115). Astronomy was Bird's hobby, and he used one of his own reflectors to observe the 1761 transit of Venus and the 1765 annular solar eclipse (Andrews 1992: 122).

Peter Dollond (1731–1820; Fig. 4.12) was part of the London 'Dollond dynasty' of scientific instrument-makers (see King 1948). He began his adult life as a silk-worker, but opened an optical business in 1750 and two years later was joined by his father, John Dollond (1707–1761). In addition to telescopes, the Dollonds manufactured micrometers, sextants and octants (King 1979: 161). Although primarily a maker of instruments, Dollond did observe the 1769 transit of Venus (Maskelyne 1768).

Jesse Ramsden (1735–1800; Fig. 4.13; Chapman 1996; McConnell 2007) was born in Halifax, Yorkshire, and after an apprenticeship to a clothworker moved to London and from 1758 worked for a mathematical instrument-maker. He learnt quickly, and in 1762 opened his own business (Beaglehole 1969; King 1979;

Fig. 4.12 An oil painting of Peter Dollond by John Hoppner and now in the British Optical Association Museum (https://en.wikipedia.org)

Mennim n.d.). He has been described as "… more a mechanical than an optical engineer, his instruments showed combined sound design and elegance." (Andrews 1995: 67). He also was known for his unreliability. Richard Lovell Edgewort, who knew Ramsden personally, once remarked: "Besides his great mechanical genius, he had a species of invention not so creditable, the invention of excuses. He never kept an engagement of any sort, never finished any work punctually, or ever failed to promise what he always failed to perform." (cited in McConnell 2007: 275).

Scottish optician, James Short (1710–1768; Fig. 4.14), was born in Edinburgh, and only began the telescope-making activities for which he became justly famous after completing an M.A. degree and qualifying as a Church of Scotland minister (Bryden 1968). Following some experimentation, Gregorian reflectors with speculum mirrors became Short's forté (Turner 1969), and in 1738 he moved his business from Edinburgh to London (King 1979: 85). Andrews (1996: 99) describes Short as "A most celebrated personality … [who] accrued a fortune by supplying excellent instruments (about 1360) to amateurs and professionals." This level of sales, despite the exorbitant prices that Short placed on his telescopes (see King 1979: 86), attests to their optical superiority. Short also made occasional Cassegrain and Newtonian reflectors. Although best-known as an instrument-maker, Short did use some of his own telescopes for astronomical observations (e.g. see Short 1761–1762); one of his programs involved the 1761

Fig. 4.13 Jesse Ramsden (https://en.wikipedia.org)

Fig. 4.14 A painting of James Short (*Courtesy* Museum of the History of Science, University of Oxford)

transit of Venus (Short 1764). Short was "... a full-faced, well-built man of medium height." (Bryden 1968: 6). A bachelor, Short died while still in his 50 s, leaving an estate reputed to have been valued a £20,000.

Beaglehole (1971: 118) has made the important point that although Cook did not invent any of the astronomical instruments assigned to the *Endeavour* (and *Resolution*), he used them "... with a skill and accuracy that few other men could match. There were men who as observers did match him, but few of them were in the navy." For the most part, they were astronomers. However, Cook was more than an astronomical observer; he had "... a passion for observation of all sorts: birds, animals, people, tides, magnetic variation, everything. His attitude to measurement and inquiry would do credit to any conventionally-trained scientist." (Badger 1970b: 30).

4.4.2 Securing Celestial Positions: The Telescopes, Quadrants and Sextants

Astronomical telescopes were taken on all three Cook voyages, and in the main were used for shore-based observations of Jovian satellite phenomena. On the First Voyage they also were employed to observe the transits of Mercury and Venus. For ease of use, the reflecting telescope was generally placed on the top of a cask, which was filled with ballast and firmly buried in the ground.

It is not clear how many different reflecting telescopes accompanied the *Endeavour*. Four different reflectors are described in the journals. Two of these were "... of two feet focus each, made by the late Mr. James Short, one of which was furnished with an object glass micrometer." (Green and Cook 1771: 398), and were supplied by the Royal Society specifically for observation of the transit of Venus (cf. Fig. 4.15). The third telescope was a reflector of 1 metre focal length owned by Solander, which was described as "... more powerful ..." than the two Royal Society telescopes (Beaglehole 1968: cxliii). The fourth is described by Cook:

> The Navy Board have been pleas'd to supply His Majestys Bark the Endeavour under my command with the Reflector Telescope that was on board the Grenville Schooner for makeing Astronomical Observations at Newfoundland; in order to make it of more general use I have got made a Micrometer for measuring the apparent magnitudes of the Heavenly Bodies, which will be of great service in the observation of the Transit Vinus ... (cited in Beaglehole 1968: 621).

In addition to these four astronomical telescopes there must have been others (see Orchiston 2015), for three different parties succeeded in observing the 1769 transit of Venus and between them had at least seven telescopes (although only the aforementioned four are described in the journals). There were three at Fort Venus itself, two at the Moorea station, and two or possibly three at the Taaupiri station (see Beaglehole 1968: 97–98, 559; Beaglehole 1963(I): 284).

Fig. 4.15 A 10.2-cm (4-in) Gregorian telescope made by Short, and similar to the two taken on the *Endeavour* during the first voyage (http://www.antiquetelescopes.org/18thc.html)

The object glass micrometer supplied with one of the Short telescopes was manufactured by John Dollond (see Fig. 4.16), and is described by him in a paper published in *Philosophical Transactions of the Royal Society* (Dollond 1754). With this ingenious device, which was developed in 1753 (Dollond 1753), astronomers could measure the angular separation of two nearby celestial objects or of two parts of the same object. Dollond (1754: 469) concluded that

> … this micrometer is a complete instrument of its kind; having many advantages above the common sort, without any of their disadvantages; and there is no doubt but, when brought into practice, it will tend much to the advancement of astronomy.

Only time would tell whether this was a realistic prognosis.

With our far from precise knowledge of the astronomical telescopes taken on the three different Cook voyages, it is not surprising that the late Derek Howse (1919–1998), who was then the world's foremost authority on Cook voyage instrumentation, experienced some difficulty correlating these with extant instruments. As a promising start, however, he does associate two Gregorian reflectors by Short lodged in the Science Museum, London, with the 1769 transit of Venus expeditions (Howse 1979: 125). Although one of these (1900-136) has a Dollond micrometer, there is no evidence that it was actually one of the two Short reflectors on the *Endeavour*. But if it was not, we are justified in assuming that it was remarkably similar—if not identical—in appearance. That being the case, it is appropriate that we describe this telescope.

Fig. 4.16 Example of a Dollond object glass micrometer and case (*Courtesy* Collection of Historical Instruments, Harvard University)

Gregorian reflector number 1900-136 is on loan to the Science Museum from the Royal Society (which supplied telescopes for the 1769 transit) and is inscribed with Short's serial number 44/1198 = 24. Short's code was deciphered by David Baxandall during the 1920s, and the numerator refers to the serial number of the telescope of that aperture, the denominator gives the total number of telescopes made to that date, and the value after the equals sign indicates the focal length in inches (King 1979: 87). This instrument therefore has a focal length of 24 inches (61cm). Being a Gregorian reflector, there is a perforated parabolic primary mirror and an ellipsoidal secondary. Both mirrors are made of speculum metal, and focussing was achieved by altering the separation of the secondary mirror from the primary. Associated with this telescope is an object glass micrometer of the type described by Dollond in his 1754 paper. If in fact this telescope is not the one that accompanied Cook and Green in 1768 it nevertheless does give us a very good indication of what the actual instrument would have looked like.

Howse also attributes a second Short reflector in the Science Museum to the 1769 transit. This instrument (number 1939-389) was presented to the Museum by the Air Ministry and the catalog entry specifies that it was "Made by James Short c.1764". It is very similar in appearance and dimensions to the aforementioned Short telescope, and has a similar serial number (42/1195 = 24). Stimson (1985) has established that originally the Royal Society and the Board of Longitude housed their scientific instruments in the same warehouse, where they were cared for by a single curator. As a result, the precise provenance of some of the instruments was lost. After the Board was abolished in 1828, what were thought to be its instruments were transferred to the Royal Navy, and during the 1840s the supposed Royal Society's instruments were relocated to the King's Observatory in Richmond Park, London. This Observatory subsequently was taken over by the British Association and then by the Air Ministry (for the Meteorological Office), which proceeded to transfer some instruments to the Science Museum (Howse, personal communication 1997). Given this historical chain of events, it is reasonable to associate this second Short reflector with the 1769 transit of Venus, but once again there is no proof that this was one of the two instruments assigned to the *Endeavour*.

Of the other two telecopes taken on the First Voyage the 3-ft instrument that was used by Solander was considerably longer than the other telescopes of this type that accompanied the various British transit of Venus expeditions in 1769, so any telescope of this length with reputed Cook First Voyage associations in a museum or private collection warrants close scrutiny. Such a telescope exists in the scientific collections of The Museum of New Zealand Te Papa Tongarewa in Wellington (Accession Number NS000010), but this has a convoluted history with supposed links to either Charles Burney or Sir Joseph Banks, both of whom were closely associated with Cook. In a paper published in the *Journal of the Antique Telescope Society* in 1999 (where the telescope features on the front cover of the journal), I examine the documentation accompanying this telescope, particularly its supposed Cook voyage provenance, and conclude that "… The only possible Cook-voyage association that we have been able to identify for this instrument is that it was … used by Dr Solander in 1769 to observe the transit of Venus. However, the evidence for this is slim and and largely circumstantial." (Orchiston 1999: 8). Later I revisited this decision (Orchiston 2005), and I now believe that this is indeed the 'Solander Telescope' (for further details see Chap. 7).

Finally, we do not know what became of Cook's own telescope, which presumable went to Mrs Cook after his death.

Let us now examine the astronomical quadrants taken on Cook's three voyages. These were used ashore to make positional observations of the Sun, Moon and selected stars, which after an intricate series of calculations produced values of longitude. Meridian observations of the Sun quickly provided the latitude. Chapman (1983) has demonstrated that quadrants underwent a rapid evolution during the late eighteenth century.

During the Cook voyages, the quadrant was usually installed in a tent observatory:

> [Near the tent] … stood the observatory, in which were set up the journeyman clock and astronomical quadrant: this last, made by Mr. Bird, of one foot radius, stood upon the head of a large cask fixed firm in the ground, and well filled with wet heavy sand. (Green and Cook 1771: 398).

Bird's First Voyage quadrant (see Fig. 4.17) came in a packing case about 45.7cm square (Beaglehole 1968: 87). In May 1769 it was damaged by the Tahitians, and although running repairs were carried out by Sporing (Beaglehole 1968: 527–529), there was some doubt about its reliability after this event.

There is considerable confusion as to the current whereabouts of the various astronomical quadrants which went on the three Cook voyages. Although only one Bird quadrant was taken on the First Voyage, there are currently two identical 30.5-cm quadrants (1900-138 and 1900-139) in the Science Museum, London, that were reputedly employed to observe the 1769 transit of Venus. Both are owned by the Royal Society and were placed on long-term loan with the Museum in 1900. Documentation held by the Museum suggests that these two quadrants were made in about 1767, used by Bailey at the North Cape and Dixon on the island of Hemmerfest, and are duplicates "… of the one provided by the Royal Society and

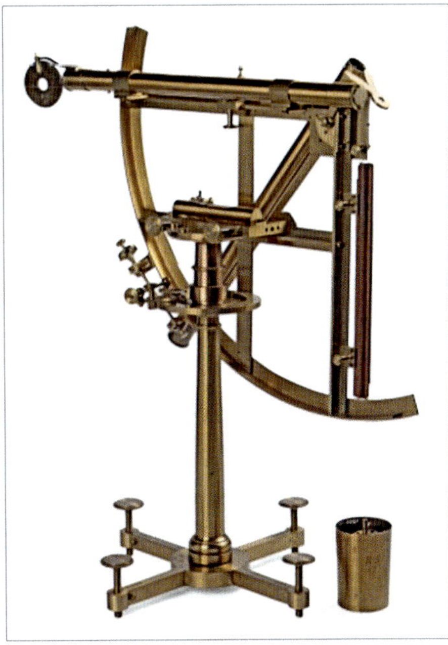

Fig. 4.17 A quadrant by Bird reputedly associated with the Cook voyages (*Courtesy* The Royal Society, London)

used by Cook for observing the transit of Venus at Tahiti." Both of the telescopes have 1.9-cm objectives of 33cm focal length, and the eyepiece end of the lower telescope is "… fitted with verniers, clamping screw and slow motion which traverses the limb which is divided into two scales one into 90° and the other 96°, according to Bird's method, each reading to 1′ of arc …" (Science Museum catalogue entry). Figure 4.17 shows one of these instruments.

While quadrants were used ashore for precise astronomical observations, sextants served the same purpose on board ship. Occasionally, they also were used on shore. On the First Voyage, Cook had his own sextant so the Royal Society had to provide just one sextant, for Green's use (ibid.). This was a 38.1-cm Hadley's sextant with edge bars, and was manufactured by Ramsden.

As of 1979, the late Derek Howse was able to identify four surviving sextants of reputed Cook voyage origin: a 45.7-cm Bird in the Science Museum, London, claimed to have been owned by Cook; a 36.8-cm Ramsden engraved "D.32", privately-owned in New Zealand; an identical Ramsden engraved "D.33" which was once in the Watt Institution, Scotland, but by 1979 had disappeared; and a 12.7-cm Ramsden in the Dixson Library, Sydney. At the time, he emphasized that "The authenticity of all of these remains to be proved." (Howse 1979: 123), and this is still the situation today. More recently, Howse (personal communication 1997) pointed out that since the appearance of his 1979 paper, "D.32" and "D.33" have surfaced in auction rooms. Both are said to have been on the *Discovery*, and the former is now owned privately in the United Kingdom and the latter is in the National Maritime Museum. If the *Discovery* association is valid, then they relate to Cook's Third Voyage, not to the First or Second Voyages.

4.4.3 Keeping Track of Time: The Clocks

During the eighteenth century, accurate determination of longitude was a major challenge, especially on long maritime voyages. Longitude could be calculated by comparing local time with Greenwich time, but the problem was that the clocks that were then available were too unreliable. What was needed was a new, precise, totally portable time-keeper, but this was still being developed when Cook's First Voyage was planned. So Cook and Green were forced to rely on 'conventional clocks' and the system of 'lunars' developed by the Astronomer Royal, Nevil Maskelyne (1732—1811; Fig. 4.18; see Howse 1989). Gould (1987: 3) clearly and succinctly describes how 'lunars' were used to derive longitude:

> If the Moon's motion be known with sufficient accuracy, tables can be drawn up forecasting her angular distance, as observed on some standard meridian (*e.g.* Greenwich), from the Sun or suitable fixed stars. These distances can also be observed (by means of the sextant)

Fig. 4.18 A mezzotint of Nevil Maskelyne (https://en.wikipedia.org)

> on board ship; and, by interpolation, the Greenwich time corresponding with such distances can be taken out of the tables.

From 1767, Maskelyne included distances of the Moon from the Sun and seven selected stars in the *Nautical Almanac*, computed for every three hours at Greenwich.

Three different clocks were taken on Cook's First Voyage: an astronomical clock, a journeyman clock and an alarum clock. The astronomical clock was the largest, most expensive and most accurate of the three, and was only used when a shore-based observatory could be set up. The usual procedure was to install the clock inside a tent that afforded protection from the elements and uninvited human interference, as the following Tahitian account dating to April 1769 indicates:

> The astronomical clock, made by Shelton and furnished with a gridiron pendulum, was set up in the middle of one end of a large tent, in a frame of wood made for the purpose at Greenwich, fixed firm and as low in the ground as the door of the clock-case would admit, and to prevent its being disturbed by any accident, another framing of wood was made round this, at the distance of one foot from it. (Green and Cook 1771: 397).

The journeyman (or assistant) clock was a smaller, less accurate clock that also generally was used on shore in a tent observatory, alongside an astronomical quadrant. Maskelyne (1764: 373) provides an excellent description of one of these clocks:

> … I fixed up a little clock there, which may be called a journeyman, or secondary clock, having a pendulum swinging seconds, which after being well adjusted would keep time very regularly for several hours. It had only a minute and second hands, and struck every minute exactly as the second hand came to sixty, which was very convenient for the counting of seconds …

The alarum clock appears to have been small, portable clock that were used when astronomical observations were made. It cost only a small fraction of the price of an astronomical clock, and seemed prone to damage and breakdown. As Howse (1969a) notes, since none of these clocks survived and written descriptions of them have not been found, we know very little about them.

The Royal Society had already purchased astronomical clocks by John Shelton and John Ellicott (costing 30 guineas and 20 guineas, respectively) for the 1761 transit of Venus, and these were available for the 1769 event. The Shelton clock (Fig. 4.19) was taken on the *Endeavour*, along with a new journeyman clock ordered from Shelton and costing £5, an alarum clock which was probably also made by Shelton, and a watch made by Geo Graham and owned by Maskelyne which was loaned for the occasion to the Royal Society (Howse 1969a, b). John Shelton (1712–1777) at the time of Cook's First Voyage was one of Britain's foremost makers of astronomical time-pieces. At the age of seven he began an apprenticeship with the London clock-maker Henry Stanbury, and in 1720 he became a member of the Clockmakers' Company. By the middle of the eighteenth century Shelton was the main person used by the noted London instrument-maker

Fig. 4.19 This Shelton astronomical clock has been identified by Howse and Hutchinson (1969b) as the one that accompanied the *Endeavour* to Tahiti (Orchiston collection)

George Graham to fabricate astronomical clocks, yet despite his obvious technical acumen and orders for transit of Venus clocks in 1761 and 1769 Shelton did not have a good business head and soon after was in financial straits (Bonhams auction catalogue 2006; Clifton 1995).

Howse (1969a) has demonstrated that none of the alarum or journeyman clocks that was taken on any of the Cook voyages can currently be identified, and he and Hutchinson (1969b) have also shown that a complicated history surrounds the five surviving Shelton astronomical clocks which are reputedly associated with James Cook:

> In the 1780s, the clocks became thoroughly mixed up, largely because the Board [of Longitude] and the [Royal] Society shared a warehouse and a storekeeper. It would not have been impossible at this time for the pendulums and even the movements of several clocks to have been cannibalised to produce one working clock. (Howse and Hutchinson 1969b: 282).

Back in 1969 Howse and Hutchinson (1969b), had difficulty identifying the Shelton Clock which went in the *Endeavour*, believing that the 'KO Clock' at the Royal Observatory, Edinburgh had the best claim. Subsequently, Howse and Murray (1997) stated that the clock in question is "… almost certainly the one now preserved in the National Museum of Scotland in Edinburgh." This is the very same 'KO Clock'.

4.5 The Astronomical Observations

4.5.1 Introduction

Cook and his colleagues visited New Zealand on six different occasions between 1769 and 1777, spending a total of almost six months ashore at a number of different locations, the most significant of which was Queen Charlotte Sound. As Table 4.2 indicates, astronomical observations were made at three different sites, Mercury Bay, Queen Charlotte Sound and Dusky Sound. In addition, solar, lunar and stellar observations were made from the ships themselves as they sailed along the coast, both for navigational purposes and in order to identify key topographical features and map the coastline.

Table 4.2 New Zealand locations where shore-based astronomical observations were made during Cook's voyages

Voyage	Date	Vessel	Location
First	1769 November	*Endeavour*	Mercury Bay
Second	1773 March–May	*Resolution*	Dusky Sound
	1773 April–June	*Adventure*	Queen Charlotte Sound
	1773 May–June	*Resolution*	Queen Charlotte Sound
	1773 November	*Resolution*	Queen Charlotte Sound
	1773 December	*Adventure*	Queen Charlotte Sound
	1774 October–November	*Resolution*	Queen Charlotte Sound
Third	1777 February	*Resolution*	Queen Charlotte Sound
	1777 February	*Discovery*	Queen Charlotte Sound

4.5.2 Coastal Mapping and the Transit of Mercury

After successfully observing the transit of Venus in June 1769 (see Herdendorf 1986; Moore 1977; Orchiston 1998a, b, 2005, 2015), Cook's secret instructions from the Admiralty required that he go in search of the postulated Great Southern Land. More specifically:

> … whereas there is reason to imagine that a Continent or Land of great extent, may be found to the Southward of the Tract lately made by Capt[n] Wallis in His Majesty's Ship the Dolphin (of which you will herewith receive a Copy) or of the Tract of any former Navigators in Pursuits of the like kind; You are therefore in Pursuance of His Majesty's Pleasure hereby requir'd and directed to put to Sea with the Bark you Command so soon as the Observation of the Transit of the Planet Venus shall be finished and observe the following Instructions.
>
> You are to proceed to the southward in order to make discovery of the Continent above-mentioned until you arrive in the Latitude of 40°, unless you sooner fall in with it. But not having discover'd it or any Evident signs of it in that Run, you are to proceed in search of it to the Westward between the Latitude before mentioned and the Latitude of 35° until you discover it, or fall in with the Eastern side of the Land discover'd by Tasman and now called New Zealand.
>
> If you discover the Continent above-mentioned either in your Run to the Southward or to the Westward as above directed, You are to employ yourself diligently in exploring as great an Extent of the Coast as you can; carefully observing the true situation thereof both in Latitude and Longitude, the Variation of the [Magnetic] Needle, bearings of Head Lands, Height, direction and Course of the Tides and Currents, Depths and Soundings of the Sea, Shoals, Rocks, &c[a] and also surveying and making Charts, and taking Views of such Bays, Harbours and Parts of the Coast as may be useful to Navigation.
>
> You are also carefully to observe the Nature of the Soil, and the Products thereof; the Beasts and Fowls that inhabit or frequent it, the fishes that are to be found in the Rivers or upon the Coast and in what Plenty; and in case you find any Mines, Minerals or valuable stone you are to bring home Specimens of each, as also such Specimens of the Seeds of the Trees, Fruits and Grains as you may be able to collect, and Transit them to our Secretary that We may cause proper Examination and Experiments to be made of them.
>
> You are likewise to observe the Genius, Temper, Disposition and Number of the Natives, if there be any, and endeavour by all proper means to cultivate a Friendship and Alliance with them, making them presents of such Trifles as they may Value, inviting them to Traffick, and Shewing them every kind of Civility and Regard; taking Care however not to suffer yourself to be surprized by them, but to be always upon your guard against any Accident.
>
> You are also with the Consent of the Natives to take possession of Convenient Situations in the Country in the Name of the King of Great Britain; or, if you find the Country uninhabited take Possession for His Majesty by setting up Proper Marks and Inscriptions, as first discoverers and possessors.
>
> *But if you should fail of discovering the Continent before-mention'd, you will upon falling in with New Zealand carefully observe the Latitude and Longitude in which that Land is situated, and explore as much of the Coast as the Condition of the Bark, the health of her Crew, and the state of your Provisions will admit of …*
>
> You will also observe with accuracy the Situation of such Islands as you may discover in the Course of your Voyage that have not hitherto been discover'd by any Europeans, and

> take possession for His Majesty and make Surveys and Draughts of such of them as may appear to be of Consequence, without Suffering yourself however to be thereby diverted from the Object which you are always to have in View, the Discovery of the Southern Continent so often Mentioned. (Cited in Beaglehole 1968: cclxxxii–cclxxxiii; my italics).

Thus it was that the *Endeavour* sailed to the south and then in a generally westward direction (see Fig. 4.20), and in October 1769 reached the New Zealand coast at Turanganui (present-day Poverty Bay). Following a brief sojourn there, the ship sailed south along the coast to Cape Turnagain then reversed direction and sailed north, stopping briefly in first Anaura Bay and then Tolaga Bay (for these localities see Fig. 4.1). All the while, Cook and Green used ship-board astronomical observations for navigational purposes and in order to map the coast of this strange new land (see Barton 1980).

After rounding East Cape and skirting the shore of the Bay of Plenty, the *Endeavour* anchored in a sheltered bay on the eastern side of the Coromandel

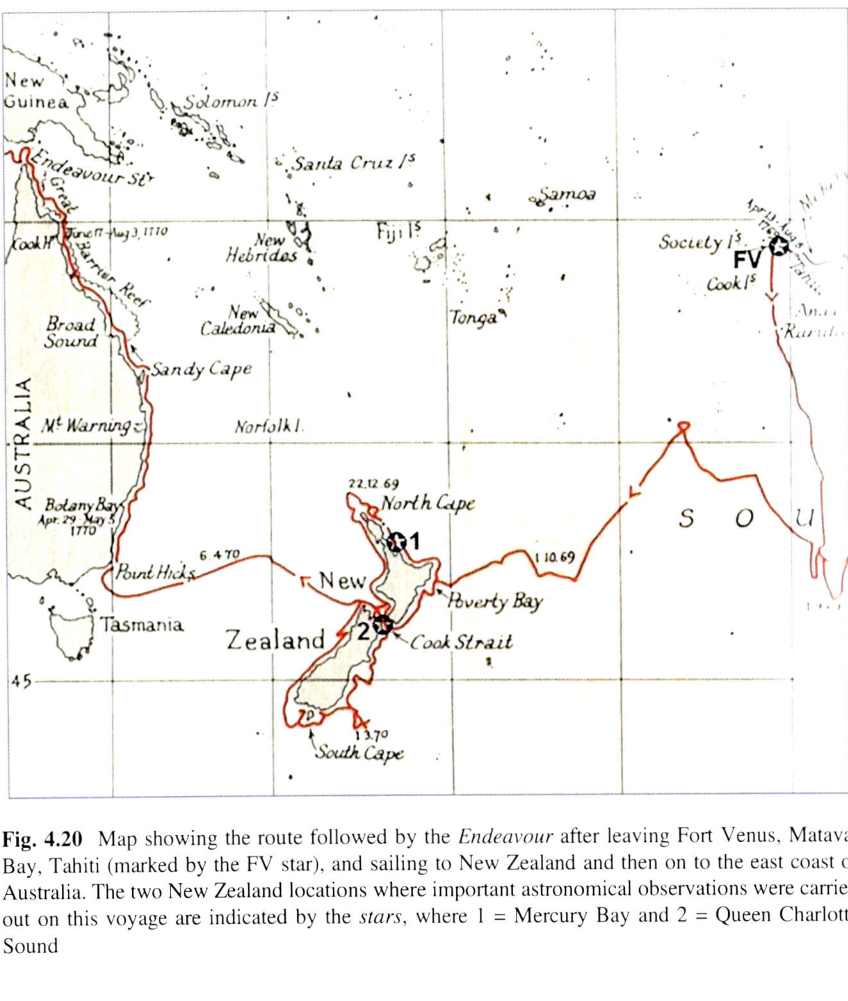

Fig. 4.20 Map showing the route followed by the *Endeavour* after leaving Fort Venus, Matavai Bay, Tahiti (marked by the FV star), and sailing to New Zealand and then on to the east coast of Australia. The two New Zealand locations where important astronomical observations were carried out on this voyage are indicated by the *stars*, where 1 = Mercury Bay and 2 = Queen Charlotte Sound

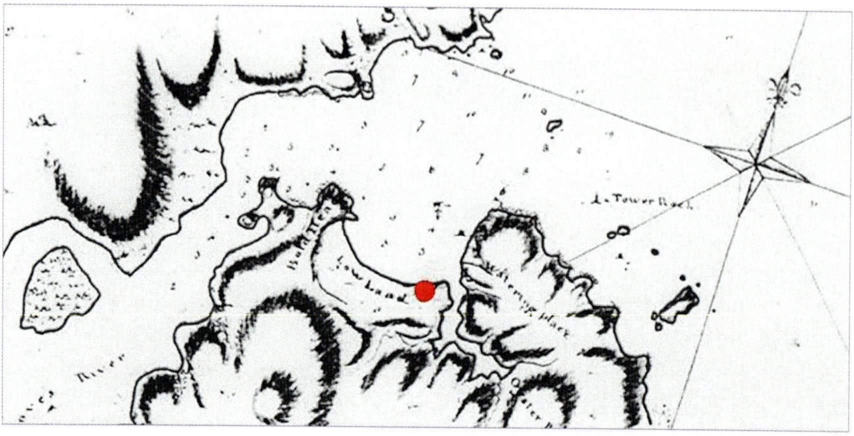

Fig. 4.21 Part of Pickersgill's 1769 map of Mercury Bay, upon which I have inserted a *red* spot to show the location of the transit of Mercury observing site, about 300 metres west of the mouth of the Oyster River

Peninsula on 3 November 1769, with the intention of observing a transit of Mercury. These transits occur when Mercury is seen as a small black spot moving across the Sun's disk. Mercury's period is just 88 days (compared with the Earth's 365 days), but because its orbit is tilted by 7° relative to the Earth's orbit, transits of Mercury do not occur every time the Earth, Mercury and the Sun (in that order) are aligned. Instead, there are about thirteen transits each century. A minimum of three years can occur between successive transits, and a maximum of thirteen years. Transits of Mercury can only occur in May or November.

The 9 November 1769 transit provided an interesting astronomical challenge, for as Cook observed,

> If we should be so fortunate as to Obtain this Observation the Longitude of this place and Country will thereby be very accurately determined. (Beaglehole 1968: 193).

At 7 o'clock on the morning of the transit Cook and Green, assisted by Hicks (Beaglehole 1963(I): 428–429), set up their telescopes on the beach adjacent to the ship's anchorage, about 300 metres west of the mouth of the Oyster River (see Fig. 4.21). Green succeeded in timing Mercury's ingress (entry onto the Sun's disk), but Cook was busy observing the Sun's altitude at the time. However, both observers recorded the egress, when Mercury moved off the Sun's disk, and their times are listed in Table 4.3 (after Wales 1788: 44).

Although their two egress timings differed respectively by 13 and 12 seconds, Green was able to use these observations to accurately calculate the longitude,[3]

[3]Keir (2010) suggests that in fact Cook and Green did not derive the longitude of Mercury Bay from these observations of the transit, but rather by extrapolating from longitude values obtained before and after the transit.

Table 4.3 Egress observations of the 3 November 1769 transit of Mercury, Mercury Bay, Coromandel Peninsula

Event	Observer	Local time
Internal contact	Cook	12 h 08 m 45 s
	Green	12 h 08 m 58 s
External contact	Cook	12 h 09 m 43 s
	Green	12 h 09 m 55 s

while Cook's measurements of the Sun's altitude provided the latitude. Their results (Beaglehole 1968: 195; Wales 1788: 44) were:

Longitude 176° 03.5′E
Latitude 36° 48′ 5.2″S

Thus, the research objectives that Cook set were readily achieved, and upon their departure this place was named "… *Mercury Bay* on acco[t] of the observ[n] be[ing] made there …" (Beaglehole 1968: 202). Mercury Bay therefore was the site of the first scientific astronomical observations conducted on New Zealand soil (see Herdendorf 1986; Orchiston 1994).

Amiable Maori-European relations were experienced while the *Endeavour* was anchored at Mercury Bay, with the crew busy trading European beads and buttons, pieces of white paper and fragments of red *tapa* cloth (from island Polynesia) for Maori clothing, weapons and ornaments, as well as fish, shellfish and crayfish (Orchiston 1974a). The Maoris here were similar to those already seen on the East Coast (at Turanganui, Anaura Bay and Tolaga Bay), except that their skins were darker, and their clothing, huts, canoes, ornaments and implements in general were less affluent in appearance. In addition, they seemed to lack the *kumara* (sweet potato) plantations that were so much in evidence on the East Coast and around the Bay of Plenty (Beaglehole 1963(I): 425–433; Beaglehole 1968: 203).

The *Endeavour* left this pleasant port on 15 November, but not before Cook claimed it for England:

> Before we left this Bay we cut out upon one of the trees near the watering place, the Ship's Name, date &c[a] and after displaying the English Colours I took formal possession of the place in the name of His Majesty. (Beaglehole 1968: 204).

Note that there is no indication that Cook first gained the approval of the local population, as stipulated in the 'secret instructions' he received from the Admiralty, before performing this ceremony.

The *Endeavour* then proceeded to near present-day Thames before calling at the Bay of Islands, rounding North Cape and sailing down the west coast of the North Island and across Cook Strait into Queen Charlotte Sound. All of the locations are indicated in Fig. 4.1.

On 16 January 1770 the *Endeavour* anchored off a small densely-wooded cove containing a freshwater stream, which was assigned the name Ship Cove (see Fig. 4.22). Nearby was a small island which contained a fortified Maori settlement, or *pa*, and was appropriately named 'Hippah Island'. The following day, the Europeans began

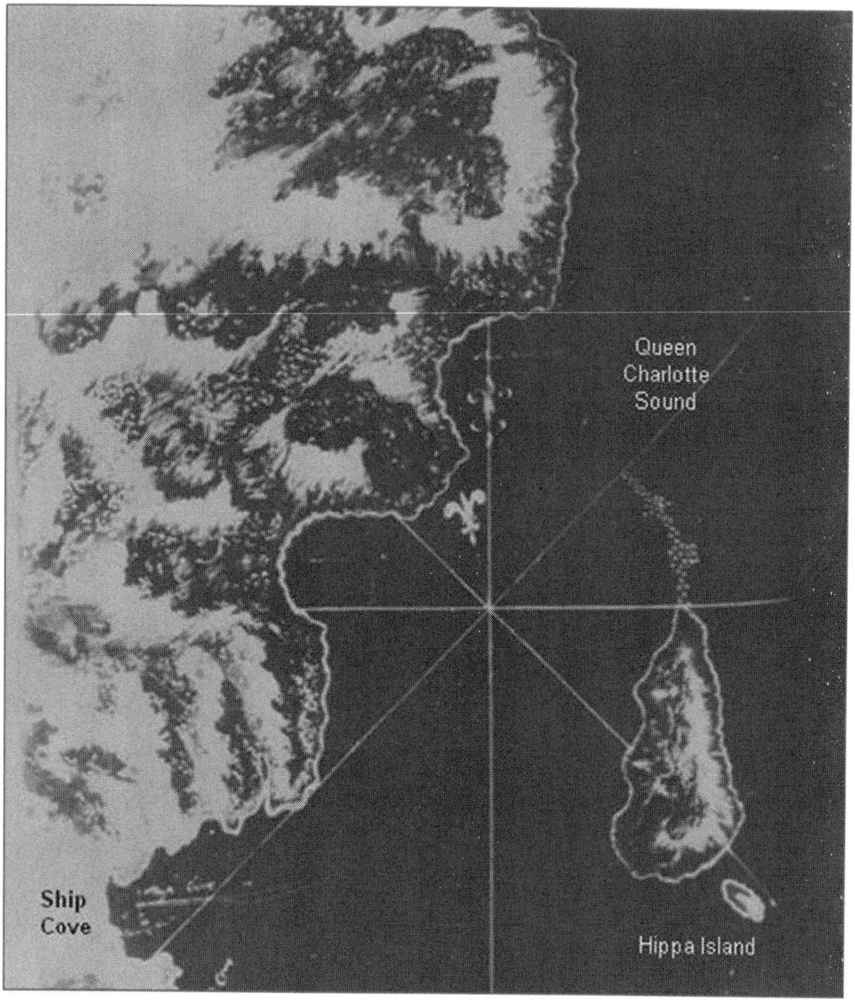

Fig. 4.22 A revised version of Pickersgill's map of Queen Charlotte Sound showing Ship Cove and Hippah Island (*Map modifications* Wayne Orchiston)

preparing the *Endeavour* for sea: the ship was careened, and while some of the crew were busily scrubbing and caulking the hull, the carpenters and coopers were employed ashore near where the forge had been set up. In the course of these operations, about one hundred Maoris came across from Hippah Island to observe proceedings and to trade fish. Meanwhile, on board ship, Green began a series of astronomical observations for latitude and longitude that was to extend over five successive days, and these provided the following mean values for Queen Charloette Sound:

Latitude 41° 05.8′S
Longitude 175° 28′ 59″E

In his general account of Queen Charlotte Sound, Cook (Beaglehole 1968: 246) gives slightly different values, of 41° 0′S and 175° 15′E, respectively. On the Second Voyage, Cook provided another figure again for the longitude: "In my Chart constructed in my former Voyage this place is laid down in ... 175° 5′ 30″ East ..." (cited in Beaglehole 1969: 579).

On 6 February 1770 Cook quit the Sound, and circumnavigated the South Island before leaving New Zealand waters. By this time, he was in a perfect position to describe the status of New Zealand vis-a-vis the mooted Great Southern Continent:

> This country, which before now was thought to be a part of the imaginary southern continent, co[n]sists of Two large Islands divided from each other by a strait or passage of 4 or 5 Leagues broad. They are Situated between the Latitudes of 34° and 48°S and between the Longitude of 181° and 194° West from the Meridion of Greenw[h] [which equals 179° and 166° East of Greenwich]. The situation of few parts of the world are better determined than these Islands are being settled by some hundreds of Observations of the Sun and Moon and one of the transit of Mercury made by M[r] Green ... (Beaglehole 1968: 274).

From New Zealand, Cook headed westward across the Tasman Sea and rediscovered the great island continent of Australia, before returning to England in July 1771.

After the voyage, Maskelyne was critical of some of Green's astronomical observations, but Cook immediately rallied to the defence of his deceased colleague pointing out that the Astronomer Royal knew full well that the quadrant had been damaged. Moreover, all of Green's observations—regardless of precision—had been handed over to Wales for analysis; "does M[r] M.," Cook asks, "publish to the world all the observations he makes good and the bad or did he never make a bad observation in his life." (cited in Beaglehole 1968: cxlv). But Cook went further, offering nothing but praise for Green's dedication and astronomical prowess:

> He was indefatigable in making and calculating these [latitude and longitude] observations. By his instructions also, several of the petty officers can make and calculate them ... It is only by this means that this method of finding the longitude at sea can be brought into universal practice – A method which we have found may be depended on to within half a degree! which is a degree of accuracy more than sufficient for all nautical purposes. (Wales 1788: viii).

This viewpoint also is vindicated if we examine maps of New Zealand made during Cook's First Voyage (e.g. see Fig. 4.23), where Green's longitude calculations led to an outline map that is easily recognizable, even if somewhat distorted in places (such as in the vicinity of the Canterbury Plains).

Partly because of Green's untimely death during the voyage, the astronomical results from Cook's First Voyage took a long time to appear in print. They were prepared by Wales (Green's brother-in-law), and were eventually published in 1788 under the title *Astronomical Observations Made in the Voyages which were Undertaken By Order of His Present Majesty, for Making Discoveries in the Southern Hemisphere* ... The full title goes on for several more lines, and is truly astronomical in length!

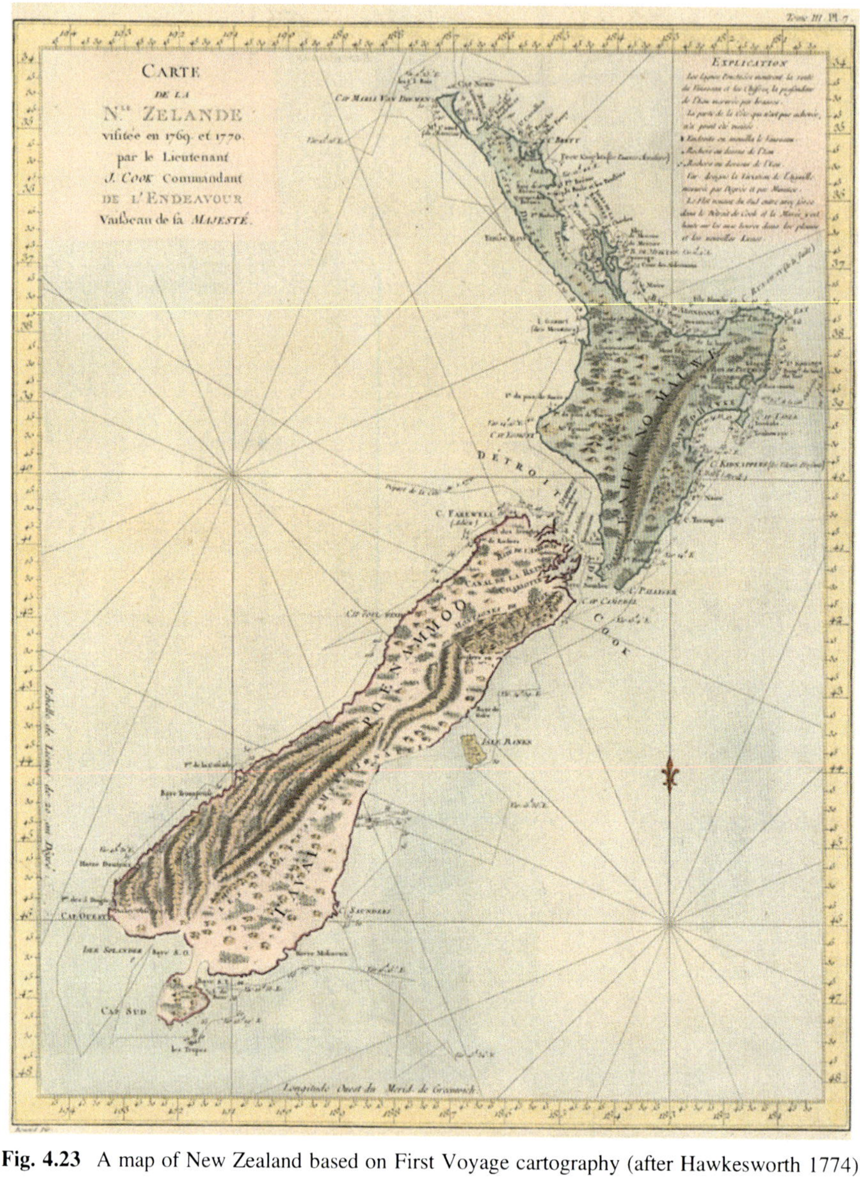

Fig. 4.23 A map of New Zealand based on First Voyage cartography (after Hawkesworth 1774)

4.6 Concluding Remarks

Cook's First Voyage to the South Seas may have started as a voyage of scientific discovery, with the 3 June 1769 transit of Venus the prime target, but after the transit it became a voyage of general discovery, with New Zealand and the east coast of Australia the principal achievements. Cartography was a primary concern, and during the mapping of the New Zealand coastline the astronomers, Cook and Green, made extensive use of astronomical observations to pin-point the changing latitude and longitude of the *Endeavour*. In addition, there were shore-based observations of a transit of Mercury, made in Mercury Bay on 3 November 1769. These were the first astronomical observations made by the visiting Europeans on New Zealand soil and triggered the start of scientific astronomy in Aotearoa/New Zealand, thereby introducing a totally different perspective on the Universe to the one enjoyed by Maori astronomers over the preceding centuries. With these endeavours, New Zealand astronomy would never be the same again.

Acknowledgements I am grateful to the following for their assistance: Dr Jonathan Betts and Dr Gloria Clifton (ex-Royal Museums, Greenwich), the late Commander Derek Howse, (Sevenoaks, England), Dr Stephen Johnston (Museum of the History of Science, Oxford), Katherine Marshall (the Royal Society, London) and the late Sir Patrick Moore (Selsey, England. I also wish to thank Gloria Clifton for reading and commenting on the first draft of this chapter, and the Museum of the History of Science, the National Maritime Museum, the Royal Society and the Scientific Instrument Collection, Harvard University (Cambridge, Massachusetts, USA) for kindly supplying Figs. 4.14, 4.16 and 4.17.

References

Andrews, A. D. (1992). Cyclopaedia of telescope makers. *Irish Astronomical Journal, 20*, 102–183.

Andrews, A. D. (1995). Cyclopaedia of telescope makers. Part 4 (O–R). *Irish Astronomical Journal, 22*, 43–96.

Andrews, A. D. (1996). Cyclopaedia of telescope makers. Part 5 (Sae–Sim). *Irish Astronomical Journal, 23*, 57–117.

Atkinson, R.d' E. (1963). Time and longitude. *Journal of the British Astronomical Association, 73*, 9–22.

Badger, G. M. (Ed.), (1970a). *Captain Cook. Navigator and Scientist*. Canberra: Australian National University Press.

Badger, G. M. (1970b). Cook the scientist. In Badger, 1970a (pp. 30–49).

Barton, P. L. (1980). The history of the mapping of New Zealand. *Map Collector, 11*, 28–35.

Baugh, D. (2011). *The Global Seven Years War, 1754–1763*. Upper Saddle River: Pearson Press.

Beaglehole, J. C. (Ed.), (1963). *The Endeavour Journal of Joseph Banks* (2nd ed.). Two Volumes. Sydney: Angus and Robertson.

Beaglehole, J. C. (Ed.), (1967). *The Journals of Captain James Cook on his Voyages of Discovery. III. The Voyage of the* Resolution *and* Discovery *1776–1780*. Two Volumes. Cambridge: Cambridge University Press.

Beaglehole, J. C. (Ed.), (1968). *The Journals of Captain James Cook on his Voyages of Discovery. I. The Voyage of the* Endeavour *1768–1771*. Cambridge: Cambridge University Press (Preprint; first published 1955).

Beaglehole, J. C. (Ed.), (1969). *The Journals of Captain James Cook on his Voyages of Discovery. II. The Voyage of the* Resolution *and* Adventure *1772–1775*. Cambridge: Cambridge University Press (Reprint; first published 1961).

Beaglehole, J. C. (1971). Cook the navigator. In J. V. S Megaw (Ed.), *Employ'd as a Discoverer* (pp. 117–134). Sydney: Reed.

Beaglehole, J. C. (1974). *The Life of Captain James Cook*. London: Hakluyt Society.

Betts, J. (2006). *Time Restored: The Harrison Timekeepers and R.T. Gould, the Man who Knew (Almost) Everything*. Oxford: Oxford University Press.

Bird, J. (1768). *The Method of Constructing Mural Quadrants*. London: Richardson & Clark.

Bliss, N. (1761). Observations of the transit of Venus over the Sun, on the 6th of June 1761 … *Philosophical Transactions of the Royal Society*, *52*, 173–177.

Bonhams auction catalogue. (2006). (Lot) 85 An important and very rare third quarter of the 18th century month going longcase (John Shelton) regulator now in a modern replica case. (http://www.bonhams.com/auctions/14228/lot/85/).

Bryden, D. J. (1968). *James Short and his Telescopes*. Edinburgh: Royal Scottish Museum.

Campbell, I. C. (1980). Savages noble and ignoble: the preconceptions of early European voyagers in Polynesia. *Pacific Studies, 4*, 45–59.

Chapman, A. (1983). The accuracy of angular measuring instruments used in astronomy between 1500 and 1850. *Journal for the History of Astronomy, 14*, 133–137.

Chapman, A. (1995). *Dividing the Circle. The Development of Critical Angular Measurement in Astronomy 1500–1850*. Chichester: Wiley.

Chapman, A. (1996). Scientific instruments and industrial innovation: the achievement of Jessie Ramsden. In A. Chapman (Ed.), *Astronomical Instruments and their Uses. Tycho Brahe to William Lassell* (pp. 418–430). Aldershot: Variorum.

Clifton, G. (1995). *Directory of British Scientific Instrument Makers 1550–1851*. London: Zwemmer and the National Maritime Museum.

Cook, J. (1767). An observation of an eclipse of the sun at the Island of New-found-land, August 5, 1766. *Philosophical Transactions of the Royal Society*, *58*, 215–216.

Cotter, C. H. (1968). *A History of Nautical Astronomy*. London: Hollis and Carter.

Deptford Yard Officers. (1768a). Letter to the Navy Board, dated March 27. Original in the Public Record Office, London (Adm 106/3315). Abstract in Beaglehole, 1968, 606.

Deptford Yard Officers. (1768b). Letter to the Navy Board, dated October 7. Original in the Public Record Office, London (Adm 106/3315). Abstract in Beaglehole, 1968, 624.

Dollond, J. (1753). A description of a contrivance for measuring small angles. *Philosophical Transactions of the Royal Society*, *48*, 178–181.

Dollond, J. (1754). Of an instrument for measuring small angles … *Philosophical Transactions of the Royal Society*, *48*, 462–469.

Dyson, F. (1929). Captain Cook as an astronomer. *Geographical Journal, 73*, 116–122.

Garnsey, E. R., (1864–1935). *Sydney Morning Herald*, May 10, 1935 (p. 17).

Gould, J. A. (1976). John Bird: the astronomer's instrument maker (1709–1776). *Journal of the British Astronomical Association, 86*, 485–486.

Gould, T. R. (1987). *John Harrison and his Timekeepers*. London: National Maritime Museum.

Green, C. (1768–1770). Log-journal. Original in the public record office, London (Adm 51/4545/151).

Green, C., & Cook, J. (1771). Observations made, by appointure of the Royal Society, at King George's Island in the South Seas. *Philosophical Transactions of the Royal Society*, *61*, 397–421.

Groube, L. (1965). *Settlement Patterns in New Zealand Prehistory*. Dunedin: Department of Anthropology, University of Otago (Occasional Papers in Archaeology No. 1).

Hawkesworth, J. (1774). *Relation des Voyages Entrepris par Ordre de Sa Majesté Britannique …* pour faires des Découvertes dans l'Hémisphère Méridional … Paris, Saillant et Nyon.

Hellman, C. D. (1932). John Bird (1709–1776) mathematical instrument-maker in the Strand. *Isis, 17*, 127–153.

Herdendorf, C. (1986). Captain James Cook and the transits of Mercury and Venus. *Journal of Pacific History, 21*, 39–55.

Higgitt, R. (2010). Charles Green, Royal Observatory Greenwich Assistant and stand-in Astronomer Royal. London, Royal Museums Greenwich. (http://blogs.rmg.co.uk/ontheline/2010/09/03/charles_green_royal_observatory/).

Hornsby, T. (1765). On the transit of Venus in 1769. *Philosophical Transactions of the Royal Society*, *55*, 326–344.

Howe, K. R. (1984). *Where the Waves Fall. A New South Sea Islands History from First Settlement to Colonial Rule*. Honolulu: University of Hawaii Press.

Howse, D. (1969a). Captain Cook's minor clocks and watches. In Howse and Hutchinson, 1969a (pp. 138–145).

Howse, D. (1969b). Captain Cook's pendulum clocks. In Howse and Hutchinson, 1969a (pp. 62–76).

Howse, D. (1979). The principal scientific instruments taken on Captain Cook's voyages of exploration, 1768–80. *Mariner's Mirror, 65*, 119–135.

Howse, D. (1983). Navigation and astronomy II: the last three hundred years. *Journal of the British Astronomical Association, 93*, 50–61.

Howse, D. (1989). *Nevil Maskelyne. The Seamen's Astronomer.* Cambridge: Cambridge University Press.

Howse, D. (1993). Green, Charles (1734–1771). In *The Dictionary of National Biography: Missing Persons* (p. 269). Oxford: Oxford University Press.

Howse, D., & Hutchinson, B. (1969a). *The Clocks and Watches of Captain James Cook 1769–1969*. London: The Antiquarian Horological Society (reprinted from four issues of Antiquarian Horology).

Howse, D., & Hutchinson, B. (1969b). The saga of the Shelton clocks. In Howse and Hutchinson, 1969a (pp. 281–298).

Howse, D., & Murray, A. (1997). Lieutenant Cook and the transit of Venus, 1769. *Astronomy & Geophysics, 38*(4), 27–30.

Jones, K. (1983). Joseph Banks' 'Fence of Poles' at Cook's Cove, East Coast, North Island, New Zealand. *Journal of the Polynesian Society, 92*, 531–536.

Jones, K. (1984). A 1761 George III medalet from the East Coast, North Island, New Zealand: Relic of Cook's first voyage? *Journal of the Polynesian Society, 93*, 315–320.

Jones, K. (1989). 'In much greater affluence': productivity and welfare in Maori gardening at Anaura Bay, October 1769. *Journal of the Polynesian Society, 98*, 49–75.

Joppien, R., & Smith, B. (1985–1997). *The Art of Captain Cook's Voyages* (4 Vols.). Melbourne: Oxford University Press.

Kaeppler, A. (1978). *'Artificial Curiosities' Being an Exposition of Native Manufactures Collected on the Three Pacific Voyages of Captain James Cook, R.N.* Honolulu: Bishop Museum Press.

Keir, B. (2010). Captain Cook's longitude determinations and the transit of Mercury—common assumptions questioned. *Journal of the Royal Society of New Zealand, 40*(2), 27–38.

Kennedy, J. (1969). *Settlement in the South East Bay of Islands, 1772*. Dunedin: Department of Anthropology, University of Otago (Studies in Prehistoric Anthropology No. 3).

King, H. C. (1948). The invention and early development of the achromatic telescope. *Popular Astronomy, 56*, 75–88.

King, H. C. (1979). *The History of the Telescope*. New York: Dover.

Kippis, A. (1788). *The Life and Voyages of Captain James Cook* (Vol. 1, pp. 176–178). London: George Newnes. (The lengthy footnote about Charles Green).

Kurtz, D. W. (Ed.). (2005). *Transits of Venus: New Views of the Solar System and the Galaxy*. Cambridge: Cambridge University Press.

Lloyd, C. (1968). *The British Seaman 1200–1860. A Social Survey*. London: Paladin.

Mackrell, B. (1985). *Halley's Comet Over New Zealand*. Auckland: Reed Methuen.

Maskelyne, N. (1764). Astronomical observations made at the island of St. Helena. *Philosophical Transactions of the Royal Society 54*, 348–349.

Maskelyne, N. (1768). Observations of the transit of Venus over the Sun, and the eclipse of the Sun, on June 3, 1769, made at the Royal Observatory. *Philosophical Transactions of the Royal Society, 58*, 355–365.

May, W. E. (1973). *A History of Marine Astronomy*. Henley-on-Thames: Foulis.

McConnell, A. (2007). *Jesse Ramsden (1735–1800): London's Leading Scientific Instrument Maker*. Aldershot: Ashgate.

McIntosh, R. A. (1957). The astronomical history of the Auckland Province. Part II: Eighteenth century. *Amateur Astronomer, 17*, 40–42.

Mead, S. M. (1969). The costume styles of the Classical Maori in New Zealand, 1642–1800 A.D. *Costume, 3*, 35–43.

Mennim, E. (n.d.). *Transit Circle. The Story of William Simms 1793–1860*. York: William Sessions.
Moore, P. (1977). Cook, Green and the transit of Venus. In J. Cook (Ed.), *The Journal of the Endeavour 1768–1771* (pp. 61–69). Guildford: Genesis Publications.
Morris, M. (1980). Charles Green—man without a face (Part 1). *Cook's Log, 3*(4), 92.
Morris, M. (1981). Charles Green—man without a face (Part 2). *Cook's Log, 4*(1), 102.
Orchiston, W. (1971). Astronomy in early New Zealand. *British Astronomical Association New South Wales Branch Bulletin, 509*, 1–2.
Orchiston, W. (1972). Maori neck and ear ornaments of the 1770s: a study in protohistoric ethno-archaeology. *Journal of the Royal Society of New Zealand, 2*, 91–107.
Orchiston, W. (1974a). Cook voyage 'trading stations' in Early Protohistoric New Zealand. *Dominion Museum Records in Ethnology, 2*, 133–156.
Orchiston, W. (1974b). *Studies in South Island New Zealand Prehistory and Protohistory*. Unpublished Ph.D. Thesis, Department of Anthropology, University of Sydney.
Orchiston, W. (1975a). Early protohistoric Maori food preparation, and application of the direct historical approach to New Zealand prehistory. *New Zealand Archaeological Association Newsletter, 18*, 18–31.
Orchiston, W. (1975b). Maori material culture change in Early Protohistoric New Zealand: the greenstone trade at Queen Charlotte Sound. *The Artefact, 39*, 40–77.
Orchiston, W. (1994). *James Cook and the 1769 Transit of Mercury*. Wellington: Carter Observatory (Information Sheet No. 3).
Orchiston, W. (1997). *Early Astronomy in New Zealand: The South Sea Voyages of James Cook*. Wellington: Carter Observatory (Information Sheet No. 10).
Orchiston, W. (1998a). From the South Seas to the Sun: the astronomy of Cook's voyages. In Lincoln, M. (Ed.), *Science and Exploration. European Voyages to the Southern Oceans in the 18th Century* (pp. 55–72). Woodbridge: Boydell and Brewer.
Orchiston, W. (1998b). *Nautical Astronomy in New Zealand: The Voyages of James Cook*. Wellington: Carter Observatory.
Orchiston, W. (1999). Cook, Banks and the Gregorian telescope in the Museum of New Zealand Te Papa Tongarewa. *Journal of the Antique Telescope Society, 18*, 4–9 (+Front Cover).
Orchiston, W. (2004). The astronomical results of Cook's voyages. In J. Robson (Ed.), *The Captain Cook Encyclopaedia* (pp. 31–35). London: Chatham Publishing.
Orchiston, W. (2005). James Cook's 1769 transit of Venus expedition to Tahiti. In Kurtz, 51–66.
Orchiston, W. (2015). Captain Cook and the 1769 transit of Venus: the Australian connection. In W. Orchiston & R. McGregor (Eds.), *Scientific Alignments: James Cook, the 1769 Transit of Venus and the Colonisation of North Queensland*. Brisbane: Queensland Museum. (in press).
Paulding, B. (2000). Charles Green astronomer. *Cook's Log, 23*, 1713.
Ritchie, G. S. (1978). Captain Cook's influence on hydrographic surveying. *Pacific Studies, 1*, 78–95.
Robertson, G. (1948). *The Discovery of Tahiti; A Journal of the Second Voyage of H.M.S. Dolphin Round the World Under the Command of Captain Wallis, R.N., in the Years 1766, 1767, and 1768, Written by her Master*. In H. Carrington (Ed.). London: Haklyut Society.
Salmond, A. (1991). *Two Worlds. First Meeting between Maori and Europeans 1642–1772*. Auckland: Viking.
Short, J. (1761–1762). An account of the transit of Venus over the Sun, on Saturday morning, 6th June 1761, at Savile-House … *Philosophical Transactions of the Royal Society, 52*, 178–182.
Short, J. (1764). Observation of the eclipse of the Sun, the 1st of April 1764, made in Surry Street, in the Strand, London. *Philosophical Transactions of the Royal Society, 54*, *107–*108, 105.
Simmons, D. R. (1981). Stability and change in the material culture of Queen Charlotte Sound in the 18th century. *Records of the Auckland Institute and Museum, 18*, 1–61.
Skelton, R. A. (1954). Captain James Cook as a hydrographer. *Mariner's Mirror, 40*, 92–119.
Skelton, R. A., (Ed.), (1969). *The Journals of Captain James Cook on his Voyages of Discovery. Charts & Views Drawn by Cook and his Officers and Reproduced from the Original Manuscripts*. Cambridge: Hakluyt Society.

Sobel, D. (1995). *Longitude: The True Story of a Lone Genius who Solved the Greatest Scientific Problem of his Time*. New York: Penguin.

Stimson, N. (1985). Some Board of Longitude instruments in the nineteenth century. In P. R. de Clercq (Ed.), *Nineteenth-century Scientific Instruments and their Makers* (pp. 93–115). Leiden: Museum Boerhaave.

Taylor, E. G. R. (1968). Navigation in the days of Captain Cook. *Journal of the Institute of Navigation, 21*, 256–276.

Turner, G. L'E. (1969). James Short, F.R.S., and his contribution to the construction of reflecting telescopes. *Notes and Records, 24* (June), 91–108.

Villiers, A. (1971). *Captain Cook, the Seaman's Seaman. A Study of the Great Discoverer*. Harmondsworth: Penguin.

Wales, W. (1788). *Astronomical Observations, Made in the Voyages which were Undertaken by Order of his Present Majesty, for Making Discoveries in the Southern Hemisphere* .... London: Elmsly.

Wales, W. (2008). Researching Charles Green after Kippis. *Cook's Log, 31*(1), 32.

Wallis, R. (2008). Hornsby, Thomas (1733–1810). In *Oxford Dictionary of National Biography*. Oxford: Oxford University Press.

Watt, J. (1979). Medical aspects and consequences of Cook's voyages. In R. Fisher & H. Johnston (Eds.), *Captain James Cook and his Times* (pp. 129–157). Canberra: Australian National University Press.

White, F. (1970). Cook the navigator. In Badger, 1970a (pp. 50–69).

Woolf, H. (1959). *The Transits of Venus: A Study in the Organization and Practice of Eighteenth-Century Science*. Princeton: Princeton University Press.

Wulf, A. (2012). *Chasing Venus. The Race to Measure the Heavens*. New York: Alfred A. Knopf.

Zuidervaart, H. J., & van Gent, R. H. (2004). A bare outpost of learned European culture on the edge of the jungles of Java. Johan Maurits Mohr (1716–1775) and the emergence of instrumental and institutional science in Dutch colonial Indonesia. *Isis, 95*, 1–33.

van Gent, R. H. (2005). Observations of the 1761 and 1769 transits of Venus from Batavia (Dutch East Indies). In Kurtz (pp. 67–73).

van de Woolley, R. (1970). The significance of the transit of Venus. In Badger, 1970a (pp. 118–135).

Chapter 5
Astronomy on Cook's Second Voyage: Dusky Sound and Queen Charlotte Sound, 1773–1774

Abstract Cook's Second Voyage to the South Seas was undertaken in a bid to locate and map the coast of the postulated Great Southern Continent, Terra Australis Incognita. Two ex-Whitby colliers were involved, the *Resolution* under the command of James Cook and the *Adventure*, which was skippered by Tobias Furneaux. Royal Observatory Greenwich astronomers William Bayly and William Wales were assigned to the *Adventure* and *Resolution* respectively. On this voyage extensive sets of observations for latitude and longitude were taken in Dusky Sound, on the far south-western tip of the South Island, and at Cook's favourite Pacific revictualling centre, Queen Charlotte Sound, near the north-eastern tip of the South Island. By the time the *Resolution* quit Charlote Sound in November 1774, after stopping at this port for a third and final time, Cook was moved to write: "... from the multitude of observations which Mr Wales took the situation of few parts of the world are better ascertained than that of Queen Charlottes Sound." (Beaglehole 1969: 580).

5.1 Introduction

Following the success of Cook's First Voyage, the Admiralty arranged Cook's Second Voyage to the South Seas, primarily to try to pin down the existence of the Great Southern Continent and map its coastline. This time two refitted ex-Whitby colliers were involved, the *Resolution* and *Adventure*, skippered respectively by Commander James Cook (promoted following the First Voyage) and Lieutenant Tobias Furneaux. Each vessel was assigned its own astronomer, William Wales on the *Resolution* and William Bayly on the *Adventure*; both came from the Royal Observatory, Greenwich.

W. Orchiston, *Exploring the History of New Zealand Astronomy*,
Astrophysics and Space Science Library 422,
DOI 10.1007/978-3-319-22566-1_5

During their exploration of the Pacific both vessels kept returning to Queen Charlote Sound (see Table 5.1), which quickly became Cooks' favourite revictualling centre with its excellent protected anchorage, fresh water, timber, native celery and 'scurvy grass', and the ingredients required for the manufacture of that antiscorbutic tonic, 'spruce beer'. There also were seemingly endless supplies of fish, shellfish and birds. As we saw in the previous chapter, Ship Cove also

> … proved an ideal setting for a makeshift 'hospital', and in this bay and nearby ones the wooden casks could be repaired and decaying bread and mouldy ship's biscuit dried out while those on board ship were busy repairing the sails, ironwork and rigging, and cleaning and caulking the hull. Revictualling stop-overs were rather hectic affairs!

Figure 5.1 shows the locations of Queen Charlotte Sound, other New Zealand anchorages listed in Table 5.1, and other places mentioned in this chapter.

Once again astronomy had a crucial role to play on this voyage as the astronomers made observations that allowed the latitude and longitude of the three anchorages and the coast of New Zealand to be determined. As we have seen, the calculation of latitude was a relatively easy matter, but longitude was a more pressing problem and required a very accurate knowledge of local time. On this occasion, however, the astronomers were assisted by the presence of four chronometers, which were trialed on the voyage.

The aim of this chapter is to describe the various astronomers involved in Cook's Second Voyage to the South Seas, their scientific instruments, and their New Zealand-based astronomical observations. Section 5.2, following, introduces us to Bayly and Wales. In Sect. 5.3 we discuss their telescopes, quadrants and sextants; their various astronomical clocks and chronometers; and their tent observatories. This leads us into Sect. 5.4 where we examine the ways in which the two astronomers put their instruments to use on New Zealand soil, while in Dusky Sound and Queen Charlotte Sound.

Although much has been written about New Zealand in 1773–1774 based on the reports of the European explorers and artifacts collected at the time (e.g. see Coutts 1969; Mead 1969; Orchiston 1972, 1974a, b, 1975a, b; Orchiston and Horrocks 1975; Salmond 1991), until the Carter Observatory published *Nautical Astronomy in New Zealand: The Voyages of James Cook* (Orchiston 1998) astronomy had been all but ignored, except for superficial references by Orchiston (1971, 1997). This chapter is based largely on material in Orchiston (1998).

Table 5.1 New Zealand anchorages during Cook's Second Voyage to the South Seas

Date	Vessel(s)	Location
March–May 1773	*Resolution*	Dusky Sound
April–June 1773	*Adventure* + *Resolution*	Queen Charlotte Sound
November 1773	*Resolution*	Tolaga Bay
November–December 1773	*Adventure*	Queen Charlotte Sound
October–November 1774	*Resolution*	Queen Charlotte Sound

Fig. 5.1 New Zealand localities mentioned in this chapter are shown in *red*

5.2 The Astronomers

5.2.1 *Introduction*

While the necessity for astronomers on the First Voyage was abundantly clear, the Second Voyage was a different matter. There was to be no transit of Venus, not even a total eclipse of the Sun or a transit of Mercury, but after the Astronomer Royal put a case to them, the Board of Longitude concurred that

> … the intended expedition to the South Seas may be rendered more serviceable to the improvement of geography and navigation than it can otherwise be if the ship be furnished with … astronomical instruments … and also some of the longitude watches and above all if a proper person could be sent out to make use of those instruments and teach the officers on board the ship the method of finding the longitudes … (Beaglehole 1969: 719).

Even though there would be officers on both vessels who already were adept at determining longitude, the Board approved as astronomers William Bayly (on the *Adventure*) and William Wales (on the *Resolution*). It was no accident that their employer, Nevil Maskelyne the Astronomer Royal, also was a prominent member of the Board (Howse 1989). He was also a member of the Council of the Royal Society, another supporter of the voyage! The Board decided that Bayly and Wales would each receive an all-expenses annual salary of £400 (Beaglehole 1969: 720).

Let us now look in turn at these two astronomers.

5.2.2 *William Bayly (1737–1810)*

William Bayly was born in Wiltshire at Bishop's Cannings or Carions in 1737 (for English localities mentioned in this chapter see Fig. 5.2). His father was a farmer, and from an early age young Bayly could be found behind the plough. However, he had an interest in mathematics, and learnt the basics from a man living in a nearby village. A man by the name of Kingston who lived in Bath heard of Bayly's achievements and gave him further assistance. As a result, Bayly was able to obtain a position as an usher at a school near Bristol, and subsequently he moved to another school nearby. During this period he carried out further mathematical studies (Hodges 1908).

Eventually Bayly came to the notice of Dr Maskelyne, and was appointed as an Assistant at the Royal Observatory, Greenwich. In 1769 the Royal Society sent him to the North Cape of Norway, together with Jeremiah Dixon, and they successfully observed the transit of Venus (Bayley 1769).

In 1772 Bayly set sail for the South Seas as astronomer in the *Adventure*, under the command of Captain Tobias Furneaux. Beaglehole (1969: xxxv) notes that "There was an incuriosity about Furneaux that made for strained relations more than once between him and Bayly … whom he could certainly have helped more …" In contrast to his attitude towards the other Second Voyage astronomer,

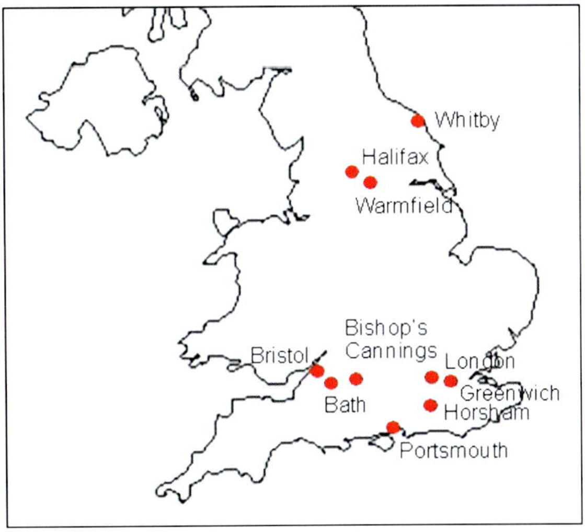

Fig. 5.2 English localities mentioned in this chapter are shown in *red*

William Wales, one of the two German natural historians on the Voyage, Johann Reinhold Forster (1729–1798), had nothing but praise for Bayly:

> M[r.] *Bailey*, the Astronomer [is] … a Man of an agreable, friendly Character, good parts, & an excellent mechanical genius, & of great learning in his profession … (Hoare 1982(II): 293).

After the voyage Bayly returned to the Observatory, and he and Wales then had the task of writing up the astronomical observations made during the Second Voyage. This was published in 1777 (Wales and Bayly 1777), while Bayly was serving as astronomer on the *Discovery* during Cook's Third Voyage. After that voyage Bayly returned once more to the Observatory, and with King's increasing ill health, the writing up of the Third Voyage astronomical observations fell to him. This volume was published in 1782, and rather generously included Cook and King as co-authors (see Cooke et al. 1782).

Beaglehole (1969: cxlii) has compared Bayly and Wales, and found the former "… a reasonably competent astronomer, but he had little of the education and none of the general abilities of Wales …"

In 1785 Bayly finally left the Royal Observatory in order to take up the post of Headmaster at the Royal Naval Academy in Portsmouth (Beaglehole 1969). Despite disciplinary and other problems, he retained this post until his Academy was transformed into the Royal Naval College in 1806 (Howse, personal communication, 1997). In 1807, at the age of 70, he was pensioned off, and he died in Portsea towards the end of 1810 (Hodges 1908). No known portraits of him have survived.

Although he was closely associated with the Royal Society through his South Sea voyages, and published a paper on his 1769 transit observations and the two volumes about the astronomy of Cook's voyages, it seems that Bayly was never made a Fellow of the Society (Rose 1853).

5.2.3 William Wales (1734–1798)

William Wales (Fig. 5.3) was yet another Yorkshireman, and was born near Warmfield to humble parents. Little is known of his childhood except that he showed an early aptitude for mathematics. During the 1760s, he contributed many examples to the *Ladies' Diary*, a magazine containing advanced mathematical problems (EIC 1909; Knight 1856), and it was probably through these that he met Charles Green, Bliss's assistant at the Royal Observatory, Greenwich. In 1765 he married Green's youngest sister, and the following year was commissioned by Maskelyne to carry out computations for the first *Nautical Almanac*. Wales was one of those who observed the 1769 transit of Venus for the Royal Society, his inhospitable observing station being on the north-west coast of Hudson's Bay in Canada. After the event, his results were published in the *Philosophical Transactions* of the Society (Wales and Dymond 1769).

In 1772 the Board of Longitude sent Wales on Cook's Second Voyage, as astronomer in the *Resolution*, and he performed his duties admirably (Moore 1981). Beaglehole (1969: cxl–cxli) clearly was impressed with him, and liked

Fig. 5.3 Pastel of William Wales by J. Russell, 1794 (*Photograph* Wayne Orchiston; courtesy: Christ's Hospital, Horsham)

> … the breadth and play of his mind, his capacity for observation, his scientific exactitude, and his integrity as a man. He is most particular in his account of physical phenomena, and of his procedure in, e.g. tidal measurements. He appreciates natural scenery, and has a poet in his mind for his more ecstatic moments. Whenever he has a chance to observe mankind he writes at length and with talent.

On the other hand, J.R. Forster, a naturalist on the *Resolution*, was not at all impressed, and an enduring antagonism developed between the two men which was to persist long after the end of the voyage (see Hoare 1976).

After returning from the voyage in 1775, Wales accepted a position as Master of the Royal Mathematical School at Christ's Hospital in London. This was founded in 1673 "… for children to be educated in Mathematics for the particular use and practice of navigation." (Plumley 1986). Wales helped turn it into a great naval seminary (Moore 1981). From all accounts, his pupils became quite attached to him (EIC 1909). One of these was Samuel Taylor Coleridge, and Frost (1972) has raised the fascinating thesis that some of his poetry may have been inspired by Wales's accounts of his South Seas experiences. Hunt, who knew Wales, described him as "… a good man, of plain simple manners, with a heavy large person and benign countenance." (cited in Beaglehole 1969: 885).

In November 1776 Wales was elected a Fellow of the Royal Society, a considerable honour (Knight 1856). During his time at Christ's Hospital, he faced the daunting task of publishing the astronomical observations from both Cook's First and Second Voyages. The Second Voyage volume, co-authored with Bayly, appeared in 1777, but because Green's papers were in such confusion (Beaglehole 1968: cxxxiv), Wales only saw the First Voyage monograph off the press in 1788. This long overdue volume also included the astronomical results deriving from the earlier voyages by Byron, Wallis and Carteret.

In 1795 Wales was appointed Secretary of the Board of Longitude, and died just three years later leaving a widow and five children. He was buried in the cloisters of Christ's Hospital, in London. Let us brings this account to a close by quoting the late Sir Patrick Alfred Caldwell-Moore (1923–2012):

> William Wales, then, was a man who earned the greatest respect for his courage, integrity and skill; but it is true to say that he will always be remembered for the part he played when, from 1772 to 1775, he travelled to the far south with Captain Cook … (Moore 1981: 92).

5.3 The Instruments

5.3.1 Introduction

A similar range of scientific instruments to those assigned to the *Endeavour* on Cook's First Voyage accompanied Bayly and Wales on the Second Voyage, but there were some interesting newcomers. In addition to the all too familiar quadrants, sextants, reflecting telescopes, astronomical clocks, journeyman clocks, alarum clocks, each vessel was supplied with a Dollond refracting telescope and with a chronometer. A single transit telescope also was provided (Beaglehole 1969: 721). There also were the usual meteorological and geomagnetic instruments.

As might be expected, the astronomical instruments taken on Cooks' Second Voyage were state-of-the-art technology, manufactured by the leading practitioners of the day, Bird, Dollond and Ramsden.

Apart from quadrants Bird made a few reflecting telescopes (King 1979: 117), and it is interesting that two of these went on Cook's Second Voyage *in lieu* of Short Gregorian reflectors. Beaglehole (1968: 87) describes how Bird "... displayed a talent for delicate and precise work which brought him European fame."

As recounted in the preceding chapter, in 1752 Peter Dollond was joined by his father, John Dollond (Fig. 5.4), who subsequently developed and then patented the achromatic refracting telescope. After his father's death in 1761, Peter Dollond developed the refracting telescope further, enjoyed the monopoly offered by the patent. This by no means endeared him to the London optical community which collectively petitioned the Privy Council in a bid to have the patent revoked. This action was unsuccessful, and Dollond successfully sued a number of his competitors who had manufactured their own achromatic telescopes. From 1764, Dollond (1765) sold a new type of refracting telescope with a triple objective, and Maskelyne bought one of these for the Royal Observatory at Greenwich and was impressed (King 1979: 159). It was this evaluation which led to the use of these instruments during Cook's Second Voyage.

Fig. 5.4 An oil painting of John Dollond by Benjamin Wilson now in the Royal Museums, Greenwich (https://en.wikipedia.org)

Hailing from Halifax in Yorkshire, Jesse Ramsden was a leading London mathematical instrument-maker. In 1765, he married one of Peter Dollond's sisters, which gave him access to the refracting telescope patent (Bennett 1987). Beaglehole (1969: 79n) believes that Ramsden was

> ... probably the greatest of eighteenth century makers of scientific instruments, renowned alike for the delicacy and scrupulous workmanship of his products, and for his unpunctuality in delivering them ...

He was a prolific inventor and manufacturer of instruments spanning the fields of optics, physics, surveying and astronomy (King 1979: 162–170). A Fellow of the Royal Society, he received their Copley Medal in 1795.

5.3.2 *Securing Celestial Positions: The Telescopes, Quadrants and Sextants*

We know of at least three reflecting telescopes that were taken on the Second Voyage. Two of these were Gregorian reflectors by Bird of 61cm focal length (Wales and Bayly 1777: vii), and the Board of Longitude supplied one for each ship (Beaglehole 1969: 721). In Cook's account of the voyage (Beaglehole 1969: 532), two different reflectors are associated with those on the *Resolution* who observed the 7 September 1774 partial eclipse of the Sun: Clerke used the Bird reflector and Cook a Watkins reflector of 45.7cm focal length. Third Voyage documentation (Beaglehole 1967(I): 243) makes it clear that the Watkins reflector (Fig. 5.5) was owned personally by Cook. Bird's credentials as a manufacturer of quality scientific instruments were widely acknowledged and have already been discussed, but the name Watkins is not so well known. In her *Directory of British Scientific Instrument Makers 1550–1851*, Clifton (1995) lists an eighteenth century London instrument-maker named Francis Watkins, who began an optical apprenticeship in 1737 and practised as an optician from 1747. He made a range of different scientific instruments, and had close links with John Dollond (for further details see Gee 2014).

One major innovation on the Second Voyage was the decision to purchase "... 2 of Dollond's last improved 3½ feet telescopes with object glass micrometers and moveable wires." (Beaglehole 1969: 721). The aforementioned optical excellence of such refractors, plus the tendency for the specula mirrors in reflecting telescopes to sometimes tarnish (e.g. see Beaglehole 1969: 835), may have motivated this decision. Whatever the actual reason(s), Dollond refracting telescopes were state-of-the-art technology, and an instrument of similar design to those taken on the voyages is shown in Fig. 5.6.

A further Second Voyage addition to the stable of astronomical instruments was the inclusion of a transit telescope, with which it was intended to observe lunar occultations and meridian transits of the Moon and selected stars in a bid to accurately determine the longitude of the Cape of Good Hope (Beaglehole 1969:

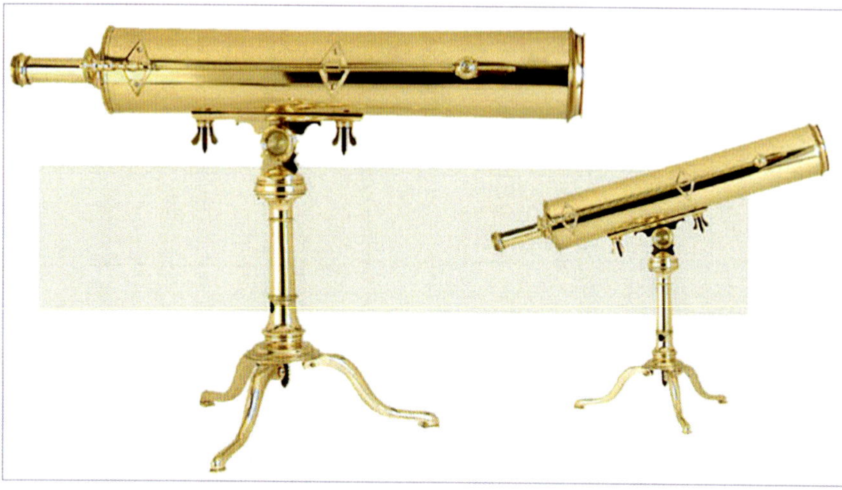

Fig. 5.5 Two different images of a brass Gregorian telescope made by Francis Watkins in about 1765 and similar to the one that Cook took on all three voyages to the Pacific (www.arsmachina.com/t-watkins5038.htm)

726–727). This instrument was manufactured by Bird (Wales and Bayly 1777: vii) and supplied by the Royal Society (Beaglehole 1969: 721). It had an object glass of 8.9cm aperture and 1.06m focal length, and

> The axis rested on two angular pieces of bell-metal, which were attached to two strong plates of brass, about six inches square; and these plates were let into two posts of Riga timber, six inches by eight, and screwed firmly to them.

Only one transit telescope was provided, and this was to be shared by Bayly and Wales.

Let us now examine the astronomical quadrants taken on Cook's Second Voyage. The Royal Society and the Board of Longitude each supplied a Bird quadrant (Wales and Bayly 1777: vii), and in addition the Society provided "... 2 common brass Hadley's quadrants." (Beaglehole 1969: 721).

We have no descriptions of any of these instruments, and as we saw in the previous chapter there is considerable confusion as to the current whereabouts of *any* of the astronomical quadrants which went on the three Cook voyages. Apart from the two quadrants in the Science Museum with a supposed First Voyage pedigree there is a third Bird pillar quadrant in the Science Museum with a Cook voyage attribution. This instrument was donated to the Royal Astronomical Society in 1873 by a Dr. W.T. Radford, and is listed in the catalogue of the Society's instrument collection as "Captain Cook's sextant [it is actually a quadrant], wooden frame; c. 1765; R [radius] = 18 in" (Howse 1986: 224). No other information is available, and the Cook attribution is just that—an attribution only. This instrument was loaned to the Science Museum in 1908 (it now has a Museum number 1908–159) and the display caption provides not further information, but it does place the date of manufacture at

Fig. 5.6 An eighteenth century Dollond achromatic refractor similar in appearance to those taken on Cook's Second Voyage (Orchiston collection)

"about 1772" rather than 1765. If this date is correct then this could be one of the quadrants that went on the Second Voyage, but it is important to stress that there are many items of reputed Cook voyage association with bogus or at best embellished histories (e.g. see Kaeppler 1972). Until additional documentation of a more persuasive nature comes to hand the imputed Cook origin of this quadrant must be treated as suspect.

Finally, to further complicate matters, Howse (1979) has reported a fourth 30.5cm Bird quadrant (catalogue number 1876–542), of unknown provenance, which is owned by the Science Museum but is on loan to the National Maritime Museum.

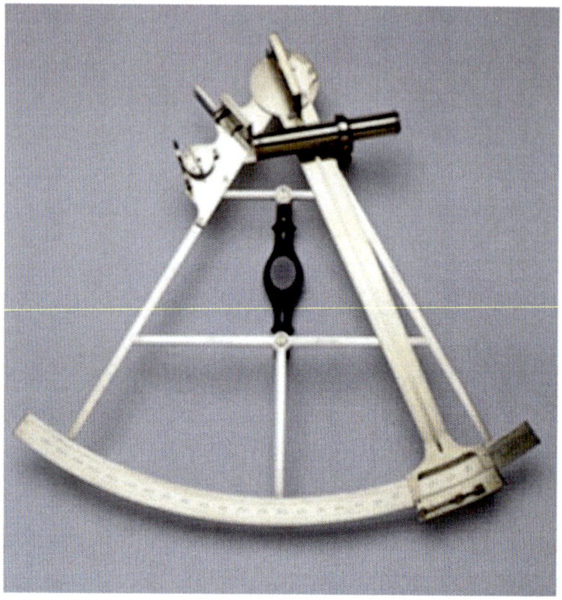

Fig. 5.7 A Hadley sextant by Jessie Ramsden, similar to the one supplied by the Royal Society for Cook's Second Voyage (natwaddell.wordpress.com/2011/12/09/the-transit-of-venus-and-the-longitude-problem-a-briefish-history/)

Sextants generally were used for ship-based astronomical observations, and the Board of Longitude assigned two Hadley's sextants—one by Ramsden and the other by Dollond—to each of the Second Voyage vessels (Wales and Bayly 1777: vii). One of these Ramsden sextants had been used on the Royal Society's transit of Venus expeditions, and was loaned by the Society. In addition, Cook took his own Bird sextant on the voyage, and there were also two other sextants owned by crew members on the *Resolution* (Beaglehole 1969: 79n). A typical Ramsden sextant is shown in Fig. 5.7. None of these Second Voyage sextants can now be traced.

5.3.3 *Keeping Track of Time: The Clocks, Watches and Chronometers*

As we saw in the previous chapter, accurately determining longitude on long maritime voyages posed a major problem during the eighteenth century, so in 1714 the British Government appointed a committee to look into the matter, and as a result of their report an Act was passed by Parliament which offered a graduated scale of financial rewards. The precise amounts payable were dependent upon the level of accuracy attained at the end of a six week voyage (Gould 1987), as detailed in Table 5.2 below.

Table 5.2 The 'longitude prize' offered by the British Parliament in 1714

Amount of error	Prize paid
41–60 milcs	£10,000
31–40 miles	£15,000
<31 miles	£20,000

The Act also set up a Board of Longitude, charged with overseeing and evaluating developments, and allocating payments, if warranted. With such large sums of money involved, it is not surprising that the Board became the target of cranks, swindlers and fanatics.

Fortunately, there were serious contenders, and these fell into two main camps: the watchmakers and the astronomers. The latter were led by Astronomer Royal, Nevil Maskelyne, who hoped to demonstrate that his system of using lunar distances was thoroughly reliable. Competing against the astronomers were the watchmakers, led by John Harrison (Fig. 5.8). Harrison (Gould 1987) manufactured

Fig. 5.8 A half-tone print of Thomas King's original portrait of John Harrison (https://en.wikipedia.org)

three increasingly more reliable chronometers, but made little progress with the Board of Longitude. He then came up with his '4th watch machine', which created such an impression that it was taken to Jamaica in 1762 and Barbados in 1764 for testing, and in the process proved to be remarkably accurate. Quill (1966: 137) reports that

> … the average error of the timekeeper was only 39.2 seconds of time, which at the latitude of Barbados (16°N) equalled 9.8 geographical miles. This was an accuracy three times better than that required to win the full £20,000 … Even John Harrison himself could hardly have dared to hope that the outcome of the trial would be such a complete success.

Despite this outstanding result, the Board of Longitude decided to pay over just half of the £20,000 prize to Harrison, and even then on the condition that he hand over all four chronometers. These developments were to impact directly on Pacific exploration, with the decision to send chronometers on Cook's Second and Third Voyages (Betts 1993).

The Board of Longitude decided that each Second Voyage vessel should take one astronomical clock (complete with instructions for setting it up—see Fig. 5.9), one journeyman clock and an alarum clock. In addition, chronometers were assigned to the voyage.

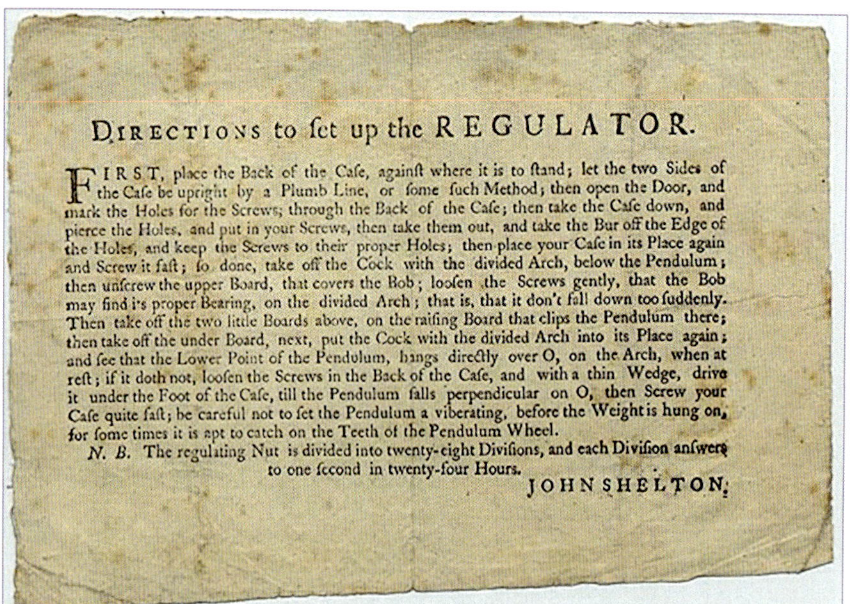

DIRECTIONS to ſet up the REGULATOR.

FIRST, place the Back of the Caſe, againſt where it is to ſtand; let the two Sides of the Caſe be upright by a Plumb Line, or ſome ſuch Method; then open the Door, and mark the Holes for the Screws; through the Back of the Caſe; then take the Caſe down, and pierce the Holes, and put in your Screws, then take them out, and take the Bur off the Edge of the Holes, and keep the Screws to their proper Holes; then place your Caſe in its Place again and Screw it faſt; ſo done, take off the Cock with the divided Arch, below the Pendulum; then unſcrew the upper Board, that covers the Bob; looſen the Screws gently, that the Bob may find i's proper Bearing, on the divided Arch; that is, that it don't fall down too ſuddenly. Then take off the two little Boards above, on the raiſing Board that clips the Pendulum there; then take off the under Board, next, put the Cock with the divided Arch into its Place again; and ſee that the Lower Point of the Pendulum, hangs directly over O, on the Arch, when at reſt; if it doth not, looſen the Screws in the Back of the Caſe, and with a thin Wedge, drive it under the Foot of the Caſe, till the Pendulum falls perpendicular on O, then Screw your Caſe quite faſt; be careful not to ſet the Pendulum a viberating, before the Weight is hung on, for ſome times it is apt to catch on the Teeth of the Pendulum Wheel.

N. B. The regulating Nut is divided into twenty-eight Diviſions, and each Diviſion anſwers to one ſecond in twenty-four Hours.

JOHN SHELTON.

Fig. 5.9 Shelton's instructions for successfully setting up his astronomical clocks (after Johnston 2005)

The Board provided its own Shelton clock which had been purchased in 1768 in order to conduct latitude and longitude observations at The Lizard, Cornwall, in 1769, and received on loan a similar clock, also by Shelton, from the Royal Society. It is not known whether this latter clock was that used on Cook's First Voyage or was one of two clocks purchased by the Society in 1768 for use during observations of the 1769 transit of Venus from near the North Cape in Norway. Howse (1969d) suspects that it was one of the two latter clocks which was loaned. Be that as it may, there is no documentation to indicate which Shelton clock was assigned to which particular Second Voyage vessel. Because one of the objectives of the voyage was to test the reliability of the chronometers, the observatories were frequently set up ashore, and so a 'clock stand' comprising an iron base plate and sturdy frame was designed to support each astronomical clock.

The Board also furnished each Second Voyage astronomer with a brand new journeyman clock by Jonathan Monk costing £7 (Howse 1969c). Wales and Bayly (1777: xviii–xix) comment on the usefulness of these timepieces:

> The loudness of the beat is of great use when the wind is high; or when, on account of any other noise or disturbance, the Astronomical Clock cannot be heard; and was particularly useful to us, whose observations stood generally on the sea-shore where the roaring of the surf seldom permitted us to hear the Astronomical Clock all the time it was going.

Fortunately the "roaring of the surf" was not often a problem in sheltered Queen Charlotte Sound where most of New Zealand's Second Voyage observatories were set up.

Each Second Voyage vessel also was assigned an alarum clock. One of these, by Shelton, had been acquired by the Board of Longitude in 1769, while a new alarum clock was purchased for two guineas from Jonathan Monk. It is not known which alarum clock went to which particular vessel (Howse 1969c). The Board of Longitude also provided Wales with a horizontal stop watch by John Ellicott, while Bayly took his own pocket watch on the *Adventure* (Howse 1969d: 64).

The Board of Longitude sent four different chronometers on the Second Voyage, two in each of the vessels. One was manufactured by a watchmaker named Larcum Kendall (1719–1790) and was an exact replica of Harrison's fourth chronometer (Howse 1969a), while the other three were designed and manufactured by Kendal's rival, John Arnold (1736–1799; Fig. 5.10; Mercer 1972). Kendall's K1 chronometer (Fig. 5.11) and Arnold's 3rd (A3) chronometers were assigned to the *Resolution* and Arnold's other two instruments (A1 and A2) to the *Adventure* (Howse 1969a, b). Tooth (1993: 37) has pointed out that Cook's Second Voyage

> … must surely be regarded as one of the most important sea voyages in British maritime history. Upon the performances of the four timekeepers, and K1 in particular, would rest the fate of the marine chronometer as the answer to the longitude problem for many years to come. As this was the first time a chronometer had been taken to sea other than on a controlled sea trial, it would be the first real test of its usefulness as a navigational aid. Many people other than the Board of Longitude and the Royal Navy would be taking much more than just a passing interest in this particular voyage.

Fig. 5.10 A painting of John Arnold in about 1767 (https://en.wikipedia.org)

As it turned out, Kendall's chronometer proved to be remarkably accurate (e.g. see Howse 1969a: Fig. 5), but Arnold's three appliances were failures. Indeed, the A2 and A3 stopped during the voyage.

Howse and Hutchinson (1969b) have tentatively identified the astronomical clock now known as RS34 (Fig. 5.12) with Cook's Second Voyage, although they point out that it is not possible to exclude either the 'Royal Society Clock' or the 'Herstmonceux Clock' as possible contenders.

In contrast to the astronomical clocks, the current whereabouts of the Kendall chronometer which went on the Second Voyage is well known: it is in the National Maritime Museum, on loan from the Navy (Howse 1969a, b). The three Second Voyage Arnold chronometers are more problematic. In addition to these Cook Voyage timekeepers, at about the same time Arnold made two others for the projected voyage of Captain Phipps towards the North Pole. All five chronometers were subsequently stored together, and in the absence of any known engraved distinguishing numbers or marks their identities quickly became blurred (see Howse 1969b). Currently two of these definitely survive, and are thought to be associated with Cook's Second Voyage. In 1968, the Royal Society loaned these to the National Maritime Museum (Howse 1969a). In addition, the Museum of the Clockmakers' Company owns an early Arnold chronometer which may be from the 1770s (Howse 1969b), but even if this chronological attribution is true there is no way of knowing whether this is of Cook Voyage origin.

Fig. 5.11 A postage stamp featuring the K1 chronometer (http://sio.midco.net/dansmapstamps/jamescook2.html)

5.3.4 Protecting the Equipment: The Tent Observatories

Following the success of the experimental prototype tent observatory taken on Cook's First Voyage (but not used in New Zealand), Bayly developed a satisfactory tent observatory for the Second Voyage (see Fig. 5.13). One was assigned to each astronomer (Beaglehole 1974: 288), on the basis that it "… will be of infinite use to them wherever they may have occasion to make observations on shore …" (Beaglehole 1969: 722).

Wales and Bayly (1777: viii) have provided a detailed description of this type of tent observatory, which I will summarise here. The framework comprised of eight vertical wooden poles, 5.1cm in diameter and 1.65m long. Each of these had a spike at one end, which was stuck into the ground. These poles supported a 2.4-m diameter circular wooden ring, which came in eight separate sections that were connected by iron plates. Around the outer edge of the wooden ring were small staples, and hooks on the top of the canvas wall covering slotted into these. A second 2.4-m wooden ring sat on top of the 'wall ring', and screwed to this were the rafters which met at a crown-piece and supported the thick conical canvas roof. When required, a section of this roof could be removed to expose a triangle of sky. Associated with the observatory was a tripod composed of three 4.7-m long wooden

Fig. 5.12 The Shelton RS34 astronomical clock (Orchiston collection)

poles, and a rope which ran along one of these was attached to the crown-piece via a hook and eye bolt. This rope was used to raise and rotate the roof when observations were to be carried out.

New Zealand gained its first genuine astronomical observatories in 1773 when tent observatories were erected at Dusky Sound and Queen Charlotte Sound.

5.4 Pinpointing the Extremities of the South Island

Cook's Second Voyage was undertaken in order to determine the existence or otherwise of Terra Australis, the postulated Great Southern Continent. Two vessels were involved, the *Resolution* under Captain James Cook and the *Adventure* under Captain Tobias Furneaux (1735–1781; Chisholm 1911).

The Board of Longitude provided Wales with specific instructions about the types of shore-based astronomical observations he was to make during the Voyage:

> Where ever you land you are to make the same observations on shore [for latitude and longitude] as you have been directed to make on Ship board, only Observing … with the Astronomical Quadrant instead of the Hadleys Sextants …

Fig. 5.13 The type of tent observatory used by the Second Voyage astronomers (after Wales and Bayly 1777: Plate II)

> Whenever you can land safely (should it be for only two or three days) Set up the Astronomical Clock and fix it very firmly to a Massey peice of wood let deep into the ground and fix the Pendulum at the same exact length as it was of when going at the Royal Observatory at Greenwich before the Voyage and take equal Altitudes of the Sun and fixed stars for determining the rate of its going …
>
> Compare the Watches with the Astronomical Clock at Noon and also about the times of equal Altitudes.
>
> Observe meridian altitudes of the Sun and also of fixed Stars some to the North and some to the South for finding the Latitude; Observe also differences of right ascension between the Moon and fixed Stars in the manner explained by the Astronomer Royal … in Order to Settle the Moons parallax in right ascension in Various Latitudes.
>
> Make Observations of Eclipses of Jupiters Satellites & Occultations of fixt Stars & Planets by the Moon and any other Observations usefull for settling the Longitude of the place … (Beaglehole 1969: 726).

Whilst *en route* to New Zealand in 1773 the *Resolution* and *Adventure* vessels were separated in the southern Indian Ocean, and the *Adventure* made straight for Queen Charlotte Sound, arriving there on 7 April 1773. Meanwhile, the *Resolution* had reached desolate inhospitable Dusky Sound, in the far south-western corner of the South Island (Fig. 5.14; Begg and Begg 1968), a week or so earlier, where Cook intended to obtain refreshments and fresh water for his crew. On his First Voyage, Cook had discovered the value of fresh food in combating scurvy (see Watt 1979),

Fig. 5.14 Cook and those on the *Resolution* found Dusky Sound a hauntingly beautiful yet secluded place, with high forested mountains that plunged directly into the sea. For Bayly, meanwhile, cloudy skies or rain meant it was difficult to make astronomical observations (*Photograph* google.earth)

and his first act upon reaching Dusky Sound was to secure a seal for fresh meat. Soon afterwards, production of 'spruce beer' began:

> … we first made a strong decoction of the leaves and small branches of the Spruce tree (*rimu*) & Tea shrub (*manuka*) by boiling them three or four hours … then take the leaves of branches out of the Copper and mix with the liquor the proper quantity of Melasses and Inspissated Juce, one Gallon of the former and three of the latter is sufficient to make a Puncheon or 80 Gallons of Beer, let this mixture just boil and then put it into the Cask and to it add an equal quantity of Cold Water … when the whole is but milk warm put in a little grounds of Beer or yeast … and in a few days the Beer will be fit to drink … (Beaglehole 1969: 137).

The resulting product was "… a very palatable pleasant drink; the Major part of the People I believe are of the same Opinion …" (Clerke 1771–1775). To Second Lieutenant Charles Clerke (1743–1779; Fig. 5.15), Dusky Sound "… for a Set of Hungry fellows after a long passage at Sea is as good as any place I've ever yet met with." (ibid.); the combination of spruce beer and abundant fresh fish, crayfish, seals and birds soon had the crew back to full health (Beaglehole 1969: 137).

However, one of the supernumeraries, the scientist Johann Reinhold Forster (Fig. 5.16) was more interested in matters of navigation:

> … I must here do Justice to our Navigators, who, when we first made the Land, coincided within a few Miles with their accounts, since the last Observation of Longitude; some were not above 3 miles out. This proves more & more the Excellence of this method of Observing the Longitude by Distances of the Moon from the Sun or some fixed Stars, which has been encouraged and introduced by the Board of Longitude & is now well understood & practised by the Gentlemen of the British Navy. The Excellence of the

Fig. 5.15 A painting of Captain Clerke by Nathaniel Dance in 1776, now in Government House, Wellington, New Zealand (https://en.wikipedia.org)

Fig. 5.16 A painting of Johann Reinhold Forster (left) and his son, Georg Forster (right), by Jean Francis Rigaud (https://en.wikipedia.org)

English Instruments, the accuracy & correctness of the Tables, the Experience & readiness in taking the most accurate Observations & calculating them with amazing expertness, all together contribute towards giving the Superiority to British Navigators in respect to astronomical Observations on board of a Ship, before those of all other Nations. (Hoare 1982(II): 239).

This was high praise indeed, both for the astronomical method of determining longitude and for the competence of British naval officers.

Wales's concerns were more with establishing his observatory, and on 29 March 1773 he settled on a site, the 15-m high promontory near where the ship was moored. This was appropriately named 'Astronomers Point' (see Fig. 5.17). Wales and two other members of the crew spent part of the next three days creating a level clearing from which celestial observations could be made. This involved an enormous amount of labour:

Every inch of the acre which was required for an open view had to be hewn laboriously with axes, and involved other heavy work in order to reclaim a space from the forest, undisturbed from the time of the Flood to this day. The trees in this wood, growing on their fallen ancestors which had rotted through the centuries and been transformed into the richest mould, reached gigantic heights ... (Sparrman 1953: 26).

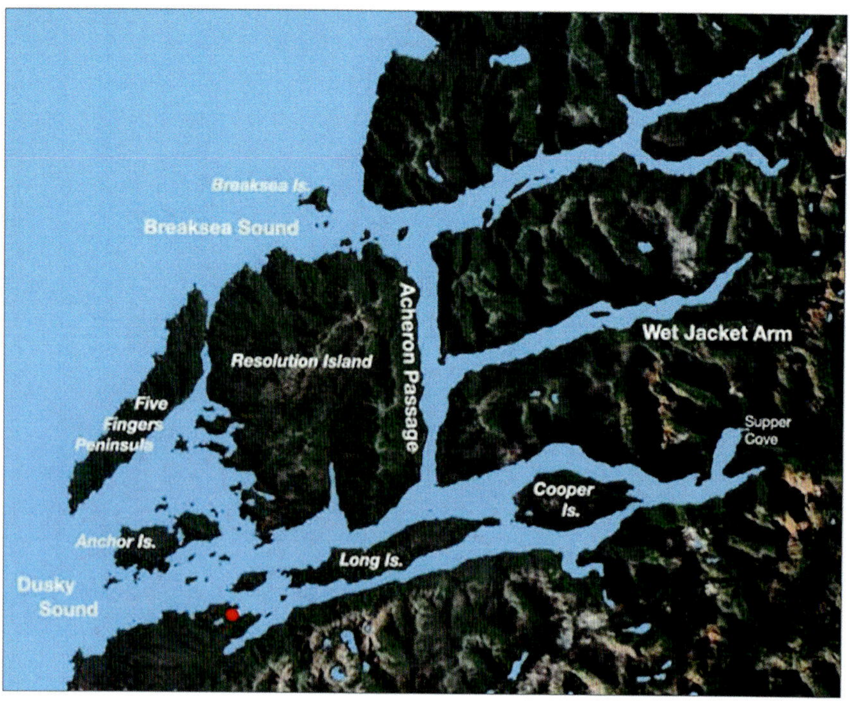

Fig. 5.17 A map of Dusky Sound, with the *red dot* showing the location of 'Astronomers Point' (*Map* scubadive.net/nz/viewtopic.php?f = 79&t = 4763)

Wales (1772–1774: 777–778) laments that on the first day alone he "… cut down and destroyed more Trees and curious shrubs and Plants, than would in London have sold for one hundred Pounds." This was exactly a quarter of his annual salary! When completed, the clearing affording excellent views of the northern sky and some of the southern sky, but it also provided a suitable shore-base close to the ship, for "… a forge, a green hut for the woodcutters and a pen for our Sheep …" (Hoare 1982(II): 266).

On 1 and 2 April, Wales (1772–1774: 778) erected his observatory:

> Set up the Clock and got a Puncheon filled with stones and gravel as a stand for the Quadrant; but had the mortification to find that neither one nor the other could be fixed with sufficient steadiness to answer any purpose, on account of the looseness of the ground: I was therefore necessitated to look out for another place & pitched on one where two large trees grew almost close together one of which I cut down almost close to the ground, and the other about 3½ feet above it … Leveled the ground for the observatory round the before-mentioned two Trees put up the Tent, & fixed the Iron frame for the clock on one; and set the Quad^{t} on the other; and found them to answer pretty well; although a smart stamp with the foot at 7 or 8 Yards distance would still make the Plumb-line Shake very plainly in the Microscope.

While there are no detailed pictorial renditions of the observatory site, a colour painting by William Hodges titled "View in Pickersgill Harbour, Dusky Bay, New Zealand" now in the National Maritime Museum, London, does provide us with a fleeting glimpse of the top of the astronomer's tent, nestled in amongst the trees (see Joppien and Smith 1985(II): Plate 18).

From his forest clearing, Wales was able to obtain some observations of the Sun, Moon, selected stars, lunar occultations and eclipses of Jovian satellites, even though the weather was "… almost continually Cloudy with Rain …" (Wales 1772–1775). This involved taking a series of measurements, determining individual latitudes or longitudes for each reading, and then calculating mean values for latitude and longitude. By way of example, while he was at Dusky Sound in April 1773 Wales observed meridian zenith distances of the Sun and four different stars in order to calculate the latitude of his observatory. Values of latitude derived from these observations are listed in Table 5.3, while longitude values derived from lunar observations alone are listed in Table 5.4. In the case of his latitude calculations, we can see that Wales utilised many different solar and stellar measurements, while his

Table 5.3 Mean values of latitude obtained by Wales at Dusky Sound, April 1773 (after Wales and Bayly, 1777: 24)

Object observed	Number of days/nights	Latitude (South) From interior arc	Latitude (South) From exterior arc
Sun	6	45° 47′ 51″	45° 47′ 29″
Beta navis	4	45° 47′ 09″	45°47′ 26.67″
Delta navis	4	45° 47′ 31.5″	45°47′ 37.75″
Gamma navis	4	45° 47′ 37.5″	45° 47′ 47″
Procyon	4	45° 47′ 13.5″	45° 46′ 53.5″

Table 5.4 Individual values of longitude obtained by Wales at Dusky Sound, April 1773 (after Wales and Bayly 1777: 18–19)

Date (1773)	Longitude (East)
April 14	166° 27′ 45″ 166° 24′ 45″
April 17	166° 03′ 18″ 166° 24′ 03″ 166° 04′ 10.5″ 166° 24′ 55.5″
April 18	165° 18′ 21″ 165° 14′ 45″

longitude values were based on two or more individual readings on each night. In addition to the sets of observations listed in these two tables, Wales obtained three double altitudes of the Sun's limb, and these gave a mean latitude of 45° 47′ 52.5″ (Wales and Bayly 1777: 24).

These various astronomical observations produced the following position for Astronomers Point:

Latitude 45° 47′ 33.3″S
Longitude 166° 02′ 41″E

Conflicting latitude values of 45° 47′ 47″S and 45° 47′S are given by Clerke and Cook respectively. Similarly, they provide the following discrepant values for the longitude: 166° 15′ 30″E and 166° 02′ 46.5″E (Beaglehole 1969: 756 and 138–139). Meanwhile, Cook was still to be convinced of the value of the chronometer:

> … uncertainty of obtaining the true rate of going of Watches [i.e. chronometers] and what is still worse their varying their rate will always render the Long^{d} deduced from them a little uncertain especially in long runs. (Beaglehole 1969: 139).

As an astronomer, he preferred to rely on co-ordinates based on standard astronomical observations.

While Wales was making his observations in Dusky Sound, Bayly was doing likewise at Queen Charlotte Sound. After the *Adventure* anchored in Ship Cove in early April 1773, Bayly's first task was to observe a partial lunar eclipse on 6 April, noting that "The moon was ⅔ of her Diam'r submerged in the shadow." (Bayly 1773: 202). He then had to select a site for his observatory. At first sight the obvious spot was somewhere on the barren ridge top above Ship Cove, but ready access was a problem given the dense vegetation on the hillsides (see Holmes 1984: 74). Although there were paths up through the forest, in reality the surrounding hilltops were out of bounds for astronomy. Another option was called for.

In mid-Sound adjacent to the anchorage was Motuara Island, and off its southern extremity Hippah Island (see Fig. 5.18). The summit of this little island was about 25 m above mean sea level, and offered a promising horizon. A Maori *pa* was located on this Island (Fig. 5.19), and had been occupied when the *Endeavour* was in the Sound three years earlier. Now it was vacant, and Bayly decided that this was

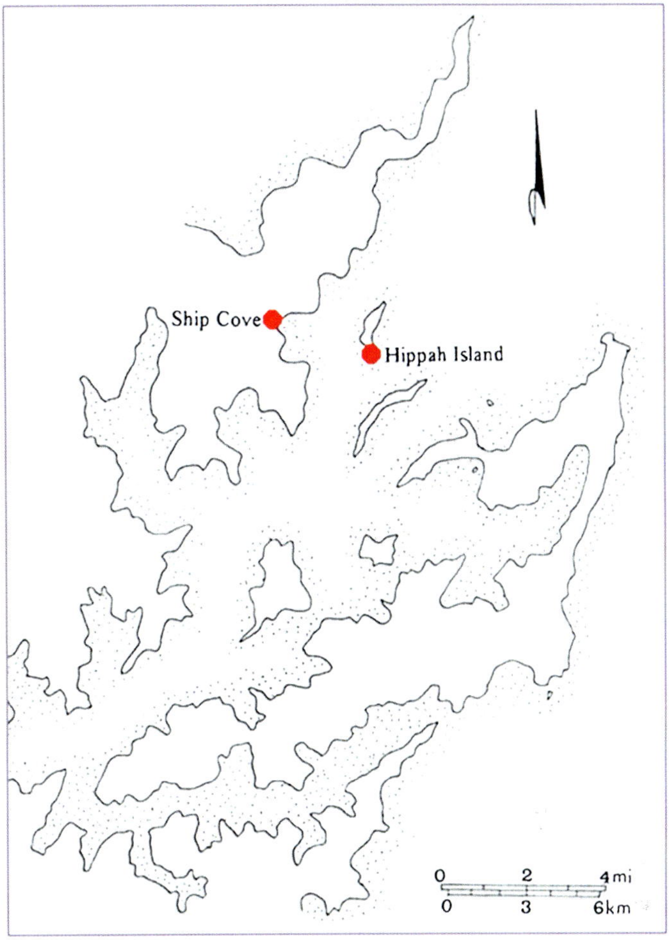

Fig. 5.18 Map showing the location of the Hippah Island observatory site relative to Ship Cove. The *Adventure* anchored in the Sound about mid-way between these two locations (*Map* Wayne Orchiston)

ideal for his observatory. He provides us with an interesting description of the Island:

> It is a rock whose sides are perpendicular in many places & indeed the whole was well fortified by nature, there being only one landing place & the passage up from it exceedingly difficult … On the top of this small island was a Town consisting of 33 houses. The most elevated part was tolerably level for about 100 yards long & 8 or 10 yards wide. This was fortified with strong posts or sticks drove into the ground & those interwoven with long sticks in a horizontal direction & then filled with small brush wood with one place two feet square where was a wooden door, so that only one man could get in at a time & that on his hands & knees & of course easily destroy'd if at war. (Bayly 1773: 207).

Fig. 5.19 Copy of a painting by John Webber of the Hippah Island settlement where Bayly set up his observatory in April 1773 (Orchiston collection)

Bayly was not on the best of terms with Captain Furneaux and other officers on the *Adventure*, and so when it came to setting up his observatory all he could rely on were his own servant and one unofficial assistant, "… a good natured Welshman, who would always work if I gave him Brandy …" (ibid.). On 17 April they transporting the astronomical instruments up to the *pa* site, but "… it was with the utmost difficulty we got them to the top of the Island." This only proved possible after they cut steps in the rock (Bayly 1772–1774).

Bayly then erected the tent observatory for the astronomical clock and a 'house' (a temporary hut) for the transit telescope. On 19 April he set up the clock:

> In one corner of the House (I had built) I sank a hole in the Rock about 12 Inches from Each Side of the House, & about 16 Inches deep; in the bottom of this hole on the Solid Rock, the Block of Iron [support for the clock] was laid as nearly Horizontal as possible, so that the bottom of the Clock-case was about 14 Inches below the floor of the House, & the Clock stood quite independent of every part of it (ibid.).

Two days later he set up the transit telescope, and he carried out further adjustments on the 24th:

> In the morning I moved the Transit Instrument very near the true Meridian by means of the Adjusting Screws, where it Accidentally cut two good marks on the tops of two Hills one to North distant about 1⅓ Mile; & the other to the South about 4½ Miles; the Instrument was constantly kept to these marks during the whole time it remained up, by examining it three or more times a day, & frequently before each observation when in the Day: as was likewise the Level, & the Posts was so well rammed that I never found it [out] more than a bredth of the Wire … (ibid.).

Two marines and a Highland piper were assigned to protect the observatory (even though two of these men were ill), and Bayly's party also included three midshipmen. Once they had excavated the floors about 30cm into the earth (Furneaux 1772–1774: 740) and got rid of "… an incredible swarm of vermin, fleas and lice …" (Sparrman 1953: 41–42), the deserted native huts provided very comfortable accommodation.

Bayly diligently observed from 'Observatory Island' until 19 May when the order came to return the instruments to the ship (Bayly 1773: 208). In order to establish the latitude of the Sound, he carried out meridian zenith distance observations of the Sun, Moon and nine different bright stars on 15 different dates between 29 April and 15 May. Meanwhile, observations of the Moon on five different evenings between 29 April and 14 May produced differing values of longitude. In addition, on 6 May he used the Dollond refractor (Bayly 1772–1774) to observe an eclipse of Jupiter's second satellite:

> At the time of this Observation, the air was very clear, and the limbs of the planet, as well as its belts, exceedingly distinct and well defined; the magnifying power used was 150 times. (Wales and Bayly 1777: 42).

Five nights later, he used the same telescope to observe a lunar occultation of the star Nu Aquarii (Bayly 1772–1774).

While Bayly was engrossed in his astronomy the crew of the *Adventure* were busy " … repairing our Sails, Casks &c & getting the Sick on shore … repairing our Rigging, wooding, Watering & other necessary things …" (Hooper 1975: 49). Several different groups of Maori from various parts of Queen Charlotte Sound visited the vessel, bringing for trade fish and fern root (a local substitute for potatoes) and great qualities of their clothing, tools and weapons. Many of the crew also succumbed to the temptations offered by the local 'heavenly bodies' of a non-celestial kind!

On 19 May 1773 the *Resolution* reached Queen Charlotte Sound, and the following day Wales erected his quadrant ashore in Ship Cove (just as Bayly was stowing his equipment away on the *Adventure*):

> I got my astronomical Quadrant on shore at a Beach near the ship, and set up a Cask filled with water for a Stand, in order to get a few equal Altitudes, & if possible see what Rates the Watches now had … (Wales 1772–1774: 788).

Wales' observatory was situated almost at sea level, and this thoroughly accessible spot was flanked by high hills to the westward which shielded the Sun from view a little after 2 pm each day. Despite the challenging access, Wales was quick to see the advantages of Bayly's former observatory site (Fig. 5.20). However, he required observations in order to rate the chronometers, not to establish the latitude and longitude of the Sound. He must have been disappointed, then, when he went to wind the chronometers on 29 May and found that

> … the middle Lock to Mr Arnold's Watch was damaged, and could not be opened, I suppose by its being opened yesterday with a wrong Key through mistake. It was proposed to open the Box by sawing off the Staples into which the Bolts of the lock shoot; but

Fig. 5.20 An aerial view of Queen Charlotte Sound looking east. In the foreground is Ship Cove where Wales set up his observatory, while Hippah Island (the site of Bayly's observatory) and part of Motuara Island are towards the top of the photograph (Orchiston collection)

> being apprehensive that the action of the saw might shake the watch too much, I proposed that the screws which fast[n] these Staples to the cover of the box might be wrenched out, by introducing the blade of a screw Driver, & turning it round; and which was accordingly put in execution. As some damage was done to the Lock by this Accident, the Lock was taken off … (Wales 1772–1774: 789).

On 5 June, Wales computed the rates for the two chronometers, and found that "… M[r] Kendall's was gaining 9″.05 P. Day & M[r] Arnold's losing 94.158 P. Day on mean time." (Wales 1772–1774: 790).

One interesting experiment that Wales and Bayly did while in Queen Charlotte Sound was to establish the longitude difference between Ship Cove and the Hippah Island observatory. By firing guns at these sites and noting the time delay between seeing the explosion and hearing the report they were able to establish a time differential of just 10 seconds (Wales and Bayly 1777: 49).

Bayly's various astronomical observations provide the following mean values for the position of Queen Charlotte Sound:

Latitude 41° 05′ 47.65″S
Longitude 173° 56′ 14.44″E

The longitude figure is after Bayly (1772–74), and values derived from lunar observations *and* those obtained from the lunar occultation and Jovian satellite observations. It is interesting that Cook gives values of 41° 05′ 35″S and 173° 48′ 55.5″E, both reputedly based on Bayly's observations (see Beaglehole 1969: 173). J.R. Forster also gives this disparate latitude figure (Hoare 1982(II): 293). Cook comments further:

> … the Latitude, [Magnetic] Variation and Tides are confirmable enough to the like observations I made when I was here in Jan[ry] 1770, but the Longitude differs considerably more than might be expected; in my Chart I have plac'd [it] in … 175° 9′ East … [which] is 1° 20′ more East than Mr Baylies which we must allow to be nearer the truth … (Beaglehole 1969: 173–174).

In another version of his journal, Cook explains why he places greater credence on Bayly's results:

> … the Longitude by M[r] Bayly appears to be nearer the truth because deduced, not only from a larger number of observations but a variety & some of them such as are less liable to error than those of the Sun & Moon; they also correspond with M[r] Wales's observations made at Dusky Bay reduced to this place by the Watches. The difference of Longitude in my Chart between these two places is agreeable enough to the differences of these gentlemen's observations … (Beaglehole 1969: 174n).

We must not forget that in 1770 the latitude and longitude of Queen Charlotte Sound were derived from ship-board observations. The significant difference in the Second Voyage longitude value reflects the merit of a shore-based observatory and firm foundations for the scientific instruments. In addition, Bayly had recourse to the added precision of a transit telescope in 1773.

The *Resolution* and *Adventure* eventually left the Sound on 7 June 1773, and visited Pitcairn Island, Tahiti and Tonga, before returning together to New Zealand. Just as they made the New Zealand coast they were separated by a long and violent storm. The *Resolution* was the first to reach Queen Charlotte Sound, on 3 November 1773, and the crew settled down to wait for the *Adventure*. Meanwhile, many local Maoris flocked to Ship Cove to trade with the visiting Europeans, and on this occasion pilfering became a major issue. Cook reports an amusing incident, in this regard:

> … one of the chiefs undertook to remove [this pilfering], and with fury in his eyes made a shew of keeping the people at a proper distance, I uploaded his conduct but at the same time kept so good a lookout as to detect him in picking my Pocket of a handkerc[h]ief, which I suffered him to put in his bosom before I seem'd to know any thing of the matter and then told him what I had lost, he seemed quite ignorant and innocent, till I took it from him, and then he put it of with a laugh, and he acted his part with so much address that it was hardly possible for me to be angry with him, so that we remained good friends and he accompanied me on board to dinner … (Beaglehole 1969: 288).

Wales records in his journal that on 3 November he and Cook went in search of a suitable site for the observatory and others to be stationed ashore, so as to "… be a mutual protection to each other …" (Wales 1772–1774: 816). They settled on Ship Cove, and the following day Wales "Finished the Observatory & set up the Astronomical and Assistant Clocks: fixed up a stand for the Quad^{t} and Adjusted it ready for Observation" (ibid.).

Almost immediately, Wales and his party of six were exposed to the very same problems those on the *Resolution* were encountering. On the morning of 6 November Wales awoke to find that a coat and a bag of dirty washing had disappeared from the tents. Cook found the missing items at a nearby native settlement and was told by the inhabitants that

> … some of them had crossed over the point of Land which separates the Cove where they were from that where we were … [and] had watched 'till we were all out of the tent and then sliped in & taken them away. We now agreed to watch by turns, two hours at a time … (Wales 1772–1774: 817).

Wales was worried about the safety of the instruments, and particularly the K1 chronometer.

Despite these security concerns, he was able to carry out a series of observations of the Sun and four bright stars in order to once more investigate the latitude of the Sound, and he conducted solar observations in order to determine the longitude. In addition, on 14 November he obtained a value of 174° 19′ 45″ from a telescopic observation of a Jovian satellite eclipse. J.R. Forster, one of the two German scientists on the *Resolution*, describes the circumstances surrounding the observation of this particular event:

> In the Evening we went to the Astronomical Observatory & set several Telescopes up, in order to observe the Emersion of one of Jupiters Satellites. We saw the Planet & three of its Satellites very finely, each in a Telescope made by different Artists. Capt Cooks was made by *Watkins*; Mr Wales observed by *Dollands* achromatic Telescope, Mr Smith by *Birds*

> Reflector & I by *Ramsdens* achromatic Telescope. The Astronomer directed us to observe on the Side where we saw two Satellites, which we all did, but this was the wrong Side, & at last when the Emersion just happened Mr Wales found he was mistaken, & every body with him saw the Satellite on the Side of Jupiter where only one Satellite before had appeared. (Hoare 1982(III): 423).

Hoare (1976) has carefully documented that there was no love lost between Forster and Wales, so Forster would have been somewhat amused by this incident.

When we examine Wales' November 1773 observations we derive the following mean values for the co-ordinates of Ship Cove at Queen Charlotte Sound:

Latitude 41° 05′ 49.7″S
Longitude 174° 25′ 41″E

In the published account of the voyage Wales gives a longitude value of 174° 17′ 4.5″E (Wales and Bayly 1777: 246), which also appears in his astronomical journal. In the latter source, Wales comments:

> I was not a little surprised to find that I made the Longit. of the Place … upwards of half a Degree more than Mr Bayley Made it. I also make the Latitude 41° 06′ 02″½S. that is 47″½ more than it ought to be by Mr Bayly's Observations. I have Computed All the Observations twice over and am certain they are done right … (Wales 1772–1775).

On 21 November, the observatory and all of the astronomical instruments were taken on board the *Resolution*, as Cook prepared the vessel for departure. When the *Adventure* still had not made an appearance by 25 November Cook determined to continue the voyage alone, and set sail for Antarctic waters in order to disprove once and for all the existence of the great Southern Continent.

Meanwhile, the *Adventure* reached Queen Charlotte Sound just six days after Cook's departure, having first sheltered at Tolaga Bay on the East Coast of the North Island. Bayly (1772–1774) reveals that the very next day (2 December 1773)

> Carried my Tent Observatory & Instruments on Shore & set all up by the side of the Garden at a little distance from the watering place [in Ship Cove] … The place where my Observatory stood when we were here before [Hippah Island] bearing E 5½° W per Compass distant …

Bayly soon found that security was a major issue, and on 14 December he was personally involved:

> I was up late observing & having taken some altitudes of Stars to the East & having set my Alarem to call me, to take them to the West, I went to bed, having nailed my old great coat at the entrance of my tent, at the inside of which I always placed the outside case of my Astronomical Quadrant … In this box I kept my Lumber, such as tools, nails, &c. … After I had been in bed & had slept some time, I was awaked by the rattling or noise of the lid of the box. I jumped up in the bed & took my gun … calling at the same time, who was there, but could neither hear nor see anything … Soon after I searched for my hat but could not find it … on going out of the Tent Obsy. I found it open & half the lid of the box at some dist. from the tent, & by feeling I found my hatchet & saw & hammer were gone out of the box. (Bayly 1773: 215–216).

After pursuing a number of natives, Bayly (1773: 216) came to a place where there were "… great quantities of things which they had stole from us, part in their canoe & part on the rocks …" This proved to be no isolated incident.

In spite of such distractions, Bayly carried on with his astronomy. He used observations of the Sun and five bright stars on four different dates to derive latitude and solar observations, also on four different days, for longitude. In addition, he obtained a longitude figure of 174° 15′ 30″E from a Jovian satellite event observed on 14 December (Wales and Bayly 1777: 71). These combined December 1773 observations produced the following mean co-ordinates for Ship Cove:

Latitude 41° 05′ 34.0″S
Longitude 174° 20′ 30″E

Throughout the stopover local Maoris continued to visit the ship, offering fish, curiosities and women, until ten of the Europeans were massacred while collecting scurvy grass in Grass Cove. This most unfortunate, yet not totally unexpected, event followed a succession of incidents and transgressions by the visiting Europeans (see Orchiston and Horrocks 1975). Luckily, Bayly was not one of those who died in what has become known as the 'Rowe Massacre'. One day later, on 18 December 1773, the *Adventure* sailed out of Queen Charlotte Sound for the last time, and returned to England.

Meanwhile, the *Resolution* was proceeding towards the Antarctic on a journey that was to disprove once and for all the existence of Terra Australis Incognita. In February 1774 the ship sailed past Easter Island, and it then proceeded via the Marquesas Islands, Tahiti, Tonga, the New Hebrides, New Caledonia and Norfolk Island back to New Zealand (Fig. 5.21). It arrived at Queen Charlotte Sound on 19 October 1774, just three months after the *Adventure* had safely reached England.

During this third and final Second Voyage stop-over in Queen Charlotte Sound Cook had the crew working in all haste to prepare the ship for the long voyage home. While some were repairing masts and sails and scrubbing and caulking the hull, others were gathering stocks of wood and fresh water ashore. Meanwhile, the Europeans soon learnt of the Rowe Massacre, but once the Maoris established that Cook would not seek revenge for those slain they flocked to Ship Cove and amiable trading relations prevailed.

The day after their arrival in the Sound Wales reports:

> … got on Shore my Observatory & Instruments, put them up in ye old Place & was employed making Observations untill November ye 8th. when I took every thing down & Carried them on board the Ship. (Wales 1772–1774).

He then completed yet another series of solar and stellar observations for latitude (see Table 5.5) and lunar measures on six different nights for longitude (Wales and Bayly 1777: 109).

On the basis of these observations we derive the following mean values for the Ship Cove observatory site:

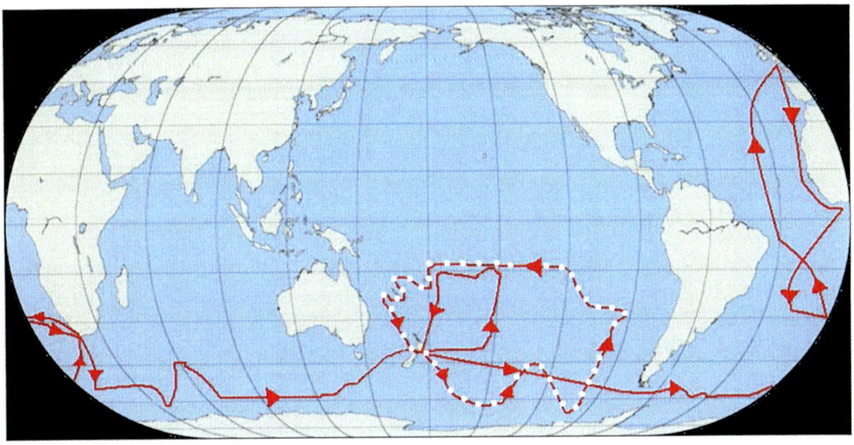

Fig. 5.21 The red and white line shows the exploration route of the *Resolution* from November 1773 until it returned to Queen Charlotte Sound for the third and final time in October 1774 (adapted from the base map in https://en.wikipedia.org)

Table 5.5 Mean values of latitude obtained by Wales at Queen Charlotte Sound, October–November 1774 (after Wales and Bayly 1777: 109)

Object observed	Number of days/nights	Latitude (South) From interior arc	Latitude (South) From exterior arc
Sun	6	41° 05′ 50.33″	41° 05′ 54.5″
Achernar	3	41° 06′ 08.5″	41° 06′ 04″
Alpha Andromedae	3	41° 05′ 57″	41° 06′ 06″
Alpha Anseris Americani	1	41° 06′ 12″	41° 05′ 55″
Alpha Gruis	2	41° 05′ 32″	41° 05′ 15.5″
Alpha Pegasi	4	41° 05′ 44.5″	41° 05′ 57.25″
Beta Gruis	3	41° 06′ 11.2″	41° 05′ 46.67″
Fomalhaut	4	41° 05′ 57.67″	41° 05′ 59″

Latitude 41° 05′ 54.4″S
Longitude 174° 30′ 11″E

When he allowed for non-lunar observations, Wales came up with a slightly different longitude:

> The mean of all the Distances taken this time & one Immer. of ♃'s 1st. Satellite gave its Longit. 11h 37′ 50″½. The mean of ye Distances taken last Year & 2 Immersions of ♃'s Sat. gave 11h 37′ 30″½. The Mean of the two is 11h 37m 40′½ = 174° 25′⅛ E. (Wales 1772–1774).

Cook also gives values that differ from the means listed above:

> M[r] Wales having from time to time communicated to me the observations he had made in this Sound, for determining the Longitude, the mean results of which gives 174° 25′ 07½″ East from the bottom of Ship Cove where the observations were made the Latitude of which is 41° 5′ 56½″ South. (Beaglehole 1969: 579).

As he prepared to leave Queen Charlotte Sound, Cook carried out a final evaluation of the available latitude and longitude values:

> In my Chart constructed in my former Voyage this place [Queen Charlotte Sound] is laid down in … 175° 5′ 30″ East, the error of this Chart is therefore 0° 40′ and nearly equal to what was found at Duskey Bay; by which it appears that the whole of *Tavai-poenammoo* [the South Island], is laid down 40′ too far East in the said Chart, as well as in the Journal of the Voyage; but the error in *Eahei-no-mauwe* [the North Island] is not more than half a degree or 30′ … (Beaglehole 1969: 579).

He also made the following important observation: "… from the multitude of observations which Mr Wales took the situation of few parts of the world are better ascertained than that of Queen Charlottes Sound." (Beaglehole 1969: 580).

On 9 November 1774, as the *Resolution* was readied for departure and just one day after Wales had dismantled his observatory, J.R. Forster went ashore in Ship Cove and made some astronomically-oriented observations of his own. He reports:

> I went after breakfast ashore, & saw … the place where Mr Bailey's Astronomical tent had been fixed. I observed the hole where the Clock had been standing, & the Pickets, where the ropes of the Tent had been fastened to. (Hoare 1982(IV): 683).

Although this observatory site was also situated in Ship Cove, Cook noted that it was "… in a place different to that where M[r] Wales had his …" (Beaglehole 1969: 570n).

Two days later the *Resolution* sailed from Queen Charlotte Sound, reaching England on the last day of July 1775.

After the voyage, Wales and Bayly began working up their astronomical observations for publication, but with Bayly's departure on Cook's Third Voyage, completion of this task fell to Wales. *The Original Astronomical Observations, Made in the Course of a Voyage Towards the South Pole, and Round the World …* (Wales and Bayly 1777) eventually made its appearance two years after the end of the Second Voyage.

5.5 Concluding Remarks

Cook's Second Voyage to the South Seas focussed on a search for the Great Southern Continent, and as such Queen Charlotte Sound played a critical role as the regular Pacific revictuallng centre. Consequently, the *Resolution* and *Adventure* visited the Sound, either individually or together, in April-June and November-December 1773 and October-November 1774. On each occasion the astronomers, William Wales on the *Resolution* and William Bayly on the *Adventure*, set up observatories and took many astronomical readings as they aimed to refine the latitude and longitude of Ship Cove. By the time the *Resolution* quit the Sound in November 1774 and headed back to England an avalanche of astronomical readings had been taken and the position of Ship Cove was better established than almost any other place on the surface of the Earth.

However, not all of Wales' astronomical observations were restricted to Queen Charlotte Sound. In April and May 1773 the *Resolution* was anchored in Pickersgill Harbour, Dusky Sound, and with considerable enterprise he created a forest clearing on a headland near the anchorage, set up his observatory, and despite persistent cloudy skies and rain still managed to pinpoint the latitude and longitude of this remote south-western corner of the South Island.

Acknowledgements I am grateful to the following for their assistance: Dr Jonathan Betts and Dr Gloria Clifton (ex-Royal Museums, Greenwich), the late Commander Derek Howse, (Sevenoaks, England) and the late Sir Patrick Moore (Selsey, England). I also wish to thank Gloria Clifton from reading and commenting on the first draft of this chapter.

References

Bayley, Mr. (1769). Astronomical observations made at the North Cape, for the Royal Society. *Philosophical Transactions of the Royal Society, 59*, 262–272.

Bayly, W. (1772–1774). (Observations Made on Board the Adventure). Records of the Board of Longitude (Vol. 45).

Bayly, W. (1773). Extract from the journal kept by William Bayly, Astronomer, on Adventure … In McNab, 1914, 201–218.

Beaglehole, J. C. (Ed.), (1967). *The Journals of Captain James Cook on his Voyages of Discovery. III. The Voyage of the* Resolution *and* Discovery *1776–1780* (Two Vols). Cambridge: Cambridge University Press.

Beaglehole, J. C. (Ed.), (1968). *The Journals of Captain James Cook on his Voyages of Discovery. I. The Voyage of* Endeavour *1768–1771*. Cambridge: Cambridge University Press (Preprint; first published 1955).

Beaglehole, J. C. (Ed.), (1969). *The Journals of Captain James Cook on his Voyages of Discovery. II. The Voyage of the* Resolution *and* Adventure *1772–1775*. Cambridge: Cambridge University Press (Reprint; first published 1961).

Beaglehole, J. C. (1974). *The Life of Captain James Cook*. London: Hakluyt Society.

Begg, A. C., & Begg, N. C. (1968). *Dusky Sound*. Christchurch: Whitcombe & Tombs.

Bennett, J. A. (1987). *The Divided Circle. A History of Instruments for Astronomy, Navigation and Surveying*. Oxford: Phaidon-Christies.

Betts, J. (1993). The eighteenth century transits of Venus, the voyages of Captain James Cook and the early development of the marine chronometer. *Antiquarian Horology, 21*, 660–669.

Chisholm, H. (1911). Furneaux, Tobias. In *Encyclopaedia Britannica* (11th ed.). Cambridge: Cambridge University Press.

Clerke, C. (1771–1775). A Log of the Proceedings of, and Occurrences onboard His Majesty's Ship Resolution … original in the Public Record Office, London (Adm 55/103).

Clifton, G. (1995). *Directory of British Scientific Instrument Makers 1550–1851*. London: Zwemmer and the National Maritime Museum.

Cooke, J. [*sic*], King, J., & Bayly, W. (1782). *The Original Astronomical Observations Made in the Course of a Voyage to the Northern Pacific Ocean, For the Discovery of a North East or North West Passage …* London: Elmsly.

Coutts, P. J. F. (1969). The Maori of Dusky Sound: a review of the historical sources. *Journal of the Polynesian Society, 78*, 178–211.

Dollond, P. (1765). An account of an improvement made by Mr. Peter Dollond in his new telescopes. *Philosophical Transactions of the Royal Society, 55*, 54–56.

EIC. (1909). Wales, William (1734–1798). In Lee, S. (Ed.), *Dictionary of National Biography* (Vol. XX, pp. 490–491). London: Smith, Elder & Co.

Frost, A. (1972). Captain James Cook and the early romantic imagination. In W. Veit (Ed.), *Captain James Cook: Image and Impact. South Seas Discoveries and the World of Letters* (pp. 90–106). Melbourne: Hawthorn Press.

Furneaux, T. (1772–74). Furneaux's narrative. In Beaglehole, 1969, 729–745.

Gee, B. (2014). *Francis Watkins and the Dollond Telescope Patent Controversy*. Farnham: Ashgate.

Gould, T. R. (1987). *John Harrison and His Timekeepers*. London: National Maritime Museum.

Hoare, M. E. (1976). *The Tactless Philosopher. Johann Reinhold Forster (1729–98)*. Melbourne: Hawthorn Press.

Hoare, M. E. (1982). *The Resolution Journal of Johann Reinhold Forster 1772–75*. Four Volumes. London: Hakluyt Society.

Hodges, W. (1908). Bayly, William (1737–1810). In L. Stephen & S. Lee (Eds.), *Dictionary of National Biography* (Vol. I, p. 1372). London: Smith, Elder & Co.

Holmes, C. (Ed.). (1984). *Captain Cook's Second Voyage: The Journals of Lieutenants Elliott and Pickersgill*. London: Caliban Books.

Hooper, B. (Ed.). (1975). *With Captain James Cook in the Antarctic and Pacific. The Private Journal of James Burney Second Lieutenant of the Adventure on Cook's Second Voyage 1772–1773*. Canberra: National Library of Australia.

Howse, D. (1969a). Captain Cook's marine timekeepers. Part I—The Kendall watches. In Howse and Hutchinson, 1969a, 190–205.

Howse, D. (1969b). Captain Cook's marine timekeepers. Part II—The Arnold chronometers. In Howse and Hutchinson, 1969a, 276–280.

Howse, D. (1969c). Captain Cook's minor clocks and watches. In Howse and Hutchinson, 1969a, 138–145.

Howse, D. (1969d). Captain Cook's pendulum clocks. In Howse and Hutchinson, 1969a, 62–76.

Howse, D. (1979). The principal scientific instruments taken on Captain Cook's voyages of exploration, 1768–80. *Mariner's Mirror, 65*, 119–135.

Howse, D. (1986). The Royal Astronomical Society instrument collection: 1827–1985. *Quarterly Journal of the Royal Astronomical Society, 27*, 212–236.

Howse, D. (1989). *Nevil Maskelyne. The Seamen's Astronomer*. Cambridge: Cambridge University Press.

Howse, D., & Hutchinson, B. (1969a). *The Clocks and Watches of Captain James Cook 1769–1969*. London: The Antiquarian Horological Society (reprinted from four issues of Antiquarian Horology).

Howse, D., & Hutchinson, B. (1969b). The saga of the Shelton clocks. In Howse and Hutchinson, 1969a, 281–298.

Johnston, S. (2005). Blast from the past. Measurement and morals in the early transit of Venus. Museum of the History of Science, Oxford University (www.physics.ox.uk/phystat05/Talks/johnston.ppt).

Joppien, R., & Smith, B. (1985–1997). *The Art of Captain Cook's Voyages*. Four Volumes. Melbourne: Oxford University Press.

Kaeppler, A. (1972). The use of documents in identifying ethnographic specimens from the voyages of Captain Cook. *Journal of Pacific History, 7*, 195–200.

King, H. C. (1979). *The History of the Telescope*. New York: Dover.

Knight, C. (1856). *Biography (The English Cyclopaedia)*. London.

McNab, R. (1914). *Historical Records of New Zealand* (Vol. II). Wellington: Government Printer.

Mead, S. M. (1969). The costume styles of the Classical Maori in New Zealand, 1642–1800 A.D. *Costume*, *3*, 35–43.

Mercer, V. (1972). *John Arnold & Son, Chronometer Makers, 1762–1843*. London: Antiquarian Horological Society.

Moore, P. (1981). William Wales. In Cook, J. *The Journal of H.M.S. Resolution 1772–1775* (pp. 86–93). Guildford, Genesis Publications.

Orchiston, W. (1971). Astronomy in early New Zealand. *British Astronomical Association New South Wales Branch Bulletin*, *509*, 1–2.

Orchiston, W. (1972). Maori neck and ear ornaments of the 1770s: a study in protohistoric ethno-archaeology. *Journal of the Royal Society of New Zealand*, *2*, 91–107.

Orchiston, W. (1974a). Cook voyage 'trading stations' in Early Protohistoric New Zealand. *Dominion Museum Records in Ethnology*, *2*, 133–156.

Orchiston, W. (1974b). *Studies in South Island New Zealand Prehistory and Protohistory*. Unpublished Ph.D. Thesis, Department of Anthropology, University of Sydney, Australia.

Orchiston, W. (1975a). Early protohistoric Maori food preparation, and application of the direct historical approach to New Zealand prehistory. *New Zealand Archaeological Association Newsletter*, *18*, 18–31.

Orchiston, W. (1975b). Maori material culture change in Early Protohistoric New Zealand: the greenstone trade at Queen Charlotte Sound. *The Artefact*, *39*, 40–77.

Orchiston, W. (1997). *Early Astronomy in New Zealand: The South Sea Voyages of James Cook*. Wellington: Carter Observatory (Information Sheet No. 10).

Orchiston, W. (1998). *Nautical Astronomy in New Zealand: The Voyages of James Cook*. Wellington: Carter Observatory.

Orchiston, W. and Horrocks, L. C. (1975). Contact and conflict: the Rowe Massacre in Early Protohistoric New Zealand. *Historical Studies*, *16*, 518–538.

Plumley, N. (1986). *Christ's Hospital. A Short History*. Horsham: Christ's Hospital.

Quill, H. (1966). *John Harrison. The Man who Found Longitude*. London: Baker.

Rose, H. J. (1853). *A New General Biographical Dictionary*. London.

Salmond, A. (1991). *Two Worlds. First Meeting Between Maori and Europeans 1642–1772*. Auckland: Viking.

Sparrman, A. (1953). *A Voyage Round the World with Captain James Cook in H.M.S. Resolution*. London: Robert Hale.

Tooth, T. (1993). *It Was All a Matter of Time*. Katoomba, printed for the author.

Wales, W. (1772–1774). Journal of William Wales. In Beaglehole, 1961, 776–869.

Wales, W. (1772–1775). Journal of a Voyage made by Order of the Commissioners of Longitude, On Board of His Majesty's Sloop Resolution … Records of the Board of Longitude (Vol. 47).

Wales, W. (1788). *Astronomical Observations, Made in the Voyages Which were Undertaken by Order of His Present Majesty, for Making Discoveries in the Southern Hemisphere* … London: Elmsly.

Wales, W., & Bayly, W. (1777). *The Original Astronomical Observations, Made in the Course of a Voyage Towards the South Pole, and Round the World* … London: Strahan.

Wales, W., & Dymond, J. (1769). Astronomical observations made by order of the Royal Society, at Prince of Wales's Fort, on the north west coast of Hudson's Bay. *Philosophical Transactions of the Royal Society*, *59*, 467–488.

Watt, Sir James. (1979). Medical aspects and consequences of Cook's voyages. In R. Fisher & H. Johnston (Eds.), *Captain James Cook and his Times* (pp. 129–157). Canberra: Australian National University Press.

Chapter 6
Astronomy on Cook's Third Voyage: Queen Charlotte Sound, 1777

Abstract Cook's Third Voyage to the South Seas was undertaken in search of an assumed 'northwest passage' linking the Pacific and Atlantic Oceans. Once again two ex-Whitby colliers were involved, the *Resolution* under the command of Captain James Cook, and the *Discovery*, skippered by Lieutenant Charles Clerke. Following up from his Second Voyage experience, Royal Observatory Greenwich astronomer William Bayly was again assigned to this voyage, and served on board the *Discovery*, while Cook and Second Lieutenant James King shared the astronomical responsibilities on the *Resolution*. This voyage only involved a single short stop-over in Queen Charlotte Sound, during which further astronomical observations were made for the calculation of latitude and longitude. These built on the very extensive series taken during the various Second Voyage stop-overs in the Sound. The final result, reported in this chapter, differs little from the currently-accepted values, and shows the level of precision these early astronomers were able to achieve with what today would be regarded as inferior instruments.

6.1 Introduction

Since the Great Southern Continent was now know to be a myth, the Admiralty turned its attention to the northwest Pacific and a possible passage leading from the Pacific Ocean to the Atlantic. Thus Cook's Third Voyage to the South Seas came into being.

Once again two vessels were involved, the *Resolution* under the commander of Captain James Cook and the *Discovery*, skippered by Captain Charles Clerke (see Fig. 5.15). This was to be the final voyage for both Captains. Cook is already well known to us, but Clerke only achieved any level of 'visibility' while a lieutenant on Cook's Second Voyage, when he became

> … three-dimensional, a positive personality of the most lively description, through both his journal and his letters to Banks … He is a man capable of systematic observation and recording; able to generalize, also, and to think … (Beaglehole 1969: xxxv).

W. Orchiston, *Exploring the History of New Zealand Astronomy*,
Astrophysics and Space Science Library 422,
DOI 10.1007/978-3-319-22566-1_6

Promoted since the Second Voyage, and with added responsibilities, Clerke was

> … always cheerful, talkative, amusing, with some of the rollicking vices as well as the rollicking virtues; a generous spirit who made friends easily; tall, long-nosed, with an eye both roving and sparkling. (Beaglehole 1974: 139).

He had "… enough mathematical ability to become a good navigator; with some interest in the scientific side of his profession …" (Beaglehole 1968: cxxxi). Following Cook's death in Hawaii Clerke took command of the expedition, and he would be called upon to exploit that navigational skill that Beaglehole writes about right up to the time of his own premature death, at sea, in 1779. He had so much to offer—he was still in his mid-30s.

Because of the northern Pacific focus of Cook's Third Voyage, New Zealand would not play a key role in this Voyage (see Fig. 6.1), although Queen Charlotte Sound would once more serve as a revictualling centre, but in this instance on a single occasion only, and for just two weeks in February 1777.

As on the previous voyage, astronomers had a crucial role to play on the Third Voyage, not just during the mapping of the northwest American coast, but also in the continued evaluation of the efficiency of the chronometers. And yet again, the latitude and longitude of Queen Charlotte Sound would receive further scrutiny.

The aim of this chapter is to describe the various astronomers involved in Cook's Third Voyage, their scientific instruments, and their astronomical observations at Queen Charlotte Sound. Section 6.2, following, introduces us to Bayly, Cook and King. In Sect. 6.3 we discuss their telescopes, quadrants and sextants; their various astronomical clocks and chronometers; and their tent observatories. This leads us into Sect. 6.4 where we examine the ways in which the three astronomers put their instruments to use on New Zealand soil, and their preoccupation with refining the longitude of Queen Charlotte Sound.

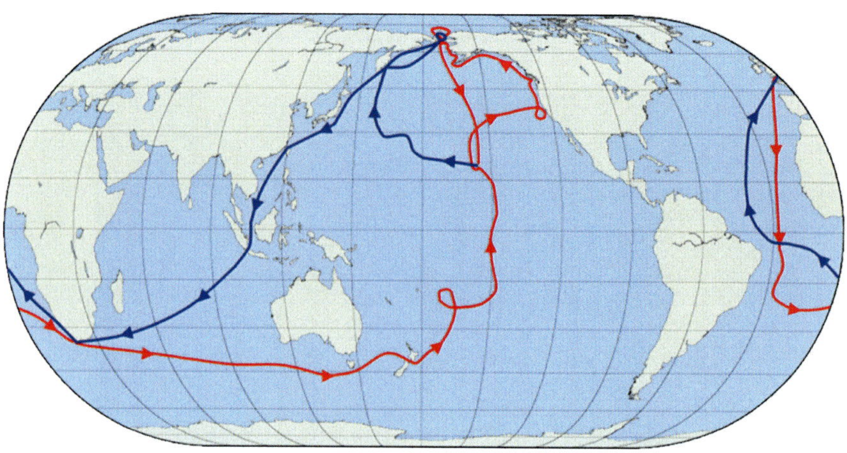

Fig. 6.1 A map showing the route of the *Resolution* and *Discovery* during the Third Voyage, prior to Cook's death, in *red*, and subsequently, in *blue* (https://en.wikipedia.org)

Although much has been written about New Zealand in 1777 based on the reports of the European explorers and artifacts collected at the time (e.g. Mead 1969; Orchiston 1972, 1974a, b, 1975a, b; Salmond 1991), until the Carter Observatory published *Nautical Astronomy in New Zealand: The Voyages of James Cook* (Orchiston 1998b) astronomy had been all but ignored, except for superficial references by McIntosh (1957), Mackrell (1985) and Orchiston (1971, 1994, 1997). This chapter, is based largely on Orchiston (1998b).

6.2 The Astronomers

6.2.1 Introduction

The two ex-Whitby colliers which took part in Cook's Third Voyage to the South Seas were naval vessels, and were manned by commissioned officers, warrant and petty officers, able seamen, servants, marines and supernumeraries. Only these last-mentioned were non-naval personnel, and they included a solitary astronomer.

However, three different astronomers were involved in Cook's Third Voyage. Once again, one of these was Captain James Cook who would serve as an astronomer on the *Resolution*, along with Lieutenant James King, who agreed to a salary of 200 guineas for the voyage itself, plus 100 guineas per year if the voyage extended beyond two years (Beaglehole 1974: 131). The astronomer on the *Discovery* was William Bayly, following on from his Second Voyage adventure on the *Adventure*.

The logic for including astronomers on the Third Voyage mirrored that of the Second Voyage: "… to make nautical and Astronomical Observations and to perform other services tending to the improvement [of] Geography and Navigation …" (Beaglehole 1967(II): 1498). But on this occasion the Board of Longitude recommended (successfully, as it turned out) that Bayly be assigned to the *Discovery*, and that

> … Capt. Cook assisted by Mr. King his lieutenant who as they are informed is well qualified for the purpose, be desired to make similar observations [to those carried out by Bayly] whenever other more necessary service will admit on board the Resolution. (Beaglehole 1967(II): 1490).

Bayly was offered an annual salary of £400, while the Board decided to finalise the salary supplement payable to Cook and King upon their return (ibid.). Since Cook was not destined to return, this proved to be a rather judicious ploy—though of course they were not to know it at the time! In the end, it was determined that Cook's widow and King should each be paid £500 (Beaglehole 1967(II): 1563).

Let us now look in turn at each of the astronomers.

6.2.2 Captain James Cook (1728–1779)

As we have already seen in Chap. 4, James Cook (Fig. 6.2) was born in Marton and went to school in nearby Great Ayton before starting an apprenticeship in Staithes (for English localities mentioned in this chapter see Fig. 6.3). In 1746 he went to sea, and he joined the Royal Navy in 1755 where his skills and dedication ensured rapid promotion. By the time he embarked on the *Endeavour* on his First Voyage to the South Seas in 1768 he was a Lieutenant and an experienced naval officer with an intimate knowledge of navigation (which included astronomy), hydrographic surveying and cartography (map-making).

After the *Endeavour* returned to London in July 1771 Cook was promoted to Commander, and at the end of August he was assigned the command of the *Scorpion* in order to re-chart the English coastline (Beaglehole 1974: 276). However, this never happened, for the following month the Admiralty decided on a second South Seas voyage, with Cook in command, and he was busy with the necessary preparations. The *Resolution* and *Adventure* sailed from England in July 1772.

Fig. 6.2 A portrait of Captain James Cook painted by John Webber in 1776, and now in the National Museum of New Zealand Te Papa Tongarewa in Wellington, New Zealand (https://en.wikipedia.org)

Fig. 6.3 English localities mentioned in this chapter are shown in *red*

Soon after the return of the *Resolution* at the end of July 1775 Cook was promoted to Captain and assigned to the *Kent*, but the very next day the ship was laid up and he was transferred to the Royal Naval Hospital at Greenwich. Cook did not take lightly to what was effectively an enforced 'retirement' and he politely protested. Soon he was to hear that a Third Voyage to the South Seas was planned, ostensibly to take Omai back to the Society Islands (where he had joined Furneaux's crew on the *Adventure* in 1773), and that he would again command the *Resolution*. Its consort this time would be the *Discovery*. As we have seen, on this occasion Cook was to share the astronomical duties with his Second Lieutenant, James King (Beaglehole 1967(II): 1490).

In March 1776, while planning for the new voyage was in train, Cook was elected a Fellow of the Royal Society, and later in the year was awarded their Copley Medal for a paper he published in their *Philosophical Transactions* on "… the methods used by him to preserve the health of his men during his second voyage …" (Beaglehole 1974: 507).

The vessels sailed for New Zealand in July 1776, and it was on this voyage —on 14 February 1779, in Hawaii, to be precise—that James Cook was slain (Beaglehole 1974). Thus ended the life of a colossus of British discovery, exploration, hydrographic surveying and nautical astronomy. Just 50 years of age, Cook was surely the "… classic example of how a man could rise [in rank] if he had exceptional talent, opportunity and patronage." (Lloyd 1968: 230). We can do no better in ending this all-too-brief sketch than to quote that doyen of Cook scholarship, the late John Cawte Beaglehole (1901–1971):

> The genius of Cook was not, in the ordinary sense, creative. Nor was it precocious. There was nothing aesthetic, nothing of early brilliance, about him. He would have been startled at the idea that anything he did might touch the emotions. His energy of mind was of that mature kind that is, on the intellectual side, critical, a sort of analytical and detective energy; on the practical, the constructive, side, it was an energy of planning, of administration and foresight. (Beaglehole 1968: xxii).

6.2.3 William Bayly (1737–1810)

As we have already seen in Chap. 5, William Bayly was born at Bishop's Cannings in 1737. He showed an early talent for mathematics, and eventually obtained a position as an usher at a school near Bristol (W.H. 1908), before joining the Royal Observatory, Greenwich. After observing the 1769 transit of Venus from Canada, he served as the astronomer on the *Adventure* during Cook's Second Voyage.

When he returned to England Bayly went back to the Royal Observatory, but soon he was off to the Pacific again on Cook's Third Voyage, this time as the astronomer on the *Discovery*. After this voyage he again returned to the Royal Observatory, finally leaving in 1785 to serve as Headmaster of the Royal Naval Academy in Portsmouth (Beaglehole 1969). He died in Portsmouth in 1810 at the age of 73. There are no known portraits of Bayly, and no descriptions of him exist that we may use to gauge what sort of man he was and what he looked like.

6.2.4 James King (1750–1784)

James King (Fig. 6.4) was born at Clitheroe, Lancashire, in 1750. He was the second son of James King, curate of Clitheroe and later dean of Raphoe. King entered the Royal Navy when just twelve years of age and subsequently served

Fig. 6.4 An oil painting of Captain James King by John Webber in 1782, now in the National Library of Australia (https://en.wikipedia.org)

under Captain Palliser in Newfoundland (at the time that Cook was surveying the coast) and later under Captain Jervis in the Mediterranean. By the age of twenty-one he was a Lieutenant (J.K.L. 1909).

Three years later King did an odd thing for a naval lieutenant: he went to Paris to study science, and then settled in Oxford. There he met Dr Thomas Hornsby, Professor of Astronomy, who was to have a lasting effect on his future. It was Hornsby who recommended King as an astronomer for Cook's Third Voyage (ibid.), and because of his naval background he was able to join the *Resolution* as a Second Lieutenant. As Beaglehole (1967(I): lxxvii) has noted:

> With enough sea service and his special training he was for Cook's purpose exceptionally well-equipped: he shared in the responsibility for the chronometer, that veteran of the second voyage – his signature shares innumerable lines of recorded observation with Cook's own – and his presence ruled out the need in the ship of a professional astronomer.

Thus, King was able to combine two different roles.

From all accounts, James King was one of the intellectuals on the voyage. He was widely read, observant, could think for himself, and was well-liked by the crew (with the notable exception of the unpopular William Bligh). For example, midshipman Trelevan [Beaglehole calls him Trevenen] described him as "… one of the best, he is one of the politest, genteelest, & best-bred men in the world." while the surgeon, Samwell, wrote: "I never in my Life knew a warmer Friend or a worthier Man than he is." (cited in Beaglehole, ibid.). Although elements of hero worship intrude here the basic sentiments are obvious, and Beaglehole (ibid.) is equally generous in his praise:

> There must have been an almost youthful charm about King, a certain refinement of mind and of body, a humanity, a kindness, a generosity and sensitivity of spirit without touch of the effeminate unusual among seamen …

During the voyage, King also was attacked by the Hawaiians, soon after Cook was killed, but he and his party succeeded in protecting themselves. When Captain Clerke died on 22 August 1779 it was King who took command of the *Discovery* and guided the expedition back to England.

After being promoted, he then took charge of the *Crocodile* attached to the Channel Fleet, and then assumed command of the *Resistence* which was in charge of a convey of 500 merchant ships *en route* to the West Indies. It is said that "… the intense anxiety of the duty … turned his hair grey …" (J.K.L. 1909). He returned from this voyage in poor health, but then had to face up to the onerous tasks of preparing a monograph on the third voyage astronomical observations and working up Cook's journal for publication. In 1783 the precarious state of his health induced him to move to Nice, and he died there in October 1784, the year that the narrative of the voyage was published. Meanwhile, the monograph on the astronomical observations (Cooke et al. 1782), which Bayly ended up editing, had appeared two years earlier. James King was buried in Nice, but a tablet in his memory was installed in the church in Clitheroe (J.K.L. 1909).

6.3 The Instruments

6.3.1 *Introduction*

The astronomical instruments assigned to the Third Voyage were almost identical to those sent on the preceding voyage: the mandatory quadrants, sextants, reflecting and refracting telescopes, chronometers, astronomical and alarum clocks, and the single transit telescope (Beaglehole 1967(II): 1499–1500), but conspicuous by their absence were journeyman clocks. Once again there also were instruments for meteorological and geomagnetic measurements, and various books, maps and charts (see Beaglehole 1967(II): 1499–1500, 1969: 724).

Most of the astronomical instruments taken on Cooks' Third Voyage were manufactured by the leading practitioners of the day, Bird, Dollond and Ramsden.

John Bird was famous for his astronomical quadrants (Chapman 1995), but he also made reflecting telescopes and transit telescopes (King 1979: 117) and these were acquired for Cook's Third Voyage. Unfortunately, Bird died in 1776 before the *Resolution* and *Discovery* returned to England, so he never got to hear how his instruments performed on the voyage.

As we saw in Chap. 5, it was Peter Dollond who developed the refracting telescope after his father's death, and these also were assigned to the astronomers on Cook's Third Voyage. Peter Dollond only retired in 1818, and died two years later (Andrews 1992: 163).

Jesse Ramsden was another leading London-based manufacturer of scientific instrument (King 1979). A Fellow of the Royal Society and recipient of their Copley Medal in 1795, he was

> … above the middle size, slender, extremely well made, and to a later period of life, possessed great activity. His countenance was a faithful index of his mind, full of intelligence and sweetness. His forehead was open and high, with a very projecting and expansive brow. His eyes were dark hazel, sparkling with animation. (cited in King 1979: 170).

6.3.2 *Securing Celestial Positions: The Telescopes, Quadrants and Sextants*

There is direct documentation for three different reflecting telescopes on the Third Voyage. The Board of Longitude once again supplied the Bird reflectors for each vessel (Cooke et al. 1782: v), and Cook brought along his Watkins reflector. Two of these reflectors are mentioned by Cook in discussing a lunar occultation of S Capricorni:

> M[r] Bailey [*sic*] observed with the Achromatic Telescope belonging to the Board of Longitude, M[r] King with the Reflector belonging to the same board and [I] Observed with the eightteen inch Reflector of my own. (Beaglehole 1967(I): 243).

The "Achromatic Telescope" referred to was one of the two Dollond refractors which, following their successful introduction on the Second Voyage, were also assigned to the Third Voyage.

The Royal Society's Bird transit telescope was also taken on the Third Voyage (Cooke et al. 1782: v), and is described by Maskelyne as "A transit instrument of 4 feet with its level and stand to Mr Bayly only." (Beaglehole 1967(II): 1500).

Unfortunately the current whereabouts is unknown of the various telescopes that went on the Third Voyage.

Documentation on the astronomical quadrants taken on the Third Voyage is sparse. Two Bird quadrants of 30.5-cm radius, supplied by the Board of Longitude for Cook and King, are mentioned in the literature (Beaglehole 1967(II): 1499; Cooke et al. 1782: v), but at least one other quadrant must have been taken on the Voyage—for Bayly's use.

On the Third Voyage the Board of Longitude assigned the *Resolution* and *Discovery* the same two Hadley's sextants—a Ramsden and a Dollond (see Wales and Bayly 1777: vii)—that went on the Second Voyage, together with two further 12.7-cm brass sextants by Ramsden. Howse (1979) records that in addition there were two further Ramsden sextants, a Dollond sextant, and Cook's Bird sextant on the *Resolution*, while one further Ramsden sextant was on the *Discovery*. He states that the additional Ramsden sextants were owned by different officers.

As we saw in Chap. 4, in 1979 Derek Howse was able to identify four surviving sextants of reputed Cook voyage origin. Two of these, a 36.8-cm Ramsden engraved "D.32" and an identical Ramsden engraved "D.33", are said to have been on the *Discovery*, and therefore on Cook's Third Voyage. The former extant is now owned privately in the United Kingdom and the latter is in the National Maritime Museum.

6.3.3 Keeping Track of Time: The Clocks, Watches and Chronometers

As we saw in the previous two chapters, during the eighteenth century, keeping track of time was a major challenge on the Cook Voyages, even following the introduction of chronometers on the Second Voyage.

Apart from chronometers and a horizontal stop watch, two different types of clocks were taken on Cook's Third Voyage: astronomical clocks and alarum clocks. Two different astronomical clocks by John Shelton, one owned by the Royal Society and the other by the Board of Longitude, were assigned to the *Resolution* and *Discovery*, but it is not clear which clock went to which vessel. It is highly likely, though, that these two clocks were the same ones that had been taken on Cook's Second Voyage (Howse 1969d). Obviously the iron supporting stands that accompanied the clocks on that voyage had not proved entirely satisfactory, for they were replaced on the Third Voyage by solid wooden tripod stands designed by Bayly, one of which is shown in Fig. 6.5.

Fig. 6.5 The RS35 Shelton astronomical clock from Cook's voyages, with a reconstructed wooden framework like those used on the Third Voyage (*Courtesy* Royal Society, London)

In *The Clocks and Watches of Captain James Cook 1769–1969* Howse and Hutchinson (1969b: 289–291) supply the following description of a typical Shelton astronomical clock, based on one now in St John's College, Cambridge, which was manufactured for one of the 1761 British transit of Venus expeditions (and therefore would have been very similar to the clock that accompanied Cook and Green to Tahiti):

> *The basic movement* consists of a pair of shaped brass plates … The average dimensions in the series are approximately 10.25 × 6.25 in. The six pillars are riveted to the back plate; the front plate is fastened by latches.
>
> *The train* has five wheels and is constructed for a month's duration. Where feasible, the pivots bear on end-plates which minimise friction and preserve the oil. One large end-plate covers all pivot holes on the back plate …
>
> *The escapement* is dead-beat, and has a 30 tooth brass wheel with relieved teeth; the anchor is steel and has a long shank and curved arms which terminate in the pallets. The collets for the escape wheel and anchor are characteristically long …
>
> *Bolt-and-shutter maintaining power* is fitted, with an additional device to prevent the clock stopping while the mechanism is being engaged. Although the movement is weight driven, stopwork (acting on a principle similar to that used in fuzee clocks) is fitted to prevent overwinding.
>
> *The motion work* is conventional; the hour-wheel pipe carries a friction-mounted hour circle instead of an hour hand. The cannon pinion and minute wheel are only partially

crossed out, and thus counterpoise the minute hand. The minute wheel and pinion are pivoted on a cock screwed to the front plate, and the arbor is extended to pivot on the back plate.

The dial is a square plate of brass - engraved, waxed, and silvered ... The dial is mounted on four pillars and can be removed by the screws ... The pillars are turned with wide feet, and are located with steady pins and screwed to brass plates which in turn are fastened to the front plate ...

The pendulum is of gridiron construction and is suspended from the back cock by a steel strip. The crutch is steel and has a brass pin which engages in a slot in the central rod ...

Shelton's punch-mark on the Cambridge movement is on the front plate ...

The movement is rigidly mounted, and is secured to the seat-board with four brass holdfasts and screws. Two more holdfasts are screwed to the back plate, and line up with brackets screwed to the back-board.

It is noteworthy that the Board of Longitude decided not to supply journeyman clocks for the Third Voyage, but each astronomer did have an alarum clock. That assigned to King on the *Resolution* was owned by Board of Longitude and was possibly the same clock that had been purchased from Jonathan Monk for the Second Voyage. Meanwhile, a brand new alarum clock made by John Arnold and costing £4 was obtained for the *Discovery* (Howse 1969c).

The same Elliott stop watch taken on the Second Voyage, was assigned to the *Resolution* for the Third Voyage, but Bayly apparently did not take his own watch on the *Discovery* (Howse 1969d: 64).

Since the accuracy of the chronometer as a portable time-piece had been adequately demonstrated on the Second Voyage only two chronometers were assigned by the Board of Longitude to the Third Voyage. Cook and King had the ever-reliable K1 while Bayly was custodian of a new Kendall chronometer, the K3 (see Fig. 6.6). In August 1779, following Clerke's death, Lieutenant Gore took over the *Resolution* and King took command of the *Discovery*, taking the K1 chronometer with him. Soon after this, Bayly and the K3 transferred to the *Resolution*. Both chronometers performed with great reliability during the Voyage (see Howse 1969a: Fig. 6), although Howse (1969a: 203) would rate the K1 as "outstanding" and the K3 as merely "excellent"!

Howse (1969c) has demonstrated that none of the alarum clocks that was taken on the Third Voyage has survived, and despite the complicated history surrounding the five surviving Shelton astronomical clocks of supposed Cook vintage he and Hutchinson (1969b) tentatively identify the astronomical clock now known as RS35 (Fig. 6.5) with Cook's Third Voyage. This attribution is partly based on the discovery of filled-in holes in the clock case which matched those required for the attachment of the style of wooden tripod used on the Third Voyage (see Howse 1969d). Indeed, in 1968 staff at the National Maritime Museum constructed and attached a replica of this tripod to RS35, and this is shown in Fig. 6.5.

In contrast to the astronomical clocks, the current whereabouts of the Kendall chronometers which went on the Third Voyages is well known: both are in the National Maritime Museum, on loan from the Navy (Howse 1969a, b).

Fig. 6.6 The K3 Kendall chronometer which was taken on the third voyage (© National Maritime Museum, Greenwich, London)

6.3.4 Protecting the Equipment: The Tent Observatories

Following the success of Bayly's tent observatories during Cook's Second Voyage, two further tent observatories were taken on the Third Voyage. One was assigned to each vessel (Beaglehole 1974: 288), and they were used at Queen Charlotte Sound in February 1777.

6.4 Refining the Location of Queen Charlotte Sound

Cook's Third Voyage to the South Seas was undertaken in search of the North-west Passage—the reputed sea link between the Pacific and Atlantic Oceans. Two vessels were involved: the *Resolution*, once more under Cook, and the *Discovery*, under Captain Charles Clerke. Yet again, New Zealand was identified as a vital revictualling centre.

On 10 June 1776 Cook recorded in his journal:

> Received on board several Astronomical & Nautical Instruments which the Board of Longitude intrusted to me and M[r] King my second Lieutenant, we having engaged to that board to make all the necessary Astronomical and Nautical observations that should accrue and to supply the place of an Astronomer which was intended to be sent out in the Ship. They also put on board the same Watch Machine that was out with me last voyage … (Clerke and King 1784).

Soon after this the two vessels sailed from England.

They arrived at Queen Charlotte Sound on 12 February 1777, and on this occasion their stay was to last just two weeks. While Cook and the crew busied themselves preparing the ships for sea the local Maoris flocked to Ship Cove and established a large makeshift settlement in order that they could engage in a lucrative trade with the ships' crews. They

> … appeared very friendly selling us plenty of fish, their Garments & other curiosities for beads, pieces of red cloth, & hatchets, in which barter they shew'd their shrewdness & indeed sense, by repressing their Eagerness, inciting ours for their curiosities, & raising the price of them, & by communicating to one another what they got. (King 1776–1778: f52).

Women were also on offer but with the Rowe Massacre still fresh in their minds, few of the sailors were prepared to take up these enticing offers, despite the privations of months at sea. In reflecting on the health and diet of the crew, Edgar (1777: 223) describes how

> Great quantity's of Fish were brought on Board during the time we lay here by the Natives and very excellent of their kind, Scurvy Grass was likewise procured – & Spruce being found we also brewed Spruce Beer for the Ships Company.

Meanwhile, the astronomers were there to achieve scientific objectives, and Bayly (1777) reports that on 13 February (the day after their arrival)

> … the Ship's tents were cary'd on Shore and set up by the side of the brook of water that runs down from the Mountains. In the afternoon I carried my Observatory and Instruments on shore & set up the Obs'y close by the Ships tents that the whole might [be] under the Centinals eye. Capt. Cook put up his observatory close by the side of mine.

Cook mentions that a place was cleared for the observatories (Beaglehole 1967(I): 59), which suggests that they were not located on precisely the same spots that Bayly and Wales used during the previous voyage. A charming painting by Webber showing the two tent observatories in Ship Cove is reproduced here as Fig. 6.7.

Ever mindful of the Rowe Massacre, Cook assigned ten marines to guard the shore camp, and from the 13th all three astronomers began taking astronomical observations in order to rate the two chronometers and refine the longitude of the Sound. They were happy to accept the value for the latitude that was adopted on the Second Voyage. Bayly used a small series of lunar observations to derive longitude values, while Cook and King used a much longer series of measures of the "Distance of the Moon and Sun from Stars" for the same purpose. Their combined observations yielded the following mean longitude for Ship Cove:

Longitude 174° 20′ 36″E

Meanwhile, Cook provides a longitude of 174° 25′ 15″E for the Ship Cove observatory site, "… a Mean of 103 sets of Observations, each set consisting of Six or more observed distances …" (Beaglehole 1967(I): 75). He adds:

Fig. 6.7 A painting by John Webber showing the two tent observatories in Ship Cove in February 1777 (after *Views in the South Seas* ... 1808)

> It will not be remiss to mention that the Longit. by Lunar Observations deffers only 6′ 45″ from what Mr Wales made it last voyage, his being so much more to the West.
>
> N.B. the Latitude is 41°6′ as found by M[r] Wales. (ibid.).

After the Third Voyage, the task of analysing and writing up the astronomical observations fell to Bayly, who produced the mandatory Board of Longitude volume in 1782 (Cooke, King and Bayly 1782). Note that Bayly generously put his name last, and that he included Cook as a posthumous author—even if his name was spelt incorrectly. Although King is listed as the second author he apparently played little if any part in the preparation of the volume for publication, and he died unexpectedly in the very same year that the volume was published.

6.5 Concluding Remarks

Cook's Third Voyage marked the end of an intensive interval of scientific astronomy at Queen Charlotte Sound, the likes of which will never again be witnessed. The stop-over also brought to an end the seeming obsession with longitude, but the outcome was excellent for once all of the 1777 results were added to those from the Second Voyage, "… the geographical co-ordinates of Ship Cove were known with greater accuracy than almost any other place on Earth, Greenwich

Table 6.1 Derived values of latitude and longitude for Queen Charlotte Sound, 1770–1777

Voyage	Date	Astronomer	Latitude (South)	Longitude (East)
First	January 1770	Green	41° 05.8′	175° 28′ 29″
Second	April–May 1773	Bayly	41° 05′ 47.65″	173° 56′ 14.44″
	November 1773	Wales	41° 05′ 49.7″	174° 25′ 41″
	December 1773	Bayly	41° 05′ 34.0″	174° 20′ 30″
	October–November 1774	Wales	41 °05′ 54.4″	174° 30′ 11″
Third	February 1777	Bayly, Cook and King		174° 20′ 36″

included!" (Orchiston 1998a: 65). In the course of the seven year period from February 1770, the accepted value for the Sound's longitude underwent increasing refinement (see Table 6.1).

In contrast, apart from Bayly's anomalous December 1773 value, there was relatively little change in the accepted latitude during this interval (Table 6.1), reflecting the comparative ease with which this parameter could be determined.

By way of comparison, the currently-accepted values of latitude and longitude for Ship Cove (G. Linnell, personal communication) are:

Latitude 41° 05′ 43″S
Longitude 174° 14′ 02″E

The Third Voyage stop-over offered the astronomers their last chance to test the chronometers on New Zealand soil. Bayly found that

> By comparing the above longitude by the Watch No 1 [the K1] with the mean of the results of 122 sets of lunar observations taken in Queen Charlotte's Sound, the Watch gave 35¾ miles too much, or East of the Lunar Observations. (Cooke et al. 1782: 102).

While this may appear to be a large discrepancy, it actually represents a small error in the context of a lengthy round-the-world voyage. The K1 chronometer certainly proved more accurate than the K3, although it showed more variability than on the preceding voyage (see Fig. 6 in Howse 1969a). But more importantly, through Cook's last two voyages the *bona fides* of chronometers as accurate time-keepers and navigational tools were established beyond any doubt.

These two voyages also proved the effectiveness of Bayly's portable tent observatory design, and as the site of all but one of the Cook voyage observatories Queen Charlotte Sound holds a special place in New Zealand astronomical history. During their various stop-overs, the astronomers erected their observatories on Hippah Island and in Ship Cove, and it is apparent that at least two different localities were utilised within Ship Cove. It is therefore appropriate that there is now a conspicuous monument in Ship Cove, to mark Cook's visits and the associated scientific investigations that were carried out in Queen Charlotte Sound between 1770 and 1777.[1]

Acknowledgements I am grateful to the following for their assistance: Dr Jonathan Betts and Dr Gloria Clifton (ex-Royal Museums, Greenwich), the late Commander Derek Howse (Sevenoaks, England), Emma Lefley (National Maritime Museum, London) and Geoff Linnell (Land Information New Zealand, Wellington). I also wish to thank Gloria Clifton for reading and commenting on the first draft of this chapter, and the National Library of Australia (Canberra), the Royal Society (London) and the National Maritime Museum (Greenwich) for kindly supplying Figs. 6.4, 6.5 and 6.6.

References

Andrews, A. D. (1992). Cyclopaedia of telescope makers. *Irish Astronomical Journal,20*, 102–183.

Bayly, W. (1777). Extract from the log of William Bayly, Astronomer, on H.M.S. Discovery … In R. McNab (Ed.), *Historical Records of New Zealandvol. II* (pp. 218–221, 1914).

Beaglehole, J. C. (Ed.), (1967). *The Journals of Captain James Cook on his Voyages of Discovery. III. The Voyage of the* Resolution *and* Discovery 1776–1780 (2 Vols.). Cambridge: Cambridge University Press.

Beaglehole, J. C. (Ed.), (1968). *The Journals of Captain James Cook on his Voyages of Discovery. I. The Voyage of* Endeavour 1768–1771. Cambridge: Cambridge University Press (Preprint; first published 1955).

Beaglehole, J. C. (Ed.), (1969). *The Journals of Captain James Cook on his Voyages of Discovery. II. The Voyage of the* Resolution *and* Adventure 1772–1775. Cambridge: Cambridge University Press (Reprint; first published 1961).

Beaglehole, J. C. (1974). *The Life of Captain James Cook*. London: Hakluyt Society.

Begg, A. C., & Begg, N. C. (1968). *Dusky Sound*. Christchurch: Whitcombe & Tombs.

Chapman, A. (1995). Dividing the circle. In *The Development of Critical Angular Measurement in Astronomy* (pp. 1500–1850). Chichester: Wiley.

Clerke, C., & King, J. (1784). *Voyage to the Pacific Ocean, Undertaken by the Command of His Majesty, for Making Discoveries in the Northern Hemisphere, Determine the Position and Extent of the West Side of North America; Its Distance from Asia; and the Practicability of a Northern Passage to Europe* (3 Vols.). London: Nicol and Cadell.

Cooke, J. [sic], King, J., & Bayly, W. (1782). *The Original Astronomical Observations Made in the Course of a Voyage to the Northern Pacific Ocean, for the Discovery of a North East or North West Passage* … London: Elmsly.

Edgar, T. (1777). Extracts from the journal of Thomas Edgar. In R. McNab (Ed.), *Historical Records of New Zealand Vol. II* (pp. 222–229, 1914).

[1]Plaques also exists at Mercury Bay and Dusky Sound (Begg and Begg 1968: 135) to mark the astronomical observations that were carried out at these two localities on the First and Second Voyages, respectively. The monument at Mercury Bay marks the site from which Cook, Green and Hicks observed the November 1769 transit of Mercury. This is the only New Zealand monument dedicated solely to Cook voyage astronomy, and the plaque, titled simply "MERCURY BAY", contains the following inscription: "NEAR THIS SPOT ON 10 NOVEMBER 1769 JAMES COOK AND CHARLES GREEN OBSERVED THE TRANSIT OF MERCURY TO DETERMINE THE LONGITUDE OF THE BAY".

Howse, D. (1969a). Captain Cook's marine timekeepers. Part I—The Kendall watches. In D. Howse & B. Hutchinson (pp. 190–205).

Howse, D. (1969b). Captain Cook's marine timekeepers. Part II—The Arnold chronometers. In D. Howse & B. Hutchinson (pp. 276–280).

Howse, D. (1969c). Captain Cook's minor clocks and watches. In D. Howse & B. Hutchinson (pp. 138–145).

Howse, D. (1969d). Captain Cook's pendulum clocks. In D. Howse & B. Hutchinson (pp. 62–76).

Howse, D. (1979). The principal scientific instruments taken on Captain Cook's voyages of exploration, 1768–80. *Mariner's Mirror,65*, 119–135.

Howse, D., & Hutchinson, B. (1969a). *The Clocks and Watches of Captain James Cook 1769–1969*. London: The Antiquarian Horological Society. (reprinted from four issues of *Antiquarian Horology*).

Howse, D., & Hutchinson, B. (1969b). The saga of the Shelton clocks (pp. 281–298). In D. Howse & B. Hutchinson.

J. K. L. (1909). King, James (1750–1884). In S. Lee (Ed.), *Dictionary of National Biography* (Vol. XI, p. 136). London: Smith, Elder & Co.

King, J. (1776–1778). Untitled log. Original in the British Library (Adm 55/116).

King, H. C. (1979). *The History of the Telescope*. New York: Dover.

Lloyd, C. (1968). *The British Seaman 1200–1860. A Social Survey*. London, Paladin.

Mackrell, B. (1985). *Halley's Comet over New Zealand*. Auckland: Reed Methuen.

McIntosh, R. A. (1957). The astronomical history of the Auckland Province—Part II: eighteenth century. *Amateur Astronomer,17*, 40–42.

Mead, S. M. (1969). The costume styles of the Classical Maori in New Zealand, 1642–1800 A.D. *Costume,3*, 35–43.

Orchiston, W. (1971). Astronomy in early New Zealand. *British Astronomical Association New South Wales Branch Bulletin,509*, 1–2.

Orchiston, W. (1972). Maori neck and ear ornaments of the 1770s: a study in protohistoric ethno-archaeology. *Journal of the Royal Society of New Zealand,2*, 91–107.

Orchiston, W. (1974a). Cook voyage 'trading stations' in Early Protohistoric New Zealand. *Dominion Museum Records in Ethnology,2*, 133–156.

Orchiston, W. (1974b). *Studies in South Island New Zealand Prehistory and Protohistory*. Unpublished Ph.D. Thesis, Department of Anthropology, University of Sydney.

Orchiston, W. (1975a). Early protohistoric Maori food preparation, and application of the direct historical approach to New Zealand prehistory. *New Zealand Archaeological Association Newsletter,18*, 18–31.

Orchiston, W. (1975b). Maori material culture change in Early Protohistoric New Zealand: the greenstone trade at Queen Charlotte Sound. *The Artefact,39*, 40–77.

Orchiston, W. (1994). *James Cook and the 1769 Transit of Mercury*. Wellington: Carter Observatory (Information Sheet No. 3).

Orchiston, W. (1997). *Early Astronomy in New Zealand: The South Sea Voyages of James Cook*. Wellington: Carter Observatory (Information Sheet No. 10).

Orchiston, W. (1998a). From the South seas to the Sun: the astronomy of Cook's voyages. In M. Lincoln (Ed.), *Science and Exploration. European Voyages to the Southern Oceans in the 18th Century* (pp. 55–72). Woodbridge: Boydell and Brewer.

Orchiston, W. (1998b). *Nautical Astronomy in New Zealand: The Voyages of James Cook*. Wellington: Carter Observatory.

Salmond, A. (1991). *Two Worlds. First Meeting Between Maori and Europeans 1642–1772*. Auckland: Viking.

Views in the South Seas from Drawings by … James [sic] Webber … London: Boydell (1808).

Wales, W., & Bayly, W. (1777). *The Original Astronomical Observations, Made in the Course of a Voyage Towards the South Pole, and Around the World …* London: Strahan.

W. H. (1908). Bayly, William (1737–1810). In L. Stephen & S. Lee (Eds.), *Dictionary of National Biography* (Vol. I, p. 1372). London: Smith, Elder & Co.

Part III
Fundamental Astronomy: Of Telescopes and Observatories

Chapter 7
The 'Cook' Gregorian Telescope in the Museum of New Zealand Te Papa Tongarewa

Abstract The Museum of New Zealand Te Papa Tongarewa in Wellington houses an eighteenth century Gregorian telescope manufactured by Heath and Wing of London which has a convoluted history reputedly involving James Cook and Sir Joseph Banks. Neither association can be verified, but we conclude that this instrument does indeed have a Cook Voyage pedigree, and was the telescope that Daniel Solander used to observe the 1769 transit of Venus from Fort Venus, Tahiti.

7.1 Introduction

In the collection of The Museum of New Zealand Te Papa Tongarewa in Wellington is an astronomical telescope (Accession Number NS000010) reputedly associated with the famous British navigator and astronomer, Captain James Cook (Fig. 7.1; Beaglehole 1974), and subsequently with Sir Joseph Banks (1743–1820; Fig. 7.2; Beaglehole 1963). This chapter, which is developed mainly from Orchiston (1999), describes the telescope and critically examines its Cook voyage provenance.

7.2 The Telescope

In a letter written in 1918, the owner of the telescope included the following 'note point' summary of its features:

> Gregorian Reflector
> Mounted as an altimuth [*sic*], with horizontal circle and vertical circle attached to the body of the Telescope, both circles graduated and having verniers attached.
> Aperture 5″ diameter.
> Made by Heath & Wing, London. No date. Two eyepieces, high and low, each having its own small concave lens.

W. Orchiston, *Exploring the History of New Zealand Astronomy*, Astrophysics and Space Science Library 422,
DOI 10.1007/978-3-319-22566-1_7

Fig. 7.1 An engraving based on a portrait of James Cook by William Hodges (after Cook and Furneaux 1777: Frontispiece)

Fig. 7.2 A portrait of Joseph Banks painted by Joshua Reynolds in 1773 (https://en.wikipedia.org)

Diameter 1½″.
Length of low eye piece 5¾″.
Length of high eye piece 4¼″.
Both fitted with sun glasses.
Length of Telescope 2ft. 10½ins.
Height of vertical column 1ft.
The instrument can be clamped so that by holding each screw with either hand the star, etc. can be kept in the field of vision.
Unclamped the instrument is free to move in any direction.
There is also a finder which is a refractor with cross wires in the eye piece.
Diameter of object glass 1″. (Relton 1918).

The material of manufacture was brass, and the telescope came with an oak travelling case into which it fitted (Ellis 1932) and a "… folding tripod of wood and iron for mounting the Telescope 6ft in height." (Relton 1918). Photographs that appeared in various newspapers in 1952 show that the tripod had a large circular top, and that the telescope sat snugly upon this.

The telescope is illustrated in Fig. 7.3, and—apart from its excessive length—in overall appearance bears a striking resemblance to the Gregorian reflectors manufactured by Short for the Royal Society's 1769 transit of Venus expeditions (e.g. see Orchiston 1998b: Fig. 6).

Being a Gregorian reflector, it has a perforated parabolic primary mirror and an ellipsoidal secondary, both made from speculum metal. The instrument was focussed by varying the separation of the two mirrors. In the standard Gregorian design, the eyepiece contains a single equi-biconvex lens (see King 1979: Fig. 33).

The name of the manufacturer, 'Heath & Wing London', is engraved on the telescope, near the eyepiece assembly. Thomas Heath (1719–1773) was one of London's most prominent scientific instrument-makers during the first half of the eighteenth century, before going into partnership with Tycho Wing (1726–1776), who was some years his junior (and with a name like that surely could be excused for making astronomical telescopes). The firm of Heath and Wing ran successfully from 1751 to 1773 as makers of mathematical, philosophical and optical instruments. Apart from astronomical telescopes, their output included barometers, protractors, sextants, sundials, theodolites and thermometers. Heath died in 1773 and Wing retired in that same year and died just three years later (Ariail, personal communication, 1999; Clifton 1995). For English localities mentioned in this chapter see Fig. 7.4.

7.3 The Chain of Ownership

Newspaper reports from 1952 when the telescope was put up for sale continually referred to it as 'Cook's telescope', or more specifically, "… the telescope used by Captain Cook on his first visit to New Zealand." (Cook's Telescope 1952). What, precisely, is the basis of this claim? Let us examine the chain of ownership.

Fig. 7.3 The 'Solander Telescope', Wellington, New Zealand (*Courtesy* Museum of New Zealand Te Papa Tongarewa, NS000010/1)

In about 1864 the Anglican Minister at Ealing, West London, the Reverend E.W. Relton (Baker and Elrington 1982) acquired the telescope (Relton 1864; cf. Ellis 1932), and at that time recorded its provenance. He

> … received it from the Widow of the late Jonathan Johnson, Esq.
> who received it from
> the late John Rickman, Esq, F.R.S. &c
> who received it from

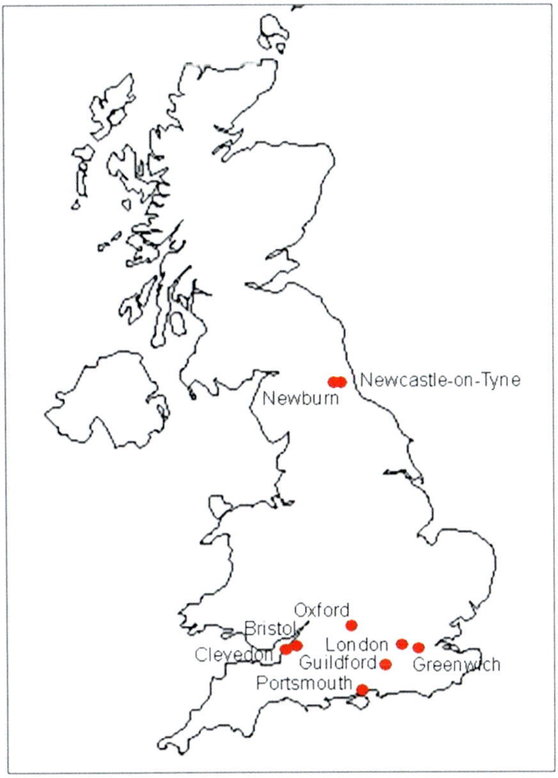

Fig. 7.4 English localities mentioned in this chapter are shown in *red*

either the late Sir J. Banks,
or — — Admiral Burney.
it is not certain which, but it is believed that it was from the former … (Relton 1864).

With the telescope was an undated note by Miss Johnson which elaborates on her father's acquisition of this historic instrument: "Presented to my father by the late John Rickman Esq[r] F.R.S. etc. Whether he received it from Sir J. Banks or Admiral Burney we cannot decidedly say, but I believe from the former—who accompanied Captain Cook." (Johnson, n.d.). John Rickman (1771–1840; Fig. 7.5) was a prominent Government official and statistician. He was born in Newburn (Northumberland) and attended Guildford Grammar School and Oxford University. Subsequently he drafted the bill that became the 1800 Census Act, and organized the first four censuses carried out in England. From 1814 he held the title of Second Clerk Assistant in the House of Commons, London (a post he occupied until his death). He also served as Secretary to two Parliamentary Commissions. He was elected a Fellow of the Royal Society in 1815 (see Williams 1912).

In 1882 the Reverend Relton gave his son (H. Relton) the telescope "… as a curiosity …" Relton (Junior) was then residing at Newcastle-upon-Tyne, and had a

Fig. 7.5 A portrait of John Rickman painted by Samuel Lane in about 1831 (published with permission of the Parliament of the United Kingdom)

local optician and scientific instrument-maker, Mr Frederick Robson (Fig. 7.6), examine it.[1] Robson reported:

> … the mirrors … were in splendid condition, the large mirror being only a little dull, and although the screw movements were all choked up with dust they could be easily cleaned and the big mirror rubbed up.
>
> Although the brass body of the Telescope had lost its polish, considering the history of the Instrument he did not advise repolishing. (Relton 1918).

Subsequently, H. Relton died, and in March 1926 the telescope was sold at the auction of his effects by the firm of 'Alonzo Dawes & Son, Auctioneers and Estate Agents' of Clevedon, Somerset (Captain Cook's Telescope, n.d.). It was described by the auctioneers as "… previously the property of Capt. Cook the explorer." (see Alonzo Dawes & Son 1926), and the purchaser was a Mr H. Stewart, acting as agent for a Mr F. Ellis of Bristol.[2]

---

[1]Web searches indicate that the firm of F. Robson & Co. Opticians was established in 1867 in the centre of Newcastle-on-Tyne, and that it has continued through to the present day.

[2]Ellis' acquisition of the telescope is in keeping with his interest in objects associated with the Cook voyages. For instance, in the Pitt Rivers Museum, Oxford, there is a bronze replica of a Maori *patu onewa* (hand club) that was bought by F. Ellis in Bristol in 1908 and donated by him to the Royal Society. In 1932 the club was loaned by the Society to the Pitt Rivers Museum and then formally donated to the Museum in 1979. This club is one of forty that Joseph Banks had made in 1772 as gift and exchange items when he was planning to accompany Cook on the Second Voyage to the South Seas (see Pitt Rivers Museum, n.d.).

Fig. 7.6 A photograph of Frederick Robson, taken in 1895 when he was about 51 years of age (www.facebook.com/opticiansnewcastle#!/opticiansnewcastle/photos/)

Nothing more is heard of the telescope until 1930, when the Great Depression may have begun to impact directly on Mr Ellis for he approached the Australian Government with a view to selling them the instrument. As a result,

> … their agent Mr Garnsey who came down to examine it told me from the evidence I possessed that he had no doubt in his own mind that it was the identical Instrument used by Dr. Solander of the British Museum on the memorable voyage with Cook & Banks & others undertaken in 1769 to observe the transit of Venus at Tahiti. (Ellis 1932).

Given the prevailing economic situation, the Australian Government was unable (or unwilling) to proceed with the sale, in spite of Garnsey's excellent assessment, but the telescope remained on the market and on 18 January 1932 we find Ellis writing to an unspecified firm with a view to their acting on his behalf:

> This fine old instrument was for a period of 60 years in the possession of the Relton family & was bought by me from their Clevedon sale; as you see by the enclosed pedigree [i.e. three original papers which accompanied the letter] they always regarded it as directly connected with Capt. Cook & as they were intellectual people they undoubtedly had good grounds for their belief. (Ellis 1932).

Precisely what happened then is not certain, but in November 1933 a reporter writing about "The 'junk' shops of Portsmouth" records an unexpected discovery:

> In a street in Portsea that boasts three or four second hand shops I came across something that made my search worthwhile. It was Captain Cook's telescope … It was obvious to the most disillusioned of collectors that the monetary value of the instrument was considerable. (FHR 1933).

Soon after this the instrument was purchased by a Mr Albert Blake, a Portsmouth antique dealer, for about £400 (Capt. Cook Relic For Sale 1952). Blake must have thought he was 'sitting on a gold-mine', for within a year of this purchase he is reported to have refused an offer of £1000 from an American dealer (FHR 1933).

Instead, shortly after this he sold the telescope to a Mr Kinch, who also was "… well-known as an antique dealer in Portsmouth in pre-war days." Kinch was apparently Blake's brother-in-law (Captain Cook's telescope 1952), which may account for the bargain basement sale price of just £300 (Captain Cook's Telescope for sale 1952), a very considerable discount on the sum offered earlier by the American dealer.

When Mr Kinch died in about Kinch 1938, the telescope passed to his daughter, Miss Edith R. Kinch (Captain Cook's telescope 1952), who promptly sent a description of it to the National Maritime Museum in Greenwich (Kinch 1938). In 1952, a local newspaper described its owner: "Miss Kinch is well-known as a concert soprano who has broadcast on numbers of occasions." (Historic Relic for New Zealand? 1952).

By the British spring of 1952 Miss Kinch was reportedly in financial difficulties and so she put the telescope up for sale (Campbell 1952; Cook's Telescope for New Zealand 1952). Although she did not specify a price, it was reported that she "… expects something in the nature of £150 or more." (Campbell 1952).

On 16 April 1952 the New Zealand Office of the High Commissioner in London sent a memorandum about the telescope to the Department of External Affairs in New Zealand, in which the writer remarks that after consulting Cook's account of the 1769 transit of Venus published in the *Philosophical Transactions of the Royal Society* he believes it is "… more than likely that this telescope came down through Sir Joseph Banks." (ibid.). Accompanying the memorandum was material on the history of the telescope provided by Miss Kinch, "… which leads us to believe there is no doubt the telescope is of genuine historical interest." (ibid.). In London, then, the association of the telescope with Cook's First Voyage was deemed proven, and the chain of ownership was seen to pass through Banks, rather than via Burney. Meanwhile, the memorandum continues:

> I would be grateful to know if there is any interest in its purchase, and if so, what offer you are prepared to make. I understand that if New Zealand is not interested, Miss Kinch will offer the telescope to the Australian Government. (ibid.).

When it reached Wellington, the memorandum and supporting documentation were forwarded to Dr (later Sir) Robert Alexander Falla (1901–1979; Fig. 7.7), Director of the Dominion Museum, for a response, and the very spectre of trans-Tasman rivalry was sufficient to guarantee a positive reaction. Falla accepted the association with Cook's First Voyage and concluded that the telescope "… is a relic that should be secured for some public institution in New Zealand, and I should like to make an offer of £150 on behalf of the Dominion Museum." (Falla 1952a).

The Department of External Affairs contacted London on 9 May, putting forward Falla's offer, and making the point that if Miss Kinch

Fig. 7.7 Sir Robert Falla and his wife Elayne in about 1971 (*Courtesy* Alexander Turnbull Library, Wellington Maritime Museum Collection, Ref. 35 mm-00272-D-F)

> … considers this offer not to be as good as she might obtain elsewhere, it is possible that she would take into consideration the fact that Captain Cook's first voyage was an important event in New Zealand history and that the Dominion Museum is an appropriate place for the permanent home of such a relic. (Secretary of External Affairs 1952).

But one other significant feature of this memorandum is that it acknowledges London's communication of 16 April, "… describing a telescope purporting to be one used by Captain Cook …" (ibid.). As we have seen, London only claimed an association with Cook's First Voyage, but the notion that this was *Cook's own telescope* immediately took root when news of the impending sale reached the English and New Zealand media (e.g., see the various newspaper reports listed in the References at the end of this chapter).

Meanwhile, despite the reassurances from Wellington, those at the Office of the New Zealand High Commission in London must have felt some disquiet for they asked Dr Henry Reginald Calvert (1904–1992) Keeper of the Department of Astronomy and Geophysics at the Science Museum, London, to comment on the telescope's authenticity, condition and value. His response is illuminating:

> He placed its date of manufacture at the middle of the 18th Century, but could express no opinion as to the likelihood of its having been on the "ENDEAVOUR". He did not consider that its condition was very good, and thought £150 was a good price, apart from any historical associations with New Zealand. (The Official Secretary 1952).

This hardly tallies with Miss Kinch's claim that the National Maritime Museum had verified the authenticity of the telescope (Captain Cook's Telescope for sale 1952)!

In their memorandum of 17 June 1952, the Office of the New Zealand High Commission went on to make further telling comments about the telescope's provenance:

> A firm of optical instrument makers, to whom the telescope was sent for minor repairs, expressed similar opinions on it [to those of Dr. Calvert].
>
> The records of the Royal Society have been searched in the hope of finding some evidence for or against the authenticity of the instrument, but with no success. The search was not exhaustive, and there may exist records at least of the makers of the telescopes used on the voyage …
>
> Enclosed are photostats of papers offered by Miss Kinch as further evidence of authenticity, but the general position seems to be that it is as difficult to prove authenticity as to disprove it. (The Official Secretary 1952).

Given their obvious concerns about authenticity, it is somewhat surprising to read that those in the Office of the New Zealand High Commissioner immediately went ahead and offered Miss Kinch £150 for the telescope (ibid.). In all likelihood, they were eager to place an offer on the table before the Australian Government did so. But they also were aware that Miss Kinch may have been hoping for a higher price, and so at the same time Wellington was asked to "… please advise me, by cable, whether a higher price is to be offered, or if you wish even to continue the offer of £150 …" (ibid.).

Falla, too, was having second thoughts about the telescope's provenance, and after reviewing the additional documentation supplied by Miss Kinch he advised External Affairs: "Considering all the circumstances and the lack of authentic proof that the instrument was used by Cook, our offer of £150 must be considered final." (Falla 1952b).

The New Zealand offer was to stand for 2 weeks (Capt. Cook Relic For Sale 1952), and we learn from a May 1952 newspaper report that it was thought likely that an offer would also be made by the Australian Government (Captain Cook's Telescope for sale 1952). Meanwhile, the £150 offered by New Zealand was a far cry from the price of above £1000 touted by the press (Historic Relic for New Zealand? 1952). By 26 May, Miss Kinch had still not determined whether the telescope would go to New Zealand or to Australia (John Citizen's Diary 1952), but she had decided to take it to London on the 27th in order to show it to the Australian Prime Minister who was then visiting England.

Several months of negotiations with both Governments followed, and on 11 August 1952 the Office of the New Zealand High Commission was pleased to be able to telegraph Wellington that Miss Kinch had finally accepted their offer (High Commissioner …, London 1952).

The sale then remained confidential while the Dominion Museum worked its way through the inevitable red-tape involved in extracting the £150 from Treasury, and it was only in October that Miss Kinch was able to announce the outcome of her negotiations. In reporting the sale, *The Cornishman* newspaper makes the following very interesting comment, which runs counter to the earlier claim that Miss Kinch put the telescope on the market because of financial hardship:

> Although Miss Kinch could have accepted offers running into over £1,000, she allowed the New Zealand Government to buy the telescope at a "very reasonable figure"– the actual amount has not been disclosed – as she feels that the telescope ought to belong to Cook's Island; moreover, Miss Kinch is thinking of emigrating to New Zealand and she would like to think that the telescope is also there. (Capt. Cook's Telescope 1952)

Another newspaper reports that the telescope was sold at "… hundreds of pounds below its value." (Cook's Telescope for New Zealand 1952). What the real value was, given the doubts about its provenance, we can only surmise, but even without a Cook association a figure of just £150 for a genuine eighteenth century telescope in fine condition would seem to be a bargain.

The actual handing over of the telescope to the New Zealand High Commissioner, Mr (later Sir) Frederick Widdowson Doidge (1884–1954; Fig. 7.8), was to be a grand affair. Miss Kirch took the telescope, travelling case and tripod from Cornwall to

Fig. 7.8 Mr and Mrs Doidge photographed on about 17 August 1951 on their way to London, where Mr Doidge was to be the New Zealand High Commissioner (*Courtesy* Alexander Turnbull Library, Evening Post Collection: Ref. 114/338/06-G)

London by train, and on 3 November 1952 at 11:45 a.m. there was an official 'ceremony' at New Zealand House. Miss Kinch could not let the occasion pass without giving a speech:

> On this momentous occasion it gives me great pleasure to hand over to Your Excellency the historic relic Captain Cook's telescope with its massive oak travelling case and the equitorial stand [i.e. tripod] together with the old and battered key.
>
> I feel convinced that if Captain Cook, Admiral Burney, Sir Joseph Banks and Dr. Solander were present here now they would be smiling down upon us, happy in the fact knowing that the telescope would find its last resting place in the islands that Cook discovered.
>
> As you are aware I was approached by the Madison Antique Show in New York to ship over the telescope which they were going to publicise widely and they predicted that the telescope would command a figure well over four figures, but I felt that it would be Captain Cook's wish for his telescope to find its last resting place in the islands he discovered and so beloved by him. (Kinch 1952).

As we see, Miss Kinch managed to display her ignorance of New Zealand history and Cook's role in the nation's discovery. Obviously she had not heard of Abel Janszoon Tasman (1603–1659), who charted part of the coast in 1642, long before Cook's arrival at Poverty Bay in October 1769.

By 26 November 1952 preparations were underway to ship the telescope to New Zealand (New Zealand Government 1952), and in early 1953 it was safely ensconced in the Dominion Museum in Wellington. After more than 180 years—if we are to believe its Cook First Voyage association—it was back on New Zealand soil.

Obviously, its presence caused considerable excitement, and it was on demand as a display item. Most notably, from about 1969 until 1991 it was among the 'props' used to dress a reconstruction of Cook's cabin on the *Endeavour*, and from 1991 it featured in another display about New Zealand history and heritage. Following the opening of the new museum building down on the Wellington waterfront it featured prominently in a short-term display titled 'The Ablest Navigator: A Look at Cook'. Since then it has been in two exhibitions (Mike Fitzgerald, personal communication, 2014). From June 2002 until January 2003 it was in "Voyagers: Discovering the Pacific", an exhibition

> … with the aim of changing many people's understanding of how our part of the world came to be explored and populated. It told the stories of how New Zealand and the rest of the Pacific was discovered and how the search for adventure in the Pacific continues today.
>
> The exhibition told the dramatic stories of four of the Pacific's most famous voyagers – Kupe, Captain Cook, Tevake, and Sir Peter Blake. Te Papa examined where these important voyagers came from, and how, where, and why they travelled. Voyagers profiled their major achievements and offered insights into the navigational techniques they used. (http://collections.tepapa.govt.nz/topic/2346).

More recently, in 2011, the 'Cook' telescope figured in the exhibition "Oceania: Early Encounters", about

> Oceania – the sea of islands in the vast Pacific. The peopling of this watery world goes back thousands of years, but only in the last 500 have Pacific and European peoples met here.

> Journey through time to explore their early encounters – some of them marked by curiosity and cooperation, others by confusion and conflict. (http://www.tepapa.govt.nz/WhatsOn/exhibitions/Oceania/EarlyEncounters/Pages/default.aspx).

7.4 The Cook Attribution

As we have seen, during the eighteenth century the telescope was traced back to one of two possible original owners, Sir Joseph Banks or Admiral Burney, and it was only in 1926, when it was about to be auctioned off, that it was assigned instead to Captain Cook. By stating that the instrument was "… previously the property of Capt. Cook …", Alonzo Dawes & Son (1926) instantly were able to enhance its profile and ensure a good sale price for by this time 'Cook memorabilia' were especially sought after.

Meanwhile, Miss Kinch innocently took Dawes' statement at face value and proceeded to reinforce the myth by proudly proclaiming in 1938 that she was "… in possession of Capt. Cook's Telescope that travelled with him in the "Endeavour" in 1768 to observe the transit of Venus." (Kinch 1938). As Dr Basil Jack Greenhill (1920–2003), the Director of the National Maritime Museum later pointed out, Miss Kinch was "… somewhat over-enthusiastic in calling it "Captain Cook's telescope"." (Greenhill 1972). But Kinch (1938) went further, and suggested that the telescope was subsequently acquired by Banks, presumably after Cook's death. This link would have appeared logical to her given Bank's presence on the First Voyage, and it offered a way of accommodating one of the nineteenth century provenance options.

Yet there are strong grounds for querying any association with either Cook or Banks. All of the reflecting telescopes supplied by the Royal Society or the Board of Longitude on Cook's three voyages were manufactured either by Short or Bird, not Heath and Wing (see Orchiston 1998b: 39–44), and although Cook is known to have owned his own astronomical telescope at that time (Beaglehole 1967(I): 243) this was "… an 18 Inch Reflector made by Watkins" (Beaglehole 1969: 532). According to Clifton (1995), Francis Watkins (d. 1791) was a London instrument-maker who began his optical apprenticeship in 1737 and practised as an optician from 1747. Apart from telescopes, he made a number of other scientific instruments. After Cook's death, his Watkins reflector was sent to his widow (Howse 1979). Vaughan and Murray-Oliver (Vaughan 1974: 32) include a photograph of a Watkins telescope that must have been very similar to the one owned by Cook.

Cook's own accounts make it clear that he did not possess a second astronomical telescope, so we cannot associate the Heath and Wing reflector with him. But even if he had owned such an instrument, there is no basis for supposing that it would have gone to Banks after his death. Banks was not listed as a beneficiary in the will, and there is no evidence that Mrs Cook, her sons, or their heirs subsequently passed relics on to him (e.g. see Beaglehole 1974: 690–695; Beddie 1970).

Nor was there any reason for them to do so, for by this time Banks already had an astronomical telescope of his own, a refractor made by Peter Dollond but with an equatorial mounting by Ramsden (see King 1979: 162). Precisely when he acquired this instrument is not recorded, but it must have been after Cook's First Voyage because he made no attempt to observe the 1769 transit of Venus (see Beaglehole 1963(I): 283–285; Beaglehole 1968: 97, 559; Orchiston 1998a; 2005) and it is hard to believe that he would have ignored so important and spectacular a celestial event had he owned the Dollond-Ramsden telescope at this time. Now, since Banks' telescope is known to have been one of four that Ramsden completed by 1773, he must have made the purchase some time between July 1771 (when he returned from the South Seas) and the end of 1772. It is interesting to speculate as to why he should have wanted to buy a telescope, given that he was not an astronomer and did not make it his practice to conduct astronomical observations. Perhaps he was anticipating Cook's Second Voyage, where his mooted scientific party was to include three astronomers, Bayly, Lind and Wales (Lyte 1980: 152–153). In the end, circumstances determined that the *Resolution* and *Adventure* would sail from England without him, although Bayly and Wales did make the voyage as the expedition's official astronomers (see Orchiston 1998b).

This only leaves Admiral Burney as the possible initial owner of the Wellington telescope, but since he was not on the *Endeavour* we can rule out any Cook First Voyage association. But he was on Cook's Second and Third Voyages (e.g. see Hooper 1975). James Burney (1750–1821; Fig. 7.9), was born in London, and went to sea at the age of 10. He joined the *Resolution* (with Cook) on the Second Voyage, but transferred to the *Adventure en route* to the Pacific. During the voyage he was promoted to Second Lieutenant (Beaglehole 1969: 877), and although

Fig. 7.9 An undated charcoal drawing of James Burney, possibly by R.H. Dyer (burneycentre.mcgill.ca/bio_james.html)

> … he sailed very little with Cook himself, is one of the most interesting of Cook's officers; a thorough seaman, and certainly one of the mainstays of the Adventure's company; in addition lively, observant, and articulate. (Beaglehole 1969: xxxviii).

On Cook's Third Voyage, Burney served as First Lieutenant on the *Discovery*, and after Clerke's death in August 1779 he transferred to the *Resolution* as second-in-command. He was promoted to Captain at the very end of the voyage, and became an Admiral in 1821, the very year of his death (Beaglehole 1969: 877; Manwaring 1931).

The only Gregorian telescopes documented on the Second and Third Voyages were two Bird reflectors supplied by the Board of Longitude for the astronomers, and Cook's trusty Watkins reflector (see Orchiston 1998b: 41–42). There is no evidence that Burney took an astronomical telescope on either voyage, and although he was familiar with nautical astronomy—as all officers had to be—unlike Bligh, Clerke, Elliott, Gilbert, Harvey, Hood, Pickersgill, Roberts, and Smith (Orchiston 1998b: 27–34), his interest in astronomy was not such that he would want to assist the astronomers in their official duties. Given these circumstances, it is hard to believe that Burney was the initial owner of the Heath and Wing telescope.

While there are no grounds, then, for assigning the telescope directly to Cook, Banks or Burney, another possibility, first raised by Garnsey in 1930, warrants further examination. This is the suggestion that the telescope was that used by Solander on Cook's First Voyage. Dr Daniel Carl Solander (1733–1782; Fig. 7.10) was a Swedish-born naturalist who studied under Linnaeus and came to London in 1759 (Beaglehole 1968: 599) where he found employment at the British Museum. In 1764 he was elected a Fellow of the Royal Society, and as "… the ablest botanist in England." (Beaglehole 1968: cxxxv) it seemed only natural that he should join Banks on the *Endeavour*. At the time he was 35 years of age,

> … short and somewhat stout, with fair hair and complexion. He [was] … a jovial sort of man, kind and obliging, with charm and humility … By some writers he has been described as lazy, dissipated, indolent, dilatory and even in one case as being nothing more than a parasite. These accusations are most unjust … (Marshall 1977: 51).

On 3 June 1769, Solander joined Cook and Green at Fort Venus, and all three succeeded in observing the transit (see Orchiston 1998a; 2005). Green and Cook (1771: 411–412) noted that Solander used a 3-ft long reflecting telescope and that it magnified more than their two Short reflectors, but nowhere is there a description of this instrument or even a mention of its maker. It is notable that Solander's telescope was very similar in length to the Heath and Wing instrument.

It is not known whether Solander actually owned 'his' telescope, or whether it was merely loaned to him for the voyage. Among the documentation in The Museum of New Zealand Te Papa Tongarewa in Wellington is a note by an anonymous writer suggesting that Banks provided Solander with this telescope (Anonymous. n.d.), but I have not managed to find any evidence of this in the extensive literature on Banks (e.g. see Beaglehole 1963, 1968; Beddie 1970; Dawson 1958; Lyte 1980). This same writer suggests that after the voyage Banks probably gave the telescope to Burney, and it is of interest to note that Banks was

Fig. 7.10 A painting (left to right) of Omai, Daniel Solander and Sir Joseph Banks now in the National Portrait Gallery, London (https://en.wikipedia.org)

known to have been in Burney's circle of friends (see Beaglehole 1969: 877). Yet the literature on Banks again fails to support of this provenance, and as such it has to be treated with suspicion.

In a paper published in the *Journal of the Antique Telescope Society* in 1999 (where the Heath and Wing telescope features on the front cover of the journal), I examined the documentation accompanying this telescope, particularly its supposed Cook voyage provenance, and concluded

> The only possible Cook-voyage association that we have been able to identify for this instrument is that it was the Gregorian reflector used by Dr Daniel Solander in 1769 to observe the transit of Venus. However, the evidence for this is slim and largely circumstantial, and the fact that Solander was part of Banks' retinue should not be seen as persuasive.
>
> Scientific instruments, memorabilia and indigenous artifacts of eighteenth century vintage and reputedly associated with Cook's voyages were eagerly sought after during the nineteenth and twentieth centuries and attracted high sale prices. As a result, many objects with bogus 'Cook' histories made their way into private collections and the world's

> museums … and it is possible that the Wellington telescope is yet another example of this trade. (Orchiston 1999: 8; cf. Kaeppler 1972).

Upon subsequently re-examining all of the available documentation I decided that perhaps I was overly cautious in my 1999 paper, so at the International Astronomical Union's Transit of Venus Conference in Preston, England, in 2004 I announced that "… the telescope used by Solander at Fort Venus is probably the Heath and Wing reflector now housed in The Museum of New Zealand Te Papa Tongarewa, in Wellington …" (Orchiston 2005: 62).

As we have seen, this idea was first promoted in 1930 by Garnsey (1864–1935), the New South Wales Agent-General in London, who had an intimate knowledge

> … of art, literature, history, and science [which] led to his being chosen from time to time to decide the authenticity or the value of documents, manuscripts, paintings, or relics alleged to have some connection with the early history of Australia. (Garnsey 1864).

7.5 Concluding Remarks

The scientific collection in the Museum of New Zealand Te Papa Tongarewa in Wellington includes an eighteenth century Gregorian reflecting telescope, and this is in remarkably good condition considering its age. This instrument was manufactured by Heath and Wing, a well-known firm of scientific instrument-makers based in London.

The principal claim to fame of this telescope is that it was apparently associated with Cook and Banks and was taken by Cook in the *Endeavour* on his First Voyage to the South Seas, but archival research does not support this provenance. Nor does a counter-claim that the instrument was originally owned by Burney—another with a Cook voyage pedigree—stand up to scrutiny. Rather, I believe that this telescope was the one that Dr Daniel Solander used in Tahiti to observe the 1769 transit of Venus.

Acknowledgements I am grateful to the following for their assistance: Bob Ariail (USA); Dr Alan Baker (former Director of the National Museum of New Zealand, Wellington); Jenni Chrisstoffels (Alexander Turnbull Library, Wellington, NZ), Dr Gloria Clifton (ex-Royal Museums, Greenwich); the late Peter Hingley (former RAS Librarian, London); the late Derek Howse (Sevenoaks, England); Ross Galbreath (Tuakau, NZ), and Dr Michael Fitzgerald, Jennifer Twist, David Riley and staff of the Hector Library (The Museum of New Zealand Te Papa Tongarewa, Wellington). Finally, I would like to thank Gloria Clifton and Michael Fitzgerald for reading and commenting on the first draft of this chapter, and the Alexander Turnbull Library, the Museum of New Zealand Te Papa Tongarewa and the Parliament of the United Kingdom for kindly supplying Figs. 7.3, 7.5, 7.7 and 7.8.

References

The following abbreviations are used:

Te Papa = MU000002/067/0002; History: Cook's Relics deposited and presented; 1914–1975; paper, Museum of New Zealand Te Papa Tongarewa, Wellington
MS = manuscript
NC = newspaper clipping
TS = typescript copy

Alonzo Dawes & Son: Note to H. Stewart, dated March 2, 1926. MS in Te Papa.

Anonymous: Untitled note about the history of the telescope, n.d. MS in Te Papa.

Baker, T. F. T. & Elrington, C. R. (Eds.), (1982). *A History of the County of Middlesex. Volume 7: Acton, Chiswick, Ealing and Brentford, West Twyford, Willesden*. Oxford: Oxford University Press.

Beaglehole, J. C. (Ed.), (1963). *The Endeavour Journal of Joseph Banks 1768–1771* (2 Vols.). Sydney: Trustees of the Public Library of New South Wales, in association with Angus and Robertson (reprint; first published 1962).

Beaglehole, J. C. (Ed.), (1967). *The Journals of Captain James Cook on his Voyages of Discovery. III. The Voyage of the* Resolution *and* Discovery *1776–1780* (2 Vols.). Cambridge: Cambridge University Press.

Beaglehole, J. C. (Ed.), (1968). *The Journals of Captain James Cook on his Voyages of Discovery. I. The Voyage of the* Endeavour 1768–1771. Cambridge: Cambridge University Press (reprint; first published 1955).

Beaglehole, J. C. (Ed.), (1969). *The Journals of Captain James Cook on his Voyages of Discovery. II. The Voyage of the* Resolution *and* Discovery *1772–1775*. Cambridge: Cambridge University Press (reprint; first published 1961).

Beaglehole, J. C. (1974). *The Life of Captain James Cook*. London: The Hakluyt Society.

Beddie, M. K. (Ed.), (1970). *Bibliography of Captain James Cook R.N., F.R.S., Circumnavigator*. Sydney: Trustees of the Public Library of New South Wales.

Campbell, A. T. (1952). Memorandum to the Secretary, Department of External Affairs, Wellington, dated April 16. MS in Te Papa.

Capt. Cook Relic For Sale. Telescope Offer. *Daily Telegraph & Morning Post*, May 26, 1952. NC in Te Papa.

Capt. Cook's Telescope For New Zealand. *The Cornishman*, October 23, 1952. NC in Te Papa.

Captain Cook's telescope. Newspaper report dating to January 1952. NC in National Maritime Museum.

Captain Cook's Telescope, n.d. (Note about the history of the telescope). TS in Te Papa.

Captain Cook's Telescope for sale. *The Cornishman*, May 22, 1952. NC in Te Papa.

Clifton, G. (1995). *Directory of British Scientific Instrument Makers 1550–1851*. London: Zwemmer and the National Maritime Museum.

Cook, J., & Furneaux, T. (1777). *A Voyage Towards the South Pole, and Round the World …* London: Strahan and Cadell.

Cook's Telescope. Newspaper report dating to 1952. NC in Te Papa.

Cook's Telescope for New Zealand. *The St. Ives Times*, October 24, 1952. NC in Te Papa.

Dawson, W. R. (Ed.), (1958). *The Banks Letters. A Calendar of the Manuscript Correspondence of Sir Joseph Banks Preserved in the British Museum, the British Museum (Natural History) and Other Collections in Great Britain*. London: British Museum.

Ellis, F. (1932). Letter to an unspecified Portsmouth antique dealer, dated January 18. MS in Te Papa.

Falla, R. A. (1952a). Memorandum to the Secretary for External Affairs, Wellington, dated May 5. TS in Te Papa.

Falla, R. A. (1952b). Memorandum to the Secretary for External Affairs, Wellington, dated July 8. TS in Te Papa.

FHR. (1933). The "junk" shops of Portsmouth. Captain Cook's telescope—an invaluable "find". *The Evening News*, November 29. NC in Te Papa.

Garnsey, E. R. (1864–1935). *Sydney Morning Herald*, May 10, 1935 (p. 17).

Green, C., & Cook, J. (1771). Observations made, by appointure of the Royal Society, at King George's Island in the South Seas. *Philosophical Transactions of the Royal Society, 61*, 397–421.

Greenhill, B. (1972). Letter to Dr R. K. Dell, dated February 2. MS in Te Papa.

High Commissioner for New Zealand, London. Telegram to the Minister of External Affairs, Wellington, dated August 11, 1952. MS in Te Papa.

Historic Relic for New Zealand? *The St. Ives Times*, May 23, 1952. NC in Te Papa.

Hooper, B. (Ed.), 1975. *With Captain James Cook in the Antarctic and Pacific. The Private Journal of James Burney Second Lieutenant of the Adventure on Cook's Second Voyage 1772–1773*. Canberra: National Library of Australia.

Howse, D. (1979). The principal scientific instruments taken on Captain Cook's voyages of exploration, 1768–80. *Mariner's Mirror, 65*, 119–135.

John Citizen's Diary. *The Evening News and Southern Daily Mail*, May 26, 1952. NC in Te Papa.

Johnson, Miss, n.d. Untitled note about the history of the telescope. MS in Te Papa.

Kaeppler, A. (1972). The use of documents in identifying ethnographic specimens from the voyages of Captain Cook. *Journal of Pacific History, 7*, 195–200.

Kinch, E. (1938). Letter to the National Maritime Museum, dated October 23. TS in National Maritime Museum.

Kinch, E. R. (1952). Speech to His Excellency Mr Doidge High Commissioners [*sic*] for New Zealand made by Miss Edith Kinch at New Zealand House Strand on November 3rd. 11.45 a. m. TS in Te Papa.

King, H. C. (1979). *The History of the Telescope*. New York: Dover.

Lyte, C. (1980). *Sir Joseph Banks. 18th Century Explorer, Botanist and Entrepreneur*. Sydney: Reed.

Manwaring, G. E. (1931). *My Friend the Admiral: The Life, Letters, and Journals of Rear-Admiral James Burney*. London: George Routledge.

Marshall, J. B. (1977). Daniel Solander. In J. Cook (Ed.), *The Journal of H.M.S. Resolution* (pp. 49–60). Guildford: Genesis Publications.

New Zealand Government Offices, London: Memorandum to The Secretary, Department of External Affairs, dated November 26, 1952. MS in Te Papa.

Orchiston, W. (1998a). From the South Seas to the Sun: the astronomy of Cook's voyages. In M. Lincoln (Ed.), *Science and Exploration in the Pacific. European Voyages to the Southern Oceans in the Eighteenth Century* (pp. 55–72). Woodbridge: Boydell Press, in association with the National Maritime Museum.

Orchiston, W. (1998b). *Nautical Astronomy in New Zealand. The Voyages of James Cook*. Wellington, Carter Observatory.

Orchiston, W. (1999). Cook, Banks and the Gregorian telescope in The Museum of New Zealand Te Papa Tongarewa. *Journal of the Antique Telescope Society, 18*, 4–9 (+Front Cover).

Orchiston, W. (2005). James Cook's 1769 transit of Venus expedition to Tahiti. In D. W. Kurtz (Ed.), *Transits of Venus: New Views of the Solar System and the Galaxy* (pp. 51–66). Cambridge: Cambridge University Press.

Pitt Rivers Museum. Objects Catalogue. 1328.86.1 (objects.prm.ox.ac.uk/pages/PRMUID14147.html).

Relton, E. W. (1864). Untitled paper documenting the history of the telescope. MS in Te Papa.

Relton, H. (1918). Letter to Major S. Maxwell, F.R.S., London, dated March. TS in Te Papa.

Secretary of External Affairs: Memorandum to The Official Secretary, Office of the High Commissioner of New Zealand, London, dated May 9, 1952. TS in Te Papa.

The Official Secretary, New Zealand Government Office, London: Memorandum to The Secretary, Department of External Affairs, dated June 17, 1952. MS in Te Papa.
Vaughan, T., & Murray-Oliver, A. A. S. C. M. (1974). *Captain Cook, R.N. The Resolute Mariner. An International Record of Oceanic Discovery*. Portland: Oregon Historical Society.
Williams, O. (1912). *Life and Letters of John Rickman*. Boston: Houghton Mifflin.

Chapter 8
Stephen Carkeek, the Wellington Time Ball, and New Zealand's Oldest Surviving Observatory

Abstract Stephen Carkeek, as the Controller of Customs in Wellington, played a key role in the founding of New Zealand's first professional observatory, the Colonial Observatory on the Wellington waterfront. After his retirement in 1867 he moved to Featherstone in the Wairarapa, where he was able to combine farming with his passion for astronomy. Accordingly, he built a commodious wooden observatory with a hexagonal dome room and an adjacent transit room and office. This observatory still exists, and despite its dilapidated condition is currently the oldest surviving example of an astronomical observatory in New Zealand. As such, it is an important part of New Zealand's astronomical heritage, and experts need to be consulted to determine whether it is still possible to preserve and restore it.

8.1 Introduction

It was long after those initial intensive forays into nautical astronomy at Queen Charlotte Sound and Dusky Sound during Cook's Second and Third Voyages (1773–1777; see Orchiston 1998; Chaps. 5 and 6 in this book) that New Zealand began to witness the emergence of scientific astronomy. During the second half of the nineteenth century central Government made a commitment to astronomy by founding the Colonial Observatory in Wellington (Eiby 1977; Chap. 9 in this book), transits of Venus in 1874 and 1882 brought teams of foreign observers to New Zealand shores (see Orchiston 2004; Chaps. 14 and 15 in this book), Christchurch's Professor Alexander William Bickerton (1842–1929; Gilmore 1982) was singing the praises of his innovative 'partial impact theory', observatories owned by amateur astronomers began to spring up across the nation in Auckland, Thames, Palmerston North, Featherston, Lower Hutt, Wellington, Nelson, Christchurch and Dunedin, and the first fledgling astronomical societies were formed in Auckland and Christchurch (see Orchiston 1998).

W. Orchiston, *Exploring the History of New Zealand Astronomy*,
Astrophysics and Space Science Library 422,
DOI 10.1007/978-3-319-22566-1_8

Fig. 8.1 Stephen James Carkeek photographed in about 1865 by an unknown photographer (*Courtesy* Alexander Turnbull Library, Ref. 1/2-038563-F)

One of the few individuals involved in both amateur and professional astronomy at this time was Stephen James Carkeek (Fig. 8.1), who has been described as "… a very prominent figure in the early days of colonisation in New Zealand." (*The Cyclopedia of New Zealand* 1897). In about 1867 Carkeek established a private observatory on his farm just south of Featherston in the southern part of the North Island (see Fig. 8.2 for localities mentioned in this chapter), and this facility is currently the oldest surviving remains of an astronomical observatory known in New Zealand.

In mid-1987 the Wellington amateur astronomer Sydney Cretney (1919–2013) heard about the existence of this observatory, and he, Anthony W. (Tony) Dodson (d. 2004) and Garry Wilmshurst (1949–2011) paid their first visit to the site. Cretney then began researching Carkeek. The following year he wrote a short account of the project for the newsletter of the Royal Astronomical Society of New Zealand (Cretney 1988), and Dodson then prepared a paper for the 1990 annual conference of the Society. In 1994 I organized a field trip to the observatory for members of the Wellington Astronomical Society, and subsequently included a brief mention of Carkeek in a book on early New Zealand astronomical history (Orchiston 1998: 98). I then published a paper in the *Journal of the Antique Telescope Society* in 2001, reporting on the status of Carkeek's observatory and speculating on the types of astronomical observations that he may have carried out there (Orchiston 2001). This chapter largely is based upon that paper, but also includes more information about Carkeek and the establishment of the time ball observatory in Wellington.

Fig. 8.2 New Zealand localities mentioned in this chapter are shown in *red*

8.2 Stephen Carkeek: A Biographical Sketch

Stephen James Carkeek (1815–1878) was born in Swansea, South Wales (Scholefield 1940). Little is known of his early life except that he went to sea, and he spent from 1838 to 1841 in Australian and New Zealand waters. In 1841 he and his family settled in New Zealand and he joined the Customs Service in Russell. The following year he transferred to Nelson, as Collector of Customs, where he was able to cater to his growing scientific interests by joining and serving on the Committee of the Nelson Literary Institution (Lash 1992). In 1849 Carkeek decided to move again, and on 14 July the following glowing report appeared in *The Nelson Examiner and New Zealand Chronicle*:

> The Government brig, when she sailed last week, took from us our late and much-respected Collector of Customs, Harbour-Master, Postmaster, and Sub-Treasurer. All these offices, and several more, has Mr. Carkeek filled in this settlement since it was first founded, or since some of the departments were established; and a more efficient Government servant, or one held in higher esteem by the public, we are sure is nowhere to the found. As it sometimes will happen to meritorious men, Mr. Carkeek has been, we should think, the hardest-worked and worst-paid, of any officer of the Government in the colony, though filling several very responsible posts.

The post Carkeek moved to was Acting Collector of Customs in Wellington, and four years later the ‘Acting’ element was dropped from his title (ibid.). He subsequently became New Zealand’s first Inspector and Commissioner of Customs (*The Cyclopedia of New Zealand* 1897). Whilst living in Wellington he was appointed to a number of other Government and public posts, and led a busy life (Scholefield 1940). Among other posts, he commanded the Wellington Militia 2nd Company, with the rank of Captain, until resigning his commission in 1864 (Presentation to Captain Carkeek 1864). Thereafter he was still widely known and referred to as ‘Captain Carkeek’.

On 23 January 1866 the *Wellington Independent* reported that on the 20th one or more burglars broke into Carkeek’s house and stole a telescope and other articles from the hall. Then on 24 April 1866 the same newspaper announced that

> S. Carkeek, Esq., Inspector of H.M. Customs for the Colony of New Zealand has resigned his appointment, and after a service of many years retires into private life. By his kindness and urbanity, Mr. Carkeek has endeared himself to many in this city, and it will be with feelings of unfeigned regret that the announcement of his retirement is learnt. For some time past severe indisposition has prevented Mr. Carkeek attending to his duties, and even now, though in a great measure convalescent, he is far from well. (H.M. Customs 1866).

This was surely a premature retirement, for he was only 50 or 51 years of age.

Carkeek then moved to Featherston in the Wairarapa, to the east of Wellington, where he purchased a farm and constructed an observatory. He continued to farm and to pursue his astronomical interests until he died in November 1878 at the relatively young age of 63. An obituary, published in the *Wairarapa Standard*, says this of him:

> Captain Carkeek, on account of his unassuming, studious, and retiring habits, was not so well-known, as he would have otherwise been, to the present generation of New Zealand colonists; but by old settlers he was both well-known and highly esteemed, on account of his many sterling qualities. He was a first-class accountant, having in that capacity few equals … (Obituary: Captain Stephen Carkeek 1878).

Meanwhile, here is an excerpt from the long obituary that appeared in Wellington's *Evening Post* newspaper:

> We yesterday announced the death of one of our oldest New Zealand colonists, Mr. Stephen Carkeek … on Wednesday evening [27 November, 1878]. Mr Carkeek … [resigned his Customs position in Wellington and] retired to a small sheep-run which he purchased near Featherston, and lived there until his death. He became tired of sheep-farming, and recently made arrangements for the sale of his Wairarapa property. He then paid a visit to his daughter … at Tauranga, and had just returned home when he was seized with an attack of *inflammation of the lungs*, which proved fatal in a very few days … Mr. Carkeek was widely known as an astronomer, and he always took a deep interest in astronomical matters. It was at his instigation that electric clocks and the time-ball were introduced by the Provincial Government … In whatever he undertook Mr. Carkeek showed an energy that eminently fitted him for a colonist, and the esteem in which he was held by successive Ministries is shown in the offices he held. (The late Mr. Stephen Carkeek 1878).

Even though both of his sons were surveyors, neither seems to have inherited an interest in astronomy (Bagnall 1976). On 17 December 1878 the furniture in Carkeek's Tirohanga homestead, plus sheep, lambs, cows and horses were auctioned off, but no mention is made of telescopes, astronomy books, or other scientific instruments (*Evening Post* 1878), and we do not know what became of Carkeek's astronomical notes, papers and instruments after his death (S. Cretney, personal communication).

8.3 The Wellington Time Ball

If Stephen Carkeek had an interest in astronomy no sign of it emerged while he was residing in Nelson, as a survey of *The Nelson Examiner and New Zealand Chronicle* failed to produce a single reported astronomical observation, or explanation or fore-warning of a current or an up-coming astronomical object or event of public interest, or even a single public lecture on astronomy or a presentation at a meeting of the Nelson Literary Institution. There may have been a latent astronomical interest there, but one gains the impression that Carkeek was simply far too busy to indulge it.

The situation was a little different once he moved to Wellington, and he was the one who in 1862 suggested that a time ball and transit observatory should be set up at the Custom House that was about to be built on reclaimed land down by the

Wellington waterfront (Eiby 1977). He was only too aware that an accurate time-service was not only of value to the local population but was essential for the captains of vessels arriving in Port Nicholson so that they could correct their clocks and chronometers. Time balls were a popular solution to this dilemma during the nineteenth century (e.g. see Bartky and Dick 1981).

The Government responded positively to Carkeek's suggestion and ordered a time ball, transit telescope and two astronomical clocks from England. There was rapid progress, and on 10 December 1863 the *Wellington Independent* announced that

> An Astronomical Clock has just arrived, and has been placed in the Custom House, in charge of S. Carkeek, Esq, whose scientific attainments are well known. A Time Ball is to be erected at the Custom House, the pole to be 28 feet above the roof. The ball will be dropped each day at noon, by an electrical discharging machine, which is connected to the Astronomical Clock. This will be the first Time Ball erected in New Zealand,[1] and we are convinced that it will be invaluable to shipmasters … (*Wellington Independent* 1863).

Soon after, on 11 January 1864 the following brief account appeared in *The Nelson Examiner and New Zealand Chronicle*, penned by their Wellington correspondent:

> Many improvements have of late been made in the way of buildings, &c. [in Wellington], and one of the chief objects that catches the eye from the wharf, is a time ball erected on the top of the Custom House [see Fig. 8.3], the pole passing through the centre of the building. It is on the same principle as the Greenwich time ball, and will fall every day at 12 o'clock. The clocks in connection with the works will be under the management of S. Carkeek, Esq., Collector of Customs …

One month later, on 10 February 1864, the local Wellington newspaper, the *New Zealand Spectator and Cook's Strait Guardian*, included a long article titled "The Time Ball", which began:

> Saturday last [6 February 1864] ought to be marked as a red letter day in all future editions of New Zealand almanacks, for at 3 o'clock the Wellington Time Ball was dropped by electricity … To Mr. Carkeek belongs the chief credit for this work as most certainly it would never have been thought of had we not possessed one so thoroughly capable of directing the setting up of the somewhat complicated gear which carries and works the ball … The ball is of zinc, weighing about two hundredweight. It is carried by an iron rod, which rod at its lower end is attached to a piston. The rod and piston are fitted into an iron cylinder resting on a foundation built up carefully from the rock below the Custom-house. The cylinder is packed at the bottom with India-rubber, forming an elastic cushion to deaden the blow of the piston when the ball is dropped. The ball when dropped falls rapidly for half-way down the mast: it then stops, leaps up again, and once more drops gradually to the bottom. The piston as it falls forcible compresses the air inside the cylinder, as the air cannot escape except through the small spaces left purposely between the pieces of India-rubber at the bottom.

[1]Subsequently, time balls were installed in Dunedin in 1868 and at the port of Lyttelton, near Christchurch, in 1876 (see Bremner and Wood 1979; Kinns 2009). The attractive and architecturally-distinguished Lyttelton Timeball Station was seriously damaged during the 4 September 2010 and 22 February 2011 major earthquakes, and collapsed during an aftershock on 13 June 2011. Plans are afoot to rebuild it.

Fig. 8.3 An 1860s view of the Wellington waterfront showing the Customs House and time ball (*Courtesy* Alexander Turnbull Library, H.N. Murray collection, Ref. PAColl-0824-1)

> It is this resistance of the air which not only stops the falling ball but forces it back a little distance up the mast.
>
> When the ball is wound up the rack work the bottom of the piston is caught by a small piece of steel, which locks it securely. This trigger forms part of a most beautiful and delicate system of levers, which work one upon the other. The last of these, when the whole are set, needs but the slightest touch to release the trigger supporting the piston, and to drop the ball. Each day, when the ball is wound up to the top of the mast, these levers must be set by the assistant. (The time ball 1864a).

The newspaper article (ibid.) stresses that this was merely a test of the time ball equipment and the transit observatory had still to be constructed beside the Customs House.

A very brief report in the *New Zealand Spectator and Cook's Strait Guardian* on 12 March 1864, indicates that the new transit observatory was completed by 8 March 1864: "On Wednesday last the time ball was brought into operation for the first time … and will continue to give the correct time at noon every day except Sunday." In the course of the next four months, Carkeek modified the construction of the ball winding and release mechanism, in order to incorporate "… an ingenious piece of mechanism … invented and manufactured by Mr. W. Davidson, gunsmith, Manners-street." (The time ball 1864b).

Carkeek subsequently carried out the appropriate astronomical observations to determine its latitude and longitude (Thomson and Jackson 1871), but after this he seems to have had no further involvement with this new facility that he helped create during the one year that he remained in Wellington prior to retiring to the Wairarapa.

Instead it was amateur astronomer and minister of religion, Archdeacon Arthur Henry Stock (1823–1901; Eiby 1977; Orchiston 1985, 1986; Chap. 9 in this book) who was responsible for the day-to-day operation of the time ball, and on 31 October 1868, the *Wellington Independent* informed its readers that

> The Time Ball will drop on Monday at 12 o'clock New Zealand mean time. This time is 9 minutes 17 seconds slower than Wellington mean time, as the longitude of the time ball is 174° 49′ 15″ This longitude differs from that generally given, but it was calculated by Mr Carkeek from several observations taken by the transit instrument of the Time Ball observatory. (The Time Ball 1868).

Meanwhile, it would seem that Carkeek's love affair with clocks went beyond those in the transit observatory, for on 22 November 1856 we read that the clock belonging to St Peter's Church "… has been lately repaired, and is now in good going order … [thanks] to the efforts of Messrs. Carkeek and Mackay …" (*The Nelson Examiner and New Zealand Chronicle* 1856).

Probably Carkeek also was able to satisfy his interest in astronomy through his membership of the Wellington Athenaeum and Mechanics Institute (William Seed 1854), which included others—like the well-known actor, singer and entrepreneur, James Henry Marriott (1799/1800–1886; Downes 1990, and Sect. 4.2 in Chap. 20 in this book)—who shared his astronomical leanings. Occasionally the meetings of the Institute featured lectures relating to astronomy. Thus, on 26 August 1852 Marriott gave a well-received lecture, "On the Telescope" (Athenaeum … 1852).

8.4 Astronomy at Featherston

The location of Carkeek's farm, Torohanga, in relation to the nearby town of Featherston is shown in Fig. 8.4, and the site of the observatory is also pinpointed, between the road leading to Featherston and the gardens fronting the homestead. Carkeek built the observatory in about 1867, and it consisted of a 4.57-m diameter octagonal room with a dome and an adjacent 4.24 × 3.45m 'transit annex' (Fig. 8.5). The entire building was constructed of rough-sawn planks, possibly of totara, a native New Zealand timber. The dome was made of wood covered with canvas (S. Cretney, personal communication), and was probably conical in shape rather than hemispherical (as is the norm today). Meanwhile, the gable roof of the transit annex contained a narrow wooden shutter that extended part way down both northern and southern walls. Steps led up from the transit room to the floor of the dome room, which was about 84cm above ground level. Access to the observatory was via a door at the southeastern corner of the transit annex.

The nineteenth century saw rapid developments in the architecture of professional observatories (see Donnelly 1973), but the basic design that Carkeek adopted of a dome room and adjacent transit annex is one that was replicated repeatedly by

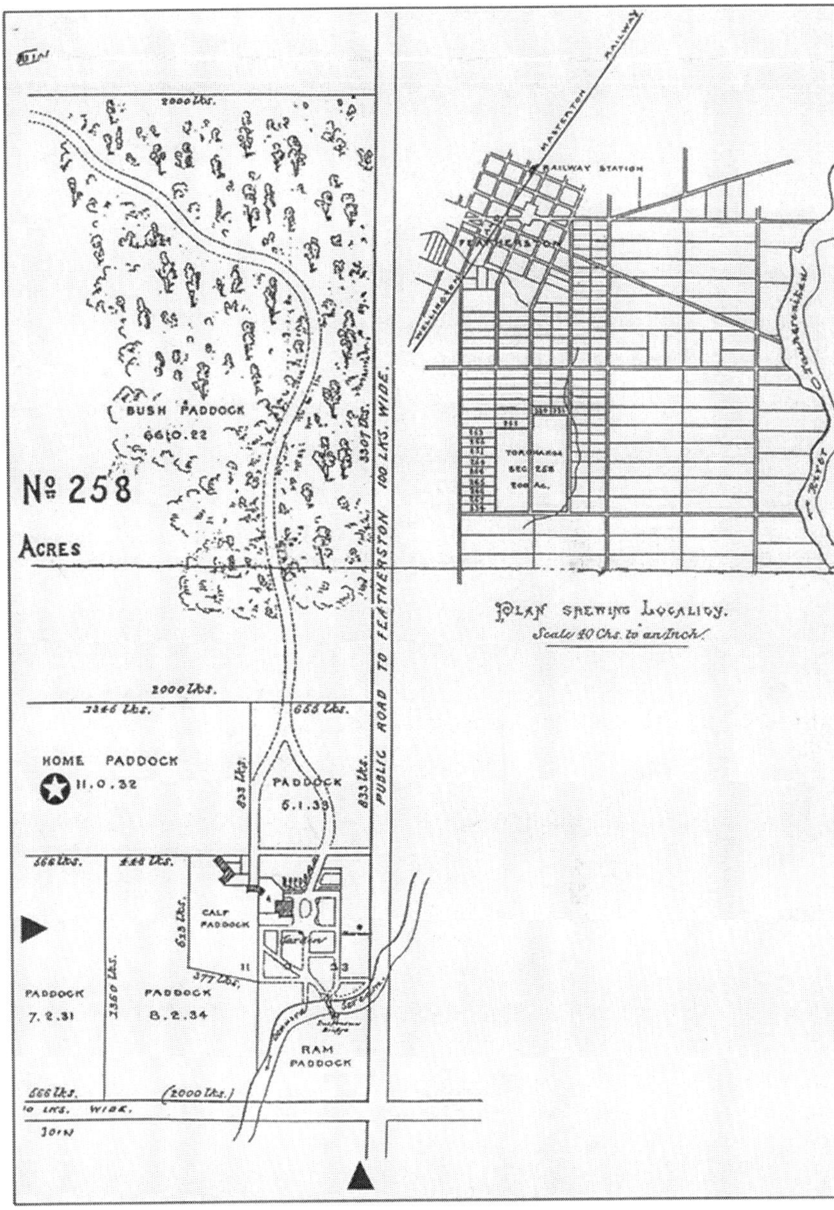

Fig. 8.4 Map showing the location of Carkeek's farm in relation to the town of Featherston. Meanwhile, the two triangles point to the site of his observatory (*Courtesy* Tony Dodson)

amateur astronomers throughout the world at this time and right through into the early twentieth century. Its simplest form was personified by the Romsey Observatory which the Reverend Edward Lyon Berthon (1813–1899) actively

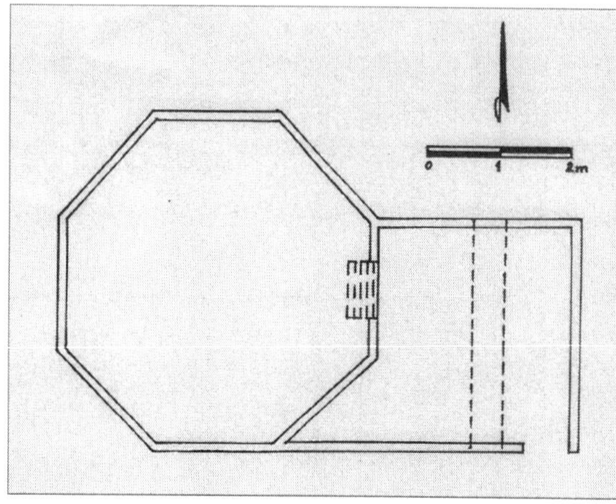

Fig. 8.5 The plan of Carkeek's observatory (*Plan* Wayne Orchiston)

promoted through the pages of the *English Mechanic and World of Science* (see Berthon 1871). However, its elements were available earlier in the form of Bedford Observatory designed by Admiral William Henry Smyth (1788–1865), which

> … directly or indirectly served as the model for nearly all the private observatories of moderate dimensions since erected in England, and it is equally certain that, whatever may be the changes which considerations of finance, or architecture, or geology, may render expedient in particular instances, no important alterations need be made in the main features of the Bedford Observatory, although upwards of half a century has elapsed since it was erected, and more than 40 years have passed away since it was pulled down. (Chambers 1890: 194).

Diagrams of and information about the Bedford Observatory appear in Smyth's book, *A Cycle of Celestial Objects* (see Smyth 1844(I): 326–327), a standard reference work which may have been in Carkeek's library.

When it was operational, Carkeek's observatory housed two quite different types of telescopes, but since none of Carkeek's records has survived and he did not publish any of his observations, we know almost nothing about these other than that there was a 10.2-cm (4-in) refractor in the 'dome room' (S. Cretney, personal communication). Clearly, a 4.57-m diameter observatory is far too large for so modest an instrument, which suggests that Carkeek originally may have entertained plans to install a much larger telescope. Although there is no documentation on the transit telescope that was in the transit annex, it most likely had an aperture somewhere in the 4.5–6.4cm range; Fig. 8.6 shows an instrument that would have been similar in general appearance to the one owned by Carkeek.

The instruments housed in the observatory were used for three different types of observations. Carkeek used the transit telescope to determine the geographical location of his observatory. Meridian transits of selected stars gave him local time and hence longitude, while zenith observations of stars of known declination provided the latitude. The transit telescope also was used to conduct regular meridian transit observations of 'clock stars', in order to rate his chronometer.

Fig. 8.6 A transit telescope similar to the one owned by Carkeek (after Chambers 1890: 123)

Carkeek may also have kept an astronomical clock, and regulated this by mean of these regular transit observations. Whether the resulting time-service was solely for his own use or was available also to the local population is not known.

In addition to observations made with the transit telescope, Carkeek used his refracting telescope for non-meridian work. Although his astronomical records have not survived, the fact that he went to the effort and expense of constructing and equipping an observatory and maintaining a local time-service would indicate that he did undertake some serious observational work, and this is reinforced by reference to his "… accumulated astronomical observations …" in the obituary that was published locally in the *Wairarapa Standard* newspaper (Obituary: Captain Stephen Carkeek 1878). Precisely what he did must remain conjectural, but Table 8.1 provides a clue by listing the sorts of programs that were undertaken by other southern hemisphere astronomers at this time, furnished with comparable equipment (see Clerke 1893).

Potentially the most important event awaiting Carkeek's attention was the 1874 transit of Venus (see Dick et al. 1998). Major Palmer, leader of the British expedition to New Zealand, was based at Burnham and he organized 'official stations' manned by his own staff or New Zealand amateur astronomers at Auckland, Thames, Wellington, Naseby and Dunedin (Airy 1881; The Transit of Venus 1874a; Tomorrow's Transit of Venus 1882). In addition, there was an American transit party at Queenstown, and other independent New Zealand observers were based at Auckland, Coromandel, Wellington, and Nelson, but there is no mention of

Table 8.1 Principal observational programs undertaken by leading amateur astronomers, 1867–1878

Transitory events
Eclipses of the Moon
Eclipses of the Sun
Lunar occultations of planets
Lunar occultations of stars
Phenomena of Jupiter's satellites
Transit of Mercury
Transit of Venus
Short-term monitoring projects
Comets (positions and appearance)
Meteors (fireballs, shower activity)
Minor planets (positions)
Planets (positions)
Long-term monitoring projects
Double stars (separation and position angle)
Planets (appearance)
Sun (surface features)
Variable stars (magnitude variations)
Search programs
New comets
New double stars
New variable stars

Carkeek in any of the published accounts (see *Lake Wakatip Mail* 1874; The Transit of Venus 1874a, b; Tomorrow's Transit of Venus 1882). As it was, inclement weather prevented most observers from seeing anything, the only notable exceptions being the Americans who succeeded in obtaining 237 photographs of the transit (see Orchiston et al. 2000; Chap. 15 in this book). If Carkeek had attempted to observe this rare event—which is very likely—then he too would have been plagued by cloudy skies.

One well-known astronomical target of special interest to southern observers during the nineteenth century was the enigmatic variable star Eta Argus (known now as Eta Carinae) and its associated nebula. During the 1860s and 1870s, observers continued to plot the star's declining magnitude (e.g. see Frew 2004; Orchiston 2000; Smith and Frew 2011), while the Tasmanian astronomer, Francis Abbott, caused considerable consternation by claiming that the nebula itself was undergoing rapid change (Orchiston 1992). Eta Argus was well placed for Carkeek to observe, and would have proved an interesting—if somewhat controversial—target.

The only major comet accessible to Carkeek during his 'Featherston years' was C/1874 H1 (Coggia), which was famous for the majesty of its tail and the 'performance' of its head when it was a northern object (see Guillemin 1877: 328–353). By the time it moved into the southern sky it was well past its best, but was still a conspicuous naked eye object during July and August 1874, and remained within telescopic range through into October. Given the media attention that it generated, it is hard to imagine that Carkeek did not observe this remarkable comet.

Finally, in addition to numerous lunar occultations of stars, several partial solar eclipses and a number of partial and total lunar eclipses, Carkeek may have been tempted by the 6 May 1878 transit of Mercury.

8.5 New Zealand's Oldest Surviving Observatory

8.5.1 Its Condition in 1994

When we visited Carkeek's observatory in 1994 I found it already to be in poor condition. For a time, the old building had been used to store farm implements, and in order to facilitate this the dome, the floor and two of the octagonal wall sections from the dome room had been removed. They were located nearby, under a large thorn bush.

The remaining dome room walls were still standing but leaned precariously, supported in part by yet another large thorn bush, and it would seem that this tree "… has provided both shelter, and something for the building to cling to, and thus ensure its lengthy survival." (Cretney 1988: 6). Upon inspecting the dome room, I noted that the interior and exterior walls of the dome room were lined with vertical planks. In reference to the plan in Fig. 8.5, even though the western and south-western walls had been removed, the horizontal wooden plate that originally ran across the tops of these two wall sections still remained in place, so the octagonal form of the dome was fully preserved (see Fig. 8.7). This wooden plate still retained the circular U-shaped iron channel upon which the dome once rotated, and immediately above this was the octagonal base-ring of the dome itself, complete with all of its large iron wheels which still sat in the U-shaped iron channel

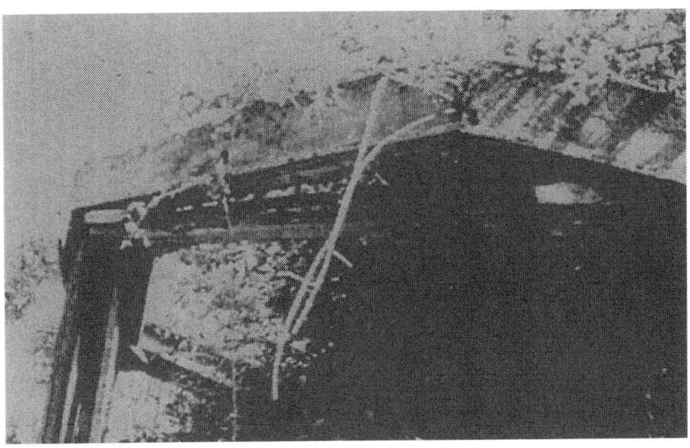

Fig. 8.7 A view of part of the dome room in 1994 showing the horizontal wall plate and directly above it the octagonal base-ring of the dome (*Photograph* Gary Wilmshurst)

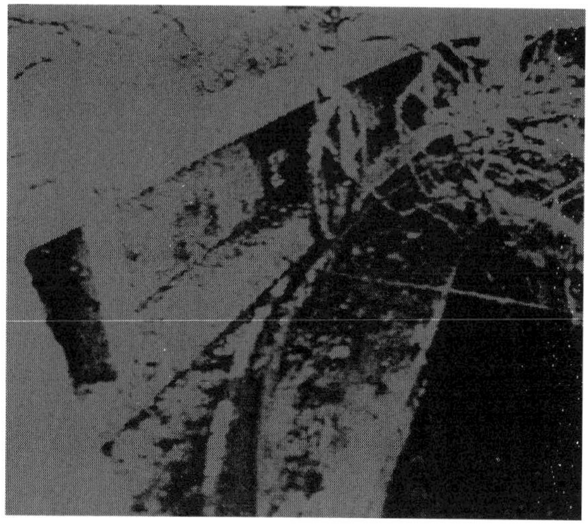

Fig. 8.8 A close up in 1994 of the horizontal wall plate, circular U-shaped channel, base-ring of the dome and one of the wheels (*Photograph* Wayne Orchiston)

(Fig. 8.8). So although the dome itself and the floor of the dome room were missing, the salient architectural features of the dome room were all there. Meanwhile, five steps leading up from the transit annex indicated where the floor of the dome room originally would have been.

The transit annex was in a much better state of preservation, and it was possible to recognize its salient features. Horizontal planks were used for the interior and exterior lining of the walls, and wooden battens for the gable roof. Offering sky access for the transit telescope was a 46-cm wide slit in the roof, which extended from the level of the roof 63.5cm down both north and south walls. Part of the reason for the continued survival of the transit annex was the mature walnut tree which was growing up through the middle of the room and out through the transit slit. Examination of the floor area failed to reveal any sign of the brickwork pier that originally would have supported the transit telescope.

8.5.2 Its Condition in 2014

In November 2014 Gordon Hudson drove to Featherston in order to determine the condition of Carkeek's observatory and assess the changes that had occurred since we both visited the site with other members of the Wellington Astronomical Society in 1994 (see Hudson and Orchiston 2015). When he first viewed the building from the east it seemed to be intact (see Fig. 8.9), but initial impressions can be deceiving and he quickly found the observatory to be in far worse condition than in 1994 (see Fig. 8.10). The key culprit was the thorn bush, which had died, thereby allowing some of the remaining walls of the dome room to collapse, along with the entire base-ring of the dome (Fig. 8.11). Only the eastern wall of the dome room and its

Fig. 8.9 View of the observatory from the east in 2014 (*Photograph* Gordon Hudson)

Fig. 8.10 View of the transit annex and dome room in 2014 (*Photograph* Gordon Hudson)

Fig. 8.11 View of the remains of the dome room from the west, in 2014 (*Photograph* Gordon Hudson)

Fig. 8.12 A collapsed section of the horizontal wall plate with the U-shaped channel, in 2014 (*Photograph* Gordon Hudson)

doorway leading to the transit annex was still intact. The horizontal plate on the top of the southeastern and southern wall sections and its U-shaped channel was still present, but the remains of this plate from other sections of the dome room lay on the ground, along with broken sections of the dome-ring and associated wheels (e.g. see Figs. 8.12 and 8.13).

Fig. 8.13 Collapsed sections of the octagonal base-ring of the dome, showing the wheels, in 2014 (*Photograph* Gordon Hudson)

By contrast, the transit annex was still in reasonable condition and the basic framework was structurally sound (see Fig. 8.14), although many of the horizontal planks used to line the walls were now detached and lying on the ground (Fig. 8.15). Thus, it too was in a more dilapidated condition than in 1994, as might be expected given the passage of the years.

8.6 Concluding Remarks

Carkeek's observatory was not the first astronomical observatory to be erected in New Zealand. As we saw in Chaps. 5 and 6, portable tent observatories (Orchiston 1998) were used at Dusky Sound and Queen Charlotte Sound on Cook's Second Voyage to the South Seas (1773–1774) and at the latter locality on the Third Voyage (1777). Meanwhile, in Wellington Archdeacon Arthur Stock (ibid.) and in Lower Hutt the province's Chief Surveyor Henry Jackson (1830–1906; Eiby 1978) are known to have erected observatories at about the same time that Carkeek built his at Featherston, and beyond Wellington there were other contemporary observatories, including one maintained by Henry Skey (1836–1914) at Dunedin (Campbell 2001).[2] But what makes Carkeek's observatory unique is that it has

[2]When I first published on Carkeek's observatory (Orchiston 2001) I also suggested that Thames' Henry Severn would have had an observatory to house his 27.9-cm (11-in) reflector, which was then the largest telescope in New Zealand. Information that has subsequently come to light shows that Severn in fact did not have an observatory and that his reflector was set up outdoors (see Chap. 21 in this book).

Fig. 8.14 A close-up showing structural details of the transit annex roof in 2014 (*Photograph* Gordon Hudson)

Fig. 8.15 A view of the transit annex showing the entrance doorway to the observatory and the mature walnut tree growing up through the transit slit in the roof of the building, in 2014 (*Photograph* Gordon Hudson)

survived, despite the passage of time and the ravages of the elements. And the elements surely have been unkind to it on occasions. For instance, in early March 1871 a severe storm hit the Wairarapa, and Featherston was one of those places that suffered considerable damage. A report in *The Nelson Examiner and New Zealand Chronicle* states:

> The stream which comes from the Rimutaka [Mountains], forms the eastern boundary of Mr. Carkeek's garden. This rose so high as to flood the house, in which there was at one time nearly three feet of water … Mr. Carkeek had just had his observatory refitted, and shelves with valuable books raised from the floor, the outsides of which became covered with mud, but fortunately the insides are very little damaged. The garden has a great deal of mud left upon it … (Effects of the late storm in Wairarapa 1871).

This raises the question of the ultimate fate of the observatory. I believe that Carkeek's astronomical work in Wellington and in the Wairarapa was of sufficient merit to make his Featherston observatory a significant part of New Zealand's astronomical heritage. However, its dilapidated condition is of great concern, and experts in architectural conservation need to assess it and determine whether it is still possible to preserve and faithfully restore it.

Acknowledgements I am grateful to the following for their assistance: Jenni Chrisstoffels (Alexander Turnbull Library, Wellington, NZ), late Syd Cretney (Wellington, NZ), the late Tony Dodson (Wellington, NZ), Dr Roger Kinns (Glasgow, Scotland), the late Gary Wilmshurst (Wellington, NZ), staff at the Alexander Turnbull Library (Wellington, NZ) and especially Gordon Hudson (Wellington, NZ) who kindly visited Featherstone in November 2014 and photographed the Carkeek Observatory. I also wish to thank Dr Richard Dodd (Martinborough, NZ), Roger Kinns and Gordon Hudson for reading and commenting on the first draft of this chapter, and the Alexander Turnbull Library, Gordon Hudson the late Tony Dodson and the late Gary Wilmshurst for kindly supplying Figs. 8.1, 8.3, 8.4, 8.7, 8.9, 8.10, 8.11, 8.12, 8.13, 8.14 and 8.15.

References

Airy, G. (1881). *Account of Observations of the Transit of Venus, 1874 December 8* … London: Her Majesty's Stationery Office.

Athenaeum and Mechanics' Institute. *New Zealand Spectator and Cook's Strait Guardian* (August 28, 1852 (p. 2).

Bagnall, A. G. (1976). *Wairarapa. An Historical Excursion*. Masterton: Hedley's Bookshop Ltd. for the Masterton Trust Lands Trust.

Bartky, I. R., & Dick, S. J. (1981). The first time balls. *Journal for the History of Astronomy, 12*, 155–164.

Berthon, E. L. (1871). On observatories. *English Mechanic and World of Science, 342*, 83–84.

Bremner, J., & Wood, F. L. W. (1979). *The Lyttelton Timeball Station: An Account of its History, Operation and Restoration*. Wellington: Historic Places Trust.

Campbell, R. N. (2001). Mr Henry Skey 1836–1914. *Southern Stars, 40*(2), 11–12.

Chambers, G. F. (1890). *A Handbook of Descriptive and Practical Astronomy* (4th ed.). Oxford: Clarendon Press.

Clerke, A. (1893). *A Popular History of Astronomy During the Nineteenth Century*. London: Adam and Charles Black.

Cretney, S. (1988). A journey 110 years into the past. *RASNZ Newsletter*, 5–6.

Dick, S., Orchiston, W., & Love, T. (1998). Simon Newcomb, William Harkness and the nineteenth century American transit of Venus expeditions. *Journal for the History of Astronomy, 29*, 221–255.

Donnelly, M. C. (1973). *A Short History of Observatories*. Eugene: University of Oregon.

Downes, P. (1990). Marriott James Henry 1799/1800?–1886. In W. H. Oliver (Ed.). *The Dictionary of New Zealand Biography. Volume One. 1769–1869* (p. 271). Wellington: Allen and Unwin New Zealand Ltd. and Department of Internal Affairs.

Effects of the late storm in Wairarapa. *The Nelson Examiner and Cook's Strait Chronicle*, March 23, 1871.

Eiby, G. (1977). The New Zealand Government time-service: an informal history. *Southern Stars, 27*, 15–34.

Eiby, G. (1978). History, astronomy and New Zealand. *Southern Stars, 27,* 118–124.

Evening Post, December 14, 1878 (p. 4).

Frew, D. J. (2004). The historical record of Eta Carinae. I: The visual light curve, 1595–2000. *Journal of Astronomical Data, 10*(6), 1–76.

Gilmore, G. (1982). Alexander William Bickerton: New Zealand's colourful astronomer. *Southern Stars, 29*, 87–108.

Guillemin, A. (1877). *The World of Comets*. London: Sampson, Low, Marston, Searle, & Rivington.

H. M. Customs. *Wellington Independent*, April 24, 1866 (p. 5).

Hudson, G., & Orchiston, W. (2015). A report on the current status of Stephen Carkeek's observatory at Featherston, the oldest surviving astronomical observatory in New Zealand. *Southern Stars* (in press).

Kinns, R. (2009). Time-keeping in the antipodes: a critical comparison of the Sydney and Lyttelton time balls. *Journal of Astronomical History and Heritage, 12*, 97–107.

Lake Wakatip Mail, November 24, 1874.

Lash, M. (1992). *Nelson Notables 1840–1940: A Dictionary of Regional Biography*. Nelson: Nelson Historical Society.

Mr. William Seed, late Honorary Secretary of the Wellington Athenaeum and Mechanics' Institute. *New Zealand Spectator and Cook's Strait Guardian*, November 18, 1854 (p. 2).

New Zealand Spectator and Cook's Strait Guardian, March 12, 1864 (p. 2).

Obituary: Captain Stephen Carkeek, J.P. *Wairarapa Standard*, November 30, 1878.

Orchiston, W. (1985). A survey of manuscript material in the Mitchell Library (Sydney) relating to the history of astronomy in New Zealand. *Southern Stars, 31*, 225–238.

Orchiston, W. (1986). Taking Stock of early astronomy books. *Southern Stars, 31*, 326–327.

Orchiston, W. (1992). The contribution of Francis Abbott to Tasmanian and Australian astronomy. *Vistas in Astronomy, 35*, 315–344.

Orchiston, W. (1998). *Nautical Astronomy in New Zealand: The Voyages of James Cook*. Wellington: Carter Observatory.

Orchiston, W. (2000). John Tebbutt of Windsor, New South Wales: a pioneer southern hemisphere variable star observer. *Irish Astronomical Journal, 27*, 47–54.

Orchiston, W. (2001). Stephen Carkeek and New Zealand's oldest surviving observatory. *Journal of the Antique Telescope Society, 20*, 19–24.

Orchiston, W. (2004). The nineteenth century transits of Venus: an Australian and New Zealand overview. *Journal of Astronomical Data, 10*, 219–308.

Orchiston, W., Love, T., & Dick, S. (2000). Refining the astronomical unit: Queenstown and the 1874 transit of Venus. *Journal of Astronomical History and Heritage, 3*, 23–44.

Presentation to Captain Carkeek. *Wellington Independent*, June 23, 1864 (p. 3).

Scholefield, G. H. (Ed.). (1940). *A Dictionary of New Zealand Biography*. Wellington: Department of Internal Affairs.

Smith, N., & Frew, D. J. (2011). A revised historical light curve of Eta Carinae and the timing of close periastron encounters. *Monthly Notices of the Royal Astronomical Society, 415*, 2009–2019.

Smyth, W. H. (1844). *A Cycle of Celestial Objects, for the Use of Naval, Military, and Private Astronomers* (Vol. 2). London: Parker.
The Cyclopedia of New Zealand. Volume 1 (1897). Wellington: The Cyclopedia Company Ltd.
The late Mr. Stephen Carkeek. *Evening Post*, November 30, 1878 (p. 3).
The Nelson Examiner and New Zealand Chronicle, July 14, 1849.
The Nelson Examiner and New Zealand Chronicle, November 22, 1856.
The time ball. *New Zealand Spectator and Cook's Strait Guardian*, February 10, 1864a (p. 2).
The time ball. *Wellington Independent*, July 2, 1864b (p. 3).
The time ball. *Wellington Independent*, October 31, 1868 (p. 5).
The transit of Venus. *Daily Southern Cross,* December 10, 1874a.
The transit of Venus. *Daily Southern Cross*, December 24, 1874b.
Thomson, J. T., & Jackson, H. (1871). *Report Relative to Measurement of Longitude Between Wellington and Otago by Means of Electric Telegraph.* Wellington: Government Printer.
Tomorrow's transit of Venus. *Evening Post,* December 6, 1882.
Wellington Independent, December 10, 1863 (p. 4).
Wellington Independent, January 23, 1866 (p. 5).

Chapter 9
The Historic Astronomical Observatories in the Wellington Botanic Garden: A Brief Introduction

Abstract Since 1869 astronomical, time-keeping, seismological, meteorological, geomagnetic and tidal studies have been carried out by staff and volunteers at five different observatories established in the Wellington Botanic Garden and the site of the Bolton Street Memorial Park. Collectively, these have played a key role in the history of New Zealand astronomy, and three of these observatories have survived to the present day.

9.1 Introduction

The history of astronomy in Aotearoa/New Zealand commenced with the arrival of the first Polynesian settlers about 800 years ago (Higham et al. 1999; Holdaway and Jacomb 2000), and over the centuries a distinctly Maori astronomical knowledge base evolved (see Best 1922; Orchiston 2000; and Chap. 2 in this book).

As we saw in Chaps. 4–6, in 1769 and during the 1770s European explorers introduced an entirely different type of astronomy to New Zealand, nautical astronomy, which was primarily concerned with establishing latitude and longitude. Then, with increasing European settlement during the nineteenth century, accurate time-keeping became a primary concern.

During the second half of the nineteenth century government astronomical observatories were established in British colonies throughout the world, but they differed from most present-day observatories in that they also took responsibility for meteorology, geomagnetism, seismology, tidal studies and trigonometrical surveys, not to mention providing a time service. Astronomy was only a part of the repertoire, even though these institutions were thought of by the public as *astronomical* observatories.

The Wellington Botanic Garden is a priceless public space located for the most part on the hills overlooking central Wellington, the New Zealand capital (see Shepherd and Cook 1988) and since 1869 it has been home to five different observatories. These are shown in Fig. 9.1, although only three (the Carter,

W. Orchiston, *Exploring the History of New Zealand Astronomy*,
Astrophysics and Space Science Library 422,
DOI 10.1007/978-3-319-22566-1_9

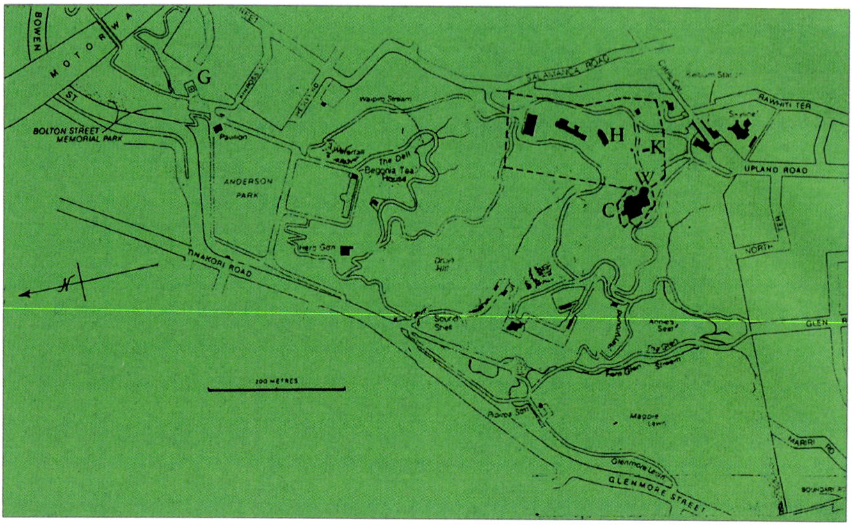

Fig. 9.1 Map of the Wellington Botanic Garden showing the locations of the Colonial (G), Hector (H), Thomas King (K), Wellington City (W) and Carter (C) Observatories and the two observatory reserves (after Orchiston 1995b: 2)

Dominion and Thomas King Observatories) currently survive. However, we will summarize the histories of all five observatories in this chapter, and examine some of the key individuals who were associated with them. As such, this chapter synthesizes and builds on material that I presented previously in the following publications and reports: Bentley et al. (1998), Orchiston (1995b, 1996), and Orchiston and Dodd (1995). New Zealand localities mentioned in this chapter are shown in Fig. 9.2.

9.2 The Colonial Observatory

Wellington's and indeed New Zealand's first professional observatory was solely involved in time-keeping, and was erected on the waterfront beside the Custom House (Eiby 1977), after the Government approved the purchase in England of a transit telescope and two astronomical clocks in 1862.

Known as the Provincial Observatory, it would hardly be recognised as an observatory by the average citizen for it comprised no more than a small wooden hut that housed the transit telescope and the two clocks. Above the telescope was a N-S shutter in the roof that could be removed to reveal a strip of sky. The transit telescope was used to observe selected stars (termed 'clock stars') as they crossed the meridian on successive nights. Since a full day (23h 56m 4s) had elapsed, the two astronomical clocks could be adjusted accordingly and a local time service maintained.

Fig. 9.2 New Zealand localities mentioned in this chapter are shown in *red*

One of the ways in which observatories around the world brought their time-services to their local populations was by means of a 'time ball' (see Bartky and Dick 1981). This normally comprised a large metal sphere which was attached to a mast that was prominently located where it could be seen by citizens of the town or city and from ships moored in the adjacent harbour. The usual procedure was to raise the ball to the top of the mast five minutes before noon each day, and drop it to the bottom of the mast precisely at midday (although some cities used 1 pm instead as their designated 'drop-time'). In Wellington, the time ball was located on the roof of the Customs House down on the waterfront, as shown in Fig. 8.3. It was made of zinc and weighed ~100 kg (The time ball 1864).

As we saw in the previous chapter, Wellington's time ball resulted from the initiative of Stephen Carkeek (Fig. 8.1), the Collector of Customs, and came into operation in 1864. What of those Wellington citizens who resided or worked at locations where the Customs House and time ball were not visible? A long article about the new Wellington time ball[1] published in the *New Zealand Spectator and Cook's Strait Guardian* newspaper on 10 February 1864 provides the answer:

> We hope that the Provincial Council will supplement the work of the ball. By voting for powder, &c., for a cannon. Which, when fired, will carry the tidings, *it is* 12 *o'clock*, to the Hutt [Valley] and to those parts of the town where the fall of the Time Ball is not visible. (ibid.).

The Wellington Botanic Garden began its long association with astronomy in 1869 when the Colonial Observatory was transferred from the waterfront to a small hilltop behind the Bolton Street Cemetery (see location G, on the far left of Fig. 9.1). A larger building was erected, and meteorological instruments also were installed. A photograph of the Observatory taken in 1906 is reproduced here in Fig. 9.3.

The nominal Director of the Colonial Observatory was Dr (later Sir) James Hector (1834–1907; Fig. 14.28; Dell 1990; Scholefield 1940(1): 374–376) who was born in Edinburgh, trained as a doctor, but changed careers and came to Dunedin, New Zealand in 1862 to conduct a geological survey of the province of Otago. Three years later he was appointed founding Director of the Geological Survey of New Zealand and moved to Wellington and supervised the construction of the Colonial Museum, where the Survey would be based. During his period as Director of the Colonial Observatory he was kept busy simultaneously serving as Director of the Colonial Museum, the Colonial Laboratory, the Wellington Botanic Garden and the Geological Survey; running the New Zealand Institute; and for a time serving as Chancellor of the University of New Zealand!

At the Observatory, fortunately Hector could rely on the assistance of Archdeacon Arthur Stock (Fig. 9.4), a keen amateur astronomer who was born in London in 1810, graduated from Cambridge University in 1845, joined the Church

---

[1]This long newspaper article (The time ball 1864) also claims that the Wellington time ball was the first one in the Southern Hemisphere, but this is not correct. By 1864, time balls had already been operational for some years at both Williamtown and Sydney Observatories in Australia, for example (see Andropoulos 2014; Orchiston 1988; Wood 1958).

Fig. 9.3 The Colonial Observatory, which was located beside the Bolton Street Cemetery from 1869 to 1906 (*Courtesy* Alexander Turnbull Library, Canterbury Times 27 June 1906, Ref. 1/2-044047-F)

Fig. 9.4 Arthur Stock, the Astronomical Observer at the Colonial Observatory (Orchiston collection)

Missionary Society, and in 1853 sailed for New Zealand. After landing in Auckland he moved to Wanganui, before being appointed Vicar of St. Peter's in Wellington in 1856 (The Venerable Archdeacon Stock 1897). Stock carried out most of the observations at the Colonial Observatory, but he did not wish to be constrained by the demands of meridian astronomy (which was all that time-keeping demanded) so he maintained a private observatory, with at first an 8-cm and from 1871 a 10.4-cm refractor. Over the years, these telescopes were used to observe the Moon and planets, lunar occultations of planets, transits of Mercury and Venus, phenomena of Jupiter's satellites, comets, southern double stars and the enigmatic Eta Carina (Orchiston 1985). Stock also independently discovered the Great Comet of 1881 just one night after the New South Welshman, John Tebbutt (Orchiston 1999).

Apart from his official and amateur observational activities, Stock was committed to popularising astronomy (see Orchiston 1986) and he published two small books on aspects of astronomy (Stock 1874a, b). One other important contribution he made was to investigate the longitude of New Zealand (Eiby 1978: 120). Nor were all Stock's astronomical thoughts directed to observation or popularisation, for in 1869 he wrote Tebbutt an interesting letter:

> Would it not be possible to see the Sun flames by inserting at the focus of the Object Glass a spot which would just cover the Sun. The Moon is such a spot outside our Atmosphere. If the place of our Atmosphere was diminished by the use of a dark glass I think then curious objects could be caught. The experiment might be worth the trial at all events. (Stock 1869).

What Stock was doing in this letter was proposing the concept of the coronograph, fully sixty years before Bernard Ferdinand Lyot (1897–1952; Danjon 1957) invented this instrument, and eleven years before William (later Sir William) Huggins (1824–1910; Becker 2011) experimented with a similar idea (see King 1979: 376).

Stock was a remarkable man: a minister of religion who preached the gospel by day and surveyed God's celestial wonders by night! Given his official role at the Colonial Observatory, I have suggested elsewhere that he "… should be seen as New Zealand's first professional astronomer—even if only in a part-time capacity." (Orchiston 1998: 99). Ill health finally forced him to retire from the Observatory in 1887, but he would live for another fourteen years. Back in 1985 I made the following assessment:

> It is apparent that Stock played a key role in New Zealand astronomy during the second half of the nineteenth century, and a detailed study of his contribution is more than overdue. (Orchiston 1985: 232).

As yet, no-one has responded to the challenge, so this is still true today.

In 1887 Stock's place at the Observatory was taken by another prominent Wellington amateur astronomer, Thomas King. Thomas King (1858–1916) was born in Glasgow, Scotland, and emigrated to New Zealand with his parents while still an infant. He was educated in Auckland and then in Wellington. Once settled in Wellington he began attending St Paul's church where Stock officiated, and developed a keen interest in astronomy. Like Stock he set up a private observatory, which housed a 14-cm (5.5-in) refractor that McKerrow (1882) used to observe the 1882 transit of Venus. In my view (Orchiston 1998: 110) King's

> … most important contribution as a part-time professional astronomer was to investigate the longitude of the Colonial Observatory, which resulted in a lengthy paper that was published in the *Transactions of the New Zealand Institute* early in the present [i.e. twentieth] century. (King 1902).

Fig. 9.5 The Seddon grave and monument at the site of the Colonial Observatory (https://en.wikipedia.org)

Sir James Hector retired in 1903 and the formal Directorship of the Colonial Observatory passed to the Director of the Museum, ethnologist and biologist Augustus Hamilton (1853–1913). The Colonial Observatory finally met an untimely end soon after, in 1906, when the Premier of New Zealand, Richard John Seddon (1845–1906), died and the site was chosen for his grave and monument. Hamilton was instructed to remove the Observatory, but this only happened on the very day of the funeral! An imposing monument (Fig. 9.5) now marks the spot.

9.3 The Dominion Observatory

With a continuing need for a national time-service the Government funded the erection of the Hector Observatory, which was completed in 1907 in Edwardian Baroque style (see Fig. 18.5). This small red brick building, sited in the upper reaches of the Botanic Garden ('H' in Fig. 9.1), initially comprised a two-storey octagonal clock room and office with adjacent transit room for the transit telescope (Fig. 9.6). In 1925 this facility was retitled the 'Dominion Observatory', and the following year an office wing was added. This attractive red brick building (Fig. 9.7) is the oldest of the surviving observatories in the Wellington Botanic Garden.

Fig. 9.6 The Hector Observatory transit telescope (Orchiston collection)

Fig. 9.7 The historic Dominion Observatory in the Wellington Botanic Garden (https://en.wikipedia.org)

The Astronomical Observer from the Colonial Observatory, Thomas King (Fig. 9.8), transferred to the Hector Observatory when this was built, but he resigned in 1911 when the Government refused to increase his annual salary from

Fig. 9.8 Thomas King (Orchiston collection)

£50 to £150 and Dr Charles Edward Adams (1870–1945; Fig. 9.9) became New Zealand's first 'Government Astronomer' when he was appointed Director of the Hector Observatory. Tides held a special fascination for Adams, who carried out a series of studies for his doctoral degree. He also was responsible for seismology (earthquake research and monitoring), and in 1920 his formal title changed to "Government Astronomer and Seismologist". Another area Adams avidly pursued

Fig. 9.9 Dr C.E. Adams, New Zealand's first 'Government Astronomer' (Orchiston collection)

was his other primary scientific interest, astronomy. At first this was not easy as the Observatory had no conventional telescopes and the small transit telescope could not be used for the observations he wished to make, but as other observatories were established nearby he was able to make use of the instruments housed in them. In this way, Adams could contribute to astronomical science in New Zealand and build a national reputation.

When the Hector Observatory was constructed in 1907 its meteorological responsibilities were handed to another Government agency, and meteorology only returned to the Wellington Botanic Garden in 1925. Three years later the Meteorological Service of New Zealand established a fenced-off observing station adjacent to the Hector Observatory (see Fig. 9.10). In the south-western corner of this triangular enclosure is a small wooden octagonal 'tower' that probably dates from 1912 (Bentley et al. 1998: 12).

By the time C.E. Adams retired as 'Government Astronomer and Seismologist' in 1936, radio time signals had largely made local time-keeping unnecessary, and Adams' departure also marked the end of the Dominion Observatory's involvement in astronomy. Soon after the opening of the nearby Carter Observatory in 1941 it was decided that all astronomical work would become the responsibility of this new facility and the Dominion Observatory would become the Seismological Observatory and restrict its charter to seismological and geophysical research.

Fig. 9.10 A view of the meteorological precinct and the nearby Dominion Observatory (*Photograph* Wayne Orchiston)

9.4 The Thomas King Observatory

This Observatory is situated on the hilltop near the old Hector Observatory ('K' in Fig. 9.1), and was erected in 1912 by the newly-formed Astronomical Section of the Wellington Philosophical Society (Seymour 1995). From the start it was known as the King Edward VII Memorial Observatory, to commemorate the King who had died in 1910. Initially a 12.7-cm (5-in) Cooke refracting telescope was installed under the dome, but this was replaced in 1918 by the 14-cm (5.5-in) Grubb refractor which was once owned and used by Thomas King, the observer at the Colonial Observatory (after Stock), and subsequently at the Hector (cum Dominion) Observatory.

During the 1920s and 1930s Dr C.E. Adams and a number of local amateur astronomers carried out a wide range of observations at the King Edward VII Memorial Observatory, and members of the public were able to attend weekly viewing nights.

From around the time of WW II the Observatory was unofficially renamed the 'Thomas King Observatory' (TKO) in honour of King's contributions to New Zealand astronomy, and it was at about this time that Ken Adams (1920–1999; Fig. 12.8) from the Carter Observatory (and no relation to Dr C.E. Adams) carried out pioneering experiments in photoelectric photometry using the Grubb telescope at the TKO (Adams 1982). His photometer is shown in Fig. 9.11, attached to the Grubb telescope.

After WWII little astronomical work was done at the TKO as the astronomers at the Carter Observatory now had their own much larger telescope. However, the daily observation of sunspots initiated by Dr Adams in 1918 was continued by Carter Observatory and Dominion Observatory staff. This program only ceased in the 1960s.

The TKO received a new lease of life in the 1970s following the founding of the Wellington Astronomical Society. The Society subsequently became the Astronomy Section of the Wellington Branch of the Royal Society of New Zealand, and assumed responsibility for the Observatory. Repairs were carried out and members used the Grubb telescope for a range of observational programs.

By the 1980s the observational interests of members had transferred elsewhere, and a combination of age, the weather and vandalism began to take its toll. In 1992 the telescope was removed following a serious act of vandalism and the future of the building was in question. However, its survival was assured in 1995 when funding for its conservation was provided by the Wellington City Council. The refurbished Observatory (Fig. 9.12) and renovated Grubb telescope (see Hudson 2003) are now included in the Carter Observatory's education and public astronomy programs.

Fig. 9.11 The photometer used by Ken Adams at the Thomas King Observatory (Orchiston collection)

9.5 The Wellington City Observatory (Aka the 'Green Tin Shed')

The fourth astronomical observatory to be erected in the Botanic Garden was opened in 1924 by the Wellington City Corporation (forerunner to the City Council). This was located on the current site of the Carter Observatory car park ('W' in Fig. 9.1), and although officially titled the 'Wellington City Observatory' was known affectionately to astronomers and public alike as the 'Tin Shed' or 'Green Tin Shed'. The reason was obvious, for the rather rustic 5.5m × 11m (18 × 36 foot) building was made of galvanised iron, painted green, and had a run-off roof (see Fig. 9.13).

Fig. 9.12 The refurbished and beautifully-restored Thomas King Observatory in 2012 (*Photograph* Wayne Orchiston)

Fig. 9.13 A photograph taken during excavation of the Carter Observatory foundations, showing the 'Tin Shed' in the background and on the hill behind it the Thomas King Observatory (Orchiston collection)

There were two square rooms of equal size, and in one of these was a 23-cm (9-in) refracting telescope which the Corporation had acquired in 1923 from the Marist Seminary at Meeanee near Napier (Orchiston 2002). The purchase price was apparently £500, a considerable sum of money in those days, but the Corporation considered it to be "… the finest instrument in New Zealand … [which] was well adapted for educational purposes, and was valuable for research work." ("Epoch marking." … 1924).

As the nation's second-largest refractor, the Wellington City Observatory telescope (Fig. 13.12) quickly attracted Adams and many of the amateurs who up until then had happily made extensive use of the King Edward VII Memorial Observatory, and throughout the 1930s the 'Tin Shed' was an observational stalwart of New Zealand astronomy (for details of this research see Chap. 13 in this book). Public nights also were transferred from the King Edward VII Memorial Observatory to the Wellington City Observatory.

9.6 The Carter Observatory

The Carter Observatory ('C' in Fig. 9.1) is the most recent of the five observatories established in the Wellington Botanic Garden, and is named after the prominent Wellington-Wairarapa pioneer, Charles Rooking Carter (1822–1896; Fig. 9.14; see Carter 1866; Orchiston 1995a). When he died, Carter left the residue of his estate to the New Zealand Institute (forerunner of the Royal Society of New Zealand)

> … to form the nucleus of a fund for the erection in or near Wellington aforesaid, and the endowment of a Professor and staff, of an Astronomic Observatory fitted with telescope and other suitable instruments for the public use and benefit of the Colony …

Fig. 9.14 C.R. Carter (Orchiston collection)

At different times unsuccessful attempts were made to secure the Carter Bequest for the Hector, King Edward VII, and Wellington City Observatories, but it was not until 1938 that the money was made available to astronomy with the passing by Government of the Carter Observatory Act. This was adopted as a 1940 'Centennial Project' commemorating the signing of the Treaty of Waitangi in 1840, but WWII intervened and the new Observatory only was opened in December 1941 with auroral expert Murray Geddes (1909–1944; Dickie 2010) as its founding Director. It occupied land in the Botanic Garden made available by the Wellington City Council, which also had agreed to hand over the historic 23-cm Cooke telescope in their Wellington City Observatory. The latter Observatory, which was quickly demolished, was located on what became the car park in front of the new Carter Observatory building. The original Carter Observatory was a modest building containing two domes and associated offices, with the Cooke telescope (Fig. 9.15) taking pride of place in the larger dome.

Over the years the original building has been expanded and renovated on various occasions, to accommodate a two storey library and office wing and a visitor centre including a planetarium (see Fig. 9.16), and a 41-cm (16-in) Boller and Chivens

Fig. 9.15 The 23-cm Cooke refractor in its new dome at the Carter Observatory (Orchiston collection)

Fig. 9.16 The Carter Observatory in 2012, as viewed from the path leading down the hill from the Thomas King Observatory. The large dome on the left houses the planetarium, the 41-cm reflector is in the central dome, while the right hand dome has been home to the historic 23-cm Cooke telescope since the observatory opened in 1941 (*Photograph* Wayne Orchiston)

reflector (Fig. 9.17), which now occupies the smallest of the three domes shown in Fig. 9.16.[2]

From the start, the Carter Observatory was involved in both research and education, with the research emphasis on solar astronomy, but with the passage of time, the appointment of professionally-trained staff engaged in astrophysical research and Wellington's deteriorating skies meant that the research emphasis changed and the Observatory's own telescopes no longer provided the bulk of the data used in research projects. Further information on Carter Observatory research projects is included here in Chaps. 13 and 23, and in successive *Annual Reports* of the Observatory.

Meanwhile, public astronomy and astronomy education have remained important aspects of the Observatory's operations ever since it opened in 1941, and are

[2]Even the 23-cm telescope has been "... expanded and renovated ..." Following the gross deterioration of Taylor's photovisual objective, Nankivell (2002) felt obliged to fabricate a substitute achromatic objective in 2001, just prior to his untimely death. This was 24.8-cm (9.75-in) in diameter, and thus the Carter Observatory was able to assume the mantle of housing New Zealand's largest operational refractor by marginally surpassing the equally-historic Cooke refractor at the Ward Observatory in Wanganui (for further details see Chaps. 11 and 13 in this book).

Fig. 9.17 The 41-cm Boller and Chivens telescope and the second Director of the Carter Observatory, Ivan Thomsen (Orchiston collection)

the mainstay of its operations today now that the research function has ceased and the Carter has become a branch museum of the Wellington City Council and a prime tourist attraction.

9.7 Concluding Remarks

For nearly 150 years the Wellington Botanic Garden has been home to a succession of observatories, which not only contributed to astronomical, geomagnetic, meteorological, seismological and tidal research but also provided, first Wellington, and then all New Zealand, with a time service. This wide-ranging charter was a feature of government observatories world-wide during the nineteenth century, but as happened elsewhere, these disparate functions were gradually assigned to different observatories and government institutes in the Wellington Botanic Garden and elsewhere during the twentieth century.

Of the five 'astronomical' observatories established in Wellington Botanic Garden in fact only three—the Edward VII Memorial Observatory (now known as the Thomas King Observatory), the Wellington City Observatory (also referred to as the 'Tin Shed') and thc Carter Observatory—were involved in astronomical research. However, the research was modest by world standards, and with the development of undergraduate and post-graduate astronomy at the University of Canterbury and the emergence of Mt John Observatory the focus of professional astronomy shifted from Wellington to Christchurch.

In addition to their research activities, the Thomas King, Wellington City and Carter Observatories all offered public viewing nights in a bid to popularise astronomy and increase the astronomical literacy of the general population, and for many years the Carter Observatory also has run a vibrant education program for school children—which continues to this day. While it is tempting for astronomers to differentiate between research and education and to place more emphasis on the former, the International Astronomical Union recognises that each is a valid and valued function of an observatory. In this light, and with the support of the

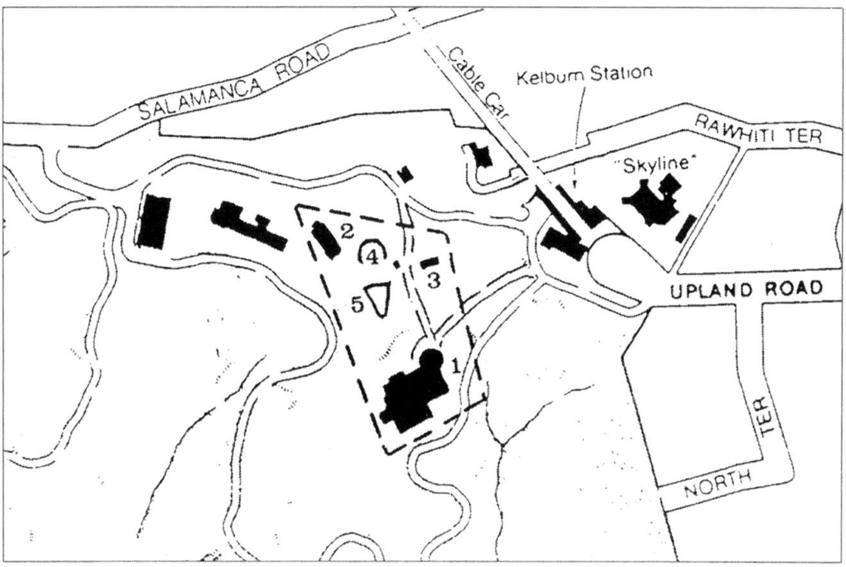

Fig. 9.18 A map of the remaining scientific and military attractions in a localised region of the Wellington Botanic Garden, close to the upper terminus of the Wellington Cable Car and the Cable Car Museum. Key: 1 Carter Observatory; 2 Dominion Observatory, 3 Thomas King Observatory, 4 gun emplacement and underground tunnels, and 5 meteorological precinct (*Map* Wayne Orchiston)

Wellington City Council, hopefully the future of the Carter and Thomas King Observatories is assured.

The Carter Observatory now boasts the largest operational refractor in New Zealand and one that is of international importance, not only because of its antiquity and association with what at the time was one of the largest photovisual objectives ever made, but also because of its connection with Edward Crossley, a famous name in the history of British astronomy. When coupled with the obvious visual appeal of the Dominion Observatory building, and the interesting histories behind the Carter, Dominion and Thomas King Observatories and the 1890s gun emplacement and tunnels near the Dominion Observatory, Wellington has a remarkable localized scientific and military precinct that is easily accessed from the upper cable car terminus (see Fig. 9.18). Back in 1998 I wrote that this exciting heritage precinct "… was intimately associated with early developments in a number of different sciences and with an important aspect of Wellington's military history, and provides a conservation and education opportunity unique to New Zealand." (Bentley et al. 1998: 6–7), while later in this same report I refer to this as "… a unique educational and recreational facility for residents of and visitors to the Wellington region." (Bentley et al. 1998: 21).

Acknowledgements I am grateful to the following for their assistance: the late Ken Adams (Auckland, NZ), Dr Grant Christie (Auckland, NZ), Jenni Chrisstoffels (Alexander Turnbull Library, Wellington, NZ), the late George Eiby (Wellington, NZ) and the late Jackie St George (Sydney). I also wish to thank Dr Richard Dodd (Martinborough, NZ) and Gordon Hudson (Wellington, NZ) for reading and commenting on the first draft of this chapter, and the Alexander Turnbull Library for kindly supplying Fig. 9.3.

References

Adams, K. D. (1982). Some aspects of the development of instrumentation for astronomical photometry [in New Zealand]. *Southern Stars, 30*, 3–11.

Andropoulos, J. (2014). *The Astronomical Publications of Melbourne Observatory*. Unpublished Ph.D. thesis, James Cook University, Townsville, Australia.

Bartky, I. R., & Dick, S. J. (1981). The first time balls. *Journal for the History of Astronomy, 12*, 155–164.

Becker, B. J. (2011). *Unravelling Starlight: William and Margaret Huggins and the Rise of the New Astronomy*. Cambridge: Cambridge University Press.

Bentley, R., Fraser, W., Orchiston, W, Siers, J., & Yardley, I. (1998). *The Observatory Reserves Heritage Park: An Interim Report. A Submission Relating to the 1998 Combined Management Plan Bolton Street Memorial Park Wellington Botanic Garden Anderson Park* (p. 23). Wellington: Carter Observatory.

Best, E. (1922). *Astronomical Knowledge of the Maori*. Wellington: Government Printer: Wellington (Dominion Museum Monograph No. 3).

Carter, C. R. (1866). *Life and Recollections of a New Zealand Colonist (Written by Himself)* (Vol. II). London, printed for the author.

Danjon, A. (1957). Funérailles de Bernard Lyot. *Académie des Sciences: Notices et Discours, 3*, 320–327.

Dell, R. K. (1990). Hector, James 1834–1907. In W. H. Oliver (Ed.), *The Dictionary of New Zealand Biography. Volume One. 1769–1869* (p. 183f). Wellington: Allen & Unwin New Zealand Ltd., and Department of Internal Affairs.

Dickie, N. (2010). Murray Geddes. *Southern Stars, 49*(2), 3–6.

Eiby, G. (1977). The New Zealand Government time-service. an informal history. *Southern Stars, 27*, 15–34.

Eiby, G. (1978). History, astronomy and New Zealand. *Southern Stars, 27*, 118–124.

"Epoch marking". The city telescope. Observatory at Kelburn. The opening ceremony. *The Evening Post*, August 27, 1924.

Higham, T., Anderson, A., & Jacomb, C. (1999). Dating the first New Zealanders: the chronology of Wairau Bar. *Antiquity, 73*, 420–427.

Holdaway, R., & Jacomb, C. (2000). Rapid extinction of moas (Aves: Dinornithiformes) in New Zealand. *Science, 287*, 2250–2255.

Hudson, G. (2003). Refurbishment of the Thomas King Telescope at Carter Observatory. *Southern Stars, 42*(4), 4–6.

King, H. C. (1979). *The History of the Telescope*. New York: Dover.

King, T. (1902). On New Zealand Mean Time, and on the longitude of the Colonial Observatory, Wellington; with a note on the Universal Time question. *Transactions of the New Zealand Institute, 35*, 428–451.

McKerrow, J. (1882). The transit of Venus, Wellington. *New Zealand Journal of Science, 1*, 330–332.

Nankivell, G. (2002). The Cooke photovisual objective and the 22.9cm refractor at the Carter Observatory, New Zealand. *Journal of the Antique Telescope Society, 24*, 4–8.

Orchiston, W. (1985). A survey of manuscript material in the Mitchell Library (Sydney) relating to the history of astronomy in New Zealand. *Southern Stars, 31*, 225–238.

Orchiston, W. (1986). Taking Stock of early astronomy books. *Southern Stars, 31*, 326–327.

Orchiston, W. (1988). From research to recreation: the rise and fall of Sydney Observatory. *Vistas in Astronomy, 32*, 49–63.

Orchiston, W. (1995a). *Charles Rooking Carter: A Brief Biography* (p. 4). Wellington: Carter Observatory.

Orchiston, W. (1995b). *The Historic Astronomical Observatories in the Wellington Botanic Garden: A Brief Introduction* (p. 8). Wellington: Carter Observatory (Information Sheet No. 5).

Orchiston, W. (1996). Observatories in the Wellington Botanic Garden. *New Zealand Historic Places*, March 9–11.

Orchiston, W. (1998). *Nautical Astronomy in New Zealand. The Voyages of James Cook*. Wellington: Carter Observatory.

Orchiston, W. (1999). C/1881 K1: a forgotten "Great Comet" of the nineteenth century. *Irish Astronomical Journal, 26*, 33–44.

Orchiston, W. (2000). A Polynesian astronomical perspective: the Maori of New Zealand. In H. Selin & X. Sun (Eds.), *Astronomy Across Cultures: A History of Non-Western Astronomy* (pp. 161–196). Dordrecht: Kluwer.

Orchiston, W. (2002). From Crossley to Carter: the life and times of an historic Cooke refractor. *Journal of the Antique Telescope Society, 24*, 9–24.

Orchiston, W., & Dodd, R. (1995). *A Brief History of the Carter Observatory* (p. 10). Wellington: Carter Observatory (Information Sheet No. 1).

Scholefield, G. H. (Ed.) (1940). *A Dictionary of New Zealand Biography. Two Volumes*. Wellington: Department of Internal Affairs.

Seymour, J. (1995). A history of the Thomas King Observatory Wellington. *Southern Stars, 36*, 102–114.

Shepherd, W., & Cook, W. (1988). *The Botanic Garden, Wellington. A New Zealand History 1840–1987*. Wellington: Millwood Press.

Stock, A. (1869). Letter to John Tebbutt, dated March 10. In *Letters to John Tebbutt*. Sydney: Mitchell Library.

Stock, A. (1874a). *Astronomy for Beginners Who Live in the Southern Hemisphere*. Wellington: Bowden.

Stock, A. (1874b) *The Transit of Venus, and How to Observe It, December 9, 1874*. Wellington: Bowden.

The time ball. *New Zealand Spectator and Cook's Strait Guardian, February 10, 1864*.

The Venerable Archdeacon Stock. (1897). In *The Cyclopedia of New Zealand (Wellington Provincial District)* (pp. 383–384). Wellington: The Cyclopedia Company.

Wood, H. W. (1958). *Sydney Observatory 1858–1958*. Sydney: Sydney Observatory.

Chapter 10
The Thames Observatories of John Grigg

Abstract In the course of a thirty year 'career' as an amateur astronomer John Grigg constructed two different observatories in Thames, and installed a small Wray refractor and a transit telescope in these. He used the refractor mainly to observe known comets and discover new ones, and to pioneer astrophotography in New Zealand. He also maintained a local time-service, and was an avid populariser of astronomer. Early in the twentieth century he was New Zealand's leading amateur astronomer.

10.1 Introduction

John Grigg is famous for his comet discoveries, and by the end of the nineteenth century was New Zealand's leading amateur astronomer. He lived in Thames, and in the course of a 30-year involvement in observational astronomy he built two different observatories. His principal instrument was an 8.9-cm (3.5-in) Wray refractor which, despite its modest size, was put to good use.

Brief general accounts of Grigg already have been published by Hughes (1991), McIntosh (1958, 1970), Mackrell (1985), Orchiston (1996a, 1998b, 2001d) and Rumsey (1974), while Orchiston (1993, 1995) has analyzed his formative role in New Zealand cometary astronomy and astrophotography. This chapter focuses on his astronomical instruments and observatories, discusses the types of observations made with these instruments and concludes with some comments on his non-observational astronomical activities.

W. Orchiston, *Exploring the History of New Zealand Astronomy*,
Astrophysics and Space Science Library 422,
DOI 10.1007/978-3-319-22566-1_10

Fig. 10.1 John Grigg, F.R.A.S. (Orchiston collection)

10.2 John Grigg: A Biographical Sketch[1]

John Grigg (1838–1920; Fig. 10.1) was born in London on 4 June 1838, the only child of James and Ruth Grigg. James was chief clerk and financial manager of Wells & Co., a successful home furnishing business in London, and he made sure that his son received a good education, including a grounding in mathematics and 'the sciences'. From an early age John Grigg showed an aptitude for science and music, interests that were to stay with him throughout his long life (Grigg 1970). In particular, he had a fascination for astronomy (Mr John Grigg elected F.R.A. Society 1906).

In 1858, at the age of 20, John married Emma Mitchell, and in 1863 they migrated to New Zealand, settling in Auckland (for New Zealand localities mentioned in this chapter see Fig. 10.2). Emma died in 1867, and this fact, coupled with news of gold strikes in the nearby Coromandel Peninsula in 1867 and 1868, probably prompted John to move to Thames in this latter year. At the time Thames was prospering and had the largest population of any North Island town (see Isdale 1967). Once there, he established a furnishing business, and later he added a music shop to this, and taught singing.

[1]The following biography is based on data provided in the unpublished *Family History* prepared by one of Grigg's sons (Grigg 1970) and in the *Monthly Notices* Obituary (Crommelin 1921), as well as two papers by McIntosh (1958, 1970) which include unreferenced data provided as personal communications to McIntosh by one of John Grigg's daughters. The Obituary published in *Monthly Notices of the Royal Astronomical Society* (Crommelin 1921) erroneously gives his place of birth as "the Isle of Thanet", and this has been repeated by some other writers in later accounts.

Fig. 10.2 New Zealand localities mentioned in this chapter are shown in *red*

While in Thames Grigg married his second wife, Sarah Allaway, in 1871, but she died suddenly in 1874. He married a third time, to Jane Henderson, in 1887. In all, Grigg had six sons and three daughters by his three wives, while he and Sarah had also adopted a son (Grigg 1970).

John Grigg died in Thames on 20 June 1920 after a long and fruitful life. He was 82 years of age (Crommelin 1921).

10.3 Telescopes and Observatories

Grigg states that the 1874 and 1882 transits of Venus were the catalysts that whetted his interest in astronomy that had been dormant since childhood (Mr John Grigg elected F.R.A. Society 1906), but I suspect that an English astronomer named Henry Severn also played a role. Severn spent just six years in Thames during the 1870s, but made a major impact at this time and interested many people in astronomy. Not only was he a brilliant public lecturer, but he also owned the largest reflecting telescope in New Zealand, and was one of an elite band of amateur observers selected by the British party based at Burnham to man 'subsidiary stations' during the 1874 December transit (see Orchiston 2001b and Chap. 14 in this book). Although McIntosh (1958: 22) maintains that "There was no great interest in astronomy in Thames … and there was therefore no one of identical interests and intellectual calibre with whom he [Grigg] could associate." Grigg would certainly have known Severn. Henry Severn is the subject of Chap. 21 in this book.

At some unspecified date prior to the 1882 transit—and possibly in anticipation of it—Grigg was loaned a 7.6-cm aperture portable altazimuth-mounted refracting telescope (Mr Grigg's observatory 1885),[2] which he would take outdoors when he wished to observe.

Then in early 1884 Grigg began thinking about purchasing a refracting telescope of his own and a transit telescope, and constructing an observatory for them, so that he could undertake astronomy on a regular basis whenever time and opportunity allowed (ibid.). Thus, on 20 March 1885 the *Thames Advertiser* informed its readers that

> Mr. Grigg recently imported a new and expensive telescope and transit instrument, with all the fittings required for accurate and important observations, and in order to make the greatest possible use of them it was necessary that a well-appointed observatory should be provided. A suitable building was therefore erected at the rear of the business premises. (ibid.).

Later in this long newspaper article we learn about the principal telescope (see Fig. 10.3), was a refractor with

[2]Nowhere does Grigg indicate who loaned him this telescope.

Fig. 10.3 The 8.9-cm Wray refractor and home-made astrocamera (*Courtesy* Ward (Wanganui) Observatory Archives)

> … an object glass of 3½ inches [8.9-cm] clear aperture, and a focal length of 54 inches [i.e. f/15], with magnifying powers ranging from 60 to 480, sun prism, diagonal eyepiece, and other adjuncts necessary for the effective use of a first-class instrument. The equatorial mounting has a latitude arc and six-inch declination and right ascension circles … (ibid.).

The telescope itself was manufactured by William Wray (1829–1885), an "… eminent English optical instrument maker and retailer …" who initially was based in London before moving to Bromley in Kent (Andrews 1997: 177), but it featured a simple home-made German equatorial mounting, with circles that could be read by verniers to 10s in right ascension and 3′ in declination (Grigg 1902e). When Grigg acquired this telescope there was no drive, so in 1893 he proceeded to construct one modeled on a design by Sparling that was published in the *English Mechanic* earlier that year (see Grigg 1902b). The equatorial mounting and drive are clearly shown in Fig. 10.4, where the telescope is set up for solar observing.

Grigg's principal observational interest was comets, and the determination of micrometric positions was one of the primary challenges of committed cometary astronomers—amateur and professional alike—during the nineteenth century. Here Grigg was severely handicapped for, as might be expected with so small

Fig. 10.4 A close-up view of the home-made equatorial mounting and drive (*Courtesy* Ward (Wanganui) Observatory Archives)

a telescope, the Wray was not supplied with a micrometer (Grigg 1902c). Nor did he have the financial means to purchase one (Grigg 1902e), and realizing it was a far from trivial task, he lacked the confidence to try and construct one himself.

Another of Grigg's interests was astronomical photography, a field which only rose to prominence late in the nineteenth century (see Lankford 1984; Norman 1938), and in order to pursue this he used two different appliances (see Orchiston 1995, and Chap. 22 in this book). By trial and error he built up a crude camera (Grigg 1902b), which was attached to the eyepiece end of the Wray telescope and probably took 4 × 5-inch glass plates. This is shown in Fig. 22.3, and was used mainly for lunar and solar photography. Meanwhile, he also constructed the simple astrocamera that is seen mounted on the telescope tube in Fig. 10.3. On the reverse of the original photograph, Grigg refers to this as a "home-made star camera".

Fig. 10.5 Grigg's first Thames Observatory, in Pollen Street, 1885–1894 (Orchiston collection)

The simple observatory that Grigg built for his new 7.6-cm Wray telescope behind his business premises in Pollen Street, Thames, is described in the *Thames Advertiser*:

> The main portion is of circular form, 8ft in diameter, with walls 6ft. 3in. high inside, carrying a revolving roof, being a segment of a sphere of somewhat larger diameter. The opening in the roof has parallel sides, giving a field of view of 30 deg., and is carried 15 deg. past the zenith. The roof is constructed of paper-mache, on a framework of pohutakawa ribs.[3] A light removable cover closes the opening when not in use. The floor and pillar are of concrete, the later having a triangular section, and being 4ft. 6in. high, this is surrounded by an iron pedestal about two feet high, carrying the equatorial mounting to which the telescope is attached. (Mr Grigg's observatory 1885).

Known as the Thames Observatory, this was personally designed by Grigg and is shown in Fig. 10.5.[4]

As can be seen in this figure, the observatory included a transit annex, which was described as "… a lean-to … [which was] covered with galvanized sheet iron. It has a meridional opening 3 inches wide, and a view arc of about 160 deg. is obtained." (ibid.). This annex housed a small transit telescope "… manufactured by A.J. Frost [of London]. Although only 13 inches in length it works very accurately, and will

[3]*Pohutakawa* was a native tree of New Zealand with excellent timber that was widely used in the building industry during the nineteenth century.

[4]Mackrell (1985: 75) errs in suggesting that this was the first observatory erected by an amateur astronomer in New Zealand. Those who maintained private observatories prior to 1885 included Thomas Cheeseman of Auckland, Stephen Carkeek of Featherstone (Orchiston 2001c and Chap. 8 in this book), Henry Jackson of Lower Hutt (Hector 1935), Thomas King of Wellington (Seymour 1995), James Townsend of Christchurch, and Arthur Beverly and Henry Skey of Dunedin.

Fig. 10.6 The transit telescope (Orchiston collection)

time a star to a fraction of a second if skillfully manipulated. The stand is mounted on a 6 inch iron pillar." (ibid.). This instrument is shown in Fig. 10.6. Grigg used his transit telescope to determine the latitude and longitude of his observatory. The American party based in Auckland for the 1882 transit of Venus determined the co-ordinates of their observing station with some precision, so with the assistance from two local surveyors Grigg was able to pinpoint the position of his Thames Observatory. It was located at:

Lat., 37° 8′ 34.06″S.
Long., 175° 32′ 50.41″E. (ibid.)

Grigg also used his transit telescope to regulate three different clocks. One of these,

> The Standard clock is a regulator, with dead beat escapement, and seconds pendulum, with zinc cylindrical compensating bob. This has not been placed in the observatory as intended, but is allowed to remain in the warehouse,[5] chiefly for the convenience of the public, who have shown their appreciation of it for some time past. It is set to New Zealand time, that is, 11 hours 30 minutes fast of Greenwich … Two common lever clocks, set to mean and sidereal times, respectively are used in the observatory, with the addition of a metronome for counting the seconds. These clock[s] are compared with standard clocks and corrected, or errors noted, just before commencing work. (ibid.).

[5]This warehouse formed part of Grigg's business premises in Pollen Street, Thames.

Fig. 10.7 Grigg's second Observatory, in Queen Street, 1894–1920 (Orchiston collection)

Small transit telescopes like Grigg's were usually portable, and during the second half of the nineteenth century they were used extensively by field surveyors engaged in trigonometrical and other surveys throughout New Zealand.

Although professional observatory architecture evolved rapidly during the nineteenth century (see Donnelly 1973), the basic design that Grigg adopted of a dome room and adjoining transit room was a standard one worldwide for modest amateur observatories throughout the century.

Even when his new observatory was operational, Grigg continued to use the 7.6-cm (3-in) portable altazimuth refractor on loan to him when surrounding buildings or trees prevented him from observing celestial objects from the observatory (Grigg 1970).

In 1894, when he was 56 years of age, Grigg retired from full-time business so that he could "… give more time to his scientific inclinations." (Grigg 1970: 25), and he relocated his observatory to Queen Street, Thames.[6] The new observatory was a larger two-storey wooden building, with the dome room and transit annex upstairs and an office and workshop downstairs (see Fig. 10.7).

[6]McIntosh (1958: 21) incorrectly states that Grigg's second Thames Observatory was constructed in "About 1884 …".

Although Grigg undoubtedly would have welcomed the significant increase in light grasp that a larger principal telescope would have brought to his comet-searching, and the precision of a micrometer in determining positions of known comets and his own discoveries, he was philosophical about his existing instruments: "… as my means are limited, I see no probability of obtaining a better outfit than I have at present so must content myself to make the best use of what I possess." (Grigg 1902e).

10.4 Astronomical Observations

Grigg's main preoccupation was comets, and he carried out systematic observations of known ones and conducted searches for new ones. He carried out morning and evening searches, using a working list of nebulae that he had compiled for reference purposes (Grigg 1902a), together with the *Nautical Almanac*, the 'Catalogue of 1500 Stars' printed in Loomis' *An Introduction to Practical Astronomy* (1861) and Proctor's *A Star Atlas for the Library, The School and the Observatory* (Grigg 1903d).

His first discovery came on 22 July 1902, when he detected an object which looked like a faint nebula near the border of Leo and Virgo (Grigg 1903a). Grigg continued to observe this object until 2 August, but he waited until 26 July to formally announce his discovery, and when the news reached Pietro Baracchi (1851–1926; Fig. 17.5; Perdrix 1979), the Director of Melbourne Observatory (which was the designated Australian and New Zealand reporting centre) and Sydney's active amateur astronomical community, they were unable make the necessary confirmatory observations because the positions Grigg supplied were so imprecise. Grigg (1902d) later realized that he had erred in not cabling details of his discovery to Baracchi, but as I have outlined elsewhere (Orchiston 1999a), Baracchi also shirked his responsibility by not immediately cabling Grigg's positions and discovery circumstances to Kiel (the international comet centre) and to regular Australian and New Zealand cometary astronomers. As a result, Grigg ended up being the only observer of this comet, and it was lost to science until its fortunate recovery by John Francis (Frank) Skjellerup (1875–1952; Fig. 10.8) in 1922 (see Orchiston 1999b, 2003a). Comet 26P/Grigg-Skjellerup is now one of the best-researched of all short period comets (e.g. see Hughes 1991).

Flush with success, Grigg continued his search program and on 17 April 1903 discovered his second comet. Vsekhvyatskii (1964: 357) provides a useful summary:

> **1903 III (1903b)**. Discovered in Eridanus by Grigg (New Zealand) on the evening of 17 April. Moved north-east (away from the Sun) through Eridanus (beginning of May), Lepus and Canis Major (end of May).
>
> Observed only in New Zealand, Australia and at the Cape [Royal Observatory, Cape of Good Hope]. Upon discovery, appeared faint through a 3-inch telescope; no nucleus (fainter than 8 to 9^{m}). Receded from the Sun (perihelion passage on 26 March) growing rapidly faint. Measured last on 28 May in Windsor [by Tebbutt] with an 8-inch refractor

Fig. 10.8 Frank Skjellerup (Orchiston collection)

> and at the Cape of Good Hope with a 7-inch refractor when hardly visible (not fainter than 12 to 13^{m}).

In this instance, Grigg followed the correct protocols and so astronomers in Australia and South Africa were able to make micrometric observations of this comet and ultimately to calculate its orbital elements (see Orchiston 1999a for details). This comet is now known as C/1903 H1 (Grigg) (Marsden and Williams 1999).

On 10 February 1906 Grigg (1906b) discovered what he thought was his third comet, only to learn that this was in fact Comet/1905 X1 which the French astronomer Michel Giacobini (1873–1938) had first detected on 8 December 1905 from Nice (Marsden and Williams 1999). Elsewhere in this book (in Chap. 17), I have suggested that "Given the long interval between the original discovery and Grigg's detection, we can hardly claim this as an independent discovery for Grigg."

Disappointment also characterized Grigg's next discovery, which took place a little over one month later. On 19 March 1906 he discovered a comet in Cetus which was "… fairly bright in my 3-in. telescope …" (Grigg 1906a). After making further observations, he sent details to Australia only to learn that David Ross (1850–1930; Fig. 17.9) of Melbourne had detected the newcomer on 14 February (Orchiston and Brewer 1990).

Grigg discovered his final comet on 8 April 1907, five days before the U.S. amateur astronomer and telescope-maker John Edward Mellish (1886–1970; Fig. 17.10). Grigg notified the New Zealand media, Joseph Thomas Ward of Wanganui, and the Australian professional observatories of the comet. Meanwhile, when he heard from Grigg, Baracchi once again vacillated on what course of action he should follow, and initially decided not to cable details of the comet to Kiel or to Australian and New Zealand cometary astronomers. Grigg (1907) was far from amused, while the whole episode disgusted Tebbutt (1907). Fortunately, Australasian astronomers were able to observe this comet up until 12 April and thereafter it was followed by northern hemisphere observatories through to 14 May (Vsekhvyatksii 1964).

Baracchi's inaction in this instance, and his earlier 'shabby treatment' of Grigg and Ross (Wright 1907), had an interesting outcome for Australian professional astronomy. Later in 1907, the New South Wales Branch of the British Astronomical Association successfully requested that Kiel transfer the Australasian 'Central Bureau' for the dissemination of astronomical information from Melbourne to Sydney Observatory (see Orchiston 1999a for details).

Precisely when Grigg decided to terminate his systematic comet searches is not known, but it was most probably in 1911 or 1912 (see Ross 1912) when he was in his 70s. By this time he had independently discovering four comets, three of which now bear his name, and received Donohoe Medals from the Astronomical Society of the Pacific for his 1902 and 1903 discoveries (Grigg 1970).

Apart from new comets, Grigg also observed comets discovered by others. Because his observing books have not survived and he published little on these objects, we have only a superficial knowledge of this aspect of his activities. Although we have no proof that he did so, it is reasonable to assume that he observed the 'Great Comets' of 1880 (C/1880 C1), 1881 (C/1881 K1, Tebbutt), 1882 (C/1882 F1, Wells and C/1882 R1), 1887 (C/1887 B1) and 1910 (C/1910 A1) as all were impressive objects in the southern sky. The only 'Great Comets' for we have direct evidence are C/1901 G1 and 1P/Halley in 1910, and Grigg obtained successful photographs of each of these (see Orchiston 1995; and Chap. 22 in this book).

The first comet that Grigg is known to have observed was C/1885 X1 (Fabry), and he recounts how he recorded its position on 33 different nights between 5 May and 25 June 1886—and it was the thrill of this experience which inspired him to embark on systematic comet-searches (Grigg 1906c). Other known comets that Grigg observed (but did not photograph) were: C/1899 E1 (Swift), C/1902 R1 (Perrine), C/1903 A1 (Giacobini), C/1907 L2 (Daniel) and the periodic comet 2P/Encke (Grigg 1903b, c, 1904, 1970; *Thames Star* 1907b; Vsekhvyatskii 1964). He observed this last-mentioned comet on three different returns, in 1895, 1898 (when he and Tebbutt were responsible for its recovery) and in 1904. Detailed accounts of Grigg's cometary work are provided in Orchiston (1993) and in Chap. 17 in this book.

Undoubtedly, the most important astronomical events that Grigg had an opportunity to observe were the 1874 and 1882 transits of Venus. These rare events were seen as offering a means of refining one of the basic astronomical yardsticks, the distance from the Earth to the Sun (see Clerke 1893; Dick et al. 1998).

There is ample evidence that Grigg intended to observe the 1874 transit (Crommelin 1921; Grigg 1970; Mr John Grigg elected F.R.A. Society 1906), and although bad weather thwarted the attempt, this transit helped launch his long 'career' as an active observational astronomer. In December 1882 the elements were somewhat more kindly, and a Mr Foy succeeded in taking two photographs of the transit using a telescope Grigg had installed in an upstairs room in his business premises. And although Grigg and his assistants did see the transit and obtain some useful results (Transit of Venus 1882), it is of interest to note that their efforts were not mentioned in the official New Zealand account of the event published by Stone (1887).

Occasionally other celestial objects and events attracted Grigg's attention, including solar eclipses, sunspots, the Moon and various planets, and lunar occultations of stars (Grigg 1970; McIntosh 1958; *Thames Star* 1900, 1905a). A particular focus was the Sun and solar-terrestrial relations, and weather permitting Grigg is reputed to have carried out daily sunspot observations (McIntosh 1958). However, given the absence of any extant observing books it is impossible to determine precisely what sorts of non-cometary observations Grigg actually carried out, and how frequently, but the lack of publications in contemporary journals does suggest he was not following in the footsteps of John Tebbutt and other leading amateur astronomers of the day and undertaking on-going systematic work (e.g. see Orchiston 1989, 2004). This is consistent with a newspaper report that "Mr Grigg's scientific efforts ... [were] necessarily limited to the few leisure hours at the disposal of a man of business." (Mr Grigg's Observatory 1885). Nor did Grigg ever take up serious double star or variable star work, as was suggested to him by the Royal Astronomical Society's Richard Henry Wesley (1841–1922) in 1906. What we do know, however, is that Grigg was interested in astronomical photography and that on 12 December 1890 he photographed a partial solar eclipse and in 1895 and 1902 he photographed the Moon (see Orchiston 1995 and Chap. 22 in this book).

In 1906, following his string of independent comet discoveries, Grigg was elected a Fellow of the Royal Astronomical Society, and in a letter to the Society he describes the modest astronomical equipment at his disposal and remarks:

> ... I am conscious of my inability to supplement the valuable & critical work done by those Fellows whose names and attainments are so familiar. My Chief work is comet seeking during evening hours, if anything more useful can be suggested, I shall be happy to consider it. (Grigg 1906d).

10.5 Other Astronomical Activities

In addition to his observational interests, Grigg ran the Thames Observatory as a *de facto* public observatory, offering a similar range of facilities and providing a comparable range of services to those typical of a larger government-funded observatory. As we have already seen, he offered a local time service which was popular with townsfolk and "Many people called daily to set their watches." (Grigg 1970: 26), while Eiby (1971; 1978) has alluded to the rumour (thus far unproven) that the Thames Observatory provided a time-service for local shipping.

Fig. 10.9 Alfred Barrett Biggs (Orchiston collection)

Like his Tasmanian contemporary, the Launceston bank officer Alfred Barrett Biggs (1825–1900; Fig. 10.9; Giordano 1995; Orchiston 1985a), Grigg was a skilled, resourceful, and innovative amateur astronomer, who also possessed a passion for music and a degree of mechanical genius. Both men successfully ran 'public observatories' in their respective cities, providing services and facilities which in larger centres were conventionally funded by colonial or provincial governments. Meanwhile, like his Melbourne-based contemporary, David Ross (Orchiston and Brewer 1990), Grigg discovered comets and experimented successfully with astronomical photography, and all three men were accomplished lecturers on astronomy. Yet all three lacked the international standing of people like John Tebbutt (Fig. 10.10), Charles James Merfield (1866–1931) and Robert Thorburn Ayton Innes (1861–1933) who were offered the opportunity to transfer from amateur to professional ranks (see Orchiston 1988, 1998a, 2001a, 2003b, 2015).

10.6 Concluding Remarks

During the second half of the nineteenth century, astronomy blossomed in New Zealand. The central Government funded the Colonial Observatory; the 1874 and 1882 transits of Venus brought teams of overseas astronomers to our shores; Christchurch's Professor Bickerton was busy promoting his novel 'partial impact theory'; amateur astronomers began to emerge across the nation; and the first fledgling astronomical societies were formed (Orchiston 1998b).

Fig. 10.10 John Tebbutt (Orchiston collection)

By the end of the century John Grigg was New Zealand's foremost amateur astronomer. With the modest equipment in his Thames Observatory he pioneered astronomical photography in New Zealand and became known nationally and internationally for his cometary discoveries. He also contributed significantly to the popularisation of astronomy. Without the Thames Observatory, nineteenth and early twentieth century New Zealand astronomy would have been much the poorer.

In 1920, after Grigg's death, the Thames Observatory was dismantled and the Wray and transit telescopes were acquired by the only one of his seven sons with an interest in science (McIntosh 1958). Eventually they passed to another relative, Mrs Beverley Angus, whose husband was an avid amateur astronomer, and he installed the Wray refractor in their observatory in Wellington and in 1990 donated the transit instrument to the Carter Observatory (Graham Blow, personal communication, 10 May 1990; Beverley Angus, personal communication, 11 May 1990). In 1993, when Mr Angus acquired a larger telescope for his observatory, he decided to donate the Wray telescope to the Carter Observatory as well. In its capacity at that time as the National Observatory of New Zealand the Carter Observatory was charged with serving as a repository for the nation's astronomical heritage, so this was a particularly appropriate final resting place for so historic an instrument.

Acknowledgements I am grateful to the following for their assistance: Mr and Mrs Paul Angus (Wellington, NZ), the late Graham Blow (Wellington, NZ), the late Peter Hingley (former RAS Librarian, London), the late Alistair Isdale (Thames, NZ), Dr Nick Lomb (ex-Sydney Observatory), John Seymour (Palmerston North, NZ), Ormond Warren (Auckland, NZ) and staff at the Mitchell Library (Sydney). I also wish to thank John Drummond (Patutahi, NZ) for reading and commenting on the first draft of this chapter, and the Ward (Wanganui) Observatory (through John Seymour) for kindly supplying Figs. 10.3 and 10.4.

References

The following abbreviation is used:

TL = Letters to John Tebbutt (1860–1915), comprising 46 volumes of bound letters and documents in the Mitchell Library, Sydney.

An Astronomical Society on the Thames. *Thames Star*, August 10, 1901 (p. 2).

Andrews, A. D. (1997). Cyclopaedia of telescope-makers. Part 7 (T–Z). *Irish Astronomical Journal, 24*, 125–192.

Austin, A. W. (1901). Letter to J. Tebbutt, dated June 3, In TL

Calder, D. (1978). Joseph Ward: pioneer astronomer and telescope maker. *Southern Stars, 27*, 104–108.

Campbell, R. N. (2001a). Arthur Beverly 1822–1907. *Southern Stars, 40*(3), 23–24.

Campbell, R. N. (2001b). Henry Skey 1836–1914. *Southern Stars, 40*(2), 11–12.

Clerke, A. (1893). *A Popular History of Astronomy During the Nineteenth Century*. London: Adam and Charles Black.

Crommelin, A. C. D. (1921). Obituary notices: John Grigg. *Monthly Notices of the Royal Astronomical Society, 81*, 258–259.

Dick, S., Orchiston, W., & Love, T. (1998). Simon Newcomb, William Harkness and the nineteenth century American transit of Venus expeditions. *Journal for the History of Astronomy, 29*, 221–255.

Donnelly, M. C. (1973). *A Short History of Observatories*. Eugene: University of Oregon.

Eiby, G. A. (1971). Astronomy in New Zealand. *Sky and Telescope, 42*(1), 18–20.

Eiby, G. A. (1977). The New Zealand Government time service: an informal history. *Southern Stars, 27*, 15–34.

Eiby, G. A. (1978). History, astronomy and New Zealand. *Southern Stars, 27*, 118–124.

Evans, R. W., & Lucas, K. J. (1989). The Skey-Ashburton College telescope. *Southern Stars, 33*, 178–187.

Gibbs, F. G. (n.d.). Notes on the history of astronomy in Nelson. Copy of a manuscript, Nelson Provincial Museum (MSS 289).

Giordano, M. (1995). *Watcher of the Skies. The Story of Alfred Barrett Biggs Pioneer in Astronomy*. Launceston: Regal Publications.

Goodman, D. (1976). A brief history of the Atkinson Observatory, Nelson. *Southern Stars, 26*, 164–166.

Grigg, J. (1902a). Über die Entdeckung eines neuen Cometen 1902c. *Astronomische Nachrichten, 159*, 389–392.

Grigg, J. (1902b). Photography with a small telescope. *Journal of the British Astronomical Association, 12*, 125–126.

Grigg, J. (1902c). Letter to John Tebbutt, dated August 2, In TL.

Grigg, J. (1902d). Letter to John Tebbutt, dated September 16, In TL.

Grigg, J. (1902e). Letter to John Tebbutt, dated December 3, In TL.

Grigg, J. (1903a). Comet in Virgo 1902 observed at Thames N.Z. *Astronomische Nachrichten, 160*, 213–214.

Grigg, J. (1903b). Comet 1902 b (Perrine). *Journal of the British Astronomical Association, 13* (169), 284–285.

Grigg, J. (1903c). Comet 1903 b. *Journal of the British Astronomical Association, 14*, 77–78.

Grigg, J. (1903d). Letter to John Tebbutt, dated June 13, In TL.

Grigg, J. (1903e). Letter to John Tebbutt, dated November 28, In TL.

Grigg, J. (1904). Letter to John Tebbutt, dated December 17, In TL.

Grigg, J. (1906a). Reports of the Observing Sections. The Comet Section. *Journal of the British Astronomical Association, 16*, 271–272.

Grigg, J. (1906b). Comet 1906 c. *Journal of the British Astronomical Association., 16*, 319.

Grigg, J. (1906c). Letter to John Tebbutt, dated April 20, in TL.

Grigg, J. (1906d). Letter to T. Lewis, dated August 16, in the Royal Astronomical Society Library.
Grigg, J. (1907). Letter to John Tebbutt, dated May 22, In TL.
Grigg, M. (Ed.), (1970). *John Grigg, FRAS of Thames, New Zealand. Together with Some Genealogical Memorials of his Family and Glimpses of Pioneer Life in New Zealand.* Unpublished typescript (copy in Orchiston Collection).
Hector, C. M. (1935). Links with the past. *Southern Stars, 1*, 115–117.
Hughes, D. W. (1991). J. Grigg, J. F. Skjellerup and their comets. *Vistas in Astronomy, 34*, 1–10.
Isdale, A. (1967). *History of the River Thames N.Z.* Thames: printed for the author.
Journal of the British Astronomical Association, 18, 55–56 (1907).
Lankford, J. (1984) The impact of photography on astronomy. In O. Gingerich (Ed.), *The General History of Astronomy* (Vol. 4). *Astrophysics and Twentieth-Century Astronomy to 1950: Part A*. Cambridge: Cambridge University Press, pp. 16–39.
Loomis, E. (1861). *An Introduction to Practical Astronomy*. New York: Harper.
McIntosh, R. A. (1958). The astronomical history of the Auckland Province Part IV: John Grigg. *Amateur Astronomer, 18*(2), 20–23.
McIntosh, R. A. (1970). Early New Zealand astronomy. *Southern Stars, 23*, 101–108.
Mackrell, B. (1985). *Halley's Comet over New Zealand.* Auckland: Reed-Methuen.
Marsden, B. G. & Williams, G. V. (1999). *Catalogue of Cometary Orbits 1999* (13th ed.). Cambridge: Minor Planet Center.
Mr Grigg's Observatory. *Thames Advertiser*, March 20, 1885 (p. 3).
Mr John Grigg elected F.R.A. Society. *Thames Star*, August 23, 1906 (p. 2).
Norman, D. (1938). The development of astronomical photography. *Osiris, 5*, 560–594.
Orchiston, W. (1983). C.J. Westland and Comet 1914IV: a forgotten episode in New Zealand cometary astronomy. *Southern Stars, 30*, 339–345.
Orchiston, W. (1985a). The contribution of Alfred Barrett Biggs to Australian and Tasmanian astronomy. *Records of the Queen Victoria Museum, 89*, 1–28.
Orchiston, W. (1985b). A survey of manuscript material in the Mitchell Library (Sydney) relating to the history of astronomy in New Zealand. *Southern Stars, 31*, 225–238.
Orchiston, W. (1986). Towards an accurate history of early New Zealand astronomy. *Southern Stars, 31*, 257–266.
Orchiston, W. (1988). Illuminating incidents in antipodean astronomy: John Tebbutt and the Sydney Observatory directorship of 1862. *Australian Journal of Astronomy, 2*, 149–158.
Orchiston, W. (1989). The role of the amateur in early Australian astronomy. *Search, 20*, 16–23.
Orchiston, W., & Brewer, A. (1990). David Ross and the development of amateur astronomy in Victoria. *Journal of the British Astronomical Association, 100*, 173–181.
Orchiston, W. (1993). John Grigg, and the genesis of cometary astronomy in New Zealand. *Journal of the British Astronomical Association, 103*, 67–76.
Orchiston, W. (1995). John Grigg, and the development of astrophotography in New Zealand. *Australian Journal of Astronomy, 6*, 1–14.
Orchiston, W. (1996a). Grigg, John 1838–1920. In C. Orange et al. (Ed.), *The Dictionary of New Zealand Biography* (Vol. 3, pp. 188–189, 1901–1920). Auckland: Auckland University Press and Department of Internal Affairs.
Orchiston, W. (1996b). Ward, Joseph Thomas 1862–1927. In C. Orange (Ed.), *The Dictionary of New Zealand Biography*. (Vol. 3, p. 555, 1901–1920). Auckland: Auckland University Press and Department of Internal Affairs.
Orchiston, W. (1998a). Mission impossible: William Scott and the first Sydney Observatory directorship. *Journal of Astronomical History and Heritage, 1*, 21–43.
Orchiston, W. (1998b). *Nautical Astronomy in New Zealand. The Voyages of James Cook.* Wellington: Carter Observatory.
Orchiston, W. (1999a). Comets and communication: amateur-professional tension in Australian astronomy. *Publications of the Astronomical Society of Australia, 16*, 212–221.

Orchiston, W. (1999b). Of comets and variable stars: the Afro-Australian astronomical activities of J.F. Skjellerup. *Journal of the British Astronomical Association, 109*, 328–338.
Orchiston, W. (2001a). From amateur astronomer to observatory director: the curious case of RTA Innes. *Publications of the Astronomical Society of Australia, 18*, 317–327.
Orchiston, W. (2001b). Henry Severn: Thames' other nineteenth century astronomer. *Southern Stars, 40*(2), 8–10.
Orchiston, W. (2001c). Stephen Carkeek and New Zealand's oldest surviving observatory. *Journal of the Antique Telescope Society, 20*, 19–24.
Orchiston, W. (2001d). The Thames observatories of John Grigg. *Southern Stars, 40*(3), 14–22.
Orchiston, W. (2002). Joseph Ward: pioneer New Zealand telescope-maker. *Southern Stars, 41*, 13–21.
Orchiston, W. (2003a). J.F. Skjellerup: a forgotten name in South African cometary astronomy. *Monthly Notices of the Astronomical Society of Southern Africa, 62*, 54–73.
Orchiston, W. (2003b). Passion and patronage: the Afro-Australian astronomy of Robert Thorburn Ayton Innes. *Monthly Notices of the Astronomical Society of Southern Africa, 62*, 4–13.
Orchiston, W. (2004). John Tebbutt and observational astronomy at Windsor Observatory. *Journal of the British Astronomical Association, 114*, 141–154.
Orchiston, W. (2015). The amateur-turned-professional (ATP) syndrome: two Australian case studies. In W. Orchiston, D. Green, & R. Strom (Eds.), *New Insights from Recent Studies in Historical Astronomy: Following in the Footsteps of F. Richard Stephenson.* (pp. 259–350). New York: Springer.
Perdrix, J. L. (1979). Baracchi, Pietro Paulo Giovanni Ernesto (1851–1926. In *Australian Dictionary of Biography* (Vol. 7, pp. 166–167). Melbourne: Melbourne University Press.
Rice, P. (1982). The story of Ward Observatory. *Southern Stars, 29*, 271–274.
Ross, D. (1912). Correspondence. *Journal of the British Astronomical Association, 23*, 101–102.
Rumsey, N. (1974). Astronomy in New Zealand. *Southern Stars, 25*, 171–207.
Scholefield, G. H. (Ed.), (1940). *Dictionary of New Zealand Biography* (Two Vols. ed.). Wellington: Department of Internal Affairs.
Seymour, J. B. (1985). Palmerston North's first observatory. *Southern Stars, 31*, 209–214.
Seymour, J. B. (1995). A history of the Thomas King Observatory, Wellington. *Southern Stars, 36*, 102–114.
Southern Stars, *5*, 93 (1939).
Southern Stars, 14, 129 (1950a).
Southern Stars, *14*, 130–131 (1950b).
Stone, E. J. (1887). *Transit of Venus, 1882*. London: Her Majesty's Stationery Office.
Tebbutt, J. (1907). Comet Grigg, 1907, April. *The Observatory, 30*, 285–286.
Thames Star, May 4, 1900.
Thames Star, November 8, 1901.
Thames Star, August 8, 1904a.
Thames Star, August 23, 1904b.
Thames Star, February 1, 1905a.
Thames Star, February 16, 1905b.
Thames Star, April 19, 1907a.
Thames Star, August 21, 1907b.
Thames Star, June 2, 1908.
Thames Star, October 13, 1909.
Thames Star, June 21, 1920.
Transit of Venus. *The Thames Advertiser & Miners News*, December 8, 1882.
Turner, E. F. (1968). History of the Atkinson Observatory. Nelson Provincial Museum: Typescript of an address to the Nelson Historical Society.
Venimore, C. (1988). *Good Viewing. (A History of the First Wanganui Astronomical Society and the Ward Observatory)*. Unpublished manuscript.

Vsekhvyatskii, S. K. (1964). *Physical Characteristics of Comets*. Jerusalem: Israel Program for Scientific Translations.
Wanganui Herald, October 19, 1904.
Wesley, W. H. (1906). Letter to J. Grigg, dated September 25, in the Royal Astronomical Society Library.
Wright, H. (1907). Letter to John Tebbutt, dated May 22, In TL.

Chapter 11
The Wanganui Refractor and Its Remarkable English Equatorial Mounting

Abstract The first all-metal English equatorial mounting of the 'Cross Axis' type was constructed in 1859 for a 24.1-cm (9.5-in) Cooke refractor owned by Isaac Fletcher of Carlisle, northern England. Over the next ten years Fletcher used this telescope for systematic observations of known double stars, and after his death it was acquired by S. Chatwood of Manchester. In 1902 the telescope was purchased by J.T. Ward for the newly-formed Wanganui Astronomical Society in New Zealand. Ward used the telescope to discover new southern double stars, and it was also the mainstay of public viewing nights. This educational function has remained through to the present day, and during the 1980s and 1990s O. Warren reactivated a micrometric double star program involving the re-measurement of the Ward and other southern double stars. Until recently, when the Carter Observatory's equally-historic Cooke refractor was renovated, the 'Fletcher Telescope' was New Zealand's largest operational refractor, and it has been maintained in excellent mechanical and optical condition.

11.1 Introduction

The English equatorial mounting has a long and distinguished history (see Gingerich 1967; King 1979). There are two principal varieties:

- the 'Through Axis' type, where the tube of the telescope "… passes between, and is pivoted from, two or more structural elements of the polar axis …" Hingley (n.d.), and
- the 'Cross Axis' type, where the telescope "… is pivoted to one side of the single polar axis and thus needs a counterweight on the other side." (ibid.).

The Palomar 'Horseshoe Mounting' should be regarded as a special variant of the 'Through Axis' type that permits direct access to the celestial pole, and its origins have been traced back to nineteenth century Australia (Orchiston 2000).

W. Orchiston, *Exploring the History of New Zealand Astronomy*,
Astrophysics and Space Science Library 422,
DOI 10.1007/978-3-319-22566-1_11

The late Hingley (n.d.) discussed the history of both types of English equatorial mounting, and showed that an important advance took place when a cast-iron 'Cross Axis' mounting was manufactured for a 24.1-cm (9.5-in) f/15.2 Cooke refractor owned by Isaac Fletcher. This was the first time that metal (rather than wood) had been used in the overall construction of an English equatorial mounting. As such, Fletcher's telescope serves, in zoological parlance, as a *type specimen*, and it is of interest to trace its chain of ownership and use. After discussing its English affiliations, this chapter focuses on the circumstances leading to the relocation of the telescope to New Zealand in 1902, and its subsequent use for research and for the popularization of astronomy. As such, this chapter builds on my paper titled "The English Equatorial Mounting and the History of the Fletcher Telescope", which was published in the *Journal of Astronomical History and Heritage* more than a decade ago (Orchiston 2001b).

11.2 The English History of the Telescope

There is some confusion surrounding the early history of the 24.1-cm refractor, and it is only possible to unravel the facts by critically examining various publications that specifically mention this telescope. At the root of the problem is King's claim (1979: 251) that in about 1851 Thomas Cooke manufactured a 24.1-cm (9.5-in) refractor for a John Fletcher Miller (1816–1856), who lived at Whitehaven on the southwestern shore of Solway Firth. One of Miller's astronomical contemporaries in far north-western England was Isaac Fletcher (1827–1879), who lived at Greysouthen, near Carlisle, and it is clear from their publications that the two men knew each other (e.g. see Miller 1853). Miller died in 1856, and it is a reasonable assumption that the 24.1-cm Cooke refractor then passed to Fletcher, as he is known to have been in possession of just such an instrument during the 1860s. This is a plausible story, but what is the truth of the matter?

Published papers by Fletcher (1853) and Miller (1852) indicate that Cooke did indeed make a refracting telescope for Miller in 1851, but the aperture was actually 10.5cm (4.14 inches) not 24.1cm (or 9.5 inches), and it is clear from Miller's obituary that he never acquired a larger instrument (Obituary 1857). Meanwhile, Fletcher had a refractor of identical aperture, which Cooke had made for him in 1847 (Fletcher 1850), and the only difference between the two instruments was in their mountings: Fletcher's featured a wooden English equatorial and Miller's a German equatorial (Miller 1853). Both men used their telescopes for micrometric observations of double stars and published their results in the *Monthly Notices* and the *Memoirs* of the Royal Astronomical Society. Where the same stars were involved, they included each other's measures in some of their publications (e.g. see Fletcher 1853, 1855; Miller 1852). Furthermore, both astronomers were elected to Fellowships of the Royal Astronomical Society in 1849, and each subsequently became an F.R.S. (see Obituary 1857, 1880).

Miller died in 1856 at the youthful age of 40 (Obituary 1857), and on 24 April in the following year Fletcher placed an order with Cooke for a 9-in (22.9-cm) refractor costing £400 (Orders … n.d.) in order to dramatically increase his available light grasp (see Smyth 1860a). The actual aperture of the completed telescope was 9.5 inches (24.1 cm), and delivery took place towards the very end of 1859 (Jacob 1860; Smyth 1860b; c.f. Crossley et al. 1879). Note that this was a new telescope that was specifically made for Fletcher by Cooke, and it was in no way associated with Miller. The parallels between the two men in relation to their instruments, observations, publications and FRAS election, and the common 'Fletcher' element in both their names, may go some way towards explaining King's garbled version of events. He may also have been misled by the statement that Miller's Cooke telescope was "… of the same size as Mr. Fletcher's instrument." (Crossley et al. 1879: 49). Crossley et al. were, of course, referring to Fletcher's smaller refractor.

Captain William Stephen Jacob (1813–1862; Director of the Madras Observatory) provides an account of Fletcher's new acquisition in an 1860 issue of *Monthly Notices of the Royal Astronomical Society*:

> The telescope in question is the property of our worthy Fellow, J. Fletcher, Esq., of Tarnbank, where it has been but recently erected by him; the optical portion by Cooke, of York; but the mounting, a long polar axis of cast-iron … has been executed under Mr. Fletcher's direction at an engineering foundry at Whitehaven, belonging to one of his brothers …
>
> The mounting seemed remarkably firm, and so smooth in its movement, that, with the telescope slightly under-counter-poised, a star could be well followed and kept pretty near the centre of the field by gentle pressure with the finger. On the whole, I should say it was creditable both to the designer and the maker, and that in such able hands it is likely to prove a most efficient instrument … (Jacob 1860: 248–249).

Accompanying the telescope were "… a double-image micrometer of varying powers, a wire micrometer with positive magnifiers to 1000, and a battery of negative eye-pieces ranging from 25 to 1500." (Smyth 1860b: 304).

W.H. Smyth's son Charles Piazzi Smyth (1819–1900; see Brück and Brück 1988) was a friend of Isaac Fletcher, and the Piazzi Smyth Collection in the Royal Observatory Edinburgh (Brück 1988) includes a diary documenting his visit to Greysouthen in April 1860:

> Isaac Fletcher describes his new equatorial. Object glass by Cooke, 9.^{i}5 sees 7 Satellites of Saturn and perhaps an 8th at least as much as Sir W. Herschel with a four foot reflector.
>
> Polar axis of home make i.e. at his brother's foundry. of English form with telescope on one side; cast in iron in one piece – 15ft long … (Smyth 1860a).

Of particular interest is the mounting, for this was the very first time that an all-metal English equatorial mounting had been used in telescope design and construction.[1] It was manufactured "… expressly at the Lowca Engine Works belonging to Messrs. Fletcher, Jennings, and Co. near Whitehaven." (Smyth 1860b: 302–305); as we have seen, Henry Allason Fletcher (1834–1884) was Isaac Fletcher's brother.[2] This innovative design, which was in part inspired by the smaller wooden mounting used by William Henry Smyth at his Hartwell Observatory (Smyth 1860b), is described in an 1865 issue of *Monthly Notices of the Royal Astronomical Society*:

> The middle portion of the axis, which is pierced for the declination axis, is a tube of 17 inches, the metal being one inch thick. From the central cube to each extremity the axis is circular in section, and tapers from 16 inches diameter at the cube to 8 inches near the ends, where it spreads out as a moulded flange to give sufficient strength to carry the pivots. The thickness of the metal in the conical portions diminishes gradually from ¾ inch to ⅜ inch. The cube is planed accurately, and the rest of the axis very carefully turned. The declination axis is of hammered iron, 4½ inches diameter; it is carried *direct* by the polar axis without the intervention of either bushes or friction rollers; a boss is cast on the *inside* of the cube on the telescope side, of sufficient depth to give a bearing 4½ inches long, and a similar boss is cast on the *outside* of the cube, on the counterpoise side, which gives a similar length of bearing. As there are no means of adjusting the declination axis, great pains were taken to bore out the bearings at right angles with the polar axis … One end of the declination axis carries a strong cradle of cast-iron, to which the tube is fixed by means of four very strong clasps; the opposite end carries the counterpoise. The pivots of the polar axis are of wrought-iron, 2½ inches diameter; they work in brass bearings without friction rollers, but have the necessary adjustments. The hour-circle is of gun metal, 42 inches diameter; the declination circle is cast from the same pattern, and reads off by opposite microscopes to single seconds …
>
> The total weight of the *moveable* portion of the mounting is upwards of 1½ tons, yet the instrument may be moved either in hour-angle or declination with a finger, and a clock-weight of 45 lbs. (acting on the clock of 22½ lbs.) is sufficient to drive it with the greatest regularity. (Fletcher 1865a: 242–243; c.f. Smyth 1860b: 303–304).

Smyth (1860a) also includes a number of interesting sketches in his diary. Fletcher (1865a: 243; his italics) was proud of this newfangled mounting, and was quick to proclaim that "In regard to *firmness*, I do not think this instrument is surpassed by any …" It was this 1865 paper which led Harper (1992; see also Harper et al. 1990) to erroneously associate this date with the actual manufacture of the telescope.

Although we have stressed the unique nature of the mounting, the telescope itself was an achievement as it was "… one of the largest telescopes yet made in

[1]However, the Fletcher refractor was not the first astronomical instrument with a part-iron English equatorial mounting. In 1775 Bird made an equatorial sector for the Radcliffe Observatory, and this featured a plywood laminated single iron axis, 7½ feet long, 3 inches square in the middle and tapering to about 2 inches at the ends (Chapman 1995). In 1993 the remains of this historic instrument were discovered by Allan Chapman and Tony Simcock in a 'store hole' at Oxford (A. Chapman, personal communication, August 2000).

[2]Fletcher, Jennings & Co. took over the business of Tulk and Ley in 1857. At the time they manufactured the telescope's equatorial mounting this engineering company was primarily involved in the production of locomotives. Lowca was the name of the village near Whitehaven where the factory was located (www.en.wikipedia.org/wiki/Fletcher_Jennings_%26_Co.).

England … and may serve to show that it is not now needful to go abroad in order to procure a first-rate instrument." (Jacob 1860: 247–248). Thomas Cooke (1807–1868) only opened the Buckingham Works at Bishopshill in 1855, and the Fletcher telescope was important for his burgeoning reputation (see King 1979; McConnell 1992; Smiles 1884). More specifically, an examination of the firm's first order book reveals that at the time it was made the Fletcher telescope ranked as Cooke's largest refractor, surpassing two 20.3-cm (8-in) instruments (for J. Nasmyth and W. Hartree) ordered in 1857 (Orders … n.d; c.f. Smiles 1884). Fletcher's telescope also rated highly on an international scale (see Table 11.1; after Crossley et al. 1879; Howse 1986; Krisciunas 1988; Welther 1984; Whitesell 1998), the equal largest operational refractors in the world at that time being the 38.1-cm (15-in) Merz and Mahler instruments at the Pulkovo and Harvard College Observatories (see van Helden 1984).

In order to house his new telescope Fletcher built a commodious new Tarn Bank Observatory (Fig. 11.1), with an 5.5-m (18-ft) diameter dome that revolved on eight railway wheels 30.5 cm (12 inches) in diameter (Smyth 1860b). Smyth's diary (1860a) contains details and sketches of the dome, and a plan of the Observatory, which housed a number of other instruments including a transit telescope by Simms and an astronomical clock by Frodsham. Fletcher believed this facility was one of "… the best and most complete observatories in private hands in existence." (Mr. Fletcher's Observatory … 1866).

Table 11.1 Some operational refractors with apertures exceeding 24.1 cm (9.5 inches) in 1859*

Aperture	Manufacturer	Observatory
39.4 cm (15.5 inches)	Merz and Mahler	Harvard College
37.9 cm (14.93 inches)	Merz and Mahler	Pulkovo
33.8 cm (13.3-inches)	Cauchoix	Colonel E.J. Cooper
32.5 cm (12.8 inches)	Merz and Mahler	Royal, Greenwich
30.5 cm (12 inches)	Merz and Mahler	Cincinnati
30.5 cm (12 inches)	Fitz	Detroit
29.5 cm (11.6 inches)	Cauchoix	Cambridge University
27.9 cm (11 inches)	Ross	Mr J.W. Grant
27.9 cm (11 inches)	Amici	Royal, Florence
25.4 cm (10 inches)	Merz and Mahler	Kazan University
24.8 cm (9.75 inches)	Fitz	West Point
24.4 cm (9.6 inches)	Fraunhofer	Dorpat
24.4 cm (9.6 inches)	Merz and Mahler	U.S. Naval
24.4 cm (9.6 inches)	Fraunhofer	Royal, Berlin

*Compiling this table was a challenge as different authors sometimes give differing apertures for the same instrument, depending on their sources of documentation, the effects of rounding (e.g. 9.6-in becomes 9.5-in and 14.93-in and 15.5-in both become 15-in), and whether total or clear apertures were involved. Where such variations occurred, I generally (but not always) opted for the dimensions given in Howse (1986).

Fig. 11.1 Fletcher's residence and his Tarn Bank Observatory (*Courtesy* Royal Astronomical Society)

Initially, Fletcher used the Cooke telescope to examine the double star, ζ Herculis (Fletcher 1865b), and his observations of solar granulation clarified the controversy over their true nature by reinforcing the views of Frederick William Herschel (1738–1822) and William Rutter Dawes (1799–1868) and conflicting with the explanation put forward by James Hall Nasmyth (1808–1890; see Fletcher 1865c, d). But he had much grander plans for the Tarn Bank Observatory, which he believed boasted "… one of the finest telescopes in this country …" (Fletcher 1865d: 25). He intended to re-observe the objects in Smyth's Bedford Catalogue, and

> … when that is done, it is his intention to bring out a new edition of Admiral Smyth's *Cycle of Celestial Objects*, for which purpose the Admiral, some time ago, made over to him his entire interest in that work. (Mr. Fletcher's Observatory … 1866; c.f. Smyth 1860b).

W.H. Smyth died in 1865, and Fletcher was left with what can only be described as a mammoth undertaking. After all, Smyth's two-volume book, published in 1844, ran to a little over 1100 pages of text, more than half of which comprised the remarkable 'Bedford Catalogue'. Chapman (1998: 80) has written an excellent evaluation of this indispensable tool of the nineteenth century observational astronomer:

> The *Cycle* provides a wealth of information about the equipping of observatories and an evaluation of instruments … But it is in the second volume of the 1844 *Cycle*, known as the *Bedford Catalogue*, that we see the celestial 'harvest' of which Smyth spoke. Here is a detailed description and analysis, occupying some 543 pages, of hundreds of binary, variable, and coloured stars in 70 constellations. No one can read these pages without appreciating the thousands of hours of sheer hard observing, undertaken purely for the 'love' of astronomy, that lay behind Smyth's book.

Smyth based his 'Bedford Catalogue' upon observations carried out with a 15-cm (5.9-in) Tully refractor (see the diagram in Smyth 1844, I: 338), a much smaller instrument than Fletcher's 24.1-cm Cooke, so different field sizes and limiting magnitudes were involved, particularly important factors when observing nebulae and star clusters. Moreover, all of the double stars needed to be re-measured as their position angles and separations would, in most instances, have changed since Smyth's initial observations in the 1830s.

Fletcher responded to the challenge and for several years he "... systematically collected materials for the revision of the work." (Smyth and Chambers 1881: vii; cf. Mr. Fletcher's Observatory 1867), although one cannot help but wonder if he completely understood the magnitude of the task before him. As it turned out, a non-astronomical factor put paid to this ambitious project once and for all, for in 1868 Fletcher "... exchanged in great part his scientific career for a political one by becoming Member of Parliament for Cockermouth." (Smyth and Chambers 1881: vii). One consequence of this was that no further papers in Fletcher's name were to appear in the *Monthly Notices* or in the *Memoirs* of the Royal Astronomical Society.

Fig. 11.2 Mr Chatwood's Observatory. The man with the sextant is the British astronomer, Dr Alfred H. Fison (*Courtesy* Ward (Wanganui) Observatory Archives)

Fletcher finally took his own life in 1879, at the age of 52 (Obituary 1880), and by that time the 'Bedford Catalogue Project' had passed to George Frederick Chambers (1841–1915). The 'Revised, Condensed, and Greatly Enlarged' Second Edition emerged from the Clarendon Press at Oxford in 1881. As a single weighty tome of about 700 pages (see Smyth and Chambers 1881), this was a worthy successor to Smyth's original Volume II.

After Fletcher's death, the 24.1-cm (9.5-in) Cooke refractor passed to Samuel Chatwood (1833–1909), the successful and "… well-known bankers' engineer …" (Baker 1902) who owned the Chatwood Lock and Safe Company (Beaumont 1963b). Chatwood does not disclose whether he acquired the telescope from the estate immediately after Fletcher's death or at a later date, but we do know that he installed it in a new observatory adjacent to his home at Worsley, near Manchester (see Figs. 11.2 and 11.3). If the singular lack of published papers in the *Journal of*

Fig. 11.3 A view of the Cooke telescope and mounting at Mr Chatwood's Observatory (*Courtesy* Ward (Wanganui) Observatory Archives)

the British Astronomical Association, *Monthly Notices of the Royal Astronomical Society* and *The Observatory* is anything to go by, we must assume that Chatwood made little if any attempt to use this excellent telescope for serious research before selling it, in 1902, to the Wanganui Astronomical Society in New Zealand.

11.3 The New Zealand History of the Telescope

In this section, we will discuss separately the transfer of the telescope to New Zealand and its subsequent role and function.

11.3.1 Transfer to New Zealand

The transfer of the telescope to New Zealand was affected by Joseph Thomas Ward (see Fig. 11.4), a prominent New Zealand amateur astronomer during the first three decades of the twentieth century and a member of the British Astronomical

Fig. 11.4 Joseph Ward (Orchiston collection)

Association (see Beaumont 1963a; Calder 1978a; Orchiston 1996b). Born in England on 7 April 1862, Ward migrated to New Zealand in about 1880 and in 1896 he and his wife settled in the North Island coastal town of Wanganui (which is about 150 km due north of the capital, Wellington). There he opened a bookshop and stationery business, and also taught the violin. From an early age he had been interested in astronomy, and in about 1899 he purchased an 11.4-cm (4.5-in) refracting telescope. When the Great Comet of 1901 (C/1901 G1) appeared, townsfolk flocked to view it through this instrument.

In this same year Richard Cockburn Maclaurin (1870–1920), Professor of Mathematics at Victoria University of Wellington, gave a public lecture on astronomy in Wanganui, and Ward used this as the catalyst to form the Wanganui Astronomical Society. As might be expected, Ward was elected to the Presidency, a post that he was to occupy until his death in 1927. One of the first decisions made by the new Society was to establish an observatory, and the committee set about fund-raising. They also obtained a suitable site in suburban Cook's Gardens from the Borough Council. Ward, meanwhile, was busy searching for "… a very large telescope capable of research as well as public observation …" (Beaumont 1963b: 6).

Thrush (1963) recounts how Ward succeeded in tracking down three suitable instruments in England, 20.3-cm (8-in) Grubb and Clark refractors and a 52.1-cm (20.5-in) Calver reflector, and by October 1901 the Society had decided to purchase this last instrument. After part of the purchase price of £400 had been dispatched to the Bolton firm of Banks & Co., a cable arrived from a Mr Chatwood advising that the payment should be stopped as he was the owner of the reflector and Banks & Co. had not been authorized to sell it!

Fortuitously, Chatwood also owned the ex-Fletcher 24.1-cm (9.5-in) Cooke refractor with the prototype metal English equatorial mounting, and he decided to make that instrument available to the Society. Negotiations on his behalf were entered into by Charles Baker of London, and in a letter dated 27 March 1902 Baker stressed the telescope's optical quality:

> I have had consultations with Mr Maunder, and with Mr W.H. Maw, Treasurer of the B.A.A. who express themselves satisfied with the instrument; and Mr Chatwood informs me that Dr Fison, Mr Antoniadi and Sir Robert Ball, who have used his instrument, all say that the object glass is a beauty. (Baker 1902).

As an added incentive, Baker offered an extremely attractive purchase price of just £400, but when the cost of constructing an observatory was added to this the Society faced a daunting task for a city with a population of just 6000 since "The [total] amount needed would be at least the equivalent of two years salary for the average citizen." (Venimore n.d.: 12). In fact, the price soon proved to be even higher, for the Society agreed to pay an additional £50 to Mr Chatwood for some improvements he made to the telescope and for some extra equipment that he added to the consignment.

Yet the all-up cost did not deter Ward and his colleagues, and at their meeting on 2 May 1902 the Committee voted to accept Baker's offer. Chatwood (1902) was quick to point out that the Society had got itself a bargain, for "Under other circumstances … I would not have sold the instrument for less than £750. It could not be replaced [today] for less than £1250 …" The instrument was then dismantled and packed ready for shipping on *S.S. Indravedi*. In December 1902 the vessel reached New Zealand, and the telescope was placed in storage pending the construction of an observatory.

The Society's 6.1-m (20-ft) diameter 16-sided wooden observatory was designed by a local architect named Atkins from drawings and a model provided by Ward, and its construction was completed in early May 1903 (Fig. 11.5) at a cost of £277-12-6. The massive supports for the telescope mounting were a special feature, and weighed heavily on the mind of a 4-year old boy who 75 years later was to recall how he "… clearly remembers the day the heavy cast-iron pillar of the

Fig. 11.5 The completed Ward (Wanganui) Observatory (*Courtesy* Ward (Wanganui) Observatory Archives)

mounting was erected. His father provided the lifting gear the foundry used … and a gang of willing helpers provided the manpower that lifted the pillar into place." (Calder 1978b: 4). This pillar and the 40-ton concrete foundation made to support the lower end of the polar axis are shown in Fig. 11.6, along with the outline of the observatory building.

Now came the installation of the telescope and mounting, in good time for the official opening of the observatory by the Premier of New Zealand, the Right Honourable Richard Seddon, on 25 May 1903. Beaumont (1963b: 7) describes what a perceptive Seddon would have seen:

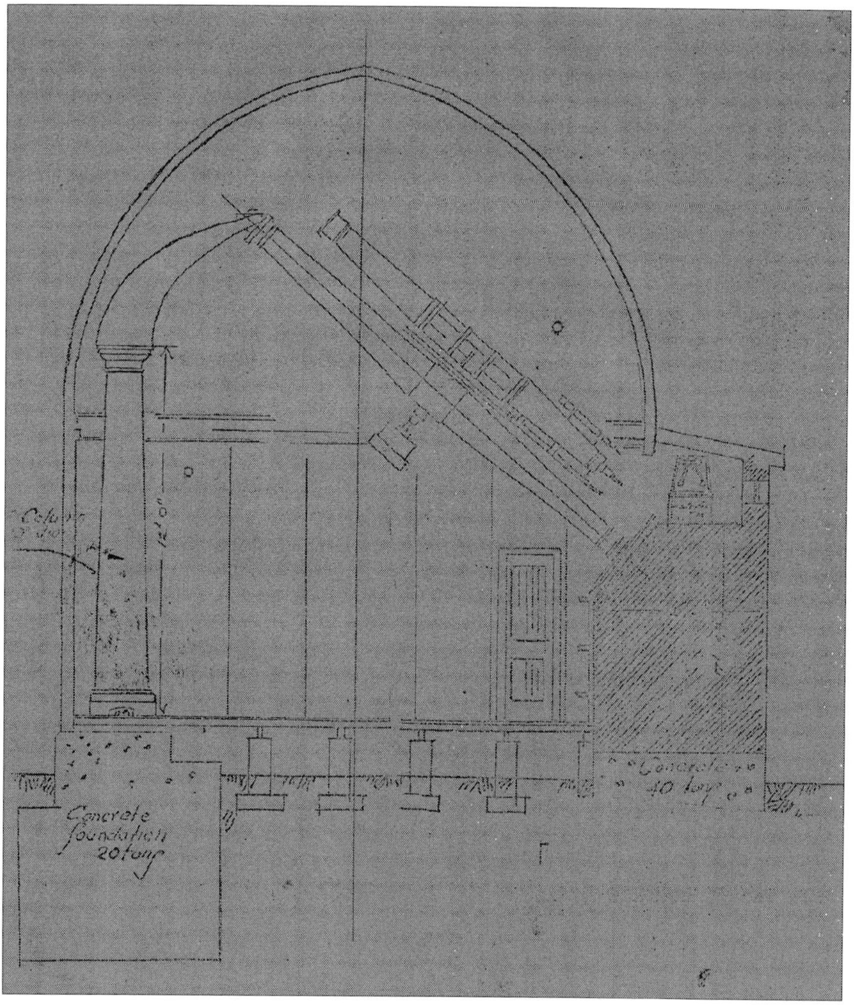

Fig. 11.6 Cut-away cross-section showing the Ward (Wanganui) Observatory, the telescope and the mounting (*Courtesy* Ward (Wanganui) Observatory Archives)

> The 9½-inch refractor has a 12-ft. focus and is equipped with a 2½-inch finder. There is a battery of eyepieces magnifying from 32 to 750 diameters; filar and position micrometer with divided heads in silver, position circle on silver, reading by opposite verniers; Cooke's two-movement adaptor for centring an object; solar and stellar diagonal eyepieces; Dawes eyepiece for observing the Sun; aluminium shutter to object glass. The telescope is mounted on what is known as the English form of equatorial. It is supported on a massive axis, weighing nearly two tons, which rests at one end on a large iron bracket, which is fastened to a cast-iron column 12 ft. high of architectural proportions. The lower support for the axis is [an] adjustable stopping-piece of brass resting on iron bearings, and moved by set screws in azimuth … A powerful driving clock is attached to the edge of the circle at the lower edge of the large axis, which is kept in motion by a weight attached to a steel cord, and working on the clock at a pressure of 22½ lbs. By this manner an object can be kept in the telescope in the centre of its field for purposes of photography, drawing or measurement.

Figure 11.7 shows a close-up view of the lower end of the polar axis, the oversize 1.07-m (42-in) diameter R.A. circle, and the eyepiece end of the main telescope tube and finder, complete with R.A. and declination clamps and slow motion controls.

Ward was appointed Honorary Director of the Observatory, a post that he was to retain until his sudden and unexpected death on 4 January 1927 at the age of 65. Nine months earlier, ownership of the Observatory had been transferred from the Astronomical Society to the Borough Council. One of Joseph Ward's sons who

Fig. 11.7 View of the lower end of the polar axis and main telescope tube (Orchiston collection)

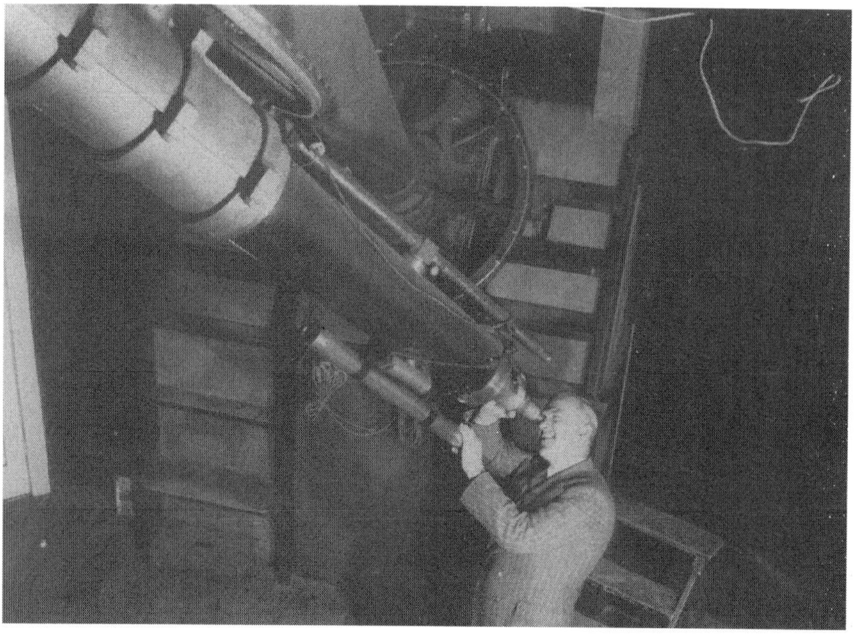

Fig. 11.8 W.H. Ward observing with the telescope (Orchiston collection)

happened to inherit a passion for astronomy was prophetically named William Herschel Ward (1900–1973), and in 1927 the local Council arranged for him to succeed his father as Honorary Director of the Observatory (see Fig. 11.8), a role that he fulfilled with distinction until 1959 (W.S.T. 1975).

The Observatory is nowadays known officially as the 'Ward (Wanganui) Observatory' in honour of its founder (Beaumont 1963b), and in spite of its antiquity the historic Cooke telescope was until recently the largest operational refractor in New Zealand when it was surpassed (but only just) by the better-known 24.8-cm (9.7-in) Cooke at the Carter Observatory in Wellington (see Orchiston 1996a, 2002a; Chap. 13 in this book).

11.3.2 Role of the Telescope in Recreation and Research

Joseph Ward was quick to recognize the potential of what was then by far the largest refractor in New Zealand, and he was keen to find a suitable research program for it. He began (Ward 1903) by canvassing the views of Australia's leading astronomer, John Tebbutt (see Orchiston 2004), and in 1904 he and his assistant, local lawyer Thomas Allison (1858–1926), commenced a search for new double stars in selected areas of the southern sky, mainly "… along the southernmost sections of the Milky Way among the constellations Triangulum Australe,

Centaurus and Musca, between −50° and −80° south declination." (Harper et al. 1990: 283). In the course of the next six years they made 212 discoveries (see Warren 1991), although unbeknown to Ward many of these had previously been detected by others (Innes 1911). In 1990, 88 of these stars were still recognized as 'Ward doubles' (Harper et al. 1990) and appeared with NZO (New Zealand Observatory) listings in international double star catalogues, thereby serving as a memorial to Ward's international contribution in this specialized field of astronomy.

In addition to double stars, Ward also observed sunspots (Venimore 1988), and 15 drawings of Mars that he made in 1905 were forwarded to the British Astronomical Association (see Antoniadi 1910). He also used the Cooke refractor to carry out numerous observations of Comet 1P/Halley in 1910 (see Mackrell 1985).

It should be mentioned that in addition to using telescopes Ward was one of that rare breed that also enjoyed manufacturing them, and over the years he made many refractors and reflectors, most of modest aperture. Undoubtedly his crowning achievement was a 52.1-cm (20.5-in) equatorially-mounted Newtonian reflector (see Fig. 12.3). This was completed in 1924 (Venimore 1988), and for 40 years remained the largest reflecting telescope manufactured by a New Zealand amateur astronomer. Further information on Ward's pioneering efforts in telescope-making is provided in Orchiston (2002b) and in the following chapter of this book.

Another of Ward's important contributions to New Zealand astronomy was in popularizing the wonders of the southern sky, and two evenings a week he ran 'public viewing nights' at the Wanganui Observatory (Venimore 1988) where he used the old Cooke telescope to eagerly share his love and knowledge of astronomy with visitors. As one writer so aptly put it, "… he loved the telescope to read the open volume of the sky." (cited by Beaumont 1963a: 14). Meanwhile, through his monthly column, "Astronomical Notes", which appeared in the *Wanganui Herald* newspaper for twenty-two consecutive years from 1904, he was able to bring astronomy to a much wider audience, and help raise the level of astronomical literacy in the general population.

Despite these various achievements, some have painted an even more glowing picture of Ward, suggesting that he proposed "… several theories concerning the movement and nature of heavenly bodies, which were accepted by leading authorities." (Beaumont 1963a: 13); that "In his time he was considered one of the world's foremost astronomers." (Beaumont 1963b: 7); and that "His writings were more widely known abroad than in New Zealand due to their publication in many British and French astronomical journals." (Rice 1982: 272). In reality, there is no basis for these exaggerated claims. Although Ward did indeed discover a number of new double stars (as, also, did other astronomers at about this time), he actually published nothing of substance in international journals (but see Venimore 1988: 156; Ward 1906, 1910). He was not even destined to publish his own double star discoveries. In 1907 he sent a list of these to the British Astronomical Association but they decided not to publish this (Harper et al. 1990), and in 1911 his catalog was eventually published in South Africa by Scottish-born R.T.A. Innes (1861–1933; Fig. 11.9), the founding Director of the Transvaal Observatory and a keen double star observer who got his grounding in this area while living in Sydney, Australia, during the 1890s (see Orchiston 2001a, 2003, 2015). Meanwhile, by this time Ward (1908) had managed to publish a half-page note about his discoveries in the *Journal of the British Astronomical Association*. Yet these comments about Ward's inability to publish prolifically should not blind us to his notable contribution to Wanganui and New Zealand astronomy.

What of the Ward (Wanganui) Observatory in more recent times? It is only fitting justice that the observing tradition begun by Fletcher in the 1860s and followed by Ward has in more recent times been perpetuated by Ormond Warren. During the 1980s and 1990s, he used the historic Cooke telescope and its original Cooke Type-A bifilar micrometer to carry out a re-examination of the 'Ward doubles' (e.g. see Warren 1991, 1995; Warren et al. 2000) and to measure the position angles and separations of other southern double stars (e.g. Warren 1992a, b, 2000). Meanwhile, the dedication to astronomical education espoused by Joseph Ward and his son (Thrush 1963) is a current priority with the Wanganui Astronomical Society (see Harper 1992), and the Cooke refractor remains the focal point of their regular public viewing nights.

As to the telescope itself, a critical examination of the 24.1-cm (9.5-in) achromatic objective in 1994 by one of New Zealand's leading professional telescope-makers, the late Garry Nankivell (1929–2001), confirmed the earlier evaluations of Sir Robert Ball (1840–1913) and Eugène Michel Antoniadi (1870–1944) as to its performance. In spite of the passage of the years, Nankivell found that it was

Fig. 11.9 R.T.A. Innes, who helped publicize Ward's double star discoveries (*Courtesy* South African Astronomical Observatory)

Fig. 11.10 A 1999 photograph of the 24.1-cm (9.5-in) Cooke refractor and 22.8-cm (9-in) Houghton-Rumsey Astrograph, along with the main telescope's distinctive English equatorial mounting (Orchiston collection)

> … in a very good state of preservation. The crown lens shows some slight incipient smearing of the front surface … [but] There are very few scratches of any consequence. The flint lens is free of any obvious tarnish patches. There are one or two "lead spots" about 2-3 mm diameter. Some localised water stains were noted close to the edge …
>
> With the provision of reasonable care, this objective should perform well for another 100 years. (Nankivell 1994: 9).

The present appearance of the telescope is impressive, with the cast-iron polar axis painted an attractive shade of blue and the declination axis (plus counterweight) and tube of the main telescope a pale grey. The natural brass finish of the eyepiece assembly and of the finder has been retained (see Fig. 11.10).

11.4 Concluding Remarks

One of New Zealand's two largest operational refractors is the 24.1-cm (9.5-in) f/15 Cooke telescope at the Ward (Wanganui) Observatory, which has some claim to fame as the first telescope in the world to be furnished with a cast-iron (as opposed to wooden) English equatorial mounting of the 'Cross Axis' type.

This instrument was manufactured in 1859 for Isaac Fletcher of Carlisle, northern England, and was later owned by Samuel Chatwood of Manchester before transferring to New Zealand and the observatory of the Wanganui Astronomical Society in 1902–1903. Fletcher and Joseph Thomas Ward (at Wanganui) both used the telescope for serious positional astronomy (including the observation of known double stars and the search for new ones) and published a number of papers on their work, while at Wanganui the telescope also was the focal point of a popular 'public viewing' program.

After a heritage of more than 150 years this historic telescope is in remarkably fine condition, and it continues to contribute to the popularization of astronomy.

Acknowledgements I am grateful to the following for their assistance: Samantha Bennett (South African Astronomical Observatory, Cape Town), Alison Brech (The Borthwick Institute of Historical Research, York), the late Dr Mary Brück (Penicuik, Scotland); the late David Calder (Wanganui, NZ); Dr Allan Chapman (Oxford University); Dr Richard Crossley (The University of York); the late Peter Hingley (former RAS Librarian, London); the late R.F. Joyce (Kaiapoi, NZ); the late Ray Lee (Wanganui, NZ), the late Garry Nankivell (Wellington, NZ); the late Sandra Ricketts (the former Anglo-Australian Observatory Librarian, Sydney); the late Jackie St. George (Sydney); the late Reverend Colin Venimore (Wanganui, NZ), Ormond Warren (Auckland, NZ), and staff at the Public Library of New South Wales (Sydney). I also am grateful to the Donovan Astronomical Trust for helping fund my archival research in England. Finally, I wish to thank John Drummond (Patutahi, NZ) and Rod Austin (New Plymouth, NZ) for reading and commenting on the first draft of this chapter, and the Royal Astronomical Society, the South African Astronomical Observatory and the Wanganui Astronomical Society (through the late David Calder) for kindly supplying Figs. 11.1, 11.2, 11.3, 11.5, 11.6 and 11.9.

References

Antoniadi, E. M. (1910). Section for the observation of Mars. Report of the Section, 1905. *Memoirs of the British Astronomical Association*, *17*(2), 35–64.

Baker, C. (1902). Letter to J.T. Ward, dated March 27. Original in Ward (Wanganui) Observatory Archives.

Beaumont, L. R. H. (1963a). J.T. Ward, founder of Wanganui's Observatory, made important contributions to science. *Journal of the Auckland Astronomical Society*, *21*(2), 13–14.

Beaumont, L. R. H. (1963b). Wanganui Observatory: the story of its foundation. *Journal of the Auckland Astronomical Society*, *21*(2), 6–7, 12.

Brück, M. T. (1988). The Piazzi Smyth collection of sketches, photographs and manuscripts at the Royal Observatory Edinburgh. *Vistas in Astronomy*, *32*, 371–408.

Brück, H. A., & Brück, M. T. (1988). *The Peripatetic Astronomer: The Life of Charles Piazzi Smyth*. Bristol: Adam Hilger.

Calder, D. (1978a). Joseph Ward: pioneer astronomer and telescope maker. *Southern Stars*, *27*, 104–108.

Calder, D. (1978b). Seventy-five years of astronomy: Ward Observatory, 1903–1978. Lecture given to the Wanganui Astronomical Society, May 24, 1978.

Chapman, A. (1998). *The Victorian Amateur astronomer. Independent astronomical research in Britain 1820–1920*. Chichester: Wiley.

Chapman, A. (1995). Out of the meridian: John Bird's equatorial sector and the new technology of astronomical measurement. *Annals of Science*, *52*, 431–463.

Chatwood, S. (1902). Letter to J. Ward, dated August 16. Copy in Ward (Wanganui) Observatory Archives.

Crossley, E., Gledhill, J., & Wilson, J. M. (1879). *A Handbook of Double Stars* ... London: Macmillan.

Fletcher, I. (1850). Brief description of a small equatoreal and dome at Tarn Bank, near Cockermouth. *Monthly Notices of the Royal Astronomical Society, 10*, 137–139.

Fletcher, I. (1853). Results of micrometrical measures of double stars made at Tarn Bank, Cumberland, from 1850.2 to 1853.4. *Monthly Notices of the Royal Astronomical Society, 13*, 256–259.

Fletcher, I. (1855). Remarks on Mr. Powell's elements of the orbit of 70 Ophiuchi. *Monthly Notices of the Royal Astronomical Society, 5*, 161–162.

Fletcher, I. (1865a). Description of an equatorial mounting. *Monthly Notices of the Royal Astronomical Society, 25*, 241–243.

Fletcher, I. (1865b). Note on ζ Herculis. *Monthly Notices of the Royal Astronomical Society, 25*, 275.

Fletcher, I. (1865c). On the photosphere of the Sun. *Monthly Notices of the Royal Astronomical Society, 25*, 231–233.

Fletcher, I. (1865d). Some remarks on the solar photosphere. *Monthly Notices of the Royal Astronomical Society, 26*, 23–26.

Gingerich, O. (1967). What is an English mounting? *Sky and Telescope, 34*(4), 293–295.

Gingerich, O. (Ed.), (1984). *The General History of Astronomy. Volume 4. Astrophysics and Twentieth-Century Astronomy to 1950: Part A*. Cambridge: Cambridge University Press.

Harper, C. T. (1992). Teaching astronomy in New Zealand: The Ward Observatory. In J. M. Pasachoff & J. R. Percy (Eds.), *The Teaching of Astronomy* (pp. 354–357). Cambridge: Cambridge University Press.

Harper, C. T., Warren, O., & Austin, R. (1990). J.T. Ward and the NZO double stars. *Southern Stars, 33*, 281–294.

Hingley, P. D. (n.d.). The Shuckburghs of Shuckburgh, Isaac Fletcher, and the history of the English mounting (unpublished manuscript).

Howse, D. (1986). The Greenwich list of observatories: a world list of astronomical observatories, instruments and clocks, 1670–1850. *Journal for the History of Astronomy, 17*.

Innes, R. T. A. (1911). Observations of southern double stars. *Transvaal Observatory Circular, 6*, 61–72.

Jacob, W. S. (1860). Account of the performance of Mr. Fletcher's new equatoreal. *Monthly Notices of the Royal Astronomical Society, 20*, 247–249.

King, H. C. (1979). *The History of the Telescope*. New York: Dover.

Krisciunas, K. (1988). *Astronomical Centers of the World*. Cambridge: Cambridge University Press.

Mackrell, B. (1985). *Halley's Comet Over New Zealand*. Auckland: Reed Methuen.

McConnell, A. (1992). *Instrument Makers to the World. A History of Cooke, Troughton & Simms*. York: William Sessions.

Miller, J. F. (1852). Micrometrical measurements of the binary star ξ Ursae Majoris, and the double star Struve 1263, made at the Observatory, Whitehaven. *Monthly Notices of the Royal Astronomical Society, 12*, 170–172.

Miller, J. F. (1853). Micrometrical measurements of double stars made at the Observatory, Whitehaven, in the years 1851, 1852 and 1853. *Memoirs of the Royal Astronomical Society, 22*, 187–188.

Mr. Fletcher's Observatory. *Monthly Notices of the Royal Astronomical Society, 27*, 131 (1867).

Mr. Fletcher's Observatory at Tarnbank. *Monthly Notices of the Royal Astronomical Society, 26*, 143–144 (1866).

Nankivell, G. R. (1994). The 9.5 inch Cooke objective of the Wanganui Observatory. *Southern Stars, 36*, 1–9.

Obituary: Isaac Fletcher. *Monthly Notices of the Royal Astronomical Society, 40*, 194 (1880).

Obituary: John Fletcher Miller. *Monthly Notices of the Royal Astronomical Society, 17*, 99–100 (1857).

Orchiston, W. (1996a). The historic 23-cm refracting telescope at the Carter Observatory. In W. Orchiston & B. Carter (Eds.), *Astronomical Handbook for 1997* (pp. 59–69). Wellington: Carter Observatory.

Orchiston, W. (1996b). Ward, Joseph Thomas 1862–1927. In C. Orange (Ed.), *The Dictionary of New Zealand Biography. Volume Three 1901–1920* (p. 555). Auckland: Auckland University Press and Department of Internal Affairs.

Orchiston, W. (2000). Illuminating incidents in Antipodean astronomy: H.C. Russell and the origin of the Palomar-type mounting. *Journal of the Antique Telescope Society, 19*, 13–15.

Orchiston, W. (2001a). From amateur astronomer to observatory director: the curious case of R.T. A. Innes. *Publications of the Astronomical Society of Australia, 18*, 317–328.

Orchiston, W. (2001b). The English Equatorial Mounting and the history of the Fletcher Telescope. *Journal of Astronomical History and Heritage, 4*, 29–42.

Orchiston, W. (2002a). From Crossley to Carter: the life and times of an historic Cooke refractor. *Journal of the Antique Telescope Society, 24*, 9–24.

Orchiston, W. (2002b). Joseph Ward: pioneer New Zealand telescope-maker. *Southern Stars, 41*, 13–21.

Orchiston, W. (2003). Passion and patronage: the Afro-Australian astronomy of Robert Thorburn Ayton Innes. *Monthly Notices of the Astronomical Society of Southern Africa, 62*, 4–13.

Orchiston, W. (2004). John Tebbutt and observational astronomy at Windsor Observatory. *Journal of the British Astronomical Association, 114*, 141–154.

Orchiston, W. (2015). The amateur-turned-professional syndrome: two Australian case studies. In W. Orchiston, D. A. Green, & R. Strom (Eds.), *New Insights From Recent Studies in Historical Astronomy: Following in the Footsteps of F. Richard Stephenson* (pp. 259–350). New York: Springer.

Orders. Astronomical & mathematical instruments. York: Order Book, Borthwick Institute (Cooke Archives).

Rice, P. (1982). The story of Ward Observatory. *Southern Stars, 29*, 271–274.

Smiles, S. (1884). *Men of Invention and Industry*. London: Murray.

Smyth, C. P. (1860a). Unpublished diary. Royal Observatory Edinburgh.

Smyth, W. H. (1844). *A Cycle of Celestial Objects, For the Use of Naval, Military, and Private Astronomers* (2 Vols.). London: Parker.

Smyth, W. H. (1860b). *The Cycle of Celestial Objects Continued at the Hartwell Observatory to 1859* …. London: Nichols and Sons.

Smyth, W. H., & Chambers, G. F. (1881). *A Cycle of Celestial Objects*. Oxford: Clarendon Press.

Thrush, W. S. (1963). Astronomy in Wanganui. *Southern Stars, 20*, 73–76.

Van Helden, A. (1984). Telescope building, 1850–1900. In Gingerich (pp. 40–58).

Venimore, C. (1988). *Good Viewing (A History of the First Wanganui Astronomical Society and the Ward Observatory)*. Unpublished manuscript.

W.S.T. (1975). Obituary. William Herschel Ward (1900–1973). *Southern Stars, 26*, 93–94.

Ward, J. T. (1903). Letter to John Tebbutt, dated November 20. Original in *Letters to J. Tebbutt 1903*. Sydney: Mitchell Library (A 3715).

Ward, J. T. (1906). The lunar eclipse of August 4. *Journal of the British Astronomical Association, 17*, 32–33.

Ward, J. T. (1908). The Wanganui Observatory. Observations of two hundred and twelve new double stars. *Journal of the British Astronomical Association, 18*, 122.

Ward, J. T. (1910). The total eclipse of the Moon, 1909, November 26. *Journal of the British Astronomical Association, 20*, 205–206.

Warren, O. C. R. (1991). The NZO double stars. *The Web Society Quarterly Journal, 85*, 1–7.

Warren, O. C. R. (1992a). Micrometrical measurements of visual double stars. *Southern Stars, 35*, 373–380.

Warren, O. C. R. (1992b). Micrometrical observations of southern double stars. *Southern Stars, 35*, 12–14.

Warren, O. C. R. (1995). Survey of NZO double star measures from 1989 to 1995. Paper presented at the Annual Conference of the Royal Astronomical Society of New Zealand.

Warren, O. C. R. (2000). Micrometric measures of southern double stars. *Southern Stars, 38*, 217–222.

Warren, O. C. R., Lee, R., & Austin, R. R. D. (2000). Micrometrical measures of NZO double stars. *Southern Stars, 39*(1), 9 12.

Welther, B. L. (1984). The world's largest telescopes, 1850–1950. In Gingerich (pp. Ai–Avi).

Whitesell, P. S. (1998). *A Creation of his Own: Tappan's Detroit Observatory*. Ann Arbor: University of Michigan.

Chapter 12
Joseph Ward: Pioneer New Zealand Telescope-Maker

Abstract During the first quarter of the twentieth century Joseph Ward was one of New Zealand's leading amateur astronomers, and he is best remembered as the founder of the Ward (Wanganui) Observatory, for his Ward double stars, and as a pioneer telescope-maker. Over the years he manufactured many reflectors and small refractors, and some of these were acquired by astronomical societies and by other well-known New Zealand astronomers (including Ronald McIntosh and Albert Jones). For many years the 52.1-cm Newtonian reflector that he completed in 1924 remained the largest telescope made by a New Zealand amateur astronomer.

12.1 Introduction

Amateur telescope-makers have a long international history of achievement, best exemplified during the nineteenth century by people like George Calver (1834–1927), Andrew Ainslie Common (1841–1901), William Lassell (1799–1880), James Nasmyth and William Parsons, the Third Earl of Rosse (1800–1867; see Chapman 1998; King 1979).

Scientific astronomy came late to New Zealand with the arrival of James Cook on the *Endeavour* (Orchiston 1998), and a local amateur telescope-making tradition only began to emerge during the 1870s through the efforts of Thomas Cheeseman in Auckland (for New Zealand localities mentioned in the text see Fig. 12.1). Very little has been written about him—largely because all of his records were destroyed after his death in 1907—but he is reputed to have made both reflectors and refractors (see McIntosh 1959).

While Cheeseman may have a legitimate claim as the 'founding father' of amateur telescope-making in New Zealand, this label is usually assigned to Joseph Ward of Wanganui, a man with "… that rare combination—the astronomer of philosophical mind and mechanical genius … [who] built many of the largest

W. Orchiston, *Exploring the History of New Zealand Astronomy*, Astrophysics and Space Science Library 422,
DOI 10.1007/978-3-319-22566-1_12

Fig. 12.1 New Zealand localities mentioned in this chapter are shown in *red*

instruments in New Zealand." (Obituary 1927; see, also, the previous chapter in this book). Back in 1978 the late George Eiby wrote:

> We certainly ought to have an adequately documented account of the extent and quality of J.T. Ward's telescope making, and of the sources of his expertise. Would it be too much to expect a *catalogue raisonné*, listing the characteristics and present ownership of at least his larger instruments? (Eiby 1978: 123).

After providing some general background information about Ward, this chapter attempts to address some of the points raised by Eiby, and as such builds on a paper I published more than a decade ago (Orchiston 2002) on Ward's telescope-making activities.

12.2 Joseph Ward: A Biographical Sketch

Joseph Thomas Ward (see Fig. 12.2) was one of New Zealand's foremost amateur astronomers during the first two decades of the twentieth century (see Beaumont 1963; Calder 1978a; Orchiston 1996). He was born in England in 1862, and after migrating to New Zealand he worked for several years as a shepherd and shearer before settled in Wanganui in 1896.

As we saw in the previous chapter, Ward was instrumental in founding the Wanganui Astronomical Society and establishing the Society's Wanganui Observatory, which was opened in 1903. He also sourced a telescope for the Observatory, a 24.1-cm (9.5-in) Cooke refractor with the first all-metal English equatorial mounting of the cross-axis type (see Orchiston 2001a). Ward and his assistant, local lawyer Thomas Allison,[1] then used the telescope to search for new double stars and in the course of the next six years they made more than 200 discoveries.

One of Ward's other loves was to popularize astronomy, and he ran twice-weekly 'public viewing nights' at the Observatory. He also took astronomy to

[1]As well as being a noted Wanganui solicitor, Thomas Allison has another claim to fame: he was associated with the introduction of the Chinese gooseberry (*Arguta deliciosa*)—better known as 'kiwifruit'—to New Zealand:

> In January 1904, Isabel (Miss M I) FRASER, who was the headmistress of Wanganui Girls' College, returned to New Zealand after visiting her sister Katie, (Miss Catherine Graham FRASER) a missionary teacher at the Scottish Mission at Yichang, China.
>
> History records that after Isabel tasted the "Ichang gooseberry" she brought seeds back with her and gave them to Thomas ALLISON, a noted Wanganui solicitor and orchardist. He passed them on to his brother, Alexander ALLISON, also a noted horticulturist. By 1910, Alexander ALLISON had plants producing fruit at his property Letham, at Marangai, State Highway Three, [about 10 km] south of Wanganui. These were probably the first kiwifruit plants grown in New Zealand. (History of the Kiwifruit Industry 2004).

Fig. 12.2 Joseph Ward posing beside a 25.4-cm (10-in) Newtonian reflector that he made (*Courtesy* Ward (Wanganui) Observatory Archives)

a much wider audience through a monthly column that appeared in the *Wanganui Herald* newspaper (Venimore 1988).

A quiet, unassuming, gentle, kindly man (ibid.), Joseph Thomas Ward died of peritonitis on 4 January 1927.

12.3 Ward Telescopes

Apart from his double star and educational achievements, Ward was a pioneer New Zealand telescope-maker (Calder 1978a), and made both reflectors and refractors. The reflectors ranged in aperture from 12.7 cm (5 inches) to 52.1 cm (20.5 inches) and the refractors from 5.1 cm (2 inches) to 10.2 cm (4 inches). Ward was so prominent in this field that in 1931 the well-known Wellington astronomer, Algernon Charles Gifford (1861–1948; Eiby 1972; Jenkinson 1940: 125–136) was convinced that most of the reflecting telescopes then in New Zealand derived from him. Over the years, Ward telescopes found their way into private and society observatories throughout the country (see Bryce 1935), where they were—and in some cases still are—used for serious observational projects or for popularizing astronomy.

There is some debate as to whether Ward taught himself the techniques of mirror-making and lens-working, or whether the rudiments were imparted to him by an optician named Harding (see Calder 1978b). Harding is known to have manufactured telescopes (but what kind is not clear) while living in England prior to emigrating to New Zealand, but there is no evidence that he actually made any telescopes whilst living in Wanganui. If Harding did teach Ward how to make telescopes then this would have happened soon after Ward's arrival in Wanganui, since Harding left the district just two years later, in 1898. On the other hand, if he was entirely self-taught, then Ward would have relied on the principles laid down by those stalwarts of the 'English School' of the nineteenth century, people like John Browning (1831–1925), George Calver, Andrew Ainslie Common (Fig. 22.9) and George Henry With (1827–1904), who documented their experiences in such magazines as *The English Mechanic* (G. Nankivell, personal communication, 2000). As the proprietor of a bookshop, Ward could easily have sourced publications relevant to his own interests.

According to one of his daughters, Ward constructed a machine

> … that polishes the … mirrors and made the delicate eyepieces too. The metal cylinders [tubes] of the telescope and the rough mirrors [blanks] were imported from England … He also made microscopes and spectroscopes … (Brown 1972).

In his brief discussion of Ward's telescope-making activities Calder (1978b) refers to "… his technical excellence … [and] the high standard which every one of his known instruments shows …", but as we shall see, this may be slightly too complimentary a statement. Optical expert, the late Garry Nankivell (personal communication, 2000), pointed out that there was a problem with many of the larger blanks that Ward imported, as

> … the glass was rather defective in composition, seemingly to have an excess of alkali oxides … The result was a premature breakdown of the surface polish. I used one such blank but was disappointed with its performance. The aluminium film tended to fail, no doubt from alkaline attack.

Smaller blanks were not affected as they were obtained from local sources and were cut from thick polished plate glass (ibid.). These concerns aside, let us now examine some of the instruments completed by Ward.

One year before he died, a paper by Ward appeared in the *Publications of the Astronomical Society of the Pacific*. This is about the history of the Wanganui Astronomical Society's observatory and its 9.5-in Cooke telescope, but Ward (1926: 17) also includes some autobiographical information about his telescope-making exploits:

> With a view of exploring the details of many southern nebulae, star clusters, clouds, etc., the writer has devoted some time during the last 18 or 20 years to the making of specula for reflecting telescopes, a 20.5-inch being the largest so far attempted.

It may be more than a coincidence that 20.5 inches was the *precise* aperture of a Calver reflector, one of three different instruments that Ward located in England and considered for the Society's observatory before the Committee settled on the Cooke refractor (see Thrush 1963).

When the 52.1-cm reflector was completed in 1924 it was installed

> … on the polar axis of the 9.5-inch refractor [at the Society's observatory], taking the place of the counterpoise which weighs about 450 pounds. Some fine views of the planet Mars were obtained with this telescope, near and at opposition, a number of visitors and friends coming to us at the time for this purpose, including Dr. C.E. Adams, Government Astronomer from Wellington, and Mr. P. O'Dea, F.R.A.S., of Hawera. *A good deal can be seen with this aperture that is quite lost in the 9.5 …* (Ward 1926: 17; my italics).

In order to facilitate easy use, Ward had furnished the 52.1-cm reflector with a rotating tube section containing the eyepiece and the finder. Later, as Fig. 12.3 illustrates, the telescope was removed from the Society's observatory, installed on a German equatorial mounting and housed in a roll-off shed observatory located at Ward's home (Venimore 1988). Although the telescope was definitely operational, there is no evidence that it was ever used for any serious observing programs.

In 1928, soon after Ward's death, the 52.1-cm reflector was acquired by Victoria University of Wellington (Brown 1972), and the optics were used by the Physics Department "… for light experiments and demonstrations …" (F. Bateson, personal communication, 2000) until they were transferred to the Carter Observatory Board

Fig. 12.3 Ward and his 52.1-cm (20.5-in) reflector (*Courtesy* Ward (Wanganui Observatory Archives)

in 1940 prior to the opening of this new national facility in Wellington (Carter Observatory Telescopes and Equipment n.d.).

The Annual Report of the Observatory for 1945–1946 expounds on the plans then entertained for this telescope:

> The 20-inch silver on glass mirror with tube in three sections, as well as a few eyepieces and finder, were collected from Victoria University College [as it was then known] and stored in the Observatory. This instrument, which was formerly in the Physics Department, was donated to the Observatory by the College ... A spare dome and pier is available for the instrument, but a complete mounting has to be designed and constructed. Since it is essential that the mounting should be good, to get the best out of an instrument of this aperture, it will incur a substantial expense, and, therefore, it is hoped to make some tests on the mirror before any recommendations are made. Previous tests some years ago in the Physics Department of Victoria University College indicated that it was a good mirror, but whether these might satisfy astronomical requirements was not certain. (Thomsen 1946: 6).

But just three years later the Board abandoned all plans to mount the Ward telescope:

> The spare dome and the 20-inch mirror were the subject of several reports to the Board by the Director. A quotation from Sir Howard Grubb, Parsons & Co., in 1931, gave the cost of mounting a 20-inch mirror efficiently, as £2,425. Rising costs since that time could easily treble the quotation for the present day, and thus places it outside the resources of the Board. Apart from the mere mounting, the cost of necessary accessories to make it a worthwhile instrument of research, together with the present man power available for its efficient use, has caused the Board to abandon all ideas of setting up such an instrument in the meantime ... The possible uses of the 20-inch mirror in conjunction with the spectrohelioscope for studies of the solar spectrum are to be investigated by the Director. (Thomsen 1949: 2–3).

At a later date, thought was given to reworking the primary mirror, constructing a new telescope of shorter focal length, and installing this in the vacant dome, but continuing funding restrictions and the relative thinness of the mirror (among other factors) prevented this scheme from materializing. Eventually, given the historical significance of the old mirror, a decision was made to simply add it to the Observatory's collection of instruments and astronomical archives. This seemed an apt solution, since one of the Carter Observatory's functions was to serve as the primary repository for New Zealand's astronomical heritage.

Before he made the 52.1-cm telescope, Ward's largest instrument appears to have been a 36.2-cm (14.25-in) reflector built sometime prior to 1923. A photograph of this telescope is in the Ward (Wanganui) Observatory Archives, and the reverse side of the photograph contains the following information:

> 14¼ inch diameter Cassegrainian Telescope made for Mr. Robinson of 149 Balmoral Rd Mt. Eden Auckland mounted on alt-azimuth stand designed by J.T. Ward (maker of the telescope) & improved by Mr. Robinson ... (Photograph 1922–1923).

The inscription is dated "1922–3". Nothing further is subsequently heard of Robinson or of his telescope.

But Robinson's was not the first instrument Ward had made in this general size range, for back in 1907 he had built a 35.6-cm (14-in) reflector specifically for the opposition of Mars:

> Your Honorary Director [i.e. J.T. Ward], with a view to taking advantage of the opportunity, had ground and figured a reflecting telescope speculum of 14 inches diameter. This mounted on the side of the counterpoise to the 9½ inch Refractor gave, at times, very good views of the broader features of the planet, with a power of 600 on the 14 inch on the night of July 5th. The detail visible on this planet was very interesting, being by far the best yet seen. The very rapid shrinkage of the south-polar cap during the month of June and early part of July was a marked feature of this opposition. A series of sketches was made by Mr. Allison and myself. (cited in Venimore 1988: 102).

Towards the end of his life, Ward made yet another 35.6cm (14-in) telescope, which in 1927 was acquired by one of Auckland's emerging amateur astronomers, Ronald Alexander McIntosh (1904–1977):

> On June 30 I ordered from W.H. Ward, of Wanganui, a 14-inch equatorially-mounted Newtonian Reflector, made by the late J.T. Ward, at a cost of £80 f.o.b. Wanganui.
>
> The deposit was paid on July 21. The telescope tube and mounting arrived on July 22 and the speculum on August 13.
>
> The mounting was set on a concrete pier at Melford Street, being quite an open-air observatory. A wooden platform surrounded this, and a canvas tent cover was suspended over a ridge-pole when not in use.
>
> Erection of the telescope was completed on August 17 and after minor alteration to the mirror cell the telescope was ready for final adjustments, which were rapidly performed. (McIntosh n.d.).

McIntosh is shown with the telescope in Fig. 12.4. Rather atypically, the primary mirror proved a poor one (perhaps because it was one of the last mirrors made by Ward), and around 1930 it was replaced by one ground by Allan Bryce (1897–1970) of Hamilton (McIntosh 1971). Shortly after this, McIntosh moved house and erected a roll-off roof observatory for the hybrid Ward-Bryce telescope (McIntosh n.d.). In 1946 the mirror cracked while being re-silvering (St. George 1987), and it was replaced by one made by W. Jackson of Auckland (see McIntosh 1952).

Ronald McIntosh turned out to be one of New Zealand's most outstanding twentieth century amateur astronomers (Bateson 1977; Orchiston 2000; see also Chap. 19 in this book), and during the 1920s, 1930s and 1940s was an international figure in meteor astronomy (Luciuk and Orchiston 2016). However, he also was committed to telescopic work, and used his Ward-Bryce reflector for regular observations of comets, Jupiter and the Moon. In 1941 he independently discovered Comet C/1941 B2 (see Orchiston 1999), and on four different occasions between 1949 and 1962 he carried out detailed studies of short-term eruptions in Jupiter's atmospheric belts. He also made an exhaustive study of the lunar crater Aristarchus. McIntosh published a succession of research papers on his telescopic work, mainly in the *Journal of the British Astronomical Association*. Given this record of achievement, his 35.6-cm reflector is perhaps the most notable Ward telescope after the 52.1-cm reflector.

Fig. 12.4 R.A. McIntosh and his 35.6-cm (14-in) Ward reflector at Melford Street, Auckland (Orchiston collection)

We know that by the time he died in 1927 Ward had completed at least three telescopes in the 35.6–36.2 cm (14–14.25 inch) aperture range. We do not know what became of the 1907 'Mars' reflector after it was removed from the refractor mounting. Presumably, it was recycled, but we can say with certainty that it was not transformed into the McIntosh telescope. Nor can the McIntosh telescope be associated with Robinson, given the slightly different apertures involved; moreover, McIntosh's telescope was a *Newtonian*, with an *equatorial* mounting.

Robinson and McIntosh certainly were not the only New Zealand astronomers with telescopes made by Ward, prior to his death, although there is surprisingly little documentation of others. In Hawera, Patrick O'Dea (1875–1935) had a 30.5-cm (12-in) reflector (Gifford 1931) and George Mortimer Townsend (1895–1954) was the proud owner of 15.2-cm (6-in) and 20.3-cm (8-in) reflectors (Obituary 1954). Both astronomers were well-known nationally, and Townsend used his telescope to carry out valuable observations following the discovery of Comet 34P/Gale in 1927 and later of Comets C/1940 R2 and 7P/Pons-Winnecke (Obituaries 1954). In addition, O'Dea and Townsend were eager to use their telescopes to popularize astronomy. O'Dea's observatory "… was free to all. In fact he was never happier than when surrounded by a group of people at his telescope …" (Gawith 1935: 114), while visitors were always welcome at Townsend's observatory, "… and there were many occasions when people came to see some special astronomical event."

(Townsend 1986). Townsend's 20.3-cm Ward reflector was also fundamental to the founding of the (first) Hawera Astronomical Society, in 1926 (Obituaries 1954). In 1958, this telescope returned to its hometown when it was sold to a Mr Arronsen of Wanganui (Townsend 1986).

Other North Island towns and cities also featured Ward telescopes at this time. In Hamilton the Reverend H. Heaslip had a 25.4cm (10-in) reflector, Eltham's T. Gawith owned a 20.3-cm (8-in) reflector (Venimore 1988) and Walter S. Thrush of Wanganui had a 6.35-cm (2.5-in) refractor (Crust 1930), but there is little evidence that any of these telescopes was used for serious astronomy.

This certainly was not the case for the 30.5-cm (12-in) equatorially-mounted Newtonian Ward reflector owned by John C. Begg (1876–1965) of Dunedin. John Begg was a founding member of the Otago Astronomical Society (in 1910),[2] and he subsequently donated his Ward telescope to the Society for its Beverley-Begg Observatory (Begg 1960; Gifford 1931), which was opened in 1922 (Fig. 12.5). For many years this telescope served as the mainstay of the Observatory's popular 'public nights', and it also was used by members for serious observations of comets, star colours and lunar occultations (Begg 1960; Campbell 1983). In 1928 Strathmore R. B. Cooke (1907–1985) constructed a photometer for the Ward telescope, and used this device to measure variable star magnitudes right up until the time of his departure for the United States the following year (Campbell n.d.). In 1973, after more than half a century's service, the Ward reflector was finally replaced by a modern new telescope of identical aperture. The old Ward telescope then passed on loan to the North Otago Astronomical Society in Oamaru.

In contrast to some of his reflectors, all of the refractors made by Ward were of very modest aperture. In all likelihood, he obtained his blanks from the English firm of Chance Bros., as they had

> … developed two glasses, a Hard Crown and Dense Flint that were suited to the making of an objective sometimes referred to as the Watson Conrady type. The crown was equi-biconvex. The interior radius of the flint lens matched the crown radius, while the last surface was near plano, depending on the dispersions of the two glasses. Arguably, this design was the easiest to fabricate, compared to the four differing radii of curvature of the aplanatic (coma-corrected) classical Fraunhofer design. (G. Nankivell, personal communication, 2000).

In order to promote his telescope-making business Ward used letterhead paper, where he was described as a "Manufacturer of Reflecting and Refracting Telescopes Altazimuth and Equatorial Mountings Etc." Also mentioned was his membership of the British Astronomical Association and the fact that he was the Honorary Director of the Wanganui Observatory. Ward was always prepared to make telescopes to order, but he seems also to have maintained a regular stock, particularly of small refractors and reflectors. The smallest reflector that he ever made (Ward 1928) had an aperture of just 12.7 cm (5 inches). Figure 12.6 shows a number of mounted and unmounted refracting telescopes that at one time he had for sale, along with a tube

[2]This later became the Dunedin Astronomical Society.

Fig. 12.5 The opening in 1922 of the Beverley-Begg Observatory, which featured a 30.5-cm (12-in) Ward reflector. Begg and Ward are fourth and fifth from the right, respectively (*Courtesy* Ward (Wanganui) Observatory Archives)

Fig. 12.6 An assortment of small refractors and a reflector made by Ward (*Courtesy* Ward (Wanganui) Observatory Archives)

for a reflecting telescope and an observing ladder. In the background is a Newtonian reflector of about 25.4-cm (10-in) aperture that may also have been for sale.

When a Mr L.R. Brakenrig of Auckland enquired about purchasing a telescope from him in 1925, Ward (1925a) was immediately able to offer him the choice of a 6.35-cm (2.5-in) refractor, a 7.6-cm (3-in) refractor and a 15.2-cm (6-in) reflector. Brakenrig selected the reflector, which was priced at £35 (including packing and freight). Ward (1925b) was quick to applaud his decision: "… I am sure you will be better satisfied with its performance than anything in the way of a refractor I could supply at twice the price." Brakenrig's reflector is the smaller of the two instruments shown in Fig. 12.3 (see Ward 1925c).

After Ward's death, his one and only astronomically-minded son, William Herschel Ward, collaborated with the afore-mentioned Allan Bryce (Fig. 12.7) of Hamilton to oversee the sale of surplus telescopes and components. Bryce was Director of the Telescope Making Section of the New Zealand Astronomical Society (later the Royal Astronomical Society of New Zealand) and he and Ward produced a list (Ward and Bryce 1927) which was sent to interested parties. This included completed, but unmounted, 30.5-cm (12-in), 25.4-cm (10-in), 20.3-cm (8-in) and 17.8-cm (7-in) reflecting telescopes; completed 35.6-cm (14-in) and 25.4-cm (10-in) mirrors and two 20.3-cm (8-in) mirrors; two 20.3-cm (8-in) glass blanks; eight secondary mirrors, and two mirror cells; a completed, unmounted, 9.5-cm (3.75-in) refracting telescope; ten different completed achromatic lenses, ranging in diameter from 2.9 cm (1.15 inches) to 5.7 cm (2 inches), some in their cells; a completed 10.2-cm (4-in) double convex lens; seven achromatic lens

Fig. 12.7 Allan Bryce (extreme right) and his family (Orchiston collection)

blanks, ranging in diameter from 4.1 cm (1.6 inches) to 7 cm (2.75 inches); two drawn brass tubes for 6.7-cm (2.65-in) refractors; five finders; four eyepieces and three eyepiece assemblies; several altazimuth mountings, and various mounting components; and other assorted "… telescopic apparatus." At the end of the two and a half page listing Bryce reminded readers that

> Practically every item on this list is much below market value and such bargains are not likely to be again offered in New Zealand. Prospective telescope makers should get in early as many useful telescope parts listed here could not be made by an expert mechanic for several times the price quoted. (Ward and Bryce 1927: 3).

We do not know what became of most of these telescopes, mirrors and lenses, although it is reasonable to assume that many were sold off between 1927 and 1930. Then the Great Depression intervened and money was scarce, and by 1935 Bryce was convinced that many Ward telescopes had "… fallen into disrepair and consequent disuse." (Bryce 1935: 110). One which had not was a 30.5-cm (12-in) reflector, for a note published in 1935 in *Southern Stars* (the journal of the New Zealand Astronomical Society) reports that "The Society has recently received as a gift from Mr. Ward of Wanganui a 12-inch reflecting telescope. It is one of several made by his late father, who was well known in astronomical circles for his careful and accurate work on mirror grinding." (Magnificent gift telescope 1935). Over the years this instrument has been loaned to a succession of Society members and has been used for serious observing, thus serving as a fitting memorial to J.T. Ward and his contribution to New Zealand telescope-making.

During the 1930s other Ward telescopes occasionally came to light. In 1933 or 1934, James Treloar (d. 1945) of Hamilton acquired a 35.6-cm (14-in) Ward Newtonian reflector. Treloar was the founder of the Hamilton Astronomical Society, and he "… subsequently altered the mounting from Altazimuth to Equatorial so that it would be more useful in any future observatory which the Society might establish." (Bryce 1945: 8). The provenance of this telescope is not stated, but it could possibly be the recycled 'Mars' telescope of 1907. Whatever the facts of the matter, Treloar does not seem to have carried out any systematic work with it, and following his death it passed to the Society in 1948 and up until 1979 was housed in the Society's observatory (Eason 1979). A photograph of this telescope was published in a 1950 issue of *Sky and Telescope* (see Holmes 1950: 146).

At about the same time that Treloar acquired his Ward telescope one of New Zealand's stand-out astronomers, the late Dr Frank Maine Bateson (1909–2007), was intent on purchasing Heaslip's 25.4-cm (10-in) Ward reflector, which "… for many years [had] … languished in a shed …" (Bryce 1935), but financial constraints eventually forced him to abandon these plans (F. Bateson, personal communication, 2000).

In about 1946 the late Albert Francis Arthur Lofley Jones (1920–2013; see Fig. 17.18) purchased one of the unsold 20.3-cm (8-in) Ward mirrors through A. Bryce, and

> … I had the steel tube made at an engineering shop while the mirror cells and focuser, finder etc. were home made. An equatorial mounting was constructed, moveable on wheels, and kept covered outside when not in use. The telescope was kept in a shed and carried out and attached to the mounting by wing nuts. (Jones, personal communication, 2000).

Initially this f/8 telescope (Nankivell, personal communication, 2000) was situated at his home in Timaru, but when Jones acquired a Hargreaves 31.75-cm (12.5-in) reflector in 1948 he decided to transfer the Ward reflector to the family's weekend fishing retreat at Rakaia Huts. At both localities the Ward telescope was used for the systematic visual observation of variable stars and comets, fields in which Jones came to achieve international eminence (see Austin 1994; Marsden 1988). Long before he died in 2013, Jones had achieved the enviable record of having made more magnitude estimates of variable stars than any other living observer (Orchiston 1990), and during the 1990s, as a Research Associate of the Carter Observatory, he was able to co-author a succession of research papers with professional astronomers. A modest, unassuming man, Jones has been dubbed "The quiet achiever" by Austin (1994), and this is a particularly apt characterization. Meanwhile, the Ward telescope had outlived its usefulness when Jones moved to Nelson in 1964, and it was purchased by a local amateur astronomer who later on-sold it. Despite its distinguished history, neither new owner used the telescope for any serious work.

During the 1970s, another old Ward mirror was converted into a functioning telescope when a partially-finished, 22.8-cm (9-in) diameter, 3.8-cm (1.5-in) thick mirror was used as the primary mirror for the so-called Houghton-Rumsey Astrograph. The late Norman J. Rumsey (1922–2007; Fig. 12.8) was then head of

Fig. 12.8 Norman Rumsey (*Photograph* Wayne Orchiston)

the optical workshop at the Department of Scientific and Industrial Research's Physics and Engineering Laboratory near Wellington, and he is known for his innovative optical designs. In this instance, he developed a new optical system based on an original design by Houghton which gave a number of advantages over the classical Schmidt camera: "Optical quality appeared to be equal, with only 0.75 of the length of an equivalent Schmidt. The primary mirror was smaller (228 mm against 245 mm for 152 mm aperture working at f3)." (Wanganui's Houghton-Rumsey Astrograph 1976). Rumsey's associate, the late Garry Nankivell, worked up the mirror, and given its Wanganui associations, they must have been very pleased when the finished instrument was mounted piggy-back on the historic 24.1-cm Cooke refractor at the Ward (Wanganui) Observatory (see Fig. 11.10). With a field of 7° and an anticipated limiting magnitude of 16 (for a 10-min exposure), a range of possible research programs was identified, including: novae patrols of the Large and Small Magellanic Clouds; minor planet astrometry; cometary astrometry; photographic photometry, using filter photography to determine blue, violet and other magnitudes; searches for new variable stars and monitoring of known ones; and studies of Hα emission in the Galaxy (ibid.). Yet, despite this obvious research potential, little effective use appears to have been made of this excellent instrument.

This completes our review of known Ward telescopes. Precisely how many different telescopes Ward made over the years is not known, but they must number in the several dozens. Unfortunately, the current whereabouts of many of these instruments is now uncertain, but as Calder (1978a: 107) has observed, "… every so often another comes to light, either a refractor up to 4 in, or a reflector of rather larger aperture." As further Ward telescopes come to the attention of astronomers it is important that every effort is made to record their histories.

12.4 The Ward Legacy

The penchant for manufacturing large reflecting telescopes initiated by Joseph Ward was continued by other twentieth century New Zealand amateur astronomers following his death. In about 1936 two Dunedin men, W. Gardner and F.M. Gillies, built and mounted equatorially a 16.5-in (41.9-cm) Newtonian reflector (Doug Berry, personal communication, 2000; Campbell 1983, personal communication, 2000; cf. Adams (1994: 2), and five years later Ken Adams (Fig. 12.9), who also was living in Dunedin at that time) completed an 18-in mirror:

> I ground and polished the mirror with considerable difficulty, to about F/6 focal ratio. This primary mirror and the associated secondary mirror were never adequately mounted, because I became much more demanding when I finally recognized my requirement for a suitable equatorial mounting, with appropriate slow motions and a drive mechanism. Indeed, I later found that a 4-inch Cooke refractor on a solid Cooke equatorial mounting with electric drive, circles etc., was really much more affordable and preferable for my purposes. (Adams 1994: 2).

Fig. 12.9 The late Ken Adams, in Auckland in 1987 (Orchiston collection)

During the 1950s and 1960s a number of individuals made 41.0–41.9cm (16–16.5-in) reflectors. One of these was Auckland's Harry Williams (1911–2008; Christie 2008), who in 1972 went on to finish a magnificently-crafted 52.1-cm (20.5-in) Dall Kirkham (Fig. 12.10) which until his death was used extensively for serious photographic projects and for variable star photoelectric photometry. Another Aucklander known for his large apertures since the 1960s was Graham Loftus (1924–2015; Fig. 12.11), and although he completed many Dobsonian telescopes with mirrors exceeding 41 cm (16 inches)—including 63.5-cm (25-in), 68.6-cm (27-in), 73.7-cm (29-in) and 91.4-cm (36-in) instruments (e.g. see Loftus 1991)—he also produced standard Cassegrainian reflectors with 57.15-cm (22.5-in) and 61-cm (24-in) mirrors.

Meanwhile, the challenge to manufacture refracting telescopes has tempted few New Zealand amateur astronomers since Ward's day, although Bruce Holmes of Hamilton did complete a 15.2-cm (6-in) telescope in the late 1940s (Holmes 1950). In addition, during the 1950s, Thomas Cole (1880–1961) of Nelson worked on a 20.3-cm (8-in) objective (Collier 1961; Some Nelson telescopes 1961), but this was only completed at a much later date when Garry Nankivell (see Fig. 13.17) reground and polished the two Chance Bros. blanks (personal communication, 2000). The objective, which by then was also owned by Holmes, was subsequently bequeathed to the Hamilton Astronomical Society, and the plan was to mount it as a guide-scope on their 61-cm (24-in) reflector (ibid.). More recently a third refractor exceeding in aperture those manufactured by Ward has been built by a New Zealand amateur astronomer. This is the 13-cm (5.1-in) f/15 instrument that was

Fig. 12.10 Harry Williams and his 52.1-cm (20.5-in) reflector and 35.6-cm (14-in) Schmidt in about 1995 (*Courtesy* Max Pow)

completed by G. Jonas of Waitara, working under the guidance of Garry Nankivell (ibid.).

Finally, we should note that while *amateur* telescope-making only achieved any level of public visibility in New Zealand during the 1870s, the first recorded *professional* telescope-maker arrived in Wellington at a much earlier date (see Orchiston 2001b). He was an Englishman named Marriott, who emigrated to New Zealand in 1842. James Henry Marriott had followed his father's calling and trained as an optician and mathematical instrument maker, but he appears to have made very few telescopes once living in New Zealand, and threw most of his energy into

Fig. 12.11 Graham Loftus (centre) in 1988 (Orchiston collection)

the theatre. Nowadays, he is known as the 'founding father' of New Zealand theatre (see Downes 1990), and the few telescopes he did make are discussed in Chap. 20.

12.5 Concluding Remarks

During the first quarter of the twentieth century, Joseph Thomas Ward of Wanganui was one of New Zealand's leading amateur astronomers, and was a pioneering telescope-maker. He manufactured small refractors, and reflectors up to 52.1 cm in aperture. For about forty years this last-mentioned instrument held the record as the largest reflecting telescope made by a New Zealand amateur astronomer.

For a number of years Ward ran a thriving business in the manufacture and sale of telescopes, and his instruments quickly acquired a pan-New Zealand distribution and were used for serious observations and for the popularisation of astronomy. Further Ward telescopes, mirrors and lenses came on the market following his death in 1927, but the financial circumstances of the 1930s hampered their immediate sale.

Reflecting telescopes made by Ward were associated with two outstanding New Zealand amateur astronomers, the late Albert Jones and the late Ronald McIntosh. Both men achieved international reputations, in part through observations made with these instruments.

Since Ward's death, a number of New Zealand telescope-makers have followed his example and completed reflectors of large aperture, but few have attempted to emulate his feats as a manufacturer of refracting telescopes.

Joseph Ward deserves to be remembered as one of the founding fathers of amateur telescope-making in New Zealand, and it is to be regretted that we know the current whereabouts of so few of his telescopes.

Acknowledgements I am grateful to the following for their assistance: the late Ken Adams (Auckland, NZ), the late Dr Frank Bateson (Tauranga, NZ), the late Doug Berry (Auckland, NZ), the late David Calder (Wanganui, NZ), the late Robert Campbell (Oamaru, NZ), Clive Collier (Richmond, NZ), the late Tony Dodson (Upper Hutt, NZ), the late Don Glass (Hawera, NZ), the late Peter Hingley (former RAS Librarian, London), the late Albert Jones (Stoke, NZ), the late Ray Lee (Wanganui, NZ), the late Graham Loftus (Auckland, NZ), the late Jackie St George (Sydney), and in particular the late Garry Nankivell (Wellington, NZ). I also wish to thank Dr Grant Christie (Auckland, NZ) and John Drummond (Patutahi, NZ) for reading and commenting on the first draft of this chapter, and Max Pow (through Grant Christie) and the Wanganui Astronomical Society (through the late David Calder) for kindly supplying Figs. 12.2, 12.3, 12.5, 12.6 and 12.10.

References

Adams, K. D. (1994). Some aspects of New Zealand astronomy in my lifetime. Typescript (copy in Orchiston collection).

Austin, R. (1994). Albert Jones—the quiet achiever. *Southern Stars, 36*, 36–42.

Bateson, F. M. (1977). Obituary: Ronald Alexander McIntosh (1904–77). *Southern Stars, 27*, 82–87.

Beaumont, L. R. H. (1963). J.T. Ward, founder of Wanganui's Observatory, made important contributions to science. *Journal of the Auckland Astronomical Society, 21*(2), 13–14.

Begg, J. C. (1960). *Historical Review*. Otago: Royal Society of New Zealand, Astronomical Section.

Brown, F. W. (1972). The Ward Family. Text of a talk given to the Porirua Historical Society, New Zealand (copy in Orchiston collection).

Bryce, A. (1935). Telescope Makers' Section. *Southern Stars, 1*, 109–110.

Bryce, A. (1945). Hamilton Astronomical Society. Annual report for year ended June 30, 1945. *Southern Stars, 11*(6), 8–9.

Calder, D. (1978a). Joseph Ward: pioneer astronomer and telescope maker. *Southern Stars, 27*(5), 104–108.

Calder, D. (1978b). Seventy-five years of astronomy: Ward Observatory, 1903–1978. Lecture given to the Wanganui Astronomical Society (copy in Orchiston Collection).

Campbell, R. N. (n.d.) Strathmore R.B. Cooke. Manuscript (copy in Orchiston Collection).

Campbell, R. N. (1983). A sketch outline astronomical history of Otago. Manuscript (copy in Orchiston Collection).

Carter Observatory Telescopes & Equipment. (n.d.). File including correspondence about the donation of the Ward 20.5-in telescope, Carter Observatory Archives.

Chapman, A. (1998). *The Victorian Amateur Astronomer. Independent Astronomical Research in Britain 1820–1920*. Chichester: Wiley.

Christie, G. W. (2008). Harry Williams (1911–2008). *Southern Stars, 47*(2), 3–4.

Collier, C. S. (1961). Obituaries. Thomas Cole. *Southern Stars, 19*, 91–92.

Crust, A. G. C. (1930). *Census of New Zealand Telescopes*. Unpublished New Zealand Astronomical Society Report (copy in RAS Archives).

Downes, P. (1990). Marriott James Henry 1799/1800–1886. In W. H. Oliver (Ed.), *The Dictionary of New Zealand Biography* (Vol. 1, pp. 1769–1869). Wellington: Allen and Unwin New Zealand Ltd. and Department of Internal Affairs, P. 271.

Eason, H. J. (1979). A brief history of the Hamilton Astronomical Society. *Royal Astronomical Society of New Zealand Newsletter*, March 9–10.

Eiby, G. A. (1972). A visit to Uncle Charlie AC Gfford, MA, FRAS (1861–1948). *Southern Stars, 24*, 109–113.

Eiby, G. (1978). History, astronomy and New Zealand. *Southern Stars, 27*, 118–124.

Gawith, F. (1935). Obituary. Patrick O'Dea, MZ, LLB, FRAS. *Southern Stars, 1*, 113–115.

Gifford, A. C. (1931). *In Starry Skies. Some Observatories*. Wellington: The 'Evening Post'.

History of the Kiwifruit Industry. (2004). (http://freepages.genealogy.rootweb.ancestry.com/~sooty/kiwihist.html).

Holmes, B. A. (1950). A report from New Zealand. *Sky and Telescope, 9*(6), 144–146.

Jenkinson, S. H. (1940). *New Zealanders and Science*. Wellington: Department of Internal Affairs.

King, H. C. (1979). *The History of the Telescope*. New York: Dover.

Loftus, G. (1991). A 36-inch telescope. *Southern Stars, 34*, 45–49.

Luciuk, M., & Orchiston, W. (2016). Ronald A. McIntosh: pioneer Southern Hemisphere meteor observer. *Journal of Astronomical History and Heritage*, 19 (in press).

McIntosh, R. A. (n.d.). Record of early astronomical work. MS (copy in Orchiston Collection).

Magnificent gift telescope. *Southern Stars, 1*, 192 (1935).

Marsden, B. (1988). Amateur astronomers and the IAU Central Bureau for Astronomical Telegrams and Minor Planet Center. In S. Dunlop, & M. Gerbaldi (Eds.). *Stargazers. The Contribution of Amateurs to Astronomy* (pp. 64–76). Berlin: Springer.

McIntosh, R. A. (1952). Members' observatories. Observatory of R.A. McIntosh, at Auckland, New Zealand. *Journal of the British Astronomical Association, 62*, 227–230.

McIntosh, R. A. (1959). Astronomical history of the Auckland Province. Part V—First astronomical society. *Amateur Astronomer Journal of the Auckland Astronomical Society, 18*, 46–48.

McIntosh, R. A. (1971). Obituary. Allan Bryce. *Southern Stars, 24*, 19.

Obituaries. George Mortimer Townsend. *Southern Stars*, *16*, 84 (1954).

Obituary. The late Mr. Joseph Thomas Ward. *Journal of the British Astronomical Association 37*, 183 (1927).

Obituary. *Hawera Star*, September 4, 1954.

Orchiston, W. (1990). Albert Jones: an interview. *Astronomy Now, 4*(9), 16–17.

Orchiston, W. (1996). Ward, Joseph Thomas 1862–1927. In C. Orange (Ed.), *The Dictionary of New Zealand Biography* (Vol. 3, p. 555, 1901–1920). Auckland: Auckland University Press and Department of Internal Affairs.

Orchiston, W. (1998). *Nautical Astronomy in New Zealand. The Voyages of James Cook*. Wellington: Carter Observatory.

Orchiston, W. (1999). Of comets and variable stars: the Afro-Australian astronomical activities of J.F. Skjellerup. *Journal of the British Astronomical Association, 109*, 328–338.

Orchiston, W. (2000). R.A. McIntosh. In C. Orange, et al. (Eds.), *The Dictionary of New Zealand biography* (Vol. 5. 1941–1955) Auckland: Auckland University Press and Department of Internal Affairs.

Orchiston, W. (2001a). The English Equatorial Mounting and the history of the Fletcher telescope. *Journal of Astronomical History and Heritage, 4*, 29–42.

Orchiston, W. (2001b). Wellington astronomers of the 1840s. *Southern Stars, 40*(4), 5–7.

Orchiston, W. (2002). Joseph Ward: pioneer New Zealand telescope-maker. *Southern Stars, 41*(2), 13–21.

Photograph of 14¼ inch Telescope. (1922–23). Original in Ward (Wanganui) Observatory Archives.

Some Nelson telescopes. *Southern Stars*, *19*, 79–80 (1961).

St. George, J. (1987). History of Auckland Astronomy. Manuscript (copy in Orchiston collection).

Thomsen, I. L. (1946). *Report of the Carter Observatory Board for the Year Ended 31 March, 1946*. Wellington: Carter Observatory (Astronomical Bulletin No. 25).

Thomsen, I. L. (1949). *Report of the Carter Observatory Board for the Year Ended 1949, March 31*. Wellington: Carter Observatory (Astronomical Bulletin No. 28).

Thrush, W. S. (1963). Astronomy in Wanganui. *Southern Stars, 20*, 73–76.

Townsend, J. I. (1986). Letter to D. Glass, dated September 5. Original in Glass Collection (copy in Orchiston Collection).

Venimore, C. (1988). *Good Viewing. (A History of the First Wanganui Astronomical Society and the Ward Observatory)*. Unpublished manuscript.

Wanganui's Houghton-Rumsey Astrograph. *Royal Astronomical Society of New Zealand Newsletter*, September 8, 1976.

Ward, J. T. (1925a). Letter to L.R. Brakenrig, dated July 29. Original in Ward (Wanganui) Observatory Archives.

Ward, J. T. (1925b). Letter to L.R. Brakenrig, dated August 8. Original in Ward (Wanganui) Observatory Archives.

Ward, J. T. (1925c). Letter to L.R. Brakenrig, dated September 21. Original in Ward (Wanganui) Observatory Archives.

Ward, J. T. (1926). The Wanganui Observatory and Astronomical Society. *Publications of the Astronomical Society of the Pacific, 38*, 314–318.

Ward, J. T. (1928). *Recent Advances in Astronomy*. Nelson: Cawthron Institute.

Ward, W. H., & Bryce, A. (1927). Telescopic apparatus for sale on behalf of the estate of the late J. T. Ward, Wanganui. Original in the Library of the Royal Astronomical Society of New Zealand, Wellington.

Chapter 13
From Crossley to Carter: The Life and Times of an Historic Cooke Refractor

Abstract The Carter Observatory in Wellington, New Zealand, houses an historic Cooke refractor that has played a role in the development of astronomy in England and New Zealand. This telescope, initially with a 23.7-cm (9.33-in) achromatic objective, was manufactured for Edward Crossley in 1867, installed in an observatory near Halifax, and used by Crossley and his astronomer Joseph Gledhill for a range of observational programs. In 1896 Crossley and Gledhill replaced the original lens with a new 22.9-cm (9-in) Cooke photovisual objective. After Crossley died in 1905 the telescope was purchased by the Reverend Dr David Kennedy and installed in an observatory near Napier, New Zealand. In 1926 the Wellington City Corporation purchased the telescope and installed it in an observatory close to the centre of the national capital. Staff at the Dominion Observatory and a number of leading amateur astronomers then used the instrument for research, but it also played a key role in promoting astronomy through regular 'public nights'. This combination of research and education continued when the telescope was transferred to the Carter Observatory in 1941, and only changed when the Observatory employed professionally-qualified staff who needed access to much larger apertures for their astrophysical research. From that time on, the historic Cooke refractor has continued to provide a public astronomy service for residents of and visitors to Wellington. By 2000, the quality of the photovisual objective had deteriorated to the point where it needed to be replaced, and a new 24.8-cm (9.75-in) achromatic objective was installed, making this now the largest operational refractor in New Zealand.

13.1 Introduction

The second half of the nineteenth century could be described as the era *par excellence* of the amateur astronomer. This was a time when positional astronomy still reigned supreme as astrophysics took its first tentative steps, and a time when gifted amateur astronomers armed with modest telescopes were able to make a significant contribution to world astronomy (see Ashbrook 1984; Chapman 1998). We see a succession of their papers in the leading astronomical journals of the day,

W. Orchiston, *Exploring the History of New Zealand Astronomy*, Astrophysics and Space Science Library 422,
DOI 10.1007/978-3-319-22566-1_13

Astronomische Nachrichten and *Monthly Notices of the Royal Astronomical Society*; we note their key roles in the formation and early development of new astronomical groups and societies; and we witness their pioneering efforts in the mass-production of moderately-priced reflecting telescopes.

Despite the increasing availability of Browning-With and Calver reflectors, most wealthy amateur astronomers wishing to carry out serious observational astronomy still opted for the precision of the equatorially-mounted professional-manufactured refracting telescope, and in favour among British astronomers were instruments by Cooke, Grubb, Troughton and Simms, and Wray.

One such telescope that was to make an impact in two widely-spaced regions of the globe was a 23.7-cm (9.33-in) Cooke refractor (Fig. 13.1) that was manufactured in 1866–1867 for the wealthy British industrialist, Edward Crossley. For more than three decades Crossley and Joseph Gledhill carried out an ambitious range of observations with this telescope and published a succession of papers in astronomical journals. During the 1890s a further advance in instrumentation occurred when the original Cooke achromatic lens was exchanged for one of the new photovisual objectives patented by Dennis Taylor. Then in 1906 the telescope was transferred to New Zealand, where it did further service in the name of observational astronomy at a private observatory and a municipal observatory before taking pride of place as the founding instrument of the Carter Observatory when this national facility was established in the nation's capital, Wellington, during World War II. Once in New Zealand it also acquired an educational role, and over the years has been used to introduce hundreds of thousands of New Zealanders and overseas visitors to the varied treasures of the southern night sky. For this combination of reasons this is an important telescope, and its history and overall contribution to astronomy deserved to be documented (but see Orchiston 1996, for an earlier, short, popular account).

In this chapter, which is based primarily on Orchiston (2002), we will discuss the successive owners of this venerable telescope and the ways in which they used it for observational astronomy and for education, before concluding with some remarks about the longevity of Cooke photovisual objectives.

13.2 The Crossley Connection

The first owner of the Cooke refractor was Edward Crossley (1841–1905; Fig. 13.2), a wealthy English industrialist who wove a fortune in the carpet-making industry while living at Halifax between the cities of Manchester and Leeds (Andrews and Budding 1992). While still a teenager, Crossley followed his father and two uncles into the successful family business. For three years he was Mayor of the local borough, and in 1885 he curtailed his involvement in the family business when he became a Member of Parliament. Indifferent health forced him to give up this calling in 1892, and he died suddenly on 21 January 1905 (Obituaries 1905; Obituary 1905).

Fig. 13.1 The 23.7-cm Cooke refractor (after Crossley et al. 1879: 10)

From an early age Crossley had a passion for astronomy and the means to indulge this interest. He started with a 7.6-cm (3-in) telescope, graduated to a 17.8-cm (7-in) refractor, and in 1867 took delivery of the 23.7-cm (9.33-in) Cooke refractor, which was installed in a new observatory. This building was "… about 18

Fig. 13.2 Edward Crossley (after Ashbrook 1984: 311)

feet square, with a dome, in a space behind his house on the edge of the town." (Obituaries 1905: 335). The telescope was an f/15.6 achromatic refractor with a conventional German equatorial mounting, and came with the usual arsenal of eyepieces, two micrometers, solar and stellar spectroscopes and a solar diagonal. Crossley et al. (1879: 32) have provided a useful description of this instrument and some of the accessories:

> Mr. Edward Crossley mounted his 9½ in. [*sic*] Cooke equatorial refractor in 1867. Its focal length is 148.5 in. The style of mounting is German. The diameter of the declination and hour circles are respectively 23½ in. and 12½ in., and they read 10″ and 2 sec.
>
> The lamp, which gives a bright field to the micrometer, swings at the end of the perforated declination axis.
>
> The aperture, and amount of colour of the light for the bright field, are regulated from the eye-end by means of rods, and a rod and cords at the same end give the observer full control over the motion of the instrument in right ascension and declination.
>
> The finder has an aperture of 2½ in., and a focal length of 2 ft. 4 in.
>
> The negative eyepieces are ten in number: powers, 60 to 1000.
>
> There are three micrometers, two filar and a double-image. The double-image and one of the filar micrometers are by Simms, and the other filar by Cooke. The eyepieces for these instruments are, in all, seventeen in number, and the powers range from 100 to 1200. The new filar micrometer by Simms is divided on the face: diameter of circle 4½ in. [cf. Fig. 13.1, here].

Note that when it was manufactured, the telescope was furnished with just the two micrometers (the third was acquired in 1877). The original Cooke drive, meanwhile, proved troublesome, and was subsequently replaced with one by Grubb of Dublin (Crossley et al. 1879) with a frictional governor, and "The driving arc is of very large radius, and is provided with a quick method of winding back ..." (The observatory ... 1907).

By the time Crossley placed an order for this telescope the firm of T Cooke & Sons was beginning to enjoy a considerable reputation for their fine instruments, culminating two years later in the construction of a 63.5-cm (25-in) refractor that was built for Robert Stirling Newall (1812–1889) and for a short time was the largest refractor in the world (see McConnell 1992; Van Helden 1984). Figure 13.3 illustrates the way in which the aperture of the world's largest refracting telescope increased in the course of the nineteenth century. Crossley's Cooke telescope is indicated by the star, and we see that by international standards it was a modest instrument, yet in an era that was initially dominated by positional and descriptive astronomy rather than astrophysics it was capable of being put to good use. But more than this, by Cooke's standards the Crossley refractor was a large telescope, for we learn from King (1979: 251–252) that up to 1863, when he began the Newall telescope, Cooke's largest refractors were in the 20.3–25.4 cm (8–10 inch) aperture range.

Crossley not only had the means to purchase the Cooke telescope and build a commodious observatory for it, but also to employ a full-time 'astronomer'. This man was Joseph Gledhill (1837–1906), who turned out to be a skilled and dedicated observer with a talent for writing up and publishing his work. Gledhill started his working life as a school teacher, but "His own tastes, however, gradually won him over to the close prosecution of scientific study, and in 1868, when Mr Edward

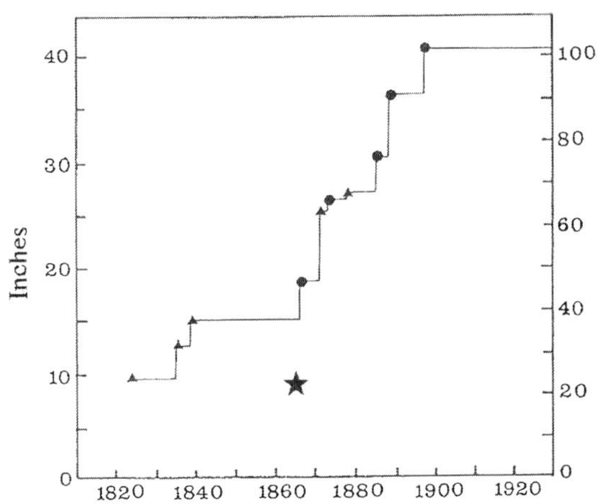

Fig. 13.3 Plot of the way in which the aperture of the world's largest refracting telescope changed during the nineteenth century. *Triangles* indicate Cooke telescopes and *circles* Alvan Clark telescopes. The Crossley refractor is represented by the *star*

Crossley established his observatory in Halifax, he asked Mr Gledhill to undertake the work of observer." (Obituaries 1907: 233). Gledhill remained in this post until Crossley's death in 1905.

As Crossley's business thrived so too did his aspirations for more opulent surroundings, and in 1872 he moved to the southern outskirts of Halifax where he had built a new palatial home, 'Bermerside', which included an impressive observatory (see Fig. 13.4). The larger 'drum' shown in this figure housed the 23.7-cm Cooke telescope, while a 17.8-cm (7-in) refractor lay under the second drum (although this instrument was later replaced by a 11.4-cm (4.5-in) Cooke photovisual refractor). The observatory also featured an 8.9-cm (3.5-in) Cooke transit telescope and a library (Obituaries 1905). During the nineteenth century astronomers were divided over the most appropriate dome design (e.g. see Donnelly 1973), and it is of interest that Crossley favoured the drum form over the more widely-used hemispherical dome. One of the most famous institutions to mimic this design was the Royal Observatory Edinburgh, when the new observatory was erected at Blackford Hill towards the end of the nineteenth century (see Brück 1983).

Back in Halifax, as soon as the 23.7-cm Cooke telescope was operational Crossley observed the lunar crater Linné and published a note on this in *Monthly Notices of the Royal Astronomical Society* (Crossley 1868). This was the first paper he submitted to an astronomical journal. He and Gledhill then began regular observations of Jupiter and Jovian satellite phenomena, and they embarked on an ambitious program of micrometric double star observations which culminated in the publication of their classic *Handbook of Double Stars* in 1879. A collaborator on this project was the Reverend James Maurice Wilson (1836–1931) from the Temple Observatory at Rugby, who three years earlier had co-authored a double star review paper with Gledhill (see Wilson and Gledhill 1876). The Crossley, Gledhill and Wilson book provides a useful historical background on double star astronomy before presenting valuable data on more than 1000 double stars, about half of which were observed from Bermerside. In Crossley's obituary published in *The Observatory*, his double

Fig. 13.4 Crossley's mansion and the Bermerside Observatory (*Courtesy* Royal Astronomical Society)

star book is described as "… a standard work." (Obituary 1905: 111), and one of its additional strengths is the 19 pages devoted to thumbnail descriptions of contemporary British and overseas observatories and telescopes. Meanwhile, Wright (1993: 433) cannot sing the praises of this book too highly:

> … a strong trio of observers [Crossley, Gledhill and Wilson] had thus been formed. The obvious next step was for them to produce a vade mecum for other observers. This would fill a long-standing need, and in 1879 the now classic *Handbook of Double Stars, with a Catalogue of Twelve Hundred Double Stars and Extensive Lists of Measures* (Crossley et al. 1879) was published. This volume deserves a historical reprinting. It is full of hints, still valid today; plus notes on orbit computation and observing practice. A tone of encouragement echoes through the book, which is complemented by some fine illustrations. A further volume of corrections was issued in 1880, which indicates the level of interest generated.

Further double star observations were made at the Bermerside Observatory right through to 1904, and the Crossley refractor certainly left its mark on British nineteenth century double star astronomy.

In addition to the Jovian and double star work, Gledhill (and initially Crossley as well) used the large Cooke refractor for a range of other observations, all in the positional and descriptive astronomy 'mould', and the main programs are listed in Table 13.1. Jupiter was a great interest at this time as astronomers tried to keep track of the ever-changing disk features: assorted short-lived black and white spots; variations in the colour, shape and size of the belts; and of course the appearance, reddening and subsequent fading of the Great Red Spot (see Hockey 1999). Likewise, Mars presented a special observational challenge during the last quarter of the nineteenth century, given the controversy surrounding Schiaparelli's 'canali' (see Sheehan 1997). Meanwhile, the Jovian and Saturnian satellites attracted observers like Gledhill as astronomers sought to accumulate the data that would later allow them to refine the orbital parameters of these bodies (e.g. see Sampson 1910), while lunar occultations offered a means of keeping the Moon's irregular motion under review (Newcomb 1903) and at the same time providing ammunition—if needs be—for fine-tuning the latitude and longitude of the observer's observatory.

Table 13.1 Main Bermerside Observatory non-meridian programs, and papers published in *Monthly Notices of the Royal Astronomical Society* and in *The Observatory*

Period	Type of observation	Number of papers	
		MNRAS	*The Obsy*
1869–1902	Jovian features (including the Great Red Spot)	11	3
1869–1904	Double stars	2	34
1873–1904	Jovian satellite phenomena	20	0
1875–1893	Saturnian satellite phenomena	7	0
1884–1894	Lunar occultations	4	0
1896–1901	Mars	3	0
	'Other'	3	3
	(Annual Reports)	(28)	0
Totals		78	40

Complimenting the principal observational programs listed in Table 13.1 were observations of Venus, the 1892 lunar eclipse (Crossley and Gledhill 1892), and variable stars. For several years Gledhill made variable star magnitude measures with a Hilger Wedge Photometer acquired in 1883 before deciding not to persevere with this line of research. Presumably he became aware that wedge photometers had their critics, and that the accuracy of the wedge photometer could not compete with the more popular comparison photometer (see Hearnshaw 1996).

Apart from the co-authored double star book, Gledhill published prolifically in *The Observatory* and *Monthly Notices of the Royal Astronomical Society* (Table 13.1), but his papers also appeared in *The Astronomical Register* and *Memoirs of the Royal Astronomical Society*. In *The Observatory* and *Monthly Notices* alone, he was sole author of 69 observational papers, and 32 of these related to Jupiter or Mars, somewhat giving the lie to the claim in his obituary that he "… contributed *several* papers on planetary observations …" (Obituaries … 1907: 233; my italics). During their association, Crossley twice joined Gledhill in writing up some of their observations, and we should note that in addition to the papers listed in Table 13.1 Crossley also published six non-observational papers.

In 1884 Edward Crossley purchased the 91.4-cm (36-in) reflector owned by Andrew Common, which at the time was the second largest reflecting telescope in the world, and for the next decade he and Gledhill had to curtail—but not totally abandon—their regular observational work while they set about housing and modifying this instrument in a bid to convert it into a viable research telescope (e.g. see Annual Reports 1888, 1890). Crossley finally gave up in frustration, "… the climate of Halifax being found quite unsuitable for so large an aperture." (Obituaries 1905: 336), and—as is now well known—in 1895 he donated this telescope to the Lick Observatory where it eventually was used by Keeler to carry out pioneering work in astrophysics (Osterbrock 1984).

Once the Common reflector had left Bermerside Gledhill was able to refocus his attention on observational astronomy and the trusty Cooke refractor, and it was at this time that he and Crossley learnt about the new photovisual lens system (Fig. 13.5) designed by Harold Dennis Taylor (1862–1943) at T Cooke & Sons (see Taylor 1894). Crossley immediately ordered an 11.4-cm (4.5-in) refractor with one of these new objectives (Annual Reports 1896), and he and Gledhill were so impressed with the superior image quality that they decided to replace the achromatic objective in the larger refractor. As a result, the original 23.7-cm two-element system was replaced by a 22.9-cm (9-in) photovisual objective in 1896. Despite its marginally smaller aperture, the new objective had a slightly longer focal length at f/16.9, which required the addition of a small extension to the top of the telescope tube. The new photovisual objective was one of the largest that Taylor had fabricated up to that point, and comprised positive crown, negative flint and positive flint elements made from Schott O-543, O-658 and O-374 glasses, respectively (Nankivell 2002). Gledhill (1897: 638) was delighted with the improvement in seeing:

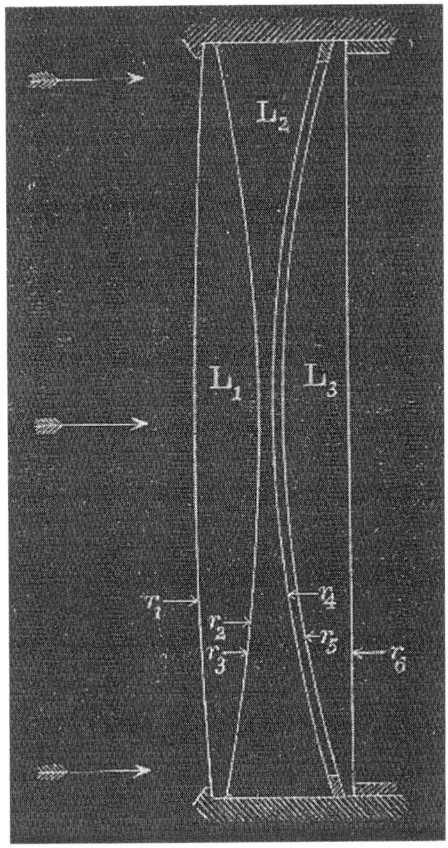

Fig. 13.5 The design of the Cooke photovisual objective (after Taylor 1894: 332)

> It was expected that with this new object-glass we should obtain a purer image and see more detail on such planets as *Jupiter* and *Mars*. Nor have our expectations been unfulfilled … the views when the definition was good certainly surpassed anything we have seen in past years with the 9⅓-inch glass; the image was as pure and soft and free from colour as that given by a good reflector.

Crossley, too, was impressed, as indicated by his testimonial, dated 21 June 1898, which appeared in Cooke's next *Catalogue of Astronomical Instruments* (1900):

> … after two years experience of your 9 inch Photovisual Objective, I have no hesitation in speaking in the very highest terms of its excellent performance. It is free from a suspicion of false colour and fully equal to a 10 inch in light grasping power. The absence of uncorrected colour on such stars as Vega and on the planets and the moon is most marked … I consider the construction a triumph of skill and patience.

But unfortunately this improvement in instrumentation came just a little too late, for Crossley was in his twilight years and had all but given up observing. He was to die suddenly soon after, on 21 January 1905, at the age of 63. Gledhill, who was then 68, was left a legacy and retired, dying just one year after Crossley, but not before the telescope was put up for sale through the London firm of W. Watson & Sons. They supplied a glowing report on the instrument for prospective buyers, even if their estimate of its age was way off the mark:

> Everything in this has been kept up to date. The telescope itself was made about 20 years ago; the Object Glass, however, has been supplied in recent years and is one of Cooke's latest designs and best quality. The observatory in which the Telescope was used was under the supervision of an astronomer who is very well-known, who took the utmost care of every part. It is, therefore, in splendid working order, and we can, with fullest confidence, recommend it to you. (Watson 1905).

It was either at this time, or possibly in 1896 when the photovisual objective was installed, that the telescope acquired a new 12.7-cm (5-in) Watson guide scope and Sun and star cameras.

The eventual new owner of the Cooke telescope (see Orchiston 1995b) was a Roman Catholic seminary at Meeanee, on the eastern coast of the North Island of New Zealand, near the city of Napier (for New Zealand localities see Fig. 13.6).

13.3 The Meeanee Years

The man behind the purchase was the Reverend Dr David Kennedy, the Superior of the Marist Seminary at Meeanee. Kennedy (Fig. 13.7) was a remarkable man (see Mackrell 1985). He was born in Lyttelton, New Zealand, but at the age of 12 went to England and Ireland to study and as a scholar at University College, Dublin, showed great brilliance. In 1891 he produced the book *Natural Philosophy for Junior Students*, which dealt with physics and mechanics and for many years was used as a text in English and Irish Catholic schools. By the time he returned to New Zealand in 1892 as the first Professor of Dogmatic Theology, Sacred Scripture and Sacred Eloquence at Meeanee, Kennedy had developed a passion for astronomy, and he brought a 15.2-cm (6-in) reflecting telescope with him. This was housed in an observatory on the grounds of the Seminary.

The long-awaited arrival of Comet 1P/Halley apparently inspired Kennedy to yearn for a larger telescope, and this is how the Cooke refractor came to New Zealand, with royalties from the sale of his popular school text book providing the necessary funding (Father David Kennedy … n.d.; Howell 1970). The purchase was finalized in late 1905 with assistance from Kennedy's friend and fellow amateur astronomer, J.T. Ward of Wanganui, making this New Zealand's second-largest refractor after the ex-Fletcher 24.1-cm (9.5-in) Cooke telescope that Ward had acquired just two and a half years earlier for the Wanganui Astronomical Society (for details of this latter instrument see the preceding chapter, or Orchiston 2001a).

Kennedy then launched a public appeal to fund a new larger observatory and "Catholic congregations and wealthy individuals responded with donations large

Fig. 13.6 New Zealand localities mentioned in this chapter are shown in *red*

Fig. 13.7 The Reverend Dr David Kennedy (Orchiston collection)

and small, and the work was soon underway." (Mackrell 1985: 84). With assistance from those at the Seminary Kennedy supervised the construction, and this facility was opened in July 1907. It featured 6-m long wooden walls, and a one tonne iron dome (Fig. 13.8). Kennedy's enthusiasm soon attracted two others from the Seminary, Ignatiss von Gottfried (1864–1951) and Brother Joseph John Cullen (1886–1963), and they became his astronomical assistants. Their first chance to test the new telescope came just one month after the official opening of the Observatory when Comet C/1907 L2 (Daniel) made an appearance in southern skies, and their initial attempts at astronomical photography were an outstanding success (for example, see Fig. 13.9). Some fine photographs of the Sun also were taken, showing solar granulation (Fig. 13.10) and details of sunspot groups and faculae (Fig. 13.11). According to Kennedy (1909), Dr George Ellery Hale (1868–1938) from Mt Wilson Observatory "… considered them very good, especially those of the Granulations." Father Aloysius Laurence Cortie S.J. (1859–1925), Director of the Solar Section of the British Astronomical Association, was also suitably impressed: "Some *beautiful photographs* of the Sun's granulated surface and of Sun-spot groups have been sent by the Rev. Dr. Kennedy." (Cortie 1909: 425; my italics).

By a cruel twist of fate, Kennedy was transferred to Wellington as the Rector of St. Patrick's College in May 1909, well before the initial appearance of Comet 1P/Halley, and it was left to Cullen and von Gottfried to continue the observational

Fig. 13.8 The new Meeanee Observatory built for the Cooke refractor (Orchiston collection)

Fig. 13.9 Photograph of Comet C/1907 L2 (Daniel) (after *Astronomical Photographs* … n.d.)

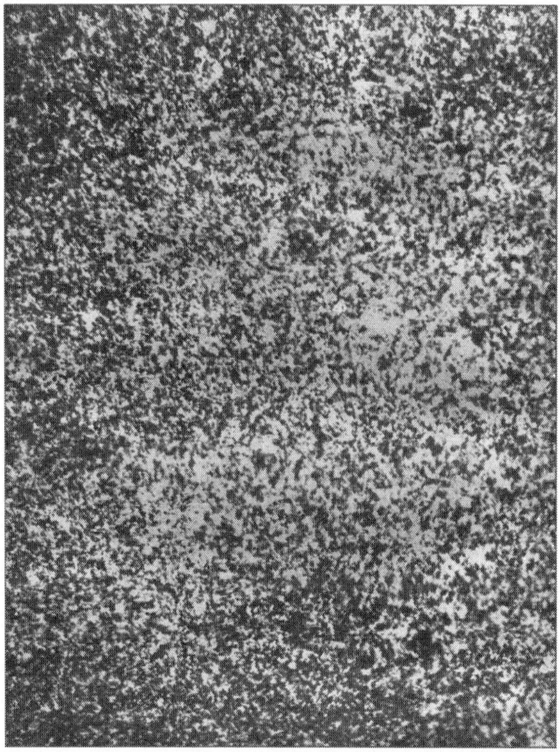

Fig. 13.10 A Meeanee photograph of solar granulation (after *Astronomical Photographs* … n.d.)

programs that he had initiated and to photograph the comet. With his guidance, they succeeded in obtaining many fine photographs, one of which is shown here in Fig. 22.20 (cf. Mackrell 1985: *passim*); some of these images were forwarded to the Royal Astronomical Society and to the British Astronomical Association. But Kennedy and his Meeanee colleagues made no attempt to write up any of this work, or their previous efforts in solar and cometary photography, and indeed not a single scientific publication emanating from the Meeanee Observatory found its way into the international astronomical journals of the day.

What is equally apparent is that during the few short years that the telescope was used at Meeanee, its research potential was largely squandered. It would appear that the Observatory lacked a transit telescope, so observational programs involving accurate timing (such as lunar occultations of stars, and Jovian and Saturnian satellite phenomena) were not an option, but there were other possibilities. At no time did Kennedy, Cullen or von Gottfried use the instrument to make visual observations of Mars or Jupiter, or to conduct magnitude estimates of southern variable stars—a field crying out for attention. The telescope came with three micrometers, and they could have emulated the efforts of Ward and made a useful contribution to southern double star astronomy, but they chose not to do so.

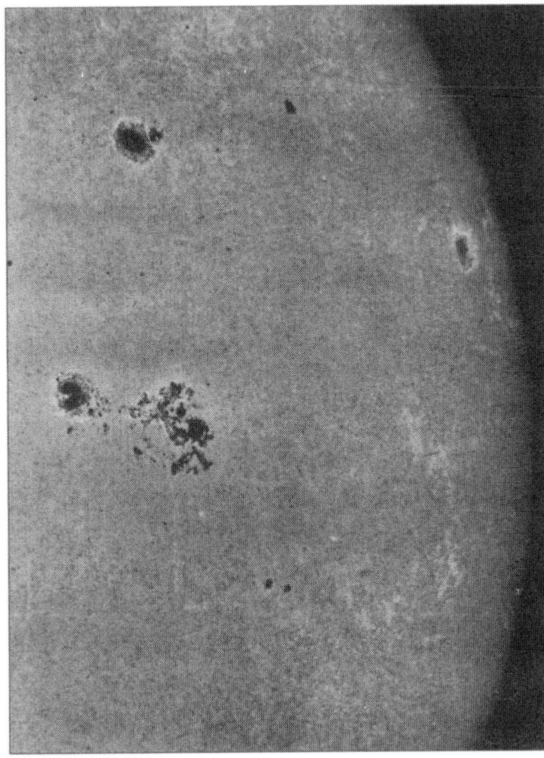

Fig. 13.11 A Meeanee photograph of sunspots and faculae (after *Astronomical Photographs …* n.d.)

Similarly, they made no attempt to obtain micrometric measures of the positions of comets, planets or minor planets. All of these were observational programs that were successfully pursued at about this time by dedicated amateur astronomers with instruments comparable to those found at Meeanee (e.g. see Orchiston 1982).

As it turned out, Kennedy's transfer to Wellington effectively spelt the end for the Meeanee Observatory, for during 1910 and 1911—soon after Comet Halley's departure—Cullen and von Gottfried also moved on. It is true that the observatory was relocated to an elevated site at nearby Greenmeadows when the Seminary moved there in 1911, but in October 1912 a violent storm lashed the district, and the dome was torn off and hurled down the hillside. The telescope was hastily dismantled and placed in storage for safekeeping and the Meeanee Observatory was closed. World War I then intervened, and although there was some talk in 1919 of remounting the telescope this did not happen. In April 1923, Reverend Father Aubrey advised the Government Astronomer, Dr C.E. Adams, that the telescope was for sale, and the Wellington City Corporation (later Wellington City Council) subsequently decided to purchase it.

13.4 The 'Green Tin Shed'

It was strange that the telescope should follow Kennedy to Wellington, but he played no part in this move for by this time he was living in Palmerston North (Father David Kennedy… n.d.). In fact, the Wellington purchase was facilitated through a deputation containing Dr Adams and J.P. Maxwell, Chairman of the Astronomical Section of the Wellington Philosophical Society, and led by the President of the recently-formed New Zealand Astronomical Society, Sir Frederick Revans Chapman (1849–1936). The purchase price of £500 was a considerable sum, and it says something of the oratory skills and persuasive powers of the delegation that it was able to convince the Finance Committee of the City Corporation that the telescope "… was a dead bargain … [and] well worth the money …" (The City Telescope … 1924). The Committee put a successful recommendation to a full meeting of the Corporation in November 1923, the deal was finalized, and the telescope arrived in Wellington the following January and was promptly placed in storage.

New Zealand's capital city was at last to have a public observatory, and the Corporation proceeded to assign land for it at Kelburn, in the Botanic Garden, near the upper terminus of Wellington's famous cable car. Just six months later, on 26 August 1924, the Mayor officially opened the new Wellington City Observatory (ibid.; "Epoch marking" 1924). Always planned as a temporary structure and with its construction hastened by the much-anticipated 1924 opposition of Mars, this rustic green-coloured 5.5 × 11-m (18 × 36-ft) corrugated iron building with a roll-off roof soon became known to locals as the 'Green Tin Shed', while those of less charitable disposition referred to it as a 'cowshed' (The City Telescope … 1924; Thomsen 1939)! Figure 9.13 shows the Wellington City Observatory in centre field, immediately behind the foreground foundations of the later Carter Observatory. On the hillside behind it is the King Edward VII (later Thomas King) Observatory which was maintained by the Wellington Philosophical Society, and just off-photo to the left of this observatory was the Dominion Observatory, which was maintained by Government and directed by Dr Adams.

Wellington City Corporation lacked the funds to employ an astronomer so its solution was to hand the public side of the Observatory's operations over to the local amateur astronomical community, which soon introduced weekly Public Viewing Nights. In this way, Wellingtonians and visitors to the city were given an opportunity to view the southern night sky from a comparatively dark site, despite the fact that the Observatory was less than ten minutes drive (back in those days) from the centre of Wellington. In 1933 it was reported that

> … the general public have always appreciated to a marked degree … the Saturday evenings when the telescope can be used by anyone … These public evenings when the wonders of the heavens can be observed through the telescope, and when interesting and instructive talks are given, are of no small educational importance, and this is recognised by the schools and colleges which make use of the city's telescope. (New Zealand Astronomical Society … 1933).

On clear nights when the telescope was not committed to public viewing, a small but dedicated band of local amateur astronomers used it for serious observational programs, and from the start they also were joined by Dr Adams (Thomsen 1947) and his assistant, Ivan Leslie Thomsen (1910–1969; Eiby 1970), from the Dominion Observatory. Although this Government facility supplied a national time service and conducted seismological and tidal research, it lacked a large telescope that could be used for non-meridian work. The construction of the Wellington City Observatory a mere 1–2 min walk away from the Dominion Observatory immediately solved this problem.

Figure 13.12 provides a view of the Cooke telescope ready for an evening observing session, and Table 13.2 indicates of the types of targets in favour at this time. This Table summarizes the full range of observations carried out during one calendar year chosen at random, in this case 1932 (Log Book … n.d.). It is of interest to note that in this particular year observations were made on 112 different nights by seven different observers, all but one of whom were amateur astronomers. Ivan Thomsen was the lone professional, and he observed on 32 different nights. However, the most prolific observer was Maxwell Stuart Butterton (1914–1982;

Fig. 13.12 View of the Cooke telescope at the Wellington City Observatory (Orchiston collection)

Table 13.2 Telescopic observations made at the Wellington City Observatory in 1932

Type of object or event	Number of days/evenings observed
Comets	36
Comet-sweeping	1
Double stars	4
Jovian satellite phenomena	10
Jupiter	37
Lunar drawings	5
Lunar eclipse	1
Lunar occultations	2
Mars	1
Minor planets	1
Planetary nebulae	1
Saturn	32
Star clusters	7
Star colours	20
Sun	10
Suspected variable stars	3
Telescopic meteors	1
Variable stars	23

see Fig. 19.4) who was active on 76 nights, and as an indication of his dedication we should note that on 22/23 July 1932 he spent the entire evening observing and between 19 h 30 m and 05 h local time measured the position of Comet C/1932 M2 (Geddes), drew Saturn, recorded 30 star colours, made 4 variable star estimates, and sketched the lunar crater Posidonius! The next most prolific amateur astronomer, with observations on 22 nights, was Frank Bateson (see Fig. 19.3), an individual who was later destined to transfer to professional ranks and built an international reputation through his variable star work (see Bateson 1989). In hindsight it could be said that Bateson carried out his observational 'apprenticeship' with the 22.9-cm Cooke refractor, as subsequently he purchased the 20.3-cm (8-in) Grubb refractor that for many years had done excellent service in the hands of Australia's leading amateur astronomer, John Tebbutt (see Orchiston 1982, 2001b).

One year later, on 23 August 1933, a newspaper article reported:

> During the last nine years this City Council telescope has more than justified the expenditure made upon it. It has been used continuously by astronomers in the observation of star clusters, double and variable stars, occultations, sunspots, and other astronomical phenomena, and it is always ready for a surprise visit on the part of a comet. The observational work which has been accomplished through its aid has been worked up, and the resulting papers circulated among those interested. (New Zealand Astronomical Society … 1933).

In addition to visual and occasional micrometric and spectroscopic observations (e.g. see Thomsen 1938a, b), the telescope was used to photograph selected comets

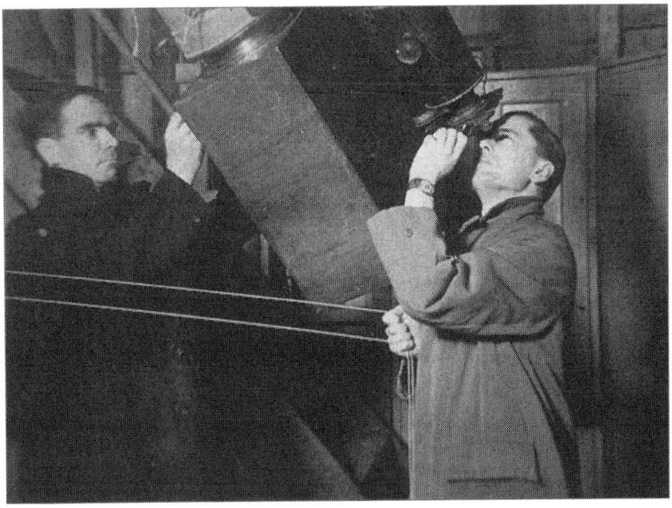

Fig. 13.13 Photographing Eros in 1931. Ivan Thomsen tracks the minor planet using the guide scope, while a local amateur named Quinnell makes minor adjustments in declination (Orchiston collection)

and other astronomical objects, including Eros (Fig. 13.13), during the 1931 campaign to refine the value of the solar parallax (Eiby n.d.). Another interesting project was the 1936 annular solar eclipse, when Thomsen and local amateur astronomer, Bruce Embleton Stonehouse (1910–1988; see Fig. 19.4), used the telescope for photographic and spectroscopic observations of the event (Stonehouse and Thomsen 1937).

From these examples, and others that could be added, it is apparent that throughout the second half of the 1920s and during the 1930s the Wellington City Observatory was one of the observational mainstays of New Zealand astronomy.

13.5 The Carter Connection

The move of the Cooke telescope to its final destination occurred in 1941 when the Carter Observatory was opened in Wellington, just a few metres away from the Wellington City Observatory.

The Carter Observatory owes its existence to the colourful local politician, farmer and businessman, Charles Rooking Carter (Orchiston 1995a), who while in Florence during the 1860s visited the Museum of Natural History. Associated with this was a Temple dedicated to Galileo, and Carter was moved to write:

> … it is small, but magnificent, and was recently erected at a cost of £36,000. In it are paintings representing the important events in Galileo's life, *a part of one of his fingers* in a

> glass bottle, and also the principal instruments with which he made his scientific discoveries, – including *his telescope*. (Carter 1866(II): 330; his italics).

Carter was a man with an innate curiosity about the natural environment, a sense of civic duty and a vision that extended far beyond his mortal years, and when he died in 1896 he left a sum of £2,240 as the nucleus of a fund to be used to erect a public astronomical observatory in or near Wellington. By the 1930s this fund grown to the point where it could be effectively utilized, and in 1938 the Government of New Zealand passed the Carter Observatory Act. This identified the Observatory as a Centennial Project to mark the signing of the Treaty of Waitangi between the British Government and the Maori chiefs of New Zealand in 1840.

What could not have been anticipated was the distraction of another world war, and as a consequence the official opening of this new national facility was delayed until 20 December 1941 with noted New Zealand amateur astronomer and auroral expert Murray Geddes as the inaugural Director. Under the terms of the Carter Observatory Act 1938, the Wellington City Council agreed to supply land for the Observatory in the Wellington Botanic Garden and to hand over the Cooke telescope in the Wellington City Observatory, while the Council and Government would provide annual funding for staffing, operations and up-keep. The City Council promptly demolished the 'Green Tin Shed'.

Since 1941 the original Carter Observatory building has been substantially extended or renovated on three different occasions (see Fig. 13.14), but the Cooke telescope has remained in its original dome. Murray Geddes died during WWII, and after the war the Observatory staff, led by Ivan Thomsen (who was Director from 1945 to 1969) and some of the more committed local amateur astronomers, used

Fig. 13.14 Carter Observatory in the late 1990s. The Cooke telescope is in the foreground left-hand dome. In the small right-hand dome, also part of the original building, is a 41-cm Boller & Chivens reflector, while the large dome at the rear, added at the end of 1991, houses a planetarium. The two-storey library-office wing on the extreme right was constructed during the 1960s (Orchiston collection)

this telescope for a range of visual, micrometric and photographic programs (see Table 13.3). Solar astronomy was a mainstay of the Observatory's early research programs (bolstered by a spectrohelioscope supplied by Mt Wilson Observatory), and the Cooke telescope came with a prime-focus camera designed specifically for solar photography. There was also a second camera, which was used to photograph comets and star fields, and this is shown in Fig. 13.15.

In addition to the range of programs indicated in Table 13.3, six projects of special importance deserve to be mentioned. In 1945 Dr Elizabeth Alexander

Table 13.3 Principal observational programs involving the Cooke telescope at the Carter Observatory

Object/event	Type(s) of observation(s)
Comets	Photographs, visual descriptions
Double stars	Micrometric measures, photographs
Jupiter	Drawings
Lunar occultation	Timings (sent to the Nautical Almanac Office)
Mars	Drawings at selected oppositions
Saturn	Drawings
Sun	Sunspot drawings and photographs; eclipse photographs
Transit of Mercury	Contact timings
Venus	Micrometric measures

Fig. 13.15 Ivan Thomsen prepares to photograph a comet (Orchiston collection)

(1908–1958) from the Radio Development Laboratory co-ordinated pioneering observations of solar radio emission using New Zealand Air Force radar antennas located at Piha near Auckland, at three other sites in Northland and on Norfolk Island (see Orchiston 1994, 2005), and sunspot data supplied by Ivan Thomsen were utilized in her paper on this work (Alexander 1946). This inspired Thomsen, and he then wrote a paper for *Nature* comparing solar radio fluxes published by Ryle and Vonberg (1946) with Carter sunspot records (Thomsen 1948). These were among the earliest attempts to examine the relationship between the radio Sun and its optical features, and Thomsen's paper was the first by a New Zealand astronomer to appear in *Nature*. This work is discussed later in this book in Chap. 23.

The Carter Observatory Cooke refractor also was involved in the early days of photoelectric photometry in New Zealand. Although the first photometer employing a photomultiplier tube was used by Albert Edward Whitford (1905–2002) and Gerald Kron (1913–2012) at Mount Wilson Observatory in the late 1930s, it was only in the mid-1940s that photoelectric photometry began to emerge as a valid astrophysical technique, and "… after 1946 many observatories acquired 1P21 photomultipliers, constructed photometers for them (often based on the advice of Kron) and undertook reliable photoelectric photometry for the first time." (Hearnshaw 1996: 413). In 1948 the Carter Observatory's newly-appointed Assistant Astronomer, Ken Adams, began experimenting with a photometer using the 22.9-cm Cooke telescope and also the 14-cm (5.5-in) Grubb refractor at the nearby Thomas King Observatory. Despite assistance from Kron, there were many difficulties, and "Regrettably, no observations of value resulted … [because] our photomultipliers were noisy, poor samples not suitable for use on such small light buckets." (Adams 1982: 4–5). At about the same time, in 1949, the late George Allison Eiby (1918–1992) from the Dominion Observatory built a photometer incorporating an RCA 1P21 tube as part of his M.Sc. program, and attached this to the Carter Observatory refractor. He and Adams then "… worked together at the telescope making a few observing runs, taking differential measurements between the components of close pairs such as Delta 1 and Delta 2 Gruis, and obtained results yielding standard deviations of the order of 0.03 magnitudes." (Adams 1982: 5). From these humble beginnings, which "… give hope for useful photometric work in the future." (Thomsen 1950), New Zealand went on to establish an international reputation for variable star photometry.

During the late 1960s the Cooke telescope was involved in one international collaboration of particular note. On 7 April 1968, Carter Observatory was one of the observatories to record an occultation of Neptune by the Moon (Occultation of Neptune 1968), and photometric and visual data provided by New Zealand, Australian and Japanese observers subsequently allowed two U.S. Naval Observatory astronomers to derive an improved value for the diameter of Neptune (see Bixby and Van Flandern 1969).

Two years later the telescope was used by amateur astronomers from the Variable Star Section of the Royal Astronomical Society of New Zealand for international flare star monitoring programs involving UV Ceti and YZ Canis Majoris (Fisher 1971).

Perhaps the last serious attempt to use the Cooke telescope for a project of research value occurred between 1973 and 1975 when local amateur astronomer, Ray Nisbet, carried out micrometric observations of selected double stars (Warren 2001), continuing an initiative begun a decade earlier by Thomsen (1965a, b).

But by this time the research days of this fine old Cooke telescope were all but over, for back in 1968 the Observatory had taken delivery of a new 41-cm (16-in) Boller & Chivens reflector and with this instrument and newly-appointed staff came a new research direction involving photoelectric studies of variable stars and multi-wavelength extragalactic astronomy. However, the Observatory's light-polluted location was hardly conducive to serious observational astronomy, and so the 41-cm telescope was moved to a new high-altitude 'outstation' at Black Birch in the South Island. This left the old Cooke refractor to continue a role that it inherited when the Observatory first opened, as the mainstay of a public astronomy program.

The 1990s witnessed a major development of the Carter Observatory's public astronomy and educational programs (e.g. see Leather et al. 1998; Orchiston 1998; Orchiston et al. 1998) when the Observatory's visitor centre was expanded to accommodate a new lecture theatre and a small Zeiss planetarium. Then, as part of the Observatory's 2001 sixtieth anniversary celebrations, new displays were installed in the 'Dome Room' housing the Cooke telescope, and the telescope itself was thoroughly refurbished and repainted.

Since the resignation of the author of this book as Executive Director of the Carter Observatory in 1999, the status of the Observatory has changed markedly. Following a review, it lost its 'national observatory' status and the Carter Observatory Act was repealed. The Observatory ceased all research functions and became a branch of the Wellington City Council, offering education and public outreach programs in astronomy and a first-class tourism destination. Nowadays, despite competition from planetarium programs, the old telescope remains popular with school groups and members of the general public making daytime visits to the Observatory, and on 'public nights'—weather permitting (see Fig. 13.16)—it provides visitors with breathtaking views of our fabulous southern skies, thereby highlighting Mullaney's (1993) conviction that given the right educational environment a planetarium *and* a telescope can enjoy a mutually beneficial symbiotic relationship.

Over the past seventy-five years the old Cooke telescope has been the mainstay of the Carter Observatory's education and public outreach programs and although the 41-cm Boller & Chivens reflector is now back at the Observatory and also in use on public viewing nights, the Cooke refractor continues to be a major draw card. Despite its more modest aperture, it is what the public thinks a 'telescope' should look like, and as such it can look forward to serving a vital educational role for many years to come.

Fig. 13.16 This newspaper cartoon from the 1960s featuring the Cooke telescope highlights the fact that this instrument is in demand on public viewing nights, notwithstanding Wellington's notoriously fickle weather (Orchiston collection)

13.6 Problems Associated with Cooke Photovisual Objectives

There is one final chapter in the history of this old Cooke telescope that remains to be told. As Dewhirst (1975: 24–25) has recounted, Dennis Taylor, the inventor of the Cooke photovisual objective,

> … was not aware that the new glass in the central component is chemically unstable, although very hard and mechanically strong. Despite all practical precautions, in a damp climate a lens made of it slowly crazes on its surface and eventually requires actual reworking, not just cleaning.

The Carter Observatory objective has suffered in this way, which is perhaps no real surprise given Wellington's notorious wet and windy weather, but the problem began long before its arrival in the nation's capital. In 1914, when the telescope was already in storage at Greenmeadows, the objective was returned to T Cooke & Sons for reworking, while the first occurrence of this problem during the 'Carter' years occurred in 1951. At that time the objective was sent to Grubb-Parsons (who had taken over the astronomical interests of the Cooke firm), and in his 1952 Annual Report, the Chairman of the Board was glad to be able to report that

> … the deterioration in the objective of the 9-inch telescope, referred to in last year's report, has been remedied. The objective was returned from England in 1951, October, after the

> repolishing of the defective surface and the subsequent performance of the telescope has been highly satisfactory. (Norman 1952: 1–2).

However, at this time Grubb Parsons stressed that "… no further restorative work would be possible on account of the increased fragility arising from the extremely small central thickness of the middle negative component." (cited in Nankivell 2001: 5).

New Zealand's leading professional telescope-maker, the late Garry Nankivell (Fig. 13.17), first examined the 22.9-cm photovisual objective in 1977 and notes made at the time

> … record the first component showing light interference colours of uniform thickness on both surfaces. The second component exhibited a slight deterioration and felt rather sandy during the cleaning operation. Some very fine pitting of the surface was also evident. The surfaces generally were in fair condition. The third component was in poor condition especially the inner surface bounding the air space. Interference films were evident on both surfaces. The inner surface was covered with a white powdery efflorescence which was removed easily with a soft cotton wool swab dampened with some distilled water. (Nankivell 1991: 7).

Nankivell (2001: 5) further notes that since Grubb-Parsons' work in 1951 the objective has continued to show signs of deterioration, and that about every five years he has had to dismantle and clean the objective, "… to remove the effects of an alkaline carbonate efflorescence on the internal surface of the third lens component."

Fig. 13.17 The late Garry Nankivell, who manufactured a new 24.8-cm aplanatic objective for the Cooke telescope in 2001 (*Photograph* Wayne Orchiston)

However, he was in for a rather unpleasant surprise when he examined the objective in January 2000, as later recounted in a letter to me:

> You will be dismayed to learn that the 23cm Refractor at Carter has developed a huge "cataract" on the first surface of the central deep flint lens of the photovisual objective. Only the outer 2 to 3 cms retain any polish. The centre is now quite matt in appearance. A comment that the images appeared quite faint and surrounded by a mist prompted an inspection … [and] I discovered with a sense of incredulity, that rain water had penetrated and wicked into the narrow air space between the first and second lenses! (Nankivell, personal communication, 23 April 2000).

As someone with an immense passion for nineteenth century optics, it was with considerable regret that Nankivell (2001) came to the conclusion that the affected element could not be reworked. His solution was to fabricate a new objective, but this differs in two important ways from the original Cooke system: firstly, instead of a 3-element photovisual replacement he substituted a 2-element aplanatic objective, and secondly, this new objective has a larger diameter, at 24.8 cm (9.75 inches), than the 22.9 cm original. By a strange twist of fate, Nankivell installed the new objective in the telescope in mid-December 2001 and a few days later he fell ill, was hospitalized, and died suddenly—a major blow for New Zealand astronomy (see Hearnshaw 2002; Rumsey 2002). Further details of this new objective are provided in a paper that was published posthumously in the *Journal of the Antique Telescope Society* (Nankivell 2002) as a companion to my own paper about this telescope (Orchiston 2002) which has formed the basis of this chapter.

13.7 Concluding Remarks

By any account the Cooke refractor at the Carter Observatory, Wellington, New Zealand, is a heritage scientific instrument. It has given continuous service to astronomy for nearly 150 years. Initially it was associated with two prominent names in nineteenth century British astronomy, and in 1896 it acquired one of Dennis Taylor's earliest large photovisual objectives. Some years after transferring to New Zealand it became the work-horse of a municipal observatory in the nation's capital, before taking pride of place as the founding telescope at the Carter Observatory when this national facility was opened in 1941. For many years it remained New Zealand's second-largest operational refractor, but with the substitution of a new 24.8-cm (9.75-in) achromatic objective at the end of 2001 it took this title from the 24.1-cm (9.5-in) Cooke at the Ward (Wanganui) Observatory. Even so, by current international standards its aperture is decidedly modest.

Astronomy world-wide has seen remarkable changes in research emphasis since this telescope was constructed. During the era of the 'grand amateur' (Chapman 1998), Gledhill and Crossley were able to use it for serious observational work, at a time when positional and descriptive astronomy reigned supreme and it was still possible for amateur astronomers to make a valuable contribution to forefront research astronomy. With the move to Meeanee its research potential was largely squandered and it did little more than provide some rather pretty pictures at a time when astrophysics was making its mark around the globe. After the transfer to Wellington, amateur astronomers continued to use the instrument in a typical nineteenth century manner, but unlike their English predecessors they rarely sought to write up their observations and publish them. However, it was at this time that the telescope first began to fulfill a major public educational role.

The founding of the Carter Observatory did little to change the types of 'research' (and I use the term loosely) carried out with this telescope, which is little wonder given the city-based location of the Observatory, the amateur backgrounds of most of the early staff, and the limitations that so modest an instrument could offer for those astrophysicists who were later employed by the Observatory. However, an equatorially-mounted instrument of this type matches almost precisely the public stereotype of what an 'astronomical telescope' is meant to look like, and most visitors *are* impressed by its appearance and size, and the excellent views it offers of the more popular celestial targets on clear moonless nights. So, its on-going educational value is unquestionable.

This raises an interesting question: given the ready availability these days of relatively cheap Schmidt-Cassegrain and other catadioptric telescopes of much larger aperture, is there a research future for nineteenth century refractors like the Carter Observatory Cooke telescope? For those of us with a sentimental attachment to such telescopes it is tempting to respond with an unqualified 'yes', but the challenge arises when we look for valid examples. While such instruments can undoubtedly be used for useful solar and planetary monitoring and for double star studies, perhaps their greatest research potential lies in variable star photometry. In this regard, it is appropriate to reflect on the fact that "… at least 22 observatories (9 in Europe and 13 in North America) engaged in photoelectric photometry at some time or other prior to 1940 … and over 200 articles were published giving observational results …" (Hearnshaw 1996: 240–241), and that some of this pioneering work was carried out with telescopes remarkably similar in aperture to the Cooke refractor at Carter Observatory. And once the 1P21 and 931-A photomultiplier tubes became available, some of the earliest serious photometric programs undertaken in Australia and South Africa were achieved with 22.9-cm (9-in) and 17.8-cm (7-in) refractors, respectively (see Hearnshaw 1996: 441–442). Given the demise of an astrophysics research program at the Carter Observatory, the Cooke telescope's future clearly lies in education and not in serious research, but this should not prevent other refractors of similar vintage and with fine optics and reliable drives from making a useful on-going contribution to research—and particularly to variable star photometry.

Given new optics and a carefully mapped out educational role the future of the Cooke refractor at the Carter Observatory looks assured, but it is to be hoped that the visiting public also learn that this telescope made a valuable contribution to research during the nineteenth century. This is an historic refractor, dating from a by-gone era, and its history provides an interesting case study of changing patterns of telescope function and technology during the nineteenth and twentieth centuries. Without this venerable instrument, British and New Zealand astronomy would have been the poorer.

Acknowledgements I am grateful to the following for their assistance: the late Ken Adams (Auckland, NZ), Alison Brecht (Cooke Telescope Archives, York), the late David Calder (Wanganui, NZ), Dr Richard Crossley (University of York), the late Peter Hingley (former RAS Librarian, London), Errol Lea-Scarlett (Riverview Archives, Sydney), the late Garry Nankivell (Wellington, NZ), the late Jackie St. George (Sydney) and staff at the State Library of New South Wales (Sydney). I also wish to thank Dr Richard Dodd (Martinborough, NZ) and Gordon Hudson (Wellington, NZ) for reading and commenting on the first draft of this chapter and the Royal Astronomical Society for kindly supplying Fig. 13.4. The initial draft of the paper upon which this chapter is based was written in 1998, prior to my resignation as Executive Director of the Carter Observatory. This chapter is dedicated to the memory of the late Garry Nankivell.

References

Astronomical Photographs Taken at the Meeanee Observatory Napier, N.Z. (n.d.). Copy in Carter Observatory Library.

Adams, K. D. (1982). Some aspects of the development of instrumentation for astronomical photometry (in New Zealand). *Southern Stars, 30*, 3–11.

Alexander, E. (1946). The Sun's radio energy. *Radio and Electronics*, *1*(1), 16–17, 20.

Andrews, F., & Budding, E. (1992). Carter Observatory's 9 inch refractor: the Crossley connection. *Southern Stars, 34*, 358–366.

Annual Reports. Mr. Crossley's Observatory, Bermerside, Halifax. *Monthly Notices of the Royal Astronomical Society, 48*, 195 (1888).

Annual Reports. Mr. Crossley's Observatory, Bermerside, Halifax. *Monthly Notices of the Royal Astronomical Society, 50*, 206–207 (1890).

Annual Reports. Mr. Crossley's Observatory, Bermerside, Halifax. *Monthly Notices of the Royal Astronomical Society, 56*, 238 (1896).

Ashbrook, J. (1984). *The Astronomical Scrapbook. Skywatchers, Pioneers, and Seekers in Astronomy*. Cambridge: Sky Publishing.

Bateson, F. M. (1989). *Paradise Beckons*. Waikanae: The Heritage Press.

Bixby, J. E., & Van Flandern, T. (1969). The diameter of Neptune. *Astronomical Journal, 74*, 1220–1222.

Brück, H. (1983). *The Story of Astronomy in Edinburgh from its Beginnings until 1975*. Edinburgh: Edinburgh University.

Carter, C. R. (1866). *Life and Recollections of a New Zealand Colonist (Written by Himself)*. Volume II. London, printed for the author.

Catalogue of Astronomical Instruments (1900). York: T Cooke & Sons.

Chapman, A. (1998). *The Victorian Amateur Astronomer. Independent Astronomical Research in Britain 1820–1920*. Chichester: Wiley.

Cortie, A. L. (1909). The Solar Section. *Journal of the British Astronomical Association, 19*, 424–425.

Crossley, E. (1868). The lunar crater Linné. *Monthly Notices of the Royal Astronomical Society, 28*, 187.

Crossley, E., & Gledhill, J. (1892). The lunar eclipse, 1892 May 11: observations made at Mr. Crossley's Observatory, Bermerside, Halifax. *Monthly Notices of the Royal Astronomical Society, 52*, 560–562.

Crossley, J., Gledhill, J., & Wilson, J. M. (1879). *Handbook of Double Stars …* London: Macmillan.

Dewhirst, D. W. (1975). A Cooke photovisual lens in a compensated cell. *Sky and Telescope, 49* (1), 24–25.

Donnelly, M. C. (1973). *A Short History of Observatories*. Eugene: University of Oregon.

Eiby, G. A. (1970). Obituaries. *Ivan Leslie Thomsen. Southern Stars, 23*, 113–116.

Eiby, G.A. (n.d.) *Jubilee Exhibition and a Brief History of Astronomy in New Zealand*. Typescript, Royal Astronomical Society of New Zealand.

"Epoch marking". The City Telescope. Observatory at Kelburn. The opening ceremony. *The Evening Post*, August 27, 1924.

Father David Kennedy S. M. Notes. MS, Marist Fathers N.Z. Archives, Wellington.

Fisher, W. J. H. (1971). *Carter Observatory Report for the Year ended 1971, March 31*. Wellington: Carter Observatory (Astronomical Bulletin No. 75).

Gledhill, J. (1897). Observations of Jupiter and Jupiter's satellites made at Mr. Crossley's Observatory, Bermerside, Halifax. *Monthly Notices of the Royal Astronomical Society, 57*, 637–654.

Hearnshaw, J. B. (1996). *The Measurement of Starlight. Two Centuries of Astronomical Photometry*. Cambridge: Cambridge University Press.

Hearnshaw, J. B. (2002). Garry Nankivell and his contribution to Mt John. *Southern Stars, 41*(1), 17–20.

Hockey, T. (1999). *Galileo's Planet. Observing Jupiter before Photography*. Bristol: Institute of Physics.

Howell, P. A. G. (1970). Father Kennedy. *Southern Stars, 23*, 121.

Kennedy, D. (1909). Letter to Sir David Gill, dated May 12. In RAS Archives.

King, H. C. (1979). *The History of the Telescope*. New York: Dover.

Leather, K., Andrews, F., Hall, R., & Orchiston, W. (1998). Coping with a new curriculum: the evolving schools program at the Carter Observatory, New Zealand. In L. Gouguenheim, D. McNally, & J. Percy (Eds.), *New Trends in Astronomy Teaching* (pp. 267–272). Cambridge: Cambridge University Press.

Log Book No. 2. Wellington City Observatory (n.d.). Original in Carter Observatory Library, Wellington.

Mackrell, B. (1985). *Halley's Comet Over New Zealand*. Reed-Methuen: Auckland.

McConnell, A. (1992). *Instrument Makers to the World. A History of Cooke, Troughton & Simms*. York: William Sessions.

Mullaney, J. (1993). The role of the telescope in the planetarium. *The Planetarian, 22*(3), 6–7.

Nankivell, G. (2002). The Cooke photovisual objective and the 22.9cm refractor at the Carter Observatory, New Zealand. *Journal of the Antique Telescope Society, 24*, 4–8.

Nankivell, G. (1991). The Cooke photovisual objective revisited. MS.

Nankivell, G. (2001). One Cooke telescope with three objectives. MS.

New Zealand Astronomical Society. City telescope. Not fully used. Chance to alter things. *The Evening Post*, August 23, 1933.

Newcomb, S. (1903). On the desirableness of a re-investigation of the problems growing out of the mean motion of the Moon. *Monthly Notices of the Royal Astronomical Society, 63*, 316–324.

Norman, E. P. (1952). *Report of the Carter Observatory Board for the Year Ended 1952, March 31*. Wellington: Carter Observatory.

Obituaries. Edward Crossley. *Monthly Notices of the Royal Astronomical Society, 65*, 335–336 (1905).

Obituaries. Joseph Gledhill. *Monthly Notices of the Royal Astronomical Society, 67*, 232–233 (1907).

Obituary. Edward Crossley. *The Observatory, 28*, 110–111 (1905).

Occultation of Neptune. Preliminary report of N.Z. observations. *Southern Stars, 22*, 178–180 (1968).

Orchiston, W. (1982). John Tebbutt and the Whakatane Eight Inch Refractor: a review of 'The Australian connection'. *Southern Stars, 29*, 215–244.

Orchiston, W. (1994). Radio waves from the Sun: the New Zealand connection. In W. Orchiston, R. Dodd, & R. Hall (Eds.), *Astronomical Handbook for 1995* (pp. 65–69). Wellington: Carter Observatory.

Orchiston, W. (1995a). *Charles Rooking Carter: A Brief Biography*. Wellington: Carter Observatory.

Orchiston, W. (1995b). The observatory. In L. Knight (Ed.), *Muskets and Marshes. The History of Meeanee. The Land the People the School* (pp. 119–123). Napier: Printed for the editor.
Orchiston, W. (1996). The historic 23cm refracting telescope at the Carter Observatory. In W. Orchiston & B. Carter (Eds.), *Astronomical Handbook for 1997* (pp. 59–69). Wellington: Carter Observatory.
Orchiston, W. (1998). Enhancing public understanding of science through astronomy: a New Zealand case study. In *Proceedings of APEC Symposium Concerning Public Understanding on Science and Technology* (pp. 63–71). Tokyo: Japan Science and Technology Corporation.
Orchiston, W. (2001a). The English Equatorial Mounting and the history of the Fletcher Telescope. *Journal of Astronomical History and Heritage, 4*, 29–42.
Orchiston, W. (2001b). 'Sentinel of the Southern Heavens': the Windsor Observatory of John Tebbutt. *Journal of the Antique Telescope Society, 21*, 11–23.
Orchiston, W. (2002). From Crossley to Carter: the life and times of an historic Cooke refractor. *Journal of the Antique Telescope Society, 24*, 9–24.
Orchiston, W. (2005). Dr Elizabeth Alexander: first female radio astronomer. In W. Orchiston (Ed.), *The New Astronomy: Opening the Electromagnetic Window and Expanding our View of Planet Earth* (pp. 71–92). New York: Springer.
Orchiston, W., Carter, B., Dodd, R., & Hall, R. (1998). Selling our southern skies: recent public astronomy developments at the Carter Observatory, New Zealand. In L. Gouguenheim, D. McNally, & J. Percy (Eds.), *New Trends in Astronomy Teaching* (pp. 214–219). Cambridge: Cambridge University Press.
Osterbrock, D. E. (1984). *James E. Keeler: Pioneer American Astrophysicist and the Early Development of American Astrophysics*. Cambridge: Cambridge University Press.
Rumsey, N. (2002). Garry Nankivell: perfectionist. *Southern Stars, 41*(1), 14–16, 20–21.
Ryle, M., & Vonberg, D. D. (1946). Solar radiation at 175 Mc/s. *Nature, 158*, 339–340.
Sampson, R. A. (1910). On the old observations of the eclipses of Jupiter's satellites. *Memoirs of the Royal Astronomical Society, 59*, 199–256.
Sheehan, W. (1997). *The Mlanet Mars. A History of Observation & Discovery*. Tucson: University of Arizona Press.
Stonehouse, B. E., & Thomsen, I. L. (1937). Results of spectroscopic observations made during the partial eclipse of the Sun at Wellington on 1936, December 13th–14th. *Southern Stars, 3*, 43–46.
Taylor, H. D. (1894). Description of a perfectly achromatic refractor. *Monthly Notices of the Royal Astronomical Society, 54*, 328–337.
The City Telescope official opening. Speeches by Mayor, Sir F. Chapman and others. First municipal telescope. *The New Zealand Times*, August 27, 1924.
The observatory and nine-inch refracting telescope recently erected at St. Mary's Seminary, Meeanee, New Zealand. *The Weekly Press*, August 21, 1907 (copy in Riverview Archives, Sydney: A2926).
Thomsen, I. L. (1938a). Measures of double stars. *Southern Stars, 4*, 40–41.
Thomsen, I. L. (1938b). Recent solar activity. *Southern Stars, 4*, 23–24.
Thomsen, I. L. (1939). Fun with Mars in the news. *Southern Stars, 5*, 87–91.
Thomsen, I. L. (1947). Obituary notices: Charles Edward Adams. *Monthly Notices of the Royal Astronomical Society, 107*, 39–40.
Thomsen, I. L. (1948). Solar radio emissions and sunspots. *Nature, 161*, 134–136.
Thomsen, I. L. (1965a). A programme of double-star observing. *Southern Stars, 21*, 43–48.
Thomsen, I. L. (1965b). Photographic measures of the double star k Puppis. *Southern Stars, 21*, 81–84.
Thomsen, I. L. (1950). Proceedings of observatories. Carter Observatory, Wellington, N.Z. *Monthly Notices of the Royal Astronomical Society, 110*, 163.
Van Helden, A. (1984). Telescope building, 1850–1900. In O. Gingerich (Ed.), *The General History of Astronomy. Astrophysics and Twentieth-Century Astronomy to 1950: Part A* (Vol. 4, pp. 40–58). Cambridge: Cambridge University Press.

Warren, O. C. R. (2001). Catalogue of double star measures made by Ray Nisbet (1973–1975). *Southern Stars, 40*(1), 19–21.

Watson, W. (1905). Letter to J.T. Ward, dated November 7. Original in Ward (Wanganui) Observatory Archives, Wanganui, New Zealand.

Wilson, J. M., & Gledhill, J. (1876). A preliminary list of binary and other interesting double stars. *Monthly Notices of the Royal Astronomical Society, 37*, 64–77.

Wright, D. C. (1993). Double star astronomy in the UK 1832–1932. *Quarterly Journal of the Royal Astronomical Society, 34*, 423–440.

Part IV
Transits of Venus: The Quest for the Astronomical Unit

Chapter 14
The 1874 and 1882 Transits of Venus: An Overview

Abstract Because of its fortuitous location, New Zealand hosted overseas expeditions, various local government-funded observing teams and a plethora of dedicated amateur astronomers intent on observing the 1874 and 1882 transits of Venus. Even though inclement weather foiled some observers in 1874 and in 1882 the ingress phase occurred before local sunrise, New Zealand was able to play an important role in the quest to determine one of astronomy's fundamental yardsticks, the mean distance from the Earth to the Sun (now known as the 'Astronomical Unit', or AU).

14.1 Introduction

The quest to accurately determine the solar parallax was one of the great issues of world astronomy during the nineteenth century, and the Venus transits of 1874 and 1882 were seen by many as the best method of approach (Clerke 1893). With a value for this parallax nailed, that basic yardstick of astronomy, the distance from the Earth to the Sun, also was solved. Ideally, the transit should be observed from sites widely separated in latitude, and, given its fortuitous location on the globe, New Zealand was to contribute in a manner that was far out of proportion to its geographical area and population.

Studies have already been published of some individual New Zealand transit stations (Dick et al. 1998; Orchiston et al. 2000; Orchiston 2001a), regional New Zealand programs (McIntosh 1958), and the entire 1874 transit program (Orchiston 2012), and in 2004 I prepared a detailed overview of the Australian and New Zealand transit observations made in 1874 and 1882 (Orchiston 2004). In 2012, William Tobin also published a valuable paper on the nineteenth century New Zealand transit observations. This chapter draws heavily on the New Zealand component of my long 2004 review paper. As is accepted practice, throughout this paper the terms 'first contact' and 'second contact' will relate to the external and internal ingress contacts respectively, and 'third contact' and 'fourth contact' to the internal and external egress contacts.

W. Orchiston, *Exploring the History of New Zealand Astronomy*, Astrophysics and Space Science Library 422,
DOI 10.1007/978-3-319-22566-1_14

14.2 The 1874 Transit

14.2.1 Introduction

As Woolf (1959) has brilliantly portrayed, the two eighteenth century transits were the world's first major international scientific projects, but they did not produce the anticipated results. Instead, 'acceptable' published values of the solar parallax deriving from the 1769 transit ranged from 8.43″ to 8.80″, corresponding to an uncertainty in the mean Earth-Sun distance of about 4.4 %—a totally unsatisfactory situation. This threw the focus onto the two nineteenth century transits, and demanded that improvements be made in instrumentation and longitude-determination of the various transit sites. One of the key developments in nineteenth century astronomy was the emergence of photography, and there was an early realization that it had special potential in astronomy (e.g. see Lankford 1984; Norman 1938). Photography was to play a key role during the 1874 transit (Lankford 1987)—as the following case studies will illustrate.

New Zealand was particularly favourably-placed for the transit of 9 December 1874, with the entire event visible throughout the nation, and this is one of the reasons that foreign powers decided to establish southern transit stations in this region of the southwest Pacific. Two different American parties, and single parties from Britain, France and Germany, found their way to the New Zealand mainland, or to offshore islands that were New Zealand territory. For mainland New Zealand localities mentioned in the text relating to the 1874 transit see Fig. 14.1.

14.2.2 Observations of the Transit

The 1874 transit of Venus promised to be the most important astronomical event in New Zealand since the European settlement of the country, and it attracted enormous public attention. Not only did the Government plan transit observations, but amateur astronomers throughout the nation readied themselves for this once-in-a-lifetime event which promised views of the entire transit and all four contacts (unlike the later 1882 transit, when only the egress phase would be visible). Members of the general public also were catered for, thanks to a booklet titled *December 9, 1874. The Transit of Venus and How to Observe It* (Stock 1874), which was penned by Archdeacon Arthur Stock (see Fig. 4.8), the talented and dedicated Astronomical Observer at the Government's Colonial Observatory in Wellington (Eiby 1977).

Because of New Zealand's unique geographical location, English (Airy 1881), French (Bouquet de la Grye 1882; Filhol 1885), German (Auwers 1887–1898) and U.S. (Newcomb 1880) transit parties were attracted to the sunny South Island and to three of New Zealand's off-shore islands: the Chatham Islands, due east of New Zealand, and to the Auckland Islands and Campbell Island, far to the south, in New

Fig. 14.1 New Zealand mainland localities relating to the 1874 transit are shown in *red*

Fig. 14.2 The French transit station on Campbell Island (after Bouquet de la Grye 1882)

Zealand's wind-swept, inhospitable sub-antarctic zone. The British, under Major H.S. Palmer, also arranged a network of official subsidiary stations, which were manned by the following New Zealand observers: Theophilus Heale in Auckland, Henry Severn in Thames, Arthur Stock in Wellington and the two surveyors J.T. Thomson and J. McKerrow in Dunedin. All were supplied with time, by telegraph, from Burnham (The Transit of Venus 1874b). In addition, a number of amateur astronomers in Auckland, Featherston, Wellington, Nelson and Dunedin prepared independently for this important event. Regrettably, inclement weather prevented the French party on Campbell Island (Fig. 14.2) from seeing the transit (Lauga 2004: 302), and this same fate befell many of those on the North and South Islands of New Zealand (The Transit of Venus 1874c, d). Yet some of the astronomers in Auckland, Canterbury and Otago, and the overseas parties on the Chatham Islands and Auckland Island, did meet with a modicum of success—as the following accounts will demonstrate.

14.2.2.1 International Expeditions. Burnham: The British Expedition

The British organised an ambitious international transit of Venus campaign (see Radcliff 2008). Leading the British expedition to New Zealand was Major Henry Spencer Palmer (1838–1893; Fig. 14.3), Royal Engineers, aided by Lieutenants Henry Crawford (Royal Navy), L. Darwin and H. Praed, Royal Engineers. The site they selected, after carefully considering seven different options (Palmer 1874a), was about 730 metres north-west of Burnham Railway Station, on the expansive Canterbury Plains (which stretch from Christchurch westwards to the South Alps, and extended ~150 km in a NE-SW direction). The railway station was on the

Fig. 14.3 Major Henry Spencer Parker (http://homepage3.nifty.com/yhiguchi/gallery1.htm)

main N-S telegraph line, which provided Palmer with a means of transmitting time signals to the 'satellite' stations throughout the nation (Airy 1881: 484). Meanwhile, acting on advice from New Zealand authorities (Thomson et al. 1874) and in order to allow for possible inclement weather at Burnham, Palmer directed Lieutenant Crawford and Captain Williams from the *Merope* to establish a separate observing station at Naseby, in dry, sunny northern Otago (The Transit of Venus 1874b). This was furnished with a 10.2-cm (4-in) Simms refractor, a portable transit telescope by Shaw, and an astronomical clock by Dent (Palmer 1874a, b).

The coordinates of the Burnham transit station were: longitude 11h 29min 13.1s E and latitude 43° 36′ 48.1″S. The main astronomical instruments used there were a 15.2-cm (6-in) Simms equatorially-mounted refractor (Fig. 14.4), a transit telescope by Simms (Fig. 14.5), a Dallmeyer photoheliograph (Fig. 14.6), a 38.1-cm (15-in) diameter Troughton and Simms azimuth instrument with a 5.7-cm (2.25-in) objective, and an astronomical clock made by Arnold but modified by Dent. Major Palmer was particularly pleased with their quality, commenting:

> The instruments throughout are but of moderate size, but are the best of their kind, and give evidence of remarkable preparation and care and forethought which the Astronomer-Royal has bestowed upon the English part of the undertaking for the actual observation of the transit. (The Transit of Venus 1874a).

With the instruments came prefabricated equatorial, photoheliograph, transit and altazimuth huts (Simmons 1880). In reporting on their preparations leading up to the transit, Palmer (1874a) heaped praise on the New Zealand and Provincial

Fig. 14.4 The Simms equatorial (*Courtesy* Royal Astronomical Society)

Fig. 14.5 The Simms transit telescope (*Courtesy* Royal Astronomical Society)

Governments: "It is impossible to speak too well of the readiness with which all facilities in their power have been afforded to us … Everything has been done to assist our operations, and to lessen our expenses."

On the morning of 9 December, unfavourable weather prevented observation of the first contact, and when the clouds thinned sufficiently Venus "… was then seen

Fig. 14.6 The Dallmeyer photoheliograph (*Courtesy* Royal Astronomical Society)

to have advanced apparently about three-eighths of her diameter on the Sun. Both Sun and planet could only be just made out through the clouds, without any coloured shade to the eye-piece." (Airy 1881: 493). Over the next 12 minutes or so, Palmer used the Simms refractor to make a dozen unsuccessful attempts to take micrometric measurements of the transit during gaps in the clouds. Immediately prior to second contact

> … the Sun showed again, when the cusps were about one-twentieth of a diameter apart, and connected by a dimly marked ligament, not nearly so sharp as the "black drop" of the model in full sunlight … [Three and a half seconds later] … the ligament seemed to undergo a change in depth of colour, but clouds prevented me from seeing whether any streak of light connecting cusps played across it. (Airy 1881: 494).

Just 3 seconds later clouds intervened, and prevented further observations of the ingress phase. Nearly 15 minutes after the estimated time of second contact, the clouds cleared sufficiently for a few photographs to be taken (e.g. see Fig. 14.7), and Palmer attempted to make further micrometric observations, but he judged the results as "… very irregular, and … of little use." The remainder of the transit—including the egress contacts—was lost owing to clouds and rain, and the Sun only emerged 10 minutes after fourth contact. As if to mock the astronomers, it then shone brilliantly until near sunset!

Palmer may have had only limited success at Burnham, but the weather conspired against Lieutenant Crawford making any observations at Naseby (Airy 1881: 484).

Fig. 14.7 A negative of one of the few successful photographs of the transit taken at Burnham (*Courtesy* STFC and J. Ratcliff)

14.2.2.2 International Expeditions. Queenstown: The U.S. Expedition

After dropping off two Australian-based transit teams in Hobart, the *U.S.S. Swatara* crossed the Tasman Sea with two parties destined for New Zealand waters, and anchored in Bluff Harbour on the extreme southern tip of the South Island of New Zealand (Dick et al. 1998; Orchiston et al. 2000). The intention, initially, was to establish a transit station at Bluff, but Dunedin astronomers recommended an inland site, in sunny Central Otago (see Thomson et al. 1874). Queenstown, on the shores of Lake Wakatipu and at longitude 11h 14m 40.4s E and latitude 45° 02′ 07″S subsequently was chosen (Newcomb 1880: 21). The Queenstown transit party comprised Dr Christian Heinrich Freidrich Peters (1813–1890; see Fig. 15.7) from Hamilton College, his Assistant Astronomer Lieutenant Edgar Wales Bass (1843–1918; from the U.S. Military Academy at West Point), and four photographers, Charles L. Phillippi (1826–1890), Israel Cook Russell (1852–1906), E.B. Pierson and Louis H. Aymé (Dick 2003: 255), and they had access to the same types of instruments as issued to all of the U.S. stations (for details see Orchiston et al. 2000), namely a photoheliograph, a refracting telescope, a transit telescope, an astronomical clock, box chronometers and meteorological instruments.

Since they wished to rely on photographs of the transit, the key instrument was the photoheliograph. The Americans had decided to use the style of horizontal photographic telescope first devised by Joseph Winlock (1826–1875; Fig. 14.8), the Director of the Harvard College Observatory. This comprised a 17.8-cm unsilvered glass heliostat mirror mounted on a solid metal pier, which tracked the Sun and directed sunlight via a 12.7-cm collimator lens onto a photographic plate mounted

Fig. 14.8 Joseph Winlock (www.kilma-luft.de/steinicke/ngcic/persons/winlock.html)

on another metal pier, situated 12 metres away, in a 'Photographic House' (see Figs. 15.3 and 15.4).

As the latter figure indicates, adjacent to the heliostat and drive was the prefabricated 'Transit House' which housed a Stackpole broken-tube transit telescope (Fig. 14.9). This instrument was used to maintain a reliable local time service and also to determine the latitude and longitude of the transit station.

Completing the principal instrumentation was a 12.7-cm (5-in) equatorially-mounted Alvan Clark refractor (see Fig. 14.10) which was in a prefabricated 'Equatorial House'. Figure 15.11 shows the plan of the Queenstown transit station and the relative positions of the various instruments and their associated 'houses'.

At 5 a.m. on 9 December the sky was cloudy and stormy, but—miraculously—the Sun appeared just 2 minutes before first contact,

> … permitting observations of Equatorial and photographs to be made without interruption for $1^2/_3$ hours about. After that only at intervals, up to within 16 minutes before computed time of 3^d contact. The Sun then remained invisible until … after 4th contact. (Peters 1874).

In all, 237 photographs of the transit were obtained, 178 of the ingress contacts and 59 of Venus superimposed on the Sun's disk (Newcomb 1881: 439–440). Peters was observing with the Clark refractor, and he specifically noted the absence of the black drop at second contact (Peters 1874). Between second and third contacts he obtained a series of micrometric measures of Venus' position relative to the Sun's limb (see Newcomb 1880).

Over all, Peters was very pleased with the outcome of the expedition: "The instruments and appliances worked admirably, and everything passed off well." (*Lake Wakatip Mail* 1874). A more detailed account of the Queenstown transit enterprise is presented in the next chapter in this book.

Fig. 14.9 One of the 'broken tube' transit telescopes at the U.S. Naval Observatory (*Photograph* Wayne Orchiston)

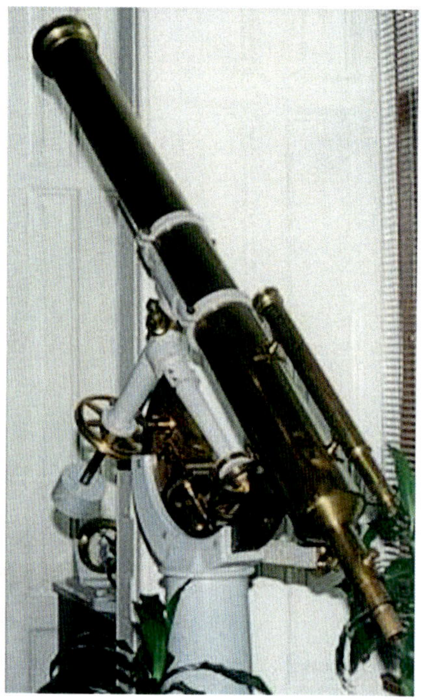

Fig. 14.10 One of the 12.7-cm Alvan Clark refractors at the U.S. Naval Observatory (*Photograph* Wayne Orchiston)

14.2.2.3 International Expeditions. Chatham Islands: The U.S. Expedition

The final Southern Hemisphere U.S. transit party was assigned to the Chatham Islands, New Zealand territory to the east of the South Island (Dick et al. 1998), and the contingent there comprised Chief Astronomer, Edwin Smith (1851–1943) from the U.S. Coast and Geodetic Survey, Assistant Astronomer, Albert H. Scott (also from the Survey), the photographers, Otto Beuhler and W.H. Rau, and an instrument-maker named Sumner Tainter (Dick 2003: 255). They were also joined by an interested local person (see Baucke 1928). Their astronomical 'houses' and equipment mirrored those found at other U.S. 1874 transit stations, and Figs. 14.11, 14.12 and 14.13 provide close-ups of the photographic telescope heliostat, the equatorial house and the broken tube transit telescope in the transit house. The transit station was set up at Whangaroa, at longitude 12h 13m 11.8s W and latitude 45° 49′ 03.2″S (Newcomb 1880: 21).

From the listing in Newcomb (1880: 135) and the account given by Baucke (1928), it is apparent that first and second contacts were observed, but only eight photographs were obtained (Dick 2003: 259; The transit of Venus. The United States Expedition … 1882e). Otherwise the observers were clouded out.

14.2.2.4 International Expeditions. Auckland Islands: The German Expedition

The German station at Port Ross on Auckland Island (longitude 166° 13′ 27.9″E; latitude 50° 32′ 14″S), was one of six set up by Germany for the 1874 transit (see Dawson and Duerbeck 2008; Duerbeck 2004). Members of the Auckland Island expedition were: two astronomers, Hugo von Seeliger (1849–1924; Fig. 14.14) and Wilhelm Schur (1846–1901); two photographers, Hermann Krone (1827–1916; Fig. 14.15) and Dr Guido Wolfram (b. 1842); and two assistants, Johannes Krone (1855–1924; he was Hermann's son) and Hermann Leyser. A comfortable station with prefabricated observatories and a house was set up on the beach at Terror Cover, Port Ross (see Fig. 14.16), some of the brick footings of which were still visible more than a hundred years later (Barberel 1975).[1] The instruments used at this transit station were: a 7.6-cm (3-in) Fraunhofer heliometer (loaned by Göttingen Observatory), used to measure the motion of Venus across the solar disk; a 10.7-cm (4.2-in) Steilheil photoheliograph (Fig. 14.17), which was designed to photographically record the transit; 11.7-cm (4.6-in), 8.25-cm (3.25-in) and 7.3-cm (2.9-in) Fraunhofer refractors (for visual observations of the ingress and egress contacts);

[1]Dawson and Duerbeck (2008: 9) note that in 2003 Kevin Jones and a team of archaeologists from the Department of Conservation mapped "… the site of the 1874 transit expedition, noting the position of instrument pillars and so on, following a brief survey done by a Lands & Survey ranger, William Barberel, in December 1973."

Fig. 14.11 The Chatham Islands horizontal telescope showing (left to right) the protective wooden framework (leading to the 'Photographic House'), the heliostat and pier, the heliostat weight drive, an engineer's level, and part of the 'Transit House' (*Courtesy* Alexander Turnbull Library, Wellington, Ref. PAColl-0058-02)

a Pistor and Martin transit telescope; and time-keeping and meteorological equipment (Auwers 1898; Duerbeck 2004).

Although light rain fell before the transit started, as it turned out all the preparations proved worthwhile, and Seeliger and his colleagues did succeed in observing much of the transit. Passing clouds made it difficult to time the two ingress contacts, but soon after second contact the clouds dispersed and the remainder of the transit was visible (although by egress the solar image was far from ideal). During the transit, Seeliger observed with the heliometer, Schur with the largest conventional refractor, Wolfram with one of the smaller refractors, while Hermann Krone exposed no less than 115 plates with the photoheliograph (Auwers 1898: 202–206). Most of the photographs showed Venus on the Sun's disk (and one of these is reproduced here in Fig. 14.18), but the last six plates were exposed immediately prior to, and during, the egress contacts.

Fig. 14.12 The Chatham Islands 'Equatorial House' (*Courtesy* Alexander Turnbull Library, Wellington, Ref. PAColl-0058-05)

Fig. 14.13 View of the Chatham Islands 'Transit House' with a wall removed, showing the broken tube transit telescope and its brick pier (*Courtesy* Alexander Turnbull Library, Wellington, Ref. PAColl-0058-03)

Fig. 14.14 Hugo von Seeliger (https://en.wikipedia.org)

Fig. 14.15 A photograph of Hermann Krone taken in 1854 (*Courtesy* Hermann-Krone-Collection, IAPP, University of Technology of Dresden)

Fig. 14.16 Engraving of the 1874 German transit of Venus camp at Ross Cove, Auckland Islands (after *The Illustrated Australian News for Home Readers*, 1875)

Fig. 14.17 The Steilheil photoheliograph at the Auckland Island transit station (*Courtesy* Hermann-Krone-Collection, IAPP, University of Technology of Dresden)

Towards third contact, Seeliger saw the black drop effect, but its formation

> … did not happen suddenly, but the two limbs (Venus and the Sun) merged slowly with one another. Then there was a moment when a noticeable intensity change in the existing very washed-out darkening took place. (Auwers 1898: 199; English translation).

Wolfram also clearly saw the back drop effect: At "5h 49m 39s a darkening occurred between both limbs, which I perceived as the moment of drop formation …" (Auwers 1898: 201). He also remarked that unlike with the transit of Venus apparatus used for practice observations, in reality when the back drop develops "… one sees that a broad band forms which is darkest at that location where the limbs

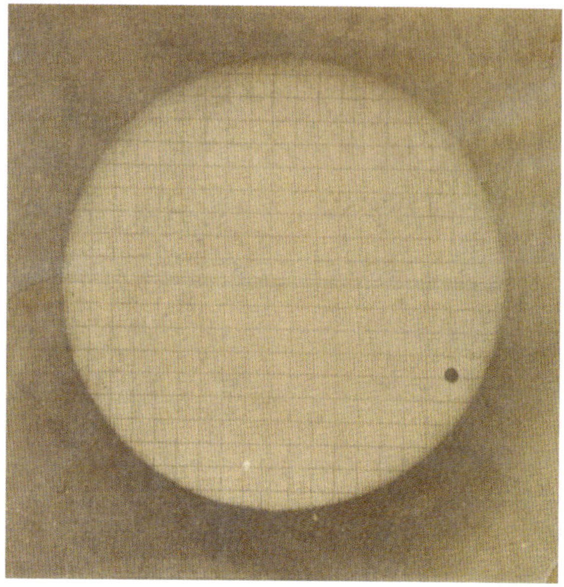

Fig. 14.18 A photograph of the transit taken at the Auckland Island transit station (*Courtesy* Hermann-Krone-Collection, IAPP, University of Technology of Dresden)

[of Venus and the Sun] are closest to each other, and this gradually fades at the upper and lower edges." (ibid.). Unfortunately, Schur did not make a successful observation of the third contact, but all three astronomers did see, and time, the fourth contact.

14.2.2.5 Independent Observers. Auckland

Although the surveyor Captain Theophilus Heale (1816–1885; Scholefield 1940(1): 371–373) was prevented from observing the transit, another transit team, located in a different Auckland suburb, had some success. Professor Samuel John Lambert, F.R.A.S., and photographers Josiah Martin (1843–1916; Maitland 1993), Pond and Redfern[2] spent four months setting up their transit station in Hobson Street, which contained telescopes, photographic equipment, and a chronometer. Time was provided via telegraphed signals (Transit of Venus 1874c). Lambert timed the first contact and 8 minutes later the black drop was observed, which was followed by second contact. Soon after this, clouds moved in and prevented any further observations of the transit. However, during the ingress phase, "A good many sun pictures of great nicety were obtained." (ibid.).

[2]Josiah Martin was a gifted Auckland school teacher and photographer (see Maitland 1993), while Redfern was either George W. Redfern (1844–1901) or Richard Redfern, who had photographic studios in Auckland at this time. In 1889 Richard Redfern moved to Melbourne, Australia.

14.2.2.6 Independent Observers. Nelson

Both ingress and egress contacts were visible from Nelson, but during the intervening period clouds and rain prevented observations being made. *The Daily Southern Cross* newspaper reports that an amateur astronomer by the name of Captain Sharp[3] succeeded in timing the egress contacts with the aid of a chronometer (The Transit of Venus 1874d).

14.3 The 1882 Transit

14.3.1 Introduction

Despite the disappointing weather conditions that greeted many New Zealand observers in 1874, the overall expectation was that observers of the 7 December 1882—the second and final transit of the nineteenth century—would be more successful.

There were, however, two factors other than the weather that might militate against this. Firstly, the 1882 transit would already be in progress at sunrise in New Zealand, so it would only be possible to observe, and time, the third and fourth contacts. And secondly, photography had not lived up to expectations in 1874 (e.g. see Airy 1881: Appendix V), and its use during the 1882 transit was therefore questioned by many (Lankford 1987). Some nations decided to rely solely on visual observations, but others, including the Americans, determined to persist with their photographic endeavours.

New Zealand localities relating to the 1882 transit are shown in Fig. 14.19.

14.3.2 Observations of the Transit

An ambitious transit program involving amateur and professional observers was planned for New Zealand, and this was designed in close collaboration with the British and U.S. parties that based themselves in Burnham and Auckland, respectively (note that on this occasion there were no French or German parties in New Zealand waters). Most observers were connected to the telegraph, and time signals were distributed to North Island and Nelson observers from an observatory set up by the Government at Mount Cook in Wellington, while other South Island observers relied on the Burnham headquarters of the British transit party for their

[3]Presumably this was John Sharp (1828–1919), who originally worked as a surveyor with the New Zealand company, was living in Nelson in 1882, and was a Captain in the Nelson Rifles (see Scholefield 1940(s): 294).

Fig. 14.19 New Zealand mainland localities relating to the 1882 transit are shown in *red*

time signals (Palmer 1875). And in order to accurately determine the geographical co-ordinates of all observing stations, exchanges of time signals were made between Burnham and Auckland, Burnham and Wellington, Wellington and New Plymouth, and Wellington and 'Bidwill's' in the Wairarapa (McKerrow 1882).

14.3.2.1 International Expeditions. Auckland: The U.S. Expedition

Frustrated by the largely unsuccessful Chatham Islands' program in 1874, the Americans on this occasion determined to base their sole New Zealand transit station in Auckland. The site selected was in the Auckland Domain, and is now covered by an extension of the Auckland War Memorial Museum. Head of the U.S. party was Edwin Smith (who was in the Chatham Islands in 1874), and assisting him was Henry Smith Pritchett (1857–1939; Fig. 14.20), Professor of Mathematics and Astronomy at Washington University, St. Louis, and two photographers, Augustus Storey and Gustav Thielkuhl (Dick 2003: 266; Transit of Venus. Arrival … 1882a). Four others completed the transit party, one of whom was the Auckland amateur astronomer, J.T. Stevenson, whose observations are discussed later.

The transit station was described in a contemporary newspaper report:

> The casual visitor to the Domain will see a small space near the old Block House enclosed with an iron wire fence. Inside this have been erected several buildings. The one nearest the Block House contains the transit instrument. Another one adjoining, with a revolving roof, contains the five-inch equatorial telescope. In front of the one containing the transit instrument is placed the stand for the heliostat, from which runs the telescopic tube connecting with the dark chamber of the photographers, at a distance of some 38 or 39 feet. But inside the Block House are many appliances of the greatest use. There is fitted up the telephonic apparatus, and the telegraphic instruments, by which Mr. Smith has been enabled to exchange signals with Colonel Tupman at Burnham. (Transit of Venus. Arrival … 1882a).

Fig. 14.20 Professor H.S. Pritchett (https://en.wikipedia.org)

The astronomical instruments associated with the transit, photographic and equatorial houses were basically unchanged from 1874 (The Transit of Venus 1882b), but apparently the astronomical clock used was different. According to Cossons (1992: 34–35) it was one of the Shelton 'regulators' used by the British during Cook's Second and Third Voyage to the South Seas, and loaned by the Royal Society of London to the U.S. Coast and Geodetic Survey from 1882 to 1884.

When the Sun rose in Auckland on 7 December the weather was not at all promising, but most of the clouds subsequently dispersed and the transit was visible. Messrs Storey and Thielkuhl then obtained a series of seventy-two photographs before further clouds caused these operations to be suspended until just before third contact when two final images were exposed (see Dick 2003: 267). Meanwhile, Smith and Pritchett successfully timed both egress contacts with the 12.7-cm refractor and a smaller (unspecified) telescope. A newspaper account mentions that

> A considerable number of people gathered outside the enclosure, and seemed to take a considerable interest in watching the various observations. Of course it will be several years before the data collected yesterday are so reduced as to give the information sought, but all must feel pleasure to know that the success of the American party of Observers at Auckland has been greater than they anticipated … (The Transit of Venus. The United States Expedition … 1882e).

14.3.2.2 International Expeditions. Burnham: The British Expedition

After an exchange of letters between Astronomer Royal Sir George Biddell Airy (1801–1892; Fig. 14.21) and Dr (later Sir) James Hector, Director of the Colonial Observatory, the Geological Survey and the Colonial Museum, Airy (1880) was moved to comment that "We are now entitled to consider New Zealand as probably the most important country of stations for observations of the Transit of Venus of 1882." Accordingly, funding was approved to support an expedition. Initially the Royal Society's British 'Transit Committee' favoured Auckland for the 1882 station, but leading politician and former Premier of New Zealand Harry (later Sir Harry) Albert Atkinson (1831–1892; Fig. 14.22) vehemently counselled against this, singing the climatic and geographical merits of Napier, if a decision was ultimately made to abandon the 1874 station at Burnham (Atkinson 1882). Eventually logic prevailed and Burnham was retained, but some of the instruments and all of the personnel assigned to New Zealand on this occasion differed. Head of the party was Lieutenant-Colonel George Lyon Tupman (1838–1922), who was assisted by G.E. Coke. Both observed from Burnham—on this occasion there was no ancillary station. For the transit, Tupman used a 15.2-cm (6-in) Cooke refractor and 'chronometer F', and Coke an 11.4-cm (4.5-in) Cooke refractor and a Molyneux chronometer. Assisting Tupman were his wife, a Mr Gill from the Telegraph Department in Wellington and Bombardier Wilson. Coke was also aided

Fig. 14.21 Sir George Biddell Airy in 1873 (https://en.wikipedia.org)

by his wife, a Mr White from the N.Z. Telegraph Service and a Mr Hamilton of the *Lyttelton Times* newspaper.

The astronomers awoke to a fine windless day on 7 December, and the roofs and eastern sides of the equatorial houses were removed entirely to facilitate easy

Fig. 14.22 Sir Harry Albert Atkinson in about 1885 (https://en.wikipedia.org)

observation. Prior to third contact, Tupman carefully examined the limb of Venus but could not detect any evidence of an atmosphere:

> The limb was absolutely sharp, of striking abruptness. There was no trace of any shadow on the Sun's surface without the planet, or of any light within the planet's disc. I assured myself of this repeatedly during the hour or more that I had little else to do. (Stone 1887: 60).

Nearly an hour before third contact was scheduled, Tupman used the Airy double-image micrometer to measure the diameter of Venus.

Just over a minute and a half before third contact was scheduled, Tupman's wife began counting seconds aloud, and then "Venus drew pretty quickly up to the Sun's limb. I expected to see the shadow between the limbs sooner than I actually did. The distance became very small without any diminution of the Sun's light … [then] A faint shadow grew slowly between the limbs." (ibid.). This faint shadow was parallel to the limbs, not many seconds of arc in length, and although apparent was certainly difficult to see. Just nine seconds later Tupman reported that "… the Sun's limb had lost its sharpness from the overlapping of Venus' atmosphere …" and seven seconds later "… the atmosphere was conspicuous, connecting the now formed cusps." (ibid.). Upon reviewing his third contact observations, Tupman stated:

> The illumination of the planet's atmosphere was, at first, almost as bright as the last glimpse of the Sun's limb, but it was different altogether. It was brightest at the sharp limb of the planet, of a different colour to the Sun's limb, and faded off at its outer border. It was less than a second of arc broad, but might, perhaps, have appeared a little broader had there been no cloud. (ibid.).

The "cloud" that Tupman alludes to was nothing more than a hazy white that spread over the Sun prior to third contact but in no way impeded his view of the transit.

Once third contact had been observed, Tupman installed the Airy micrometer and measured the distance between the cusps, and upon replacing the 210 power eyepiece

> … again saw the illumination in Venus' atmosphere, but only in the upper or south portion of that part of the planet which was outside the Sun … It seemed to start from the Sun's limb, and to extend 60° or 70° along the south limb of Venus, about half a second of arc broad, distinctly visible, though faint. (ibid.).

In all, visual evidence of Venus' atmosphere lasted for about fifteen minutes (cf. Shiltsev 2014). About 5 minutes later, Tupman obtained a clear view of fourth contact.

Coke also successfully observed the egress contacts. As third contact approached he noticed a "shadow" between the limbs of Venus and the Sun, and although this subsequently became more decided, even at its darkest "… it never approached in the slightest degree to the darkness of the planet; it was never more than a hazy shadow." (Stone 1887: 61). But at the time he estimated third contact to take place, Coke could see "… no appearance of an atmosphere of Venus …" (ibid.). He also observed fourth contact without incident, but did not place much faith on the time he recorded.

14.3.2.3 Government Observers. New Plymouth

At New Plymouth, Thomas Humphries, F.R.A.S. (1841–1928; Fig. 14.23; Scholefield 1940(1): 420), the Chief Surveyor for Taranaki, observed the transit with a 10.2-cm (4-in) Cooke refractor and a chronometer, assisted by T.K. Skinner. The longitude of the site was 11h 36m 17.55s E of Greenwich and the latitude 39° 04′ 0.8″S. Shortly before third contact Humphries observed that "A dark haze appeared between the limbs, which I thought was going to merge into "black drop", but five seconds [later] … clear light appeared." (Stone 1887: 68).

14.3.2.4 Government Observers. Wairarapa

When the Government was planning its transit stations, a decision was made to "… also observe in the Wairarapa district, so as to secure a chance of clear weather on the other side of the Rimutaka range, in case clouds prevail on the Wellington side of the range and obscure the view." (The Transit of Venus 1882a). A site referred to as "Bidwill's", with a longitude of 11h 41m 41.47s E and a latitude of 41° 11′ 29″S, was selected just south of the town of Martinborough. The Chief Surveyor of Wellington Province, John William Allman Marchant (1841–1920; Fig. 14.24; Scholefield 1940(2): 54), was assigned to this temporary transit station, and he used a 10.2-cm (4-in) altazimuth-mounted Browning refractor (loaned by David Gray of Wellington) and a Barraud chronometer. He was assisted by G. Struthers and a

Fig. 14.23 Thomas Humphries (after Ward 1928: Fig. 295)

Fig. 14.24 J.W.A. Marchant (after Ward 1928: Fig. 294)

Mr Wright. The times of both egress contacts were recorded, and prior to third contact Marchant noted that the limb of Venus "… appeared to be surrounded by a haze." (Stone 1887: 54). Just after third contact he observed that "… the rim of the Sun appeared to advance before the outwards limb of Venus, that is to say, it appeared to be pushed out by the black disk of the planet, and the aureole was distinctly visible round the outward rim of Venus." (ibid.). This feature lasted for some minutes and then disappeared.

The other observer at "Bidwill's" was Captain J.D. Hewitt, R.N., a Palmerston North farmer and amateur astronomer (Seymour 1985) who used a 21.6-cm (8.5-in) Browning reflector loaned by James Henry Pope (1837–1913; Fig. 14.25;

Fig. 14.25 J.H. Pope (https://en.wikipedia.org)

Scholefield 1940(2): 179), Inspector of Native Schools. This historic telescope was later owned by the Wellington Astronomical Society (see Dodson 1996). Hewitt timed the fourth contact, but provided no other information about the transit (Stone 1887).

14.3.2.5 Government Observers. Wellington

In Wellington, 'Mount Cook Observatory' was specifically set up for the transit on what was until recently the site of the National Museum, at longitude 11h 39m 05.91s E, and latitude 41° 18′ 0.6″S (Stone 1887: 55). It was "… a plain wooden building externally, but internally is very complete and elaborately fitted up." (Tomorrow's Transit of Venus 1882). Leading the transit party was Charles William Adams (1840–1918; Fig. 14.26; Scholefield 1940(1): 4) from the Survey Department (see Adams 1993), assisted by Mrs. Adams and W. Holmes, and at their disposal was a 10.6-cm (4.17-in) Solomons refractor (loaned by Mr Barnard of Wellington), a sidereal clock by Dent, and a chronograph. In addition, there was a 7.6-cm (3-in) Troughton and Simms transit telescope (Tomorrow's Transit of Venus 1882) and a "… new zenith telescope of American invention, for determining latitude, has been erected … and has enabled the latitude also to be precisely ascertained …" (The Transit of Venus 1882c). Transit day was bright and clear, and

Fig. 14.26 C.W. Adams (*Courtesy* Collection of Toitū Otago Settlers Museum)

the entire egress phase was seen. Since no suitable solar filter could be obtained in time, Adams observed the transit by projecting the image onto a screen. He successfully recorded the times of the third and fourth contacts but reported that "… there was an almost entire absence of the incidental phenomena that I had been led to expect." (Stone 1887: 55). Perhaps this was because of his less than satisfactory observing procedure.

The transit also was observed by James McKerrow (1834–1919; Fig. 14.27), the Surveyor-General of New Zealand (Obituaries 1920), from Thomas King's observatory in downtown Wellington (with a longitude of 11h 39m 05.57s E, and a latitude of 41° 17′ 14.3″S). At the time King was a prominent amateur astronomer (Orchiston 1998), but he later succeeded Arthur Stock as the Astronomical Observer at the Colonial Observatory. His observatory housed a sidereal clock and a recently-acquired 14-cm (5.5-in) equatorially-mounted Grubb refractor which later ended up in the King Edward VII Observatory in the Wellington Botanic Garden (see Seymour 1995). Assisting McKerrow were Alexander Barron and Thomas Grant, both officers of the Survey Department. McKerrow obtained a "most satisfactory" view of the transit and succeeded in timing both egress contacts. No black drop was observed at third contact, and the fourth contact "… was not marked by any phenomena …" (Stone 1887: 57). In conclusion, McKerrow was moved to comment that "… in the absence of clouds, wind, or other disturbing cause of weather, I cannot conceive of more favourable conditions for the observation." (ibid.). Soon after his observation of the transit, McKerrow was elected a Fellow of the Royal Astronomical Society (Obituaries 1920).

Fig. 14.27 James McKerrow (jamesmckerrowsurveyor.blogspot.com/)

14.3.2.6 Government Observers. Clyde

Wellington-based Dr James Hector (Fig. 14.28), Director of the Colonial Observatory, the Geological Survey and the Colonial Museum—among other posts (see Dell 1990)—selected Clyde in Otago for his transit site because of its

> … having an almost continental climate. Rain rarely falls, and a bright still atmosphere is almost invariably experienced at early morning. It was one of the few places in New Zealand that enjoyed a clear sky throughout the whole day of the last transit in 1874. It is also conveniently circumstanced as regards the telegraph system, being the junction point of two separate lines by which communication could be maintained with Burnham observatory. (Hector 1882: 326).

From his viewpoint, it was the ideal site, and circumstances subsequently vindicated this choice.

Hector was furnished with a 12.7-cm (5-in) f/13 altazimuth-mounted refractor by Cooke (loaned by G.V. Shannon of Wellington), a Russell chronometer, a stop watch and a chronograph. By 22 November the observing station was operational, and on the 24th time signals were exchanged with the British station at Burnham for the first time. Assisting Hector during the transit were E. Ashcroft; Mr. Henry, the local telegraphist at Clyde; Mr McKay, the district surveyor; and Major Jackson Kedell (1831–1910; Scholefield 1940(1): 452–453), the Resident Magistrate of Central Otago.

The day of the transit dawned cloudy, but cleared prior to third contact. At this time, Hector (1882: 328) noted: "As soon as the last shred of cloud passed, all boiling of the sun's edge cleared, and it was more sharp and steady than I had ever

Fig. 14.28 Sir James Hector in about 1900 (https://en.wikipedia.org)

before seen it." Venus was then near the solar limb, and "No trace of a halo or indefinite outline was seen round the planet …" (ibid.). Unfortunately a small dense cloud intervened and prevented observation of third contact, but soon after the Sun was again visible Hector noticed that

> … the outline of the emerged limb of the planet suddenly became apparent against the background of space as a delicate violet-tinted streak, having its concave edge sharp but the convex edge discontinuous. I brought this appearance out more distinctly by cutting off the sun's limb with a dense glass. Its extreme width I estimate at about 1-50th of its distance from the sun, which was about the semi-diameter of the planet. Suddenly, with a twinkle, this phenomenon disappeared, and I called time at 7h. 42m. This twinkle made a most distinct impression on me. There was not the least vibration at the time and my eye was not fatigued, as I still saw the "rice grains" [=granulation] on the sun's surface. (Hector 1882: 329).

Nearly nine minutes later, Hector successfully timed the fourth contact.

14.3.2.7 Independent Observers: Introduction

In 1875, Lieutenant Colonel Tupman (head of the 1874 British station at Burnham) observed that

> There is already a good deal of private scientific enterprise in New Zealand … The last Transit will undoubtedly have given a stimulus in the direction of astronomy, and it may be expected that by 1882 there will be many good private telescopes, and amateur observers willing to help in an extensive scheme of observation.

This certainly proved to be the case, with private observers scattered the length of the nation. These are discussed below, city by city, starting with Auckland in the north.

14.3.2.8 Independent Observers. Auckland

In order to observe the transit, the former Chief Surveyor of Auckland Province, Captain Theophilus Heale (Fig. 14.29), used a 12.7-cm (5-in) equatorially-mounted Cooke refractor (loaned for the occasion) and a chronometer, which he installed in an observatory that he built near his house in the suburb of Parnell. The co-ordinates of this observatory were: longitude 11h 39m 9.40s E; latitude 36° 51′ 9.5″S (Stone 1887: 72). At sunrise the transit was visible through thin misty clouds, which thickened and prevented him from timing third contact. Soon after, however, the clouds dispersed and "The external contact was, therefore, observed with the greatest accuracy." (The Transit of Venus. The United States Expedition … 1882e).

As we have seen, John Torrens Stevenson (1854–1914; Fig. 17.16) observed the transit from the U.S. compound in the Auckland Domain, thanks to the intervention of Lieutenant-Colonel Tupman. Irish-born Stevenson, FRAS, was an accountant with a passion for astronomy, and prior to the transit had published a paper on the

Fig. 14.29 Theophilus Heale (http://timespanner.blogspot.com/2010/07/island-called-motuketekete.html)

Great Comet of 1882, C/1882 R1, in *Monthly Notices of the Royal Astronomical Society* (Stevenson 1882). During the transit, while he had free access to the time service maintained by the Americans, Stevenson used his own telescope, a 16.5-cm (6.5-in) equatorial-mounted reflector with a mirror by Calver (The Transit of Venus 1882c). Light clouds covered the Sun during the egress phase, and he timed the third contact with difficulty. Although observing conditions were far from ideal, he reported that "There was no appearance of the black drop observed, neither was there any appearance of the light surrounding the limb of the planet which observers at previous transits have seen." (The Transit of Venus. The United States Expedition … 1882e). By the end of the transit the Sun was largely free of clouds, and he obtained a reliable time for fourth contact, which was included in Stone's monograph (1887: 73).

Professor S.J. Lambert from Auckland University College and local amateur astronomer Samuel Stuart (see Orchiston 1998: 110) were located at Stuart's house on New North Road, and supplied with 7.6-cm (3-in) and 6.35-cm (2.5-in) refractors and an "… instrument fitted up for photographing the transit." Lambert does not describe this last-mentioned instrument, other than to state that its lenses were by Dallmeyer (The Transit of Venus. The United States Expedition … 1882e); it may have been an astrograph, or a Dallmeyer refractor fitted out for prime focus photography. During the transit, Stuart and Lambert successfully obtained a good series of photographs. Prior to third contact, "… nothing peculiar was noticed, except a faint ring of light surrounding the planet … but [there were] no traces of anything appertaining to a satellite …" (ibid.). Both astronomers observed the third contact, but neither of them saw the black drop. Clouds prevented them from seeing fourth contact.

Fig. 14.30 Thomas Cheeseman (https://en.wikipedia.org)

Thomas Cheeseman (Fig. 14.30) was another Auckland amateur who observed the transit, using the "large reflecting telescope" in his observatory in the suburb of Remuera. He did not attempt to obtain accurate times for the egress contacts, but like Stuart and Lambert he reported the presence of a Venusian atmosphere:

> … as the time for internal contact approached it appeared to Mr. Cheeseman as if the outer edge of the sun was bulged out a little just before the contact was actual. The phenomenon may have various explanations, but the more reasonable is probably that as Venus neared the sun's edge the glowing light behind the dark body of the planet shone through the dense atmosphere of Venus, and thus appeared to swell the illuminated edge in front of the dark body of the planet. (The Transit of Venus. The United States Expedition … 1882e).

14.3.2.9 Independent Observers. Thames

During the second half of the nineteenth century, John Grigg (see Fig. 10.1) was one of New Zealand's most distinguished amateur astronomers (see Grigg 1970). Not only did he later discover a number of comets (Orchiston 1993), but he was one of those who pioneered astronomical photography in New Zealand (Orchiston 1995). Accounts of Grigg appear in Chaps. 10, 17 and 22 in this book.

Although Grigg built two different observatories during his lifetime in the wealthy gold-mining centre of Thames (Orchiston 2001b), both of these post-dated the 1882 transit, for which he fitted out a temporary observing station. A contemporary newspaper reports that

> At the Thames very general interest was taken in the transit and smoked glass and good telescopes were at a premium. At sunrise until 6am Old Sol very modestly concealed his

> face behind a bank of clouds but about the hour mentioned he came distinctly into view the transit then being well advanced. The atmosphere was tolerably clear, the view being occasionally intercepted by light clouds, but for about an hour and a half Venus was distinctly visible upon the face of the sun. The planet was well defined appearing as a jet black dot crossing the sun. Towards half past seven the sun entered a dense bank of clouds and except during a momentary break at 7.45 was not again visible until after the transit was past. Mr. Grigg of Pollen St had perfected arrangements for the observation of the transit on more scientific principles than any others we have seen in the district *and had fitted up a temporary observatory in one of the rooms of his establishments*. The apparatus was provided by Mr. Grigg and adjusted accurately. Mr. Foy[4] was in attendance to photograph the sun during the transit and Mr. & Mrs. Neil recorded the time of the different occurrences by means of a seconds pendulum in the lower room. One photograph was taken soon after the sun became visible viz at 6h. 1m. 11sec and another at 6h. 37m. 22sec. These are all that could conveniently be made, though it was hoped that several more would have been perfected by means of which the path of the planet could have been accurately determined. The actual egress was not visible at Thames … though the Rev. Mr. Neil who was watching in Mr. Grigg's garden for a break in the clouds to afford an opportunity for another photograph saw the planet on the sun's disc at 7h. 45m. 20sec through the slight break in the clouds and at this time it appeared to be just about at internal contact, so that this time may be set down approximately though not with certainty. (Transit of Venus 1882b; my italics).

The "apparatus" referred to above presumably was a prime focus camera that was attached to the 7.6-cm (3-in) refractor Grigg had on loan at this time (see Chap. 10 in this book; cf. Orchiston 2001b). The newspaper article also explains the reason for this interest in photographically recording the transit: "The pictures were taken in the hope that they might be of use should the operations in other parts fail through any mishap or unfavourable weather …" (ibid.).

Another Thames observer was County Clerk Edwin Wise Hollis (d. 1906; *Thames Star* 1888; 1906), who viewed the transit through what is only described as "… a powerful telescope …" (ibid.). He saw

> … an exceedingly beautiful phenomena … which bears upon the disputed point as to whether Venus possesses an atmosphere. Several times during the transit there appeared on the western edge of the planet a pretty roseate light such as would be deflected only through an atmosphere medium apparently shining between the equator and the north pole. (ibid.).

When he died, in 1906, The *Thames Star* newspaper (1906) reported that Hollis "… had been a good and faithful servant of the Council for years and had taken the keenest and most intelligent interest in all matters pertaining to the welfare of the district."

[4]From 1872, the Foy Brothers, James Joseph Foy (1844–1890) and Joseph Michael Foy (1847–1923) ran a successful photographic studio in Pollen Street, in the same street as John Grigg's Thames Observatory (Early New Zealand Photographers …, n.d.). It is not known which of the brothers photographed the transit.

14.3.2.10 Independent Observers. New Plymouth

In New Plymouth at a site with a longitude of 11h 36m 17.55s E of Greenwich and a latitude of 39° 03′ 57″S—about one hundred yards away from the observatory of the Chief Surveyor, T. Humphries—a Mr A.O.N. O'Donoghue carried out independent observations of the third contact with a 6-cm (2.375-in) Elliott refractor and a watch. He also failed to detect the black drop, or "… haze of any description at the time of contact." (Stone 1887: 69; cf. The Transit of Venus. The United States Expedition … 1882e).

14.3.2.11 Independent Observers. Wellington

The only independent Wellington observers mentioned by the newspapers (The Transit of Venus. The United States Expedition … 1882e) as reporting "… thoroughly satisfactory observations of both internal and external contact." were Arthur Stock (see Fig. 9.4) and Thomas King (see Fig. 9.8) who were based at Stock's observatory in Thorndon, and Messrs Shannon and Littlejohn (who were located on The Terrace), yet for some reason their observations were not included in Stone's New Zealand analysis. This is particularly remarkable in the case of Stock and King, as both were experienced observers (see Orchiston 1998); at the time, Stock also doubled as the part-time Astronomical Observer at the Government's Colonial Observatory in the Wellington Botanic Garden (Eiby 1977). Immediately before the transit, Stock wrote:

> What is proposed to be done at the Wellington Observatory, Thorndon, is as follows:- Two observers, Mr T. King and Archdeacon Stock, will be stationed there, with two 4-inch telescopes. At the side of each will be two assistants – one with a chronometer, the other to write down what is said by the observer.
> (Tomorrow's Transit of Venus 1882).

The only other Wellington observation of note is the report that William Brickell Gibbs (1844–1898; Early New Zealand Photographers … n.d.) "… has also taken several excellent photographs of the phenomenon in its different stages." (The Transit of Venus. The United States Expedition … 1882e). A professional photographer, Gibbs moved from Nelson to Wellington in 1878 (*Evening Post* 1878; *Nelson Evening Mail* 1878), but in 1885 decided to close down his photography business and move back to Nelson. Accordingly, on 2 February the entire contents of his photography premises were auctioned off (*Evening Post* 1885). It is of some regret that the current whereabouts of his photographs of the transit are unknown, as they constitute an important contribution to the early development of astronomical photography in New Zealand (see Orchiston 1995: 9–12).

14.3.2.12 Independent Observers. Nelson

Nelson's leading amateur astronomer was the talented lawyer, Arthur Samuel Atkinson (Fig. 14.31; Porter 1993), who was furnished with a 10.2-cm (4-in) altazimuth-mounted Browning refractor, a Porthouse chronometer and a stop watch. He was aided by two assistants, one of whom, Maurice William Richmond (1860–1911; Scholefield 1940(2): 242), was at the chronometer. The co-ordinates of their observing station, a small shed erected for the occasion, were: longitude 11h 33m 08.82s E; latitude 41° 17′ 01.9″S. Figure 14.32 shows this shed, the telescope, and from left to right, Richmond, Atkinson and the second (unnamed) assistant who Tobin (2012: 2) suggests may be Atkinson's middle daughter, Ruth. On the day of the transit the sky was clear, but "... the Sun's limb [was] considerably agitated, and the planet also, though to a less extent." (Stone 1887: 62). However, Atkinson was convinced that this situation did not interfere with his observations of the egress. He wrote that when third contact was imminent

> ... I saw that dark waves, as it were, were carried from her [Venus] across the narrow intervening portion of the Sun's disc to his limb; narrow symmetrical portions of the preceding limb of the planet seemed to detach themselves and advance to the Sun's limb, and then pass away. (Stone 1887: 63).

Atkinson judged third contact to occur when these dark waves "... suddenly adhered, as it were, to the Sun's limb ..." instead of disappearing. He then found—to his great surprise—that although Venus' image was clear and undistorted,

Fig. 14.31 A.S. Atkinson (https://en.wikipedia.org)

Fig. 14.32 Maurice W. Richmond (left) and A.S. Atkinson (centre) with the 5-in telescope (*Courtesy* Harry H. Atkinson)

> … it was not apparently the whole planet that was adherent to the Sun's limb, but a large detached segment of it … I turned my attention to the dividing light at its thinnest part. The *apparent planet* seemed to press upward and curve in; the chord of the segment and the two bodies were apparently touching along a considerable part of the segment, but for a very faint white line between them, though it looked more as if this line were marking their junction than separating them. (ibid.).

The faint white line persisted for more than a minute before disappearing. Atkinson then saw "… the upper limb of the segment clearly protuberant above the line of the Sun's limb, its contour marked by a very faint line of grey light, and separating the Sun's cusps by a considerable interval …" (ibid.), which he estimated at about one-third of the planet's diameter. He noted that when Venus was about one-third of her diameter off the Sun, "… the part still on the Sun was evidently distorted, being somewhat pear-shaped …" (ibid.). He subsequently watched the fourth contact.

14.3.2.13 Independent Observers. Christchurch

Mr James Townsend's observatory was the site of Christchurch's assault on the transit, with the Lands and Survey Office surveyor, Walter Kitson (1835–1914) making use of the 15.2-cm (6-in) equatorially-mounted Cooke refractor (Fig. 14.33)[5] and a chronometer, while Townsend himself used his 8.6-cm (3.375-in) Dallmeyer equatorial and a stop watch. During the morning the sky was covered by light cirrus cloud, but in spite of this both observers saw the transit and timed the two egress contacts.

Kitson specifically looked for signs of a Venusian atmosphere, but "… no halo nor any light at all different from the light of the Sun was visible round the planet, nor was there any shadow round the planet's limb." (Stone 1887: 58). What he did notice, though, was that Venus "… appeared to have a lighter tinge towards its limb. This light tinge began to displace the blackness of the centre at about one-tenth of the planet's diameter from the edge. The tinge was of a bluish colour, the centre being almost black." (ibid.). About ten seconds before third contact, Kitson recorded "… the first appearance to me of a shadow or darkening of the Sun's face between the limbs of the planet and Sun near the point of contact." (ibid.), and third contact subsequently occurred, without any sign of the notorious 'black drop'. About twenty minutes later Kitson observed fourth contact.

Townsend provides an extremely brief account of his observation of the transit, although this does include times for the third and fourth contacts. He noted that "Part of outline of Venus seen till 12 minutes after internal contact, outside limb of Sun. Centre of planet when on the Sun quite black, margin for one-third diameter indigo … No markings or light seen on body of Venus." (Stone 1887: 59).

[5]In 1891, three years before he died, Townsend donated his 1864 Cooke telescope to Canterbury College and in 1896 it was installed in a newly-erected tower observatory on campus (see Tobin 1996). This campus precinct subsequently became the Christchurch Arts Centre with the establishment of the new University campus at Ilam, but the Townsend Observatory remained under the control of the University's Department of Physics and Astronomy, and was used for regular 'public viewing nights'. Unfortunately, the observatory tower was damaged during the 2010 earthquakes and collapsed during the 2011 earthquakes. Although the telescope was badly damaged, surprisingly the objective was found intact.

Fig. 14.33 Townsend's 15.2-cm Cooke refractor (www2.phys.canterbury.ac.nz/~jhe25/townsend/townsend.html)

14.3.2.14 Independent Observers. Timaru

The Venerable Archdeacon Henry W. Harper of Timaru is listed as one of the New Zealand observers organized by Tupman to view the transit (The Transit of Venus 1882c), but no details are provided of the instrumentation available to him or whether his observations were successful (see *Evening Post* 1882). We do know, however, that on 29 November more than a week before the grand event, he gave a public lecture on the transit in the Oddfellows' Hall (*Timaru Herald* 1882). In this lecture he mentioned having successfully observed the 1874 transit from Hokitika on the West Coast of the South Island "… together with Mr O'Connor, the Resident Engineer, and since then, *having become possessed of a good telescope* … in preparation for the transit of 1882." (Transit of Venus 1882c; my italics). A subsequent newspaper article mentions his "… fine equatorial telescope … an excellent chronometer .. [loaned by] Mr A.J.H. Bower, who was to assist in making the observations." (The Transit of Venus 1882d), and the fact that the sky was cloudy and they did not see the transit. Consequently, "A good deal of regret is felt that Archdeacon Harper was disappointed, it being well known that he is an ardent astronomer." (ibid.). Fortunately, this was one of the very few New Zealand transit stations where the observers did not view the transit—though this was no consolation for Henry Harper.

14.3.2.15 Independent Observers. Dunedin

Three different astronomers carried out successful observations of the transit from Dunedin. One of these was the talented Arthur Beverly (Fig. 14.34) who so impressed Lieutenant-Colonel Tupman that he was moved to write John Tebbutt in the following terms:

> Since I have been in this wonderful country I have discovered "a truly bright light under a bushel" in the person of Mr. Arthur Beverley [*sic*] of Dunedin. Like many distinguished Astronomers he is a mechanic of a high order, making his own microscope objectives on his own formulae; inventing ingenious apparatus and possessing high mathematical attainments. In Europe he would be in the first rank among physicists & Astronomers. In New Zealand scarcely anyone knows his name not one soul knows his merits. A watchmaker, I think, by trade, from an obscure part of Scotland, with consumption and another about as deadly disease, he made achromatic doublets of surpassing excellence for Sir David Brewster, Prof Thwaites & other distinguished men until he was able to pay his passage to New Zealand. There he soon saved as much as he required, or acquired enough somehow, chiefly by plying his trade, and at once gave up business to live & revel in botany, physics & astronomy. He quickly gets out the orbit of any comet that he sees and probably was the first to find that of the present comet[6] …
>
> It is right that you should know what a neighbour you have. He only possesses a three inch achromatic, but it is very good and very well mounted equatorially, with circles graduated to 3′ … [and] an excellent chronometer … (Tupman 1883; cf. Campbell 2001; Gillies 1881).

Beverly's observatory was located at longitude of 11h 22m 02.18s E and latitude 45° 52′ 20″S, and apart from the afore-mentioned small equatorially-mounted telescope housed an Arnold and Dent chronometer. Beverly used these facilities to observe the transit and record the two egress contacts, and as third contact approached he noted that "… the thread of light separating the limbs of the Sun and Venus, having become excessively slender, appeared to part or darken rather suddenly at the point of contact. No appearance of ligament or anything anomalous." (Stone 1887: 65). Seven seconds later "… a very faint ruddy line begins to appear between the cusps …" (ibid.), and 18 seconds later he noticed that the ruddy line had become "… more distinct, longer and sensibly arched." (ibid.). Fourth contact was observed without incident, and after the transit Beverly summarised his impressions:

> There was nothing of a prolonged or doubtful character about the phenomena at internal contact such as we were led to expect … The luminous thread became gradually thinner and finer as the planet advanced, until it was clearly obliterated … Possibly the pink line joining the cusps might have been a second or two sooner had I been on the look out for it. (ibid.).

---

[6]Tupman was referring to the Great Comet of 1882, C/1882 R1, which was still a naked eye object in southern skies in mid-January 1883 when Tupman wrote to Tebbutt.

Fig. 14.34 Arthur Beverly (Orchiston collection)

But elsewhere, he presents a somewhat different perspective:

> In about ten seconds after [third] contact the only other phenomena worth mentioning began to appear. The part of Venus which projected beyond the sun's disc showed a very fine pink outline, caused no doubt by sun light refracted through the atmosphere, which continued visible until the disc of Venus projected about a fifth of its diameter beyond the solar disc. It then gave way at the north side, but continued visible at the south side until Venus was half off, when it appeared like a minute pink hair standing perpendicular to the sun's margin at the edge of the same circular notch. (The Transit of Venus. The United States Expedition … 1882e).

The second Dunedin amateur observer was Robert Gillies (1836–1886; Scholefield 1940(1): 296–297), a successful businessman and former trigonometrical and geodetical surveyor (Gillies 1881; The late Mr Robert Gillies 1886), and he was assisted during the transit observations by a Mr. Keys. Gillies' observatory was sited near Beverly's home at effectively the same latitude and longitude, and it housed an astronomical clock, a chronometer, a chronograph and a 15.2-cm (6-in) equatorially-mounted refractor that Tupman examined and found to exhibit considerable spherical aberration (Stone 1887: 67). Despite this shortcoming, Gillies observed the third contact under clear skies, and about two seconds later he noticed that "… there was a faint grey light on Venus." (Stone 1887: 66). As fourth contact approached, he noted that Venus "… seemed to appear to draw out towards the limb into a pear shape, and I drew Mr. Keys' attention to it …" (ibid.). By the time of fourth contact the viewing conditions were not so favourable, and Gillies had some doubts about the contact time he recorded. After the transit, he wrote that during his observations

Fig. 14.35 Henry Skey (after *The Cyclopedia of New Zealand* … 1905)

> … there always was a sort of radiance or faint light in front of the planet, which now, I have no doubt, was the atmosphere of Venus … it was this faint light that I saw distinctly beyond the line of the disk of the Sun, and which in my report I call the limb of Venus. At the time I thought it was the planet itself, slightly illuminated. I think now it must have been its atmosphere … (ibid.).

Dunedin's third observer, Henry Skey (Fig. 14.35), was one of New Zealand's most experienced amateur astronomers (Campbell 2001), and he observed the transit from his observatory (at longitude 11h 21m 58.08s E; latitude 45° 52′ 11″S) with the 23.5-cm (9.25-in) Browning reflector that is now at Ashburton College (Evans and Lucas 1989). Contact times were recorded by local politician and optician Archibald Hilson Ross (1821–1900; Scholefield 1940: 258) using a Porthouse chronometer and by J.K. Logan with reference to the observatory's astronomical clock. At 08 h 00 min 50.9 s GMT, immediately before third contact, Skey observed that "… the limbs of the Sun and Venus seemed to make a vibratory approach towards each other, for the extremely thin thread of light broke, and then joined again four or five times; in other words, thin pointed and well-defined cusps kept forming and then joining." (Stone 1887: 67). This lasted about three seconds, then "… a straw coloured shade was observed on the previously white disc of the Sun between the limbs of the Sun and Venus, which rapidly passed through a brown to a nearly black shade, and completely obscured the Sun's limb." (ibid.). Skey noted that

> The ligament arising from this contact [i.e. third contact] was of the same shade of darkness as the body of Venus and the background of the sky, that is, nearly black, and no change in this colour was afterwards observed in any part of the ligament. The cusps at this time were now blunted, and, when the leading limb of Venus was well off the Sun's disc, the points of the cusps were off by straight shaded lines. *No "black drop" was observed, nor did Venus ever assume a "pear shape."* (ibid.; my italics).

Nearly 21 minutes later fourth contact was observed. At the end of his account of the transit Skey pointed out that "There was a slight haze over the Sun during these observations … This, however, had the effect of wonderfully improving the definition of the limbs of Venus and the Sun …" (Stone 1887: 67).

14.4 Discussion

Organizing the various New Zealand transit programs in 1874 and 1882 was a major logistical exercise, but after the transit came the equally challenging task of reducing the various observations and deriving meaningful values for the solar parallax.

The end result of these endeavours for the 1874 transit was a paper penned by Lieutenant-Colonel G.L. Tupman that appeared in an 1878 issue of *Monthly Notices of the Royal Astronomical Society*, where he examined separately each of the ingress and egress contact timings. It is notable that because of cloudy skies New Zealand is hardly represented. Tupman concludes his long paper with comments on previously-published papers and reports about the British 1874 transit observations, where he is openly critical of Airy's "Parliamentary Report", other reports, and a paper published by Edward James Stone (1831–1897).[7] He cautions:

> In commencing an investigation of the parallax from these and similar observations, some selection of phases, depending on the observer's language and on the mutual agreement of observers not very far apart, must be made. Any selection of times made *after an investigation of the effects of parallax*, such as that I have now made, will always expose the result to the suspicion of having been "doctored." (Tupman 1878: 456; his italics).

In the final analysis, the British transit observations of 1874, "… only enable us to determine that the solar parallax probably lies between the values of 8.82″ and 8.88″ … (ibid.).

While this must have been disappointing news for Airy, who was almost certainly hoping for a more definitive result, it was consistent with the values obtained by the Americans and the Germans, in part on the basis of observations made in New Zealand.

[7]At the time London-born Edward James Stone was Director of the Royal Observatory at the Cape of Good Hope, but by the time he published his monograph on the British 1882 Transit of Venus observations (Stone 1887) he had returned to England and succeeded the Reverend Robert Main as the Radcliffe Observer at Oxford University.

The Americans relied for their result on the photographic observations, and the task of measuring the plates obtained at the eight different 1874 transit stations fell to U.S. Naval Observatory astronomer William Harkness (1837–1903; Fig. 14.36). Although these plates yielded "excellent results" for the interval when Venus was on the Sun's disk, photographs of the ingress and egress were of "no value", because of the black drop effect (Harkness 1883). Measurements of all of the American photographs were completed by the end of 1877, and then came the laborious task of establishing the longitudes of the transit stations. When this was accomplished the official report of the 1874 American transit program was to have been published in a succession of volumes, but funding restrictions only allowed the appearance of the first of these (Newcomb 1880). Unfortunately, this contained none of the results, as these were planned for subsequent volumes. Further delays ensued, and in the end it was David Peck Todd (1855–1839; Fig. 14.37) from the Nautical Almanac Office who published a provisional American value of 8.883 ± 0.034″ (Todd 1881). Australian and New Zealand transit stations played an important role in contributing to this result: 46 % of photographs used in deriving this solar parallax came collectively from Hobart, Campbell Town, Queenstown and the Chatham Islands (see Table 14.1). Nonetheless, Todd's result remained contentious, because of certain concerns about the quality of the photographic images. These showed some limb-darkening, and there was also a difficulty in establishing plate scales (see Lankford 1984).

The German results for the 1874 transit rested upon photographic observations made at six different locations, of which the Auckland Island station was but one. In contrast to the American result, the overall New Zealand contribution to the German parallax values was not so prominent. Duerbeck (2004: 15) has recounted how German Transit of Venus Commission President, Georg Friedrich Julius Arthur von Auwers (1838–1915), was responsible for editing six different volumes, totalling 3600 pages, which were published between 1887 and 1898. These contained

Fig. 14.36 William Harkness (https://en.wikipedia.org)

Fig. 14.37 David Peck Todd (https://en.wikipedia.org)

> … instructions, reports, measurements, reductions, and finally the overall results of the project. Auwers painstakingly presented the complete material, even including uncomplimentary remarks by expedition members … [He] painstakingly analysed all observations, visual contacts, photographic positions, and heliometer settings.

The final result? There were two of them: a parallax of 8.810 ± 0.120″ based on photographic observations made at four stations (one of which was on Auckland Island), and a value of 8.8796 ± 0.0320″ based on a detailed analysis of all heliometer observations (Auwers 1898). Duerbeck (2004: 16) notes that the more accurate heliometer observations "… suffered from uncorrected systematic errors (which were overlooked even by diligent Auwers), the cause [of which] is difficult to establish today."

Unlike in 1874, in 1882 New Zealand played a vital role in providing observations that yielded further solar parallax values. One local newspaper reported

Table 14.1 Photographs from American 1874 transit stations used in deriving a value for the solar parallax (adapted from Dick et al. 1998)

Station	Number of photographs	% of total
Vladivostok (Russia)	13	3.7
Nagasaki (Japan)	60	17.1
Peking (China)	90	25.7
Kerguelen Island (Indian Ocean)	26	7.4
Campbell Town (Australia)	55	15.7
Hobart (Australia)	39	11.1
Queenstown (New Zealand)	59	16.9
Chatham Islands (New Zealand)	8	2.3
Total	350	99.9

with considerable pride that "… now that the transit is past, it is pleasant to reflect that a very large measure of success has been achieved, especially in New Zealand …" (The Transit of Venus. The United States Expedition … 1882e).

All of the 'British' observations were brought together in a monograph by Stone (1887), where solar parallaxes of 8.827 ± 0.051″ and 8.882 ± 0.045″ were derived for the first egress contact. These values were based on observations by thirty different observers, thirteen of whom (43 %) were based in New Zealand (see Table 14.2). Stone also obtained a figure of 8.942 ± 0.047″ for the solar parallax based upon observations of the second egress contact. Of the thirty different astronomers who contributed observations, twelve (40 %) were based in New Zealand, and these are also listed in Table 14.2.

In 1882, only one US transit station was sited New Zealand (in Auckland), and this contributed in a relatively minor way to the resulting solar parallax value derived from all transit stations. Of the 345 measurable plates taken at the four US and four foreign transit stations, only 31 (a mere 2.7 %) came from the New Zealand transit station (Table 14.3).

Primarily because of the better weather encountered worldwide there were large numbers of successful photographs, so "If ever the transits of Venus could be made successfully to yield the solar parallax, it was now." (Dick et al. 1998: 245). This task fell to William Harkness, and by August 1884 all the photographs had been reduced, but this was merely the first step in a long and tedious process that involved measurements of time and latitude for each station, determination of plate scales, and many other calculations. Finally, on 11 October 1888 Harkness (1888: 17–18) announced a figure of 8.847 ± 0.012″, and the following year he revised this

Table 14.2 New Zealand observers featuring in Stone's 1882 first and second egress contact analyses

First egress contact observer	Residuals 8.827″	Residuals 8.882″	Second egress contact observer	Residuals 8.942″
Atkinson	+0.303	−0.444	Atkinson	−0.101
Beverly	−0.101	+0.503	Beverly	+0.914
Coke	−0.302	−0.619	Gillies	−0.820
Gillies	−0.351	−1.098	Heale	−0.134
Humphries	+0.592	−0.155	Hewitt	−0.549
King	−2.230	+1.373	King	−1.041
Kitson	−0.782	−0.377	Kitson	−0.554
McKerrow	−0.203	−0.152	McKerrow	+0.558
Marchant	+0.592	+0.418	Marchant	−0.078
O'Donahoe	+0.420	−0.327	Skey	+0.154
Skey	−0.937	−1.684	Townsend	+0.561
Townsend	+0.016	+0.047	Tupman	+0.915
Tupman	−0.587	−0.861		

Table 14.3 Photographs from American 1882 transit stations used in deriving a value for the solar parallax*

Station	Number of photographs	% of total
Auckland (New Zealand)	31	2.7
Cedar Keys (USA)	165	14.5
Cerro Roblero (USA)	216	18.9
San Antonio (USA)	121	10.6
Santa Cruz (Patagonia)	204	17.9
Santiago (Chile)	152	13.3
Washington (USA)	53	4.6
Wellington (South Africa)	200	17.5
Total	1142	100.0

*Adapted from Dick et al. (1998)

to 8.842 ± 0.0118″ (Harkness 1889: 424–425), which corresponds to a mean Earth-Sun distance of 92,455,000 miles.

Part of the problem faced by those who reduced the 1874 and 1882 transit observations was to determine which of the reported ingress and egress phenomena corresponded to the different contacts. Many observers had spent long hours religiously observing dry runs of the up-coming transits with the aid of artificial transit devices, and when it came to the notorious black drop and other anticipated phenomena they knew precisely what to expect. But for many the 'real thing' proved somewhat of a challenge and some of the phenomena observed were totally unexpected and ran counter to their well-drilled pre-transit practice observations. Some observers also had false expectations based upon what they had read of earlier transits, whilst after the event others went so far as to query their own observations because these apparently did not tally with the 'real' sequence of ingress and egress events. Bias in one form or another was obviously incorporated into the records of some observers.

With this caveat in mind, precisely what phenomena were seen by New Zealand observers at ingress and/or egress in 1874 and 1882? The most widely-reported phenomenon was the illumination of that section of the Venusian limb that was off the Sun during ingress or egress. Sometimes the entire limb was lit, while at other times only an arc of it was illuminated. This feature was reported by many of the astronomers, including Beverly, Cheeseman, Gillies, Hector, Lambert, Marchant and Tupman, all of whom were experienced observers. Both reflectors and refractors were implicated, and these ranged widely in aperture. Analysis of the various observations and observers revealed no obvious patterns or trends. Several observers attributed this limb illumination to an atmosphere of Venus, which we now know to be the correct interpretation (see Shiltsev 2014).

Despite the uncooperative weather in New Zealand in 1874, the black drop was reported (or accurately described) by Lambert, Seeliger and Wolfram, all experienced observers. Instruments used in making these observations were refractors ranging in aperture from 8.2 to 10 cm, and a 7.6 cm heliometer. No consistent pattern or trend is apparent, but the modest apertures involved may give some credence to Ashbrook's suggestion (1984: 230) that the black drop tended to be associated with small telescopes and low magnifications. The cause of the black drop effect has been hotly debated for more than a century, but it was only comparatively recently that Schaefer (2001: 334) critically evaluated the competing explanations and came up with the correct interpretation:

> … the ideal image (a circular Venusian disk silhouetted against the Sun) will suffer smearing (from many physical mechanisms) that will produce a somewhat fuzzy image with contour lines (i.e., what is perceived as the edge) that are shaped like the Black Drop. The primary causes of smearing are the usual astronomical seeing (associated with small angle scattering in our Earth's atmosphere) and the usual diffraction in the telescope (the Airy pattern). Other contributing smearing mechanisms that generally do not dominate are imperfections in the telescope's optics, imperfections in the observer's eyes, the finite angular resolution of the detector, and even the physical size of the telescope's aperture.

Subsequent investigations by Pasachoff et al. (2005) have supported this interpretation.

14.5 Concluding Remarks

The two nineteenth century transits of Venus marked the end of one astronomical era and the birth of another. They were the last transits used in a serious attempt to resolve the solar parallax problem, before other methods gained favour. They also marked the first internationally-co-ordinated assault on a major astronomical problem using the emerging technology of photography. Photography and its investigative 'handmaiden', spectroscopy, would quickly become the indispensable tools of the 'new astronomy', astrophysics.

Because of its fortuitous longitudinal position on the globe and its southerly latitude, New Zealand played a role in the international transit efforts of 1874 and 1882, but much more so during the latter transit. However, the resulting solar parallax values (see Table 15.1) differ significantly from the value of 8.794,148 ± 0. 000007″, which is based upon radar observations of Venus and was approved by the International Astronomical Union (IAU) in 1976. It was only in the 1890s that solar parallax figures that were consistent with the current value emerged, and only when Newcomb and Harkness revisited all of the eighteenth and nineteenth century transit results and also considered an alternative approach to the problem involving the system of constants.

Now the solar parallax no longer features in discussions of the mean Earth-Sun distance (the Astronomical Unit), which in 2009 was set by the IAU at 149,597,870.700 km, with an uncertainty of just 3 metres (Tobin 2012). As a matter of interest, this "… corresponds to a solar parallax of 8.7941433″, with the uncertainty being in the last decimal place and being due to the uncertainty of the terrestrial equatorial radius." (William Tobin, pers. comm., November 2014).

Acknowledgements I am grateful to the following for their assistance: the late Robert Campbell (Oamaru, NZ), Dr Grant Christie (Auckland, NZ), Jenni Chrisstoffels (Alexander Turnbull Library, Wellington, NZ), Dr Steve Dick (USA), the late George Eiby (Wellington, NZ), Kiri Griffin (Toitū Otago Early Settlers Museum, Dunedin, NZ), Andreas Heine (The Krone-Sammlung-Collection, Dresden, Germany), the late Peter Hingley (former RAS Librarian, London), the late Jackie St George (Sydney), Dr William Tobin (Vannes, France) and staff at the Alexander Turnbull Library (Wellington, NZ). Finally, I also wish to thank Steve Dick, Dr Richard Dodd (Martinborough, NZ) and William Tobin for reading and commenting on the first draft of this chapter, and the Alexander Turnbull Library, Harry Atkinson (England), The Krone-Sammlung-Collection, Toitū Otago Settlers Museum, Royal Astronomical Society (London), and the STFC and J. Radcliff (England) for kindly providing Figs. 14.4, 14.5, 14.6, 14.7, 14.11, 14.12, 14.13, 14.15, 14.17, 14.18, 14.26 and 14.32.

References

Adams, R. D. (1993). Adams, Charles William (1840–1918). In C. Orange (Ed.), *The Dictionary of New Zealand Biography*, Volume Three: *1901–1920* (p. 1f). Auckland: Auckland University Press and Department of Internal Affairs.

Airy, Sir G. B. (1880). Letter to Dr J. Hector, dated November 1. In Airy Papers. Cambridge University Library (RGO 6/283).

Airy, Sir G. B. (1881). *Account of Observations of the Transit of Venus, 1874, December 8, Made Under the Authority of the British Government: And of the Reduction of the Observations.* London: Her Majesty's Stationery Office.

Ashbrook, J. (1984). *The Astronomical Scrapbook. Skywatchers, Pioneers, Seekers in Astronomy.* Cambridge, MA: Sky Publishing.

Atkinson, H. A. (1882). Memorandum to Sir A. Gordon (Governor of New Zealand), dated February 22. In Airy Papers. Cambridge University Library (RGO 6/283).

Auwers, A. (Ed.), (1898). *Die Venus-Durchgänge 1874 und 1882. Bericht uber die Deutschen Beobachtungen im Auftrage der Commission für die Beobachtung des Venus-Durchgangs herausgegeben.* Berlin.

Barberel, W. (1975). Report on the Victoria Tree and the German 1874 transit of Venus expedition camp. In J. C. Yaldwyn (Ed.), *Preliminary Results of the Auckland Islands Expedition 1972–73: From Reports of Participants to the Director-General of Lands* (pp. 288–291). Wellington: Department of Lands and Survey.

Baucke, W. (1928). *Where the White Man Treads Across the Pathway of the Maori.* Auckland: Wilson and Horton.

Bouquet de la Grye, A. (1882). *Recueil de Mémoires, Rapports et Documents Relatifs à l'observation du Passage de Venus sur le Soleil* (Vol. III, Part 1). Paris: Academy of Science.

Campbell, R. N. (2001). Henry Skey 1836–1914. *Southern Stars, 40*(2), 11–12.

Clerke, A. (1893). *A Popular History of Astronomy During the Nineteenth Century.* London: Adam and Charles Black.

Cossons, N. (Ed.), (1992). *Making of the Modern World: Milestones of Science and Technology.* London: John Murray.

Dawson, E. W., & Duerbeck, H. W. (2008). *The German Transit of Venus Expedition at the Auckland Islands 1874–1875.* Eastbourne: Hutton Foundation at the Wallypug Press.

Dell, R. K. (1990). Hector, James 1834–1907. In W. H. Oliver (Ed.), *The Dictionary of New Zealand Biography, Volume One: 1769–1869* (p. 183f). Wellington: Allen & Unwin New Zealand Ltd., and Department of Internal Affairs.

Dick, S. J. (2003). *Sky and Ocean Joined. The U.S. Naval Observatory 1830–2000.* Cambridge: Cambridge University Press.

Dick, S., Orchiston, W., & Love, T. (1998). Simon Newcomb, William Harkness and the nineteenth century American transit of Venus expeditions. *Journal for the History of Astronomy, 29*, 221–255.

Dodson, T. (1996). The With-Browning telescope at Pauatahanui. *Southern Stars, 37*, 45–51.

Duerbeck, H. W. (2004). The German transit of Venus expeditions of 1874 and 1882: Organization, methods, stations, results. *Journal of Astronomical History and Heritage, 7*, 8–17.

Early New Zealand Photographers and their Successors. Gibbs, William Brickell (n.d.). (canterburyphotograhpy.blogspot.com/2013/02/gibbs-william-brickell.html).

Eiby, G. A. (1977). The New Zealand Government time-service: an informal history. *Southern Stars, 27*, 15–34.

Evans, R. W., & Lucas, K. J. (1989). The Skey-Ashburton College telescope. *Southern Stars, 33*, 178–187.

Evening Post, September 18, 1878 (p. 2).

Evening Post, December 7, 1882 (p. 2).

Evening Post, January 31, 1885 (p. 3).

Filhol, H. (1885). *Recueil de Mémoires, Rapports et Documents Relatifs à l'observation du Passage de Venus sur le Soleil* (Vol. III, Part 2). Paris: Academy of Science.

Gillies, R. (1881). Letter to Sir George Airy, dated February 20. In Airy Papers. Cambridge University Library (RGO 6/283).

Grigg, M. (Ed.), (1970). *John Grigg, F.R.A.S. of Thames, New Zealand. Together with some Genealogical Memorials of his Family and Glimpses of Pioneer Life in New Zealand.* Unpublished typescript (Orchiston collection).

Harkness, W. (1883). Address by William Harkness. *Proceedings of the American Association for the Advancement of Science 31st Meeting*. Salem: AAAS.

Harkness, W. (1888). *Annual Report*. Washington, U.S. Naval Observatory.

Harkness, W. (1889). *Annual Report*. Washington, U.S. Naval Observatory.

Hector, J. (1882). The transit of Venus. Clyde. *New Zealand Journal of Science, 1*, 326–330.

Lake Wakatip Mail, December 11, 1874.

Lankford, J. (1984). The impact of photography on astronomy. In O. Gingerich (Ed.), *The General History of Astronomy. Astrophysics and Twentieth-Century Astronomy to 1950* (Part A, Vol. 4, pp. 16–39). Cambridge: Cambridge University Press.

Lankford, J. (1987). Photography and the 19th-century transits of Venus. *Technology and Culture, 28*, 648–657.

Lauga, J. (2004). Les missions Françaises pour le passage de Vénus de 1874. *L'Astronomie, 118* (Mai), 296–305.

Maitland, G. (1993). Martin, Josiah 1843–1916. In C. Orange (Ed.), *The Dictionary of New Zealand Biography. Volume Two. 1870–1900* (pp. 313–314). Wellington: Bridget Williams Books Ltd. and Department of Internal Affairs.

McIntosh, R. A. (1958). Astronomical history of the Auckland Province. Part III: Nineteenth century. *Amateur Astronomer* (Journal of the Auckland Astronomical Society), *18*(1), 3–6.

McKerrow, J. (1882). The transit of Venus, Wellington. *New Zealand Journal of Science, 1*, 330–332.

Nelson Evening Mail, January 19, 1878 (p. 2).

Newcomb, S. (1880). *Observations of the Transit of Venus, December 8–9, 1874 Made and Reduced under the Direction of the Commission Created by Congress. Part I.* Washington: Government Printing Office.

Newcomb, S. (Ed.). (1881). *Observations of the Transit of Venus, December 8–9, 1874 Made and Reduced under the Direction of the Commission Created by Congress. Part II.* Washington: Government Printing Office (unpublished typescript, USNO Library).

Norman, D. (1938). The development of astronomical photography. *Osiris, 5*, 560–594.

Obituaries. James McKerrow. *Monthly Notices of the Royal Astronomical Society, 80*, 348 (1920).

Orchiston, W. (1993). John Grigg, and the genesis of cometary astronomy in New Zealand. *Journal of the British Astronomical Association, 103*, 67–76.

Orchiston, W. (1995). John Grigg, and the development of astrophotography in New Zealand. *Australian Journal of Astronomy, 6*, 1–14.
Orchiston, W. (1998). *Nautical Astronomy in New Zealand. The Voyages of James Cook*. Wellington: Carter Observatory.
Orchiston, W. (2001a). Henry Severn: Thames' other nineteenth century astronomer. *Southern Stars, 40*(2), 8–10.
Orchiston, W. (2001b). The Thames observatories of John Grigg. *Southern Stars, 40*(3), 14–22.
Orchiston, W. (2004). The nineteenth century transits of Venus: an Australian and New Zealand overview. *Journal of Astronomical Data, 10*, 219–308.
Orchiston, W. (2012). The 1874 transit of Venus: New Zealand's first foray into international astronomy. *Journal of the Royal Society of New Zealand, 42*, 145–152.
Orchiston, W., Love, T., & Dick, S. (2000). Refining the astronomical unit: Queenstown and the 1874 transit of Venus. *Journal of Astronomical History and Heritage, 3*, 23–44.
Palmer, H. (1874a). Letter to Sir George Airy, dated October 20. In Airy Papers. Cambridge University Library (RGO 6/273).
Palmer, H. (1874b). Letter to Sir George Airy, dated November 22. In Airy Papers. Cambridge University Library (RGO 6/273).
Palmer, H. (1875). *Transit of Venus. Recommendations for 1882, by Major H. Palmer, R.E., Observer at New Zealand in 1874*. London: Her Majesty's Stationery Office.
Pasachoff, J., Schneider, G., & Golub, L. (2005). The black-drop effect explained. In D. W. Kurtz (Ed.), *Transits of Venus: New Views of the Solar System and the Galaxy* (pp. 242–253). Cambridge: Cambridge University Press.
Peters, C. H. F. (1874). General Journal New Zealand party. Manuscript, U.S. Naval Observatory (Queenstown transit box).
Porter, F. (1993). Atkinson, Arthur Samuel 1833–1902. In Orange, C. (Ed.), *The Dictionary of New Zealand Biography. Volume Three. 1901–1920* (pp. 13–14). Auckland: Auckland University Press and Department of Internal Affairs.
Radcliff, J. (2008). *The Transit of Venus Enterprise in Victorian Britain: Science and Culture in the Nineteenth Century*. London: Pickering & Chatto.
Schaefer, B. E. (2001). The transit of Venus and the notorious black drop effect. *Journal for the History of Astronomy, 32*, 325–336.
Scholefield, G. H. (Ed.), (1940). *Dictionary of New Zealand Biography* (2 Vols.). Wellington: Department of Internal Affairs.
Seymour, J. (1985). Palmerston North's first observatory. *Southern Stars, 31*, 209–214.
Seymour, J. (1995). A history of the Thomas King Observatory, Wellington. *Southern Stars, 36*, 102–114.
Shiltsev, V. (2014). The 1761 discovery of Venus' atmosphere: Lomonsov and others. *Journal of Astronomical History and Heritage, 17*, 85–112.
Simmons, J. (1880). List of transit of Venus 1874 huts. In Airy Papers. Cambridge University Library (RGO 6/279).
Stevenson, J. T. (1882). Observations of the Great Comet (b) 1882, made at Auckland, New Zealand. *Monthly Notices of the Royal Astronomical Society, 43*, 54–55.
Stock, A. [Arthur] 1874. *December 9, 1874. The Transit of Venus and How to Observe It*. Wellington: Bowden.
Stone, E. J. (1887). *Transit of Venus, 1882*. London: Her Majesty's Stationery Office.
Thames Star, February 21, 1888 (p. 3).
Thames Star, May 2, 1906 (p. 2).
The Cyclopedia of New Zealand [*Otago & Southland Provincial Districts*]. (1905). Christchurch: The Cyclopedia Company.
The Illustrated Australian News for Home Readers, April 19, 1875.
The late Mr Robert Gillies. *Otago Witness*, June 18, 1886 (p. 9).
Timaru Herald, November 29, 1882 (p. 4).

Thomson, J. T., McKerrow, J., & Skey, H. (1874). *Report of a Meeting of the Committee of the Otago Institute Appointed to Consider the Best Climate Within the Province for Observing the Transit of Venus*. Copy in RGO Archives, Cambridge University Library (RGO 6/275).

The Transit of Venus. *The Daily Southern Cross*, October 10, 1874a.

The Transit of Venus. *The Daily Southern Cross*, December 8, 1874b.

The Transit of Venus. *The Daily Southern Cross*, December 10, 1874c.

The Transit of Venus. *The Daily Southern Cross*, December 24, 1874d.

The Transit of Venus. *The New Zealand Herald*, October 4, 1882a.

The Transit of Venus. *The New Zealand Herald*, December 6, 1882b.

The Transit of Venus. *The New Zealand Herald*, December 7, 1882c.

The Transit of Venus. *Timaru Herald*, December 28, 1882d.

The Transit of Venus. The United States Expedition in the Domain. *The New Zealand Herald*, December 8, 1882e.

The New Zealand Herald, December 8, 1882.

Tobin, W. (1996). The Townsend telescope—A century of university stargazing. In W. Tobin & G. M. Evans (Eds.), *Stars in a Cluster* (pp. 78–83). Christchurch: Department of Physics & Astronomy, University of Canterbury.

Tobin, W. (2012). A transit of Venus primer. *Southern Stars, 51*(1), 3–18.

Tomorrow's transit of Venus. *The Evening Post*, December 6, 1882.

Transit of Venus. Arrival of the American Expedition. *The New Zealand Herald*, October 16, 1882a.

Transit of Venus. *The Thames Advertiser & Miners News*, December 8, 1882b.

Transit of Venus. *Timaru Herald*, December 2, 1882c.

Tupman, G. L. (1878). On the mean solar parallax as derived from the observations of the transit of Venus, 1874. *Monthly Notices of the Royal Astronomical Society, 38*, 429–457.

Tupman, G. L. (1883). Letter to John Tebbutt, dated January 16. In *Letters to J. Tebbutt 1883*. Sydney: Mitchell Library (A 3689).

Ward, L. E. (1928). *Early Wellington*. Auckland: Whitcombe and Tombs.

Woolf, H. (1959). *The Transits of Venus: A Study of Eighteenth Century Science*. Princeton: Princeton University Press.

Chapter 15
Refining the Astronomical Unit: Queenstown and the 1874 Transit of Venus

Abstract The 1874 transit of Venus was regarded as a major event which promised to produce an improved value for the solar parallax, and hence the Astronomical Unit. As a result, the United States dispatched eight different observing parties to far northern and southern hemisphere locations. This chapter documents the activities at the Queenstown transit station in the South Island of New Zealand and examines the scientific outcome of the overall American 1874 and 1882 transit programmes. It also traces the early development of astronomical photography in New Zealand at a time when this innovative methodology was emerging internationally as a valid tool of the 'new astronomy', astrophysics.

15.1 Introduction

One of the fundamental yardsticks of astronomy is the 'Astronomical Unit' (AU), the mean distance from the Earth to the Sun. During the eighteenth century, the opportunity to use transits of Venus to determine the AU prompted a number of nations to dispatch expeditions to the far corners of the globe (see Orchiston 1998, 2005; van Helden 1995; Woolf 1959). The 1761 and 1769 transits produced conflicting results, with solar parallax figures ranging from 8.28″ to 10.60″, and although a smaller spread characterised the latter transit (see Table 15.1), greater precision was required. This led to a focus on the next pair of transits, in 1874 and 1882 (see Dick et al. 1998; Forbes 1874; Grant 1874; Janiczek and Houchins 1974; Meadows 1974; and Proctor 1874), and "Every country which had a reputation to keep or to gain for scientific zeal was forward to co-operate in the great cosmopolitan enterprise ..." (Clerke 1893: 289).

As a consequence, in 1874 the United States, England, France, Germany and Russia dispatched observing parties to selected sites that usually were widely separated in latitude and longitude. In addition, Italy also funded three observing parties, and Holland one. The expeditions represented one of the first major international scientific collaborations, and occurred at a time when photography

W. Orchiston, *Exploring the History of New Zealand Astronomy*,
Astrophysics and Space Science Library 422,
DOI 10.1007/978-3-319-22566-1_15

Table 15.1 Solar parallax values (after Orchiston et al. 2000)

Astronomer(s)	Date	Method	Parallax (″)
Pingré	1770	1769 transit of Venus	8.88 ± 0.05
Lalande	1771	1769 transit of Venus	8.55–8.63
Encke	1835	1761 and 1769 transits of Venus	8.57116 ± 0.0371
Gillis and Gould	1856	Meridian parallax of Mars	8.495
Powalky	1864	1761 and 1769 transits of Venus	8.83
Hall	1865	Meridian parallax of Mars	8.842
Newcomb	1867	Meridian parallax of Mars	8.855
Stone	1868	1761 and 1769 transits of Venus	8.91
Todd	1881	U.S. 1874 transit of Venus programme	8.883 ± 0.034
Obrecht	1885	French 1874 transit of Venus programme	8.81 ± 0.06
Harkness	1889	U.S. 1882 transit of Venus programme	8.842 ± 0.0118
Newcomb	1891	1761 and 1769 transits of Venus	8.79 ± 0.051
Harkness	1894	System of constants	8.809 ± 0.0059
Newcomb	1895	System of constants	8.800 ± 0.0038
Spencer Jones	1941	Eros campaign	8.790 ± 0.001
IAU	1976	Radar	8.794148 ± 0.000007

was beginning to make an impact on astronomy (Langford 1984, 1987; Norman 1938), but they were plagued by controversy over instrumentation and methodology.

Positional astronomy had long been the research focus of the U.S. Naval Observatory, and it is only natural that it was charged with overseeing the two nineteenth century American transit programmes. The 1874 programme involved eight different stations. As Fig. 15.1 indicates, these were located at Vladivostok in far eastern Russia; Nagasaki in Japan; Peking in China; Queenstown in the South Island of New Zealand; the Chatham Islands (New Zealand territory in the Pacific Ocean, to the east of the South Island); Hobart and Campbell Town in Tasmania; and Kerguelen Island in the southern Indian Ocean. This chapter (based largely on Orchiston et al. 2000) is about the Queenstown station, its contribution to the overall American programme, and the place of the 1874 transit in the evolution of New Zealand astronomy.

15.2 Instrumentation

The American transit of Venus program was planned, initiated and implemented by a special Commission which was created by Congress in 1871. Dominating the Commission were U.S. Naval Observatory personnel, one of the most influential of

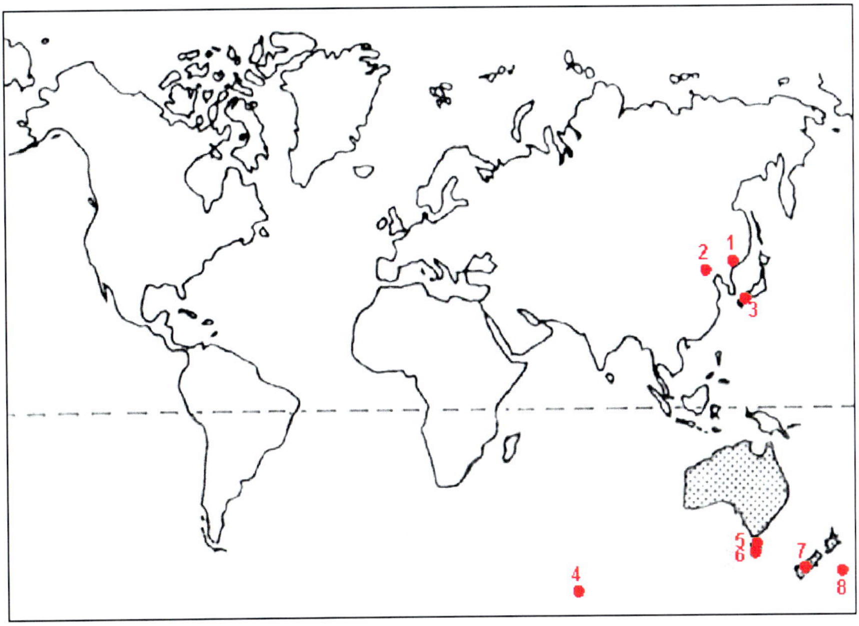

Fig. 15.1 Locations of the U.S. 1874 transit of Venus observing stations. Key: 1 Vladivostok; 2 Peking; 3 Nagasaki; 4 Kerguelen Island; 5 Campbell Town; 6 Hobart; 7 Queenstown; 8 Chatham Islands

whom was Simon Newcomb (1835–1909; Fig. 15.2), Professor of Mathematics at the Observatory. Newcomb was destined to become a leading figure in American astronomy (see Archibald 1924; Newcomb 1903).

The first challenge of the Commission was to decide on the instrumentation, and for this they had to choose between visual and photographic observations. Although they decided to adopt both, photography was to play the leading role. It was then a matter of selecting the type of photographic telescope, and two basic designs were considered: the equatorially-mounted heliograph developed by the talented British amateur astronomer, Warren De la Rue (1815–1889; Fig. 22.7) and used by the British transit parties, and the fixed horizontal solar telescope first conceived by Joseph Winlock (see Fig. 14.8), the Director of the Harvard College Observatory, and used successfully by that Observatory. The Commission chose the latter, and Newcomb decided that two discrete series of photographs should be taken: of the ingress and egress contacts, and of Venus superimposed on the disk of the Sun during the transit.

The commission then finalised the instrumentation to be issued to each observing party. It comprised: the photographic telescope, a standard refracting telescope, a transit telescope, a sidereal clock, three box chronometers, a chronograph, five thermometers, a barometer and equipment for measuring terrestrial magnetism.

Fig. 15.2 Simon Newcomb (https://en.wikipedia.org)

The principal research instrument was the photographic telescope, or photoheliograph (see Newcomb 1880, for details). Vital was the 20.3-cm square brass photographic plate-holder (Fig. 15.3), which was mounted on a solid metal pier. Permanently cemented into the plate-holder was a transparent sheet of plate glass about 7.6 mm thick,

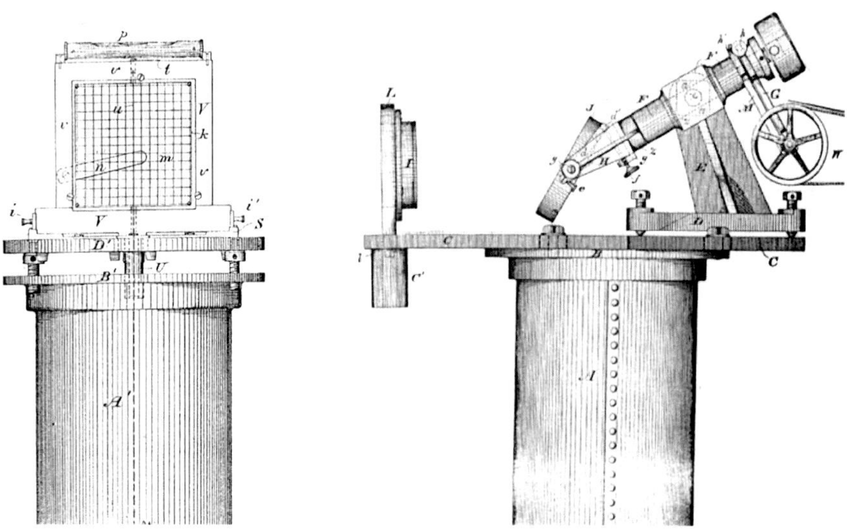

Fig. 15.3 Details of the design of the horizontal photographic telescope, showing the plate-holder (left) and heliostat (right) (after Newcomb 1880)

> … divided into small squares by very fine [etched] lines about one-five-hundredth of an inch thick … the sensitive plate goes into the other side of the frame, and when in position for taking the photograph, there is a space of about one-eighth of an inch between the ruled lines and the plate. The former are, therefore, photographed on every picture of the sun which is taken, and serve to detect any contraction of the collodion film on the glass plate. (Newcomb 1887: 189).

Also attached to the plate-holder was a plumb line, which was used to indicate vertical on each of the photographs. A simple, pre-fabricated flat-roofed, wooden 'Photographic House' protected the plate-holder, and contained facilities for preparing and developing the photographic plates.

Twelve metres away from the 'Photographic House', in the open air, was a second metal pier supporting a 17.8-cm unsilvered glass heliostat mirror (which captured the rays of the Sun) and a 12.7-cm objective (which focussed the light on the photographic plate). The heliostat tracked the Sun with the aid of a weight drive (Fig. 15.3). If it was deemed necessary, a temporary cover could be erected to shelter the heliosat and drive. Between the heliostat and the plate-holder was a measuring rod which was used to accurately determine the focal length of the telescope, a key element in establishing the plate scale. The measuring rod was supported by a wooden framework, and was protected by a narrow 'roof'. This roof also covered a tube which extended about 4 metres from the Photographic House and through which the light from the heliostat passed. An image of the Sun about 10.8 cm in diameter was formed. Four thermometers and a barometer were generally kept in the Photographic House to monitor variations in environmental conditions. The overall arrangement of the photographic telescope is shown in Fig. 15.4.

In order to visually monitor the transit and obtain precise micrometric measures of Venus on the Sun's disk, each transit party was supplied with a 12.7-cm (5-inch) f/14 refracting telescope manufactured by Alvan Clark (1804–1887; Fig. 15.5),

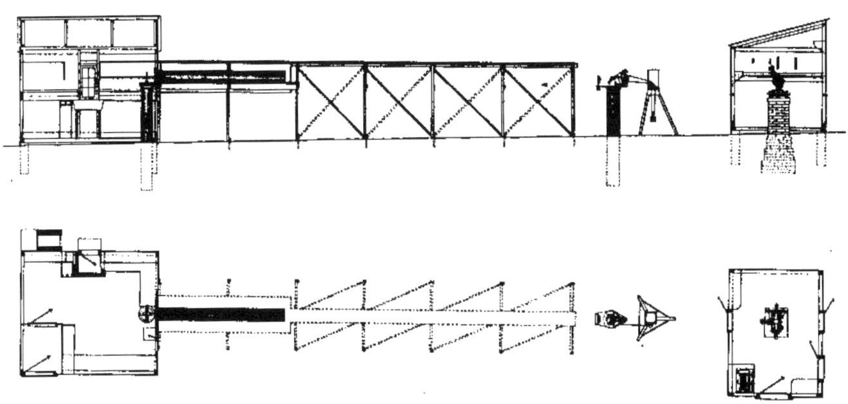

Fig. 15.4 The overall arrangement of the horizontal photographic telescope and 'Transit House'. Top side elevation; bottom plan (after Orchiston et al. 2000: 27)

Fig. 15.5 A portrait of Alvan Clark (after Gerry 1891)

America's leading nineteenth century telescope-maker (see Warner and Arial 1996). This instrument (see Fig. 14.10) was supported by a solid metal pier rather than a tripod, and was supplied with a German equatorial mounting, circles, a weight drive, slow-motion controls and the all-important double-image micrometer. Protecting the refractor from wind and inclement weather was a prefabricated, wooden, octagonal observatory 3-m diameter, with a rotating conical roof containing a single hinged shutter. Those involved in the transit expeditions generally referred to this observatory as the 'Equatorial House'.

Accurate time-keeping was paramount for the success of the venture, and so each transit station was supplied with an astronomical clock. All of the clocks were built by the Howard Clock Company of Boston, and because they were designed for long and difficult journeys to the observing stations they were not particularly elegant (Newcomb 1880: 14–15).

Each station also was provided with a transit instrument, which was used to regulate the clock. These transit telescopes had an aperture of 6.35 cm, and were manufactured by Stackpole Brothers. They were of the 'broken tube' construction,

> … a prism being placed in the center of the tube, by interior reflection from which the pencil rays is [*sic*] thrown along the axis; and the image is thus formed at the end of the latter … This form of instrument has the great advantages of convenience in observing and rapid and easy manipulation, but is still subject to the disadvantages of collimation varying with zenith-distance of the object observed. (Newcomb 1880: 14).

A schematic diagram of this unusual type of transit telescope is shown in Fig. 15.6. This instrument was mounted on a solid brick foundation capped by a stone slab, and protecting it from the elements was a simple, prefabricated, wooden 'Transit House' with a sloping roof. An observing shutter which was aligned north-south could be readily removed to reveal a strip of sky. This building was located near the heliostat so that it could also be used to accurately align the photographic telescope north-south (see Fig. 15.4). One thermometer was normally kept in the Transit House. In addition to its time-keeping role, the transit telescope was used for latitude and longitude determinations.

In order to accommodate these assembled instruments, a transit station required a nearly level north-south site about 25 m in length offering a firm foundation, protection from the winds, and a clear view of the Sun for the entire duration of the transit (Commission on the Transit of Venus 1874).

15.3 Queenstown, and the New Zealand Observing Party

Each American transit party normally consisted of a Chief Astronomer and Assistant Astronomer, who were responsible for visual observations and the time-service, respectively, and three or four photographers (Herman 1984: 38). The Chief Photographer was always a professional photographer, while the Assistant Photographers were typically "… young gentlemen of education, recent graduates of different colleges, who had been practised in chemical and photographic manipulation." (Newcomb 1880: 16).

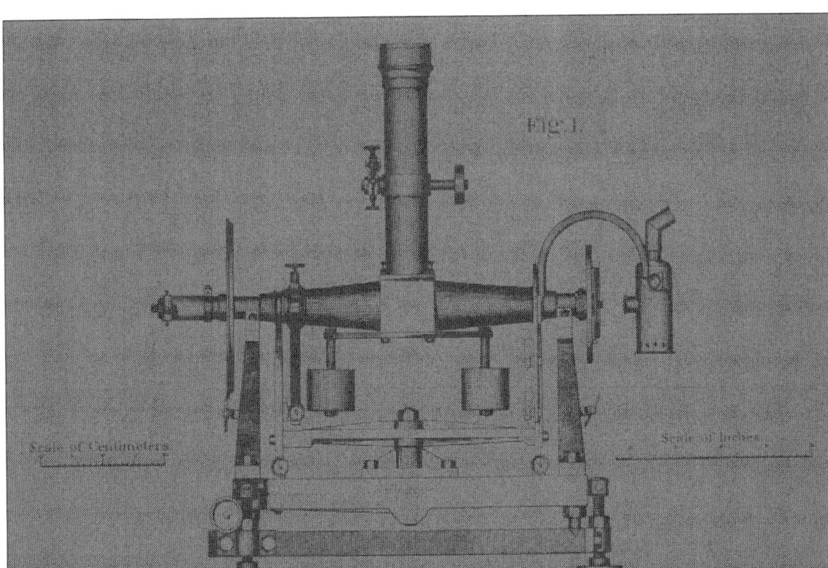

Fig. 15.6 A schematic of the Stackpole 'broken tube' transit telescope (after Orchiston et al. 2000: 28)

Fig. 15.7 C.F.H. Peters (https://en.wikipedia.org)

Leading the New Zealand party was C.H.F. Peters (Fig. 15.7), Professor of Astronomy at Hamilton College, New York, and Director of the College's Litchfield Observatory (Sheehan 1998). Christian Heinrich Friedrich Peters was born at Coldenbüttel, Schleswig, in Denmark (now Germany) on 19 September 1813. He studied mathematics and astronomy at the University of Berlin, completing his Ph.D. in 1836. In 1854 he emigrated to the United States, and after working with the U.S. Coast Survey and at the Dudley Observatory he was appointed to the Hamilton College post in 1858. One of Peters' interests was solar astronomy, and he also searched—with considerable success—for new minor planets (see Ashbrook 1984; Baum and Sheehan 1997).

The Assistant Astronomer, Lieutenant E.W. Bass, was Assistant Professor of Natural and Experimental Philosophy the U.S Military Academy at West Point.[1] Bass was born in Wisconsin but moved to St. Paul, Minnesota, at the age of four. In 1864 he entered the West Point, and four years later graduated fourth in a class of 54. He then joined the U.S. Corps of Engineers and in 1869 returned to West Point

> … at an unusually early age … Upon his return [from New Zealand] in September 1875, Captain Bass commanded the Engineer Company and then was the Battalion Adjutant at Willet's Point, New York. In September 1876, he returned to USMA [West Point, again] as

[1]The following biographical account of Bass differs from that published in Orchiston et al. (2000), and is based on documentation that has become available since that paper was researched and written.

Assistant Professor of Natural and Experimental Philosophy. On 2 May 1878 he was appointed Professor of Mathematics, Head of the Mathematics Department …

One of Professor Bass' significant contributions to the field of mathematics was the text books that he wrote …

In October 1898, due to the severe worsening of his eyesight, Professor Bass retired as the Head of the Mathematics Department and from the Army. He resided for the next twenty years at Bar Harbor, Maine and in New York City. On 6 November 1918, Edgar Wales Bass, age 75, died from the illness of pneumonia. (Edgar Wales Bass n.d.).

At Queenstown, in 1874, Peters was extremely happy with Bass' performance, as the following concluding remarks in his report indicate:

… it is my pleasant duty to mention still another circumstance that contributed not a little to the good work that the New Zealand party can show. The companionship of Lieutenant E.W. Bass (now lieutenant-colonel and professor of higher mathematics at West Point) was a most happy one. It is impossible to separate what in the work belonged to each of us, and when in the foregoing pages I have spoken in the plural I include the assistant astronomer. Our judgement agreed almost even in the minute particulars, so that we formed one head with four hands, and, if I may say so, a double brain. (Peters 1881: 548).

Completing the Queenstown party were four photographers: Charles L. Phillippi (Chief Photographer), Israel Cook Russell (Fig. 15.8; First Assistant Photographer), E.B. Pierson (Second Assistant Photographer) and Louis H. Aymé (Third Assistant Photographer).[2] Phillippi was a professional photographer (from Philadelphia); Russell and Aymé were college students. Assisting the photographers was a local volunteer named Beckett, who just happened to be from the U.S.A. (Peters 1881). Figure 15.9 shows members of the Queenstown party (minus Mr Beckett) assembled at the U.S. Naval Observatory prior to their departure for New Zealand.

On 8 June 1874 the U.S. man-of-war *S.S. Swatara* sailed from New York with the five southern hemisphere transit parties, and proceeded to the Indian Ocean, via Bahia in Brazil and Cape Town. The following newspaper report describes this vessel:

The Swatara is a screw steamer, ship-rigged, and is constructed entirely of wood, live oak being the timber used. She has an extreme clipper entrance, but is rather full aft, and it must be confessed that much handsomer models are to be found in the American mercantile marine. She was built at the Government Navy-yard, New York, and launched in June last, and got ready with all despatch for her present mission … Her screw is driven by compound engines of 1,400-horse power indicated, and the consumption of fuel is about 25 tons per diem, and with this it is said she can steam 12 knots. (The United States S.S. Swatara 1875).

[2]Of these four photographers Israel Cook Russell would go on to achieve prominence as a pioneering geomorphologist and glacial geologist. After returning from the New Zealand expedition he added a geology qualification to the B.S. and C.E. degrees he already possessed, and after serving as an Assistant Professor in the School of Mines at Columbia University in 1880 joined the United States Geological Survey and explored Alaska. In 1890 he was appointed Professor of Geology at the University of Michigan, and at the time of his death—in 1906—was the President of the Geological Society of America. In honour of his achievements (including numerous publications), a fiord, two glaciers and three mountains have been named after him. For further details see Aalto (2009) and Sylvestre (2008).

Fig. 15.8 Israel C. Russell in about 1900 (https://en.wikipedia.org)

Fig. 15.9 The Queenstown transit of Venus observing party and *ex officio* members party at the U.S. Naval Observatory. Seated (left to right): Mr. Phillippi, Professor Peters, Admiral Davis (President U.S. Transit of Venus Commission) and Professor H. Draper. Standing (left to right): Messrs Russell, Pierson and Amyé, and Lieutenant Bass (*Courtesy* U.S. Naval Observatory)

On 7 September the *Swatara* reached Kerguelen Island, and the first of the observing parties was dropped off. Subsequently, two parties disembarked at Hobart (see Orchiston and Buchanan 1993, 2004; and Orchiston 2004, for accounts of these), and the *Swatara* then headed for New Zealand, anchoring at Bluff Harbour on 16 October 1874.

There Peters was met by Bass, who had arrived in New Zealand several weeks earlier in order to finalise a suitable site for the transit station. Initially, the Americans had planned to install their South Island transit station at Bluff, a convenient coastal location, but the surveyor J.T. Thompson (1821–1884; President of the Otago Academy) and James McKerrow (Chief Surveyor of the Province of Otago) counselled Bass against this. Instead, they suggested a site in central Otago, an elevated inland area known in summer for its warm temperatures and clear sunny skies (Peters 1881: 439). After considering various options, Bass selected Queenstown on the shores of Lake Wakatipu, 320 m above sea level (see Fig. 15.10 for New Zealand localities mentioned in the text). In hindsight, this was to prove a particularly astute decision.

The New Zealand Government supported the U.S. transit project by freighting the boxes of equipment free of charge to Winton. After being off-loaded from the train, they were taken to Kingston by wagon while the transit party travelled by stage coach. The last short leg of the journey from the southern tip of Lake Wakatipu to Queenstown was by steamer, and Peters' party reached its final destination on 23 October (Peters 1874a), six and a half weeks before the transit. This may appear to be an unreasonably long lead-in period but a great deal of preparation was necessary before the 'grand event' (see Bass 1874; Newcomb 1880; Peters 1874a, 1881).

On 24 October the astronomers pinpointed a suitable observing site on an old terrace surface 408 m above the lake and about 0.8 km north-east of the centre of town. There were already a few houses scattered about in the general vicinity, and Peters (1881: 516) noted that streets and sections had been laid out, anticipating the future growth of Queenstown. The transit party proceeded to mark out a 24.4 × 12.2 m enclosure on Melbourne Street, and erect a perimeter fence.

Then began the busy schedule of setting up a functioning transit station. This involved assembling the prefabricated buildings they had brought; installing and adjusting the various scientific instruments; conducting observations for latitude, longitude and time; and practising the photographic procedures to be followed on the day of the transit. Fortunately Newcomb had provided the necessary detailed instructions for all of the observing parties (see Commission on the Transit of Venus 1874). On 26 October the *Otago Daily Times* (1874) reported that the station had been connected to the local telegraph. The first observations with the transit telescope were obtained on 30 October, the Alvan Clark telescope came into operation on 5 November, and the first experimental plates were exposed with the all-important photographic telescope on 14 November. By 24 November the photographers were ready to commence their 'practice drill', in preparation for the

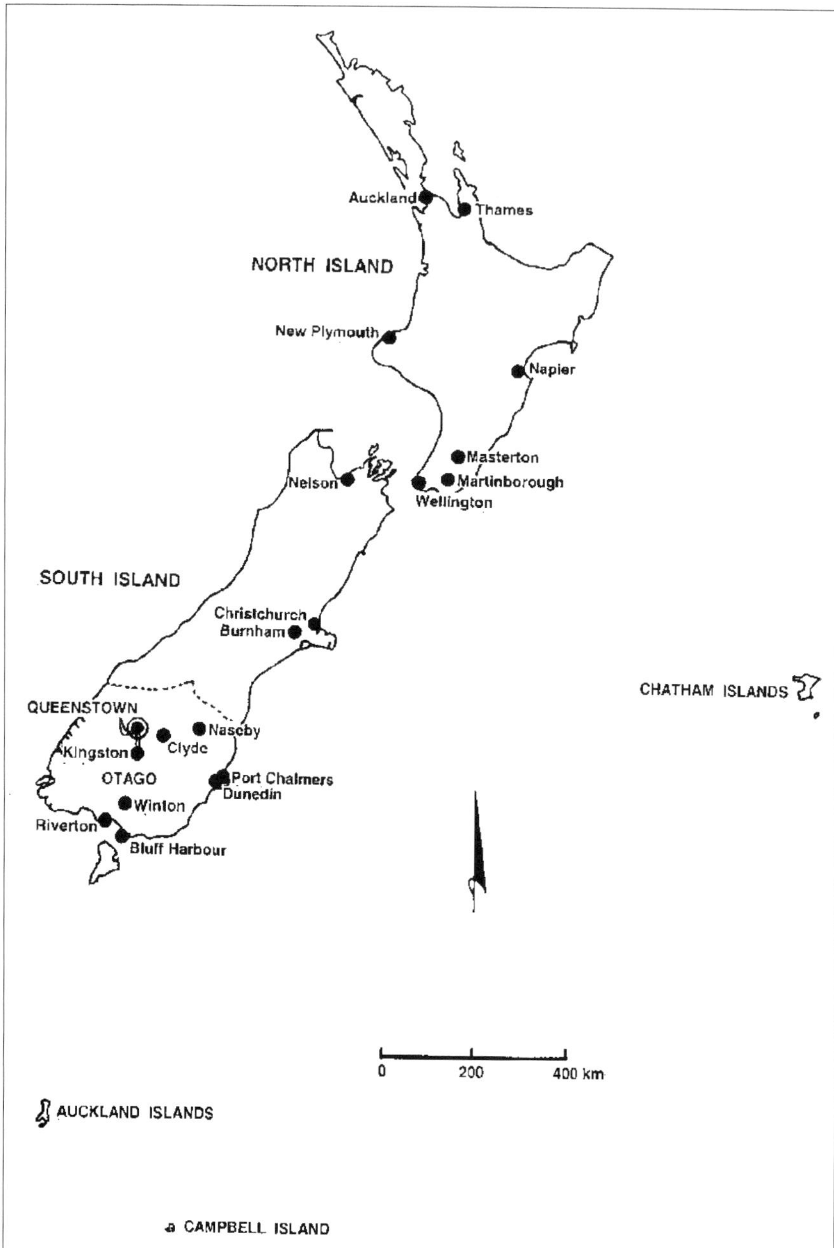

Fig. 15.10 New Zealand localities mentioned in the text

9 December transit. Peters (1881: 472) noted that in the course of these preparations daytime temperatures were rather pleasant, but at night "… during observing hours, the air was sometimes very chilly."

A plan of the Queenstown transit station is reproduced in Fig. 15.11, and the specific instruments assigned to the transit party are listed in Table 15.2.

Knowledge of the precise location of the transit station was critical, and twenty pairs of stars almost equal distances from the zenith were used to determine the

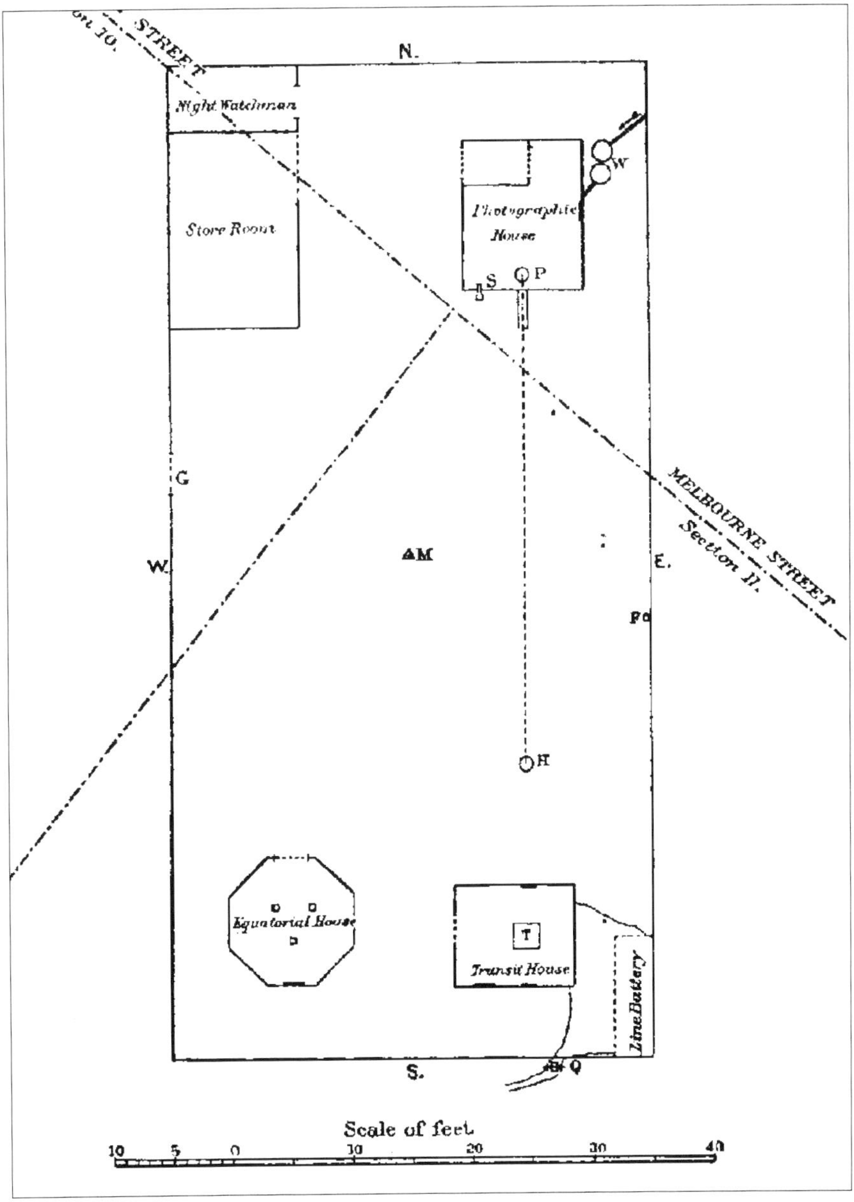

Fig. 15.11 Plan of Queenstown transit station. Key: T transit telescope; H heliostat; P photographic plate-holder (after Newcomb 1880)

Table 15.2 Instruments used at the Queenstown transit station (after Newcomb 1880; Peters 1881)

Instrument/component	Manufacturer	Number
Sidereal clock	Howard Clock Company	626
Mean time chronometer	T.S. & J.D. Negus	994
Sidereal chronometer	T.S. & J.D. Negus	1470
Sidereal chronometer	Bond	335
Transit telescope	Stackpole Bros.	1504
127-mm refracting telescope	Alvan Clark & Sons	858
Heliostat mirror	Alvan Clark & Sons	II
Photographic objective	Alvan Clark & Sons	IV
Measuring rod	Alvan Clark & Sons	VIII
Photographic plate-holder	Alvan Clark & Sons	8
Engineer's level	Alvan Clark & Sons	1489

latitude. The first observations were made on 12 November, and further observations followed on 13, 14, 20, 21, 28, 29 and 30 November (Peters 1881: 500–509). Meanwhile, the clock had been set up on 18 November, and on the 25th time signals were exchanged with the British transit party at Burnham, near Christchurch, in order to establish the longitude of the Queenstown site. Further exchanges with Burnham took place on 27 and 30 November, and on 2 December. In addition, on 5, 8 and 28 December signals were exchanged with the Rockyside Observatory near Dunedin where McKerrow and Thompson were based; with the *Swatara* on 2 November and 27 December, while it was temporarily anchored at Port Chalmers; and with the *Alexandrine* (the German Auckland Islands transit party's vessel) on 17 and 19 December, while it was in Bluff Harbour (see Peters 1881: 479–499). Apart from the various telegraphic exchanges, Peters (1881: 496) lamented that "… we were able to observe only a few occultations and no more than two moon culminations." The derived co-ordinates of the site (after Peters 1881: 499–550) were:

Latitude 45° 02′ 08.10 ± 0.14″S
Longitude 11h 14m 40s 4E

15.4 Observations and Results

On 9 December the long-awaited morning of the transit, Peters (1874a) wrote in his journal: "The eventful day at 5 o'clock in the morning looked very ugly and black. The forenoon was cloudy and inclined to be stormy." Despite this far from promising situation, all the final adjustments were made to the instruments, and

About 2 minutes before computed time of 1st contact the Sun came out, permitting observations of Equatorial and photographs to be made without interruption for 1¾ hours about. After that only at intervals, up to within 16 minutes before computed time of 3^{d} contact. The Sun then remained invisible until 30^{s} to 40^{s} after 4th contact. Attempts were made for reversed photographs at 6^{h}p.m. but invain [*sic*]. (ibid.).

During the transit, Bass was in charge of the Photographic House. Phillippi exposed the different plates, Russell and Pierson developed them, and Aymé entered information on each plate in the photographic journal (see Bass 1874). Becker took care of the heliostat and its drive (Peters 1881: 530).

In all, 237 photographs of the transit were obtained, 178 of the first and second contact, and 59 of Venus superimposed on the disk of the Sun (Peters 1874b). The final photograph was taken 16 minutes before the commencement of egress (Peters 1881). Regrettably, none of the Queenstown photographs has survived, but one of the images obtained at the U.S. transit station at Campbell Town, Tasmania, is shown in Fig. 15.12 for reference purposes.

While Bass and the photographers were busy in the Photographic House, Peters was observing the first and second contacts with the 12.7-cm refractor, and he specifically noted the absence of the notorious 'black drop effect' (Peters 1874a). He succeeded in obtaining fourteen measures of chords and cusps while Venus was on the Sun's limb (see Newcomb 1880: 135), and twenty-one measures of the distance of the planet from the limb while the transit was in progress. He also made ten determinations of the diameter of Venus. He remarked that during the transit

… perfect silence reigned in and around the station. Though my door [of the Equatorial House] was shut, whenever a picture registered itself in the photographic house the distant click was audible, and gave encouraging satisfaction and delight. (Peters 1881: 543).

All in all, Peters was pleased with the outcome of the program and that "The instruments and appliances worked admirably, and everything passed off well."

Fig. 15.12 Photograph of the 1874 transit of Venus obtained by the U.S. party at Campbell Town, Tasmania (*Courtesy* Queen Victoria Museum and Art Gallery)

(*Lake Wakatip Mail* 1874c). Despite the intermittent cloud cover, Queenstown proved to be one of the few 'successes' of the American transit program. Global weather conditions were generally poor at the time (Janiczek 1983) and although they were far from ideal in Queenstown, Peters' party did succeed in observing two of the four transit contacts and obtaining more photographs than at any of the other southern transit stations (see Dick et al. 1998: Table 2). Peters (1881: 548) was quick to acknowledge that their success at Queenstown was in part due to following the advice proffered by Thompson and McKerrow and selecting "… an inland station". Most other regions of New Zealand were clouded out at the critical time.

After the transit, Peters and his party dismantled the transit station and journeyed to Bluff Harbour where the *Swatara* was waiting at anchor. After loading the boxes of equipment, they sailed for the United States on 20 January 1875. Their route took them first to Hobart and Melbourne (Newcomb 1880: 19). Soon after their return Peters continued his regular research, and on 3 June 1875 he discovered two new minor planets which he promptly named after the Roman goddesses of journeyings and homecomings to commemorate his Queenstown success (see Ashbrook 1984).

For the Commission, the challenge now was to analyse the entire suite of observations from the expedition and produce a meaningful result. Newcomb believed that the visual observations from all the different transit parties (not just those of the Americans) would, when combined, "… give a value of the solar parallax of which the probable error will lie between 0.02″ and 0.03″." (*Annual Report* 1875). He also felt that the American photographs alone would produce a result of comparable, if not better, accuracy.

As we saw in the previous chapter, the task of measuring these plates was assigned to William Harkness (see Fig. 14.36) from the U.S. Naval Observatory. The first of the volumes reporting on the American campaign appeared in 1880 and comprised 157 quarto pages; the first 117 of which discussed the photographic work (without providing any conclusions), while the final 40 pages related to the visual observations (Newcomb 1880). The planned Part II, in two volumes, reached the proof stage in 1881 and today exists as only a single copy in the Library of the U.S. Naval Observatory in Washington, D.C. In the course of 564 pages, Part II provides detailed accounts of the eight different transit stations, including Queenstown (Peters 1881), but does not report any results. Parts III and IV were supposed to contain the results, but because of funding problems did not even reach the proof stage.

Because of this unfortunate situation, David Peck Todd (Fig. 14.37), an Assistant at the Nautical Almanac Office, decided to analyse some of the successful 1874 photographs, and he obtained a value of 8.883 ± 0.034″ for the solar parallax. This result was published in 1881 (Todd 1881), and as we have seen it differed significantly from some of the values obtained during previous transits (see Table 1). Also, there was concern about the use of the photographic method to investigate the AU (see Lankford 1984), and some astronomers decided to try other methods (including the night-time observation of minor planets, which appeared as point-sources).

Following what appeared to be a far from satisfactory outcome of the 1874 transit observations, scientists from 14 nations met in Paris in October 1881 to plan for the 1882 transit, and after some debate most nations decided to reject the use of photography. The United States was one of the few countries which determined to persevere with a photographic investigation, but for a time there was even doubt as to whether an international program would be mounted. In the end it was agreed that eight different stations would be established, four in the United States itself and four at overseas localities. Parties were subsequently dispatched to South Africa, South America and New Zealand. As we have seen, the single New Zealand station was located in Auckland, under the leadership of Edwin Smith from the United States Coast and Geodetic Survey.

This time it was Harkness, another U.S. Naval Observatory astronomer, who was responsible for the overall U.S. program:

> Although Newcomb had initiated American interest [in the nineteenth century transits], it was Harkness who had designed much of the equipment and personally led one of the parties, and it was Harkness who after Newcomb's abdication in 1882 would not only be the driving force behind the 1882 expeditions but also (in sharp contrast to Newcomb) produce a final result. Bringing the transit of Venus observations to fruition became a major goal of Harkness and a landmark in his career. (Dick et al. 1998: 242).

Six years after the 1882 transit Harkness reported a value of 8.847 ± 0. 012″ for the solar parallax based on his analysis of 1475 photographs of the event (Harkness 1888: 17–18), and just four months later he revised this to 8.842 ± 0. 0118″ (Harkness 1889: 424–425).

15.5 Discussion

Investigation of the Astronomical Unit did not end with the publication of the 1882 transit results. Harkness insisted that the solar parallax was intricately entwined with lunar parallax, the constants of precession and nutation, the parallactic inequality of the Moon, the masses of the Earth and Moon, and the velocity of light. By treating these constants as a system, he was able to produce what has been described as "... the crowning achievement of a lifetime of work ..." (Dick et al. 1998: 247), a monograph tilted "The solar parallax and its related constants" (Harkness 1891). Three years later, Harkness (1894) published a "best estimate" for the solar parallax of 8.809 ± 0.0059″, and Newcomb (1895) then produced a value of 8.800 ± 0.0038″, which was adopted as the international standard at the Conference Internationale des Étoiles Fondamentales in 1896. By way of comparison, the value of the solar parallax ratified by the International Astronomical Union (IAU) in 1976 was 8.794148 ± 0.000007″. In 2009 the IAU approved a value of 149,597,870.700 km for the mean distance from the Earth to the Sun.

As one of the most visually-appealing of all sciences, astronomy has long enjoyed a high public profile, and the importance of the nineteenth century transit of

Venus received world-wide attention (e.g. see Cottam et al. 2011, 2012). Certainly the importance of the 1874 transit was not lost on the good citizens of Queenstown. Throughout their stay, amiable relations persisted between the Americans and the townsfolk, and when Peters gave a public lecture about the transit program on 26 November this was well attended and was reported in the local newspaper, the *Lake Wakatip Mail* (1874b). The account ran to two and a half columns of detail, and summarised the history of astronomical distance measurement from antiquity through to the nineteenth century transits of Venus. Given Cook's intimate association with New Zealand, Peters made a point of labouring his exploits, to the obvious delight of the audience:

> To this astronomer, who had fought his way up, the highest credit was due, considering the instruments he had to work with. He had no time-piece that would exceed an ordinary American clock … yet his efforts were, to this day, appreciable (Applause). He did not intend to flatter, but these were facts well known, and they were much indebted to Captain Cook (Renewed applause). He took his observations at Cape Venus, Otaheite. The observations of this memorable explorer were made in the year 1769. They should be proud of him who afterwards met his death in the South Seas Islands he loved so well (Applause). And to him, more than anyone else, was due the presence of the present expedition (Cheers).

On 17 December, after the 1874 transit, there was a dinner in the Queen's Arms Hotel to farewell the visiting scientists, and the coverage in column inches assigned to this event in the *Lake Wakatip Mail* (1874d) outstrips by far the account of the transit itself! Yet the speeches, which were reported in detail, hardly touched on the transit or on science. Instead, they focussed on the glories of New Zealand in general and Queenstown in particular (to the frequent accompaniment of enthusiastic applause). In a diplomatic yet prophetic moment, Peters saw

> … in the stars a large city here [at Queenstown] in the future; railways converging upon it; people coming from many other parts of the earth to enjoy its beautiful climate, and behold its grand scenery (Loud Cheers).

Finally, on Boxing Day 1874 the transit party attended a picnic and dance given in their honour by the Masonic fraternity. All in all, the Americans were made to feel very welcome during their two-month stay in Queenstown.

Nor did the Americans restrict their non-astronomical activities to socialising. On 2 and 3 December the photographers went on 'an excursion', in order to photograph Queenstown and the magnificent scenery offered by Lake Wakatipu and the surrounding countryside. Pierson and Phillipi subsequently published an album of New Zealand photographs which presents a unique record of Otago in the 1870s. Janiczek (1983: 66) has reproduced some of the photographs from this album in one of his papers.

The transit party also extended its scientific interest beyond astronomy. In an interesting seismological diversion, Peters (1881: 469, 474) wondered whether some of the problems associated with the sidereal clock and the chronometers were caused by minor earthquakes—not that members of the transit party noticed any during their sojourn at Queenstown. Meanwhile, Russell (one of the photographic team) was always on the look-out for local 'curiosities', and in the course of their stay he succeeded in collecting seven Maori adzes, together with native birds, plants, insects, crustaceans, small vertebrates and moa bones. This collection found its way into the Smithsonian Institution, and the adzes, together with ethnographic specimens sourced from Riverton and the Chatham Islands, are documented by Keyes (1967). It is of interest to note that the adzes are stylistically and petrologically typical of those found in early prehistoric southern South Island New Zealand (Orchiston 1974), and they reflect the important role that Central Otago (and the Queenstown area) played in early Maori life.

One of those living in Queenstown during the transit was a 10-year old girl, Sarah Cockburn (1864–1956), who had a passion for astronomy. In later years (under her married name of Salmond) she lobbied relentlessly for a monument to be erected on the site of the transit station, to mark what is arguably the most important venture in international science ever carried out at Queenstown. Her endeavours were finally rewarded in 1953 when at the age of 88 she had the honour of unveiling the monument (see Salmond 1993). The plaque (Fig. 15.13) reads:

> From this site a transit of the planet Venus across the solar disc was observed on 1874 December 9 by an American scientific expedition which came to Otago in the ship "Swatara".

In 1995 the monument was incorporated into the new multi-storey international Millennium Hotel which was constructed on the site of the transit station.

As the previous chapter in this book has shown, from an astronomical point of view, the 1874 transit was important because it also attracted British, French and German parties to New Zealand and its outlying islands, Campbell Island and the Auckland Islands, respectively (see Airy 1881; Auwers 1898; Bouquet de la Grye

Fig. 15.13 The commemorative plaque at the Millennium Hotel in Queenstown (after Orchiston et al. 2000)

1882; Filhol 1885), and Major Palmer at Burnham even had one of his party set up a 'satellite station' at Naseby, 300 km to the southwest (see Fig. 15.10). As in Australia (see Orchiston 2004; Orchiston and Buchanan 1993), the transit served as a catalyst for the development of local astronomy as it was the first opportunity for local astronomers to become involved in a project of international importance. Observers in Auckland, Thames, Wellington, Nelson and Dunedin, among other centres, prepared for the event (see *Lake Wakatip Mail* 1874a; The transit of Venus 1874a, b; Tomorrow's transit of Venus 1882), and as we have already noted one of the Wellington astronomers, Archdeacon Arthur Stock, produced a small popular book about the transit (Stock 1874). Unfortunately, cloudy skies prevented most observers from viewing the transit.

There was even more New Zealand interest in the 1882 transit (e.g. see Hector 1882; Orchiston 2004; Stone 1887; Tomorrow's transit of Venus 1882), with observers located in Auckland, Thames, New Plymouth, Martinborough, Wellington, Nelson, Christchurch, Dunedin and Clyde.

15.6 Concluding Remarks

The 1874 U.S. transit of Venus program was an expensive venture designed to refine that fundamental yardstick of Solar System astronomy, the Astronomical Unit. It was a major logistical exercise, involving the transportation of extensive equipment (including prefabricated buildings) to eight different transit stations, five of which were in the southern hemisphere. One of these was situated at Queenstown, in the South Island of New Zealand.

While most other New Zealand sites were clouded out during the transit, those at the Queenstown station succeeded in observing both ingress contacts and obtaining a succession of photographs of these contacts and of Venus while it was visible on the disk of the Sun. Todd used the latter photographs and those from other U.S. transit stations to produce a value for the solar parallax that differs only 0.089″ from the currently-accepted figure, but because the reliability of photography was called into question at the time this result was not seen as improving on the value of the astronomical unit.

Despite this disappointing outcome, from a national perspective the 1874 transit did serve to introduce astronomical photography to New Zealand, and it provided an important impetus for the development of local astronomy.

Acknowledgements I wish to thank the following for their assistance: Dr Steve Dick (ex-U.S. Naval Observatory, Washington); the late Peter Hingley (former RAS Librarian, London), Tom Love (ex-Carter Observatory, Wellington, NZ), Gregory Sheldon (U.S. Naval Observatory, Washington), Dr William Sheehan (USA), and staff at the National Archives (Washington). Finally, I wish to thank Dr Steven Dick (USA), Dr Richard Dodd (Martinborough, NZ) and Dr William Tobin (Vannes, France) for reading and commenting on the first draft of this chapter, and the U.S. Naval Observatory and the Queen Victoria Museum and Art Gallery (Launceston, Australia) for kindly supplying Figs. 15.9 and 15.12.

References

Aalto, K. R. (2009). Rock stars: Israel Cook Russell (1852–1906). *GSA Today*, February, 14–15.

Airy, S. G. B. (Ed.), (1881). *Account of Observations of the Transit of Venus, 1874, December 8, Made Under the Authority of the British Government: And of the Reduction of the Observations*. London: Her Majesty's Stationery Office.

Annual Report. Washington, U.S. Naval Observatory (1875).

Archibald, R. C. (1924). Simon Newcomb 1839–1909. Bibliography of his life and work. *Memoirs of the National Academy of Sciences, 17.*

Ashbrook, J. (1984). The adventures of C.H.F. Peters. In J. Ashbrook (Ed.), *The Astronomical Scrapbook. Skywatchers, Pioneers, and Seekers in Astronomy* (pp. 56–66). Cambridge, MA: Sky Publishing.

Auwers, A. (Ed.), (1898). *Die Venus-Durchgänge 1874 und 1882: Bericht über die Deutschen Beobachtungen im Autrage der Commission für die Beobachtung des Venus-Durchgangs*. Berlin.

Bass, E. W. (1874). Photographic record and journal New Zealand party. Manuscript, U.S. Naval Observatory (Queenstown transit box).

Baum, R., & Sheehan, W. (1997). *In Search of Planet Vulcan. The Ghost in Newton's Clockwork Universe*. New York: Plenum.

Bouquet de la Grye, A. (1882). *Recueil de Mémoires, Rapports et Documents Relatifs à l'observation du Passage de Venus sur le Soleil. Volume III. Part 1*. Paris: Academy of Sciences.

Clerke, A. M. (1893). *A Popular History of Astronomy During the Nineteenth Century*. Edinburgh: Adam and Charles Black.

Commission on the Transit of Venus. (1874). *Instructions for Observing the Transit of Venus, December 8–9, 1874*. Washington: Government Printing Office.

Cottam, S., Orchiston, W., & Stephenson, R. (2011). The 1874 transit of Venus, and the popularisation of astronomy in the U.S.A. as reflected in *The New York Times*. In W. Orchiston, T. Nakamura, & R. Strom (Eds.), *Highlighting the History of Astronomy in the Asia-Pacific Region* (pp. 225–241). New York: Springer.

Cottam, S., Orchiston, W., & Stephenson, F. R. (2012). The 1882 transit of Venus and the popularization of astronomy in the USA as reflected in *The New York Times*. *Journal of Astronomical History and Heritage, 15*, 183–199.

Dick, S., Orchiston, W., & Love, T. (1998). Simon Newcomb, William Harkness and the nineteenth century American transit of Venus expeditions. *Journal for the History of Astronomy, 29*, 221–255.

Early New Zealand Photographers and their successors. (n.d.). (canterburyphotography.blogspot.com/).

Edgar Wales Bass. (n.d.). (fredrickey.info/dms/DeptHeads/Bass.html).

Filhol, H. (1885). *Recueil de Mémoires, Rapports et Documents Relatifs à l'observation du Passage de Venus sur le Soleil. Volume III. Part 2*. Paris: Academy of Sciences.

Forbes, G. (1874). *The Transit of Venus*. London: Macmillan.

Gerry, S. L. (1891). Old masters of Boston. *New England Magazine, 3*(6), 682–695.

Grant, R. (1874). *The Transit of Venus in 1874*. Glasgow: Maclehose.

Harkness, W. (1888). *Annual Report*. Washington, U.S. Naval Observatory.

Harkness, W. (1889). *Annual Report*. Washington, U.S. Naval Observatory.

Harkness, W. (1891). The solar parallax and its related constants. *Washington Observations for 1885.*

Harkness, W. (1894). On the magnitude of the solar system. *Astronomy and Astro-physics, 13*, 605–626.

Hector, J. (1882). The transit of Venus. Clyde. *New Zealand Journal of Science, 1*, 326–330.

Herman, J. K. (1984). *A Hilltop in Foggy Bottom. Home of the Old Naval Observatory and the Navy Medical Department*. Washington: Department of the Navy.

Janiczek, P. M. (1983). Remarks on the transit of Venus expedition in 1874. In S. J. Dick & L. E. Doggett (Eds.), *Sky and Ocean Joined* (pp. 52–73). Washington: U.S. Naval Observatory.

Janiczek, P. M., & Houchins, L. (1974). Transits of Venus and the American expedition of 1874. *Sky and Telescope, 48*, 366–371.

Keyes, I. W. (1967). New Zealand artifacts from the United States "transit of Venus expedition" 1874–1875. *Smithsonian Contributions to Anthropology, 2*(2), 21–27.

Lake Wakatip Mail, November 24, 1874a.

Lake Wakatip Mail, December 4, 1874b.

Lake Wakatip Mail, December 11, 1874c.

Lake Wakatip Mail, December 18, 1874d.

Lankford, J. (1984). The impact of photography on astronomy. In O. Gingerich (Ed.), *The General History of Astronomy* (Vol. 4, pp. 16–39). *Astrophysics and Twentieth Century Astronomy to 1950: Part A*. Cambridge: Cambridge University Press.

Lankford, J. (1987). Photography and the 19th-century transits of Venus. *Technology and Culture, 28*, 648–657.

Meadows, A. J. (1974). The transit of Venus in 1874. *Nature, 250*, 749–750.

Newcomb, S. (1887). *Popular Astronomy*. New York: Harper.

Newcomb, S. (1895). *The Elements of the Four Inner Planets and the Fundamental Constants of Astronomy*. Washington: Government Printing Office.

Newcomb, S. (1903). *The Reminiscences of an Astronomer*. London: Harper.

Newcomb, S. (1880). *Observations of the Transit of Venus, December 8–9, 1874 … Part I. General Discussion of Results*. Washington: Government Printing Office.

Norman, D. (1938). The development of astronomical photography. *Osiris, 5*, 560–594.

Orchiston, W. (1974). *Studies in South Island New Zealand Prehistory and Protohistory*. Unpublished Ph.D. Thesis, Department of Anthropology, University of Sydney.

Orchiston, W. (1998). From the South Seas to the Sun: the astronomy of Cook's voyages. In M. Lincoln (Ed.), *Science and Exploration in the Pacific. European Voyages to the Southern Oceans in the Eighteenth Century* (pp. 55–72). Woodbridge: Boydell Press, in association with the National Maritime Museum.

Orchiston, W. (2004). The nineteenth century transits of Venus: an Australian and New Zealand overview. *Journal of Astronomical Data, 10*, 219–308.

Orchiston, W. (2005). James Cook's 1769 transit of Venus expedition to Tahiti. In D. W. Kurtz (Ed.), *Transits of Venus: New Views of the Solar System and Galaxy* (pp. 52–66). Cambridge: Cambridge University Press.

Orchiston, W., & Buchanan, A. (1993). Illuminating incidents in Antipodean astronomy: Campbell Town, and the 1874 transit of Venus. *Australian Journal of Astronomy, 5*, 11–31.

Orchiston, W., & Buchanan, A. (2004). 'The Grange', Tasmania: survival of a unique suite of 1874 transit of Venus relics. *Journal of Astronomical History and Heritage, 7*, 34–43.

Orchiston, W., Love, T., & Dick, S. J. (2000). Refining the astronomical unit: Queenstown and the 1874 transit of Venus. *Journal of Astronomical History and Heritage, 3*, 23–44.

Otago Daily Times, October 26, 1874.

Peters, C. H. F. (1874a). General Journal New Zealand party. Manuscript, U.S. Naval Observatory (Queenstown transit box).

Peters, C. H. F. (1874b). Letter to Rear-Admiral C.H. Davis, dated December 11. In Peters, 1881.

Peters, C. H. F. (1881). Section VII. Observations made at Queenstown … In S. Newcomb (Ed.), *Observations of the Transit of Venus, December 8–9, 1874* … Part II. Washington: Government Printing Office (unpublished typescript U.S. Naval Observatory Library).

Popular Science Monthly, 86 (1903).

Proctor, R. (1874). *Transits of Venus: A Popular Account of Past and Coming Transits*. London: Longmans Green.

Salmond, J. A. (1993). Salmond, Sarah 1864–1956. In C. Orange (Ed.), *Dictionary of New Zealand Biography. Volume Two. 1870–1900* (pp. 439–440). Wellington: Bridget Williams Books Ltd. and Department of Internal Affairs.

Sheehan, W. (1998). Christian Heinrich Peters. *Biographical Memoirs of the National Academy of Sciences, 76*.

Stock, Archdeacon [Arthur] (1874). *December 9, 1874. The Transit of Venus, and How to Observe It*. Wellington: Bowden.

Stone, E. J. (1887). *Transit of Venus, 1882*. London: Her Majesty's Stationery Office.

Sylvestre, P. D. (2008). *The Art and Science of Natural Discovery: Israel Cook Russell and the Emergence of Environmental Exploration*. Unpublished M.A. Thesis, History Department, Colorado State University.

The Transit of Venus. *Daily Southern Cross*, December 10, 1874a.

The Transit of Venus. *Daily Southern Cross*, December 24, 1874b.

The United States S.S. Swatara. *The Argus*, February 20, 1875.

Todd, D. P. (1881). The solar parallax as determined from the American photographs of the transit of Venus, 1874, December 8–9. *American Journal of Science, 21*, 491–493.

Tomorrow's transit of Venus. *Evening Post*, December 6, 1882.

van Helden, A. (1995). Measuring solar parallax: the Venus transits of 1761 and 1769 and their nineteenth-century sequels. In R. Taton & C. Wilson (Eds.), *The General History of Astronomy. Volume 2. Planetary Astronomy from the Renaissance to the Rise of Astrophysics. Part B: The Eighteenth and Nineteenth Centuries* (pp. 153–168). Cambridge: Cambridge University Press.

Warner, D. J., & Arial, R. B. (1996). *Alvan Clark & Sons: Artists in Optics*. Richmond: Willmann-Bell.

Woolf, H. (1959). *The Transits of Venus: A Study of Eighteenth Century Science*. Princeton: Princeton University Press.

Part V
Stunning Spectacles: Eclipses, Comets and Meteor Showers

Chapter 16
The 1885 Total Solar Eclipse: An Amazing Public Spectacle

Abstract The second half of the nineteenth century saw a blossoming of interest in solar eclipses as astronomers tried to establish whether the corona was a solar, lunar or terrestrial phenomenon, and as they investigated the nature of the corona, the chromosphere and prominences. Critical in these investigations were astronomy's newest allies: photography and spectroscopy. Photography was used with great effectiveness throughout the half century, but spectroscopy was first applied during the 'Indian eclipse' of 1868. Thereafter, almost every total solar eclipse was subjected to scrutiny, the intensity of which depended upon the duration of the eclipse and the location of its path of totality. The first total solar eclipse visible from New Zealand since the introduction of scientific astronomy occurred on 9 September 1885, and attracted the attention of professional scientists, amateur astronomers and—overwhelmingly—the general public. The centre of the path of totality extended from West Wanganui Inlet on the far northern reaches of the west coast of the South Island to Castlepoint on the Wairarapa Coast in the North Island, and a total eclipse was visible from population centres like Collingwood, Nelson, Wellington, Feilding and Dannevirke. In this chapter we briefly examine the scientific investigation of this eclipse before looking at the ways in which the nation's newspapers portrayed the eclipse and used it as a vehicle to promote a popular interest in astronomy among the general public.

16.1 Introduction

A total solar eclipse can be an unforgettable spectacle and one of the most remarkable astronomical events open to naked eye observation, yet for centuries these eclipses were viewed by many as portends of death or destruction (Zirker 1984), or were interpreted in terms of terms of heroic conflicts between major mythological characters. To the Maori, the Sun was *Te Ra*, one of the grandchildren of *Ranginui* (the Sky Father) and *Papanuanuku* (the Earth Mother), and an eclipse of the Sun (*Ra Kutia*) occurred when *Te Ra* was "… being attacked and devoured

W. Orchiston, *Exploring the History of New Zealand Astronomy*, Astrophysics and Space Science Library 422,
DOI 10.1007/978-3-319-22566-1_16

by demons, from which attacks, however, it invariably recovers." (Best 1955: 20; cf. Orchiston 2000).

During the last eight hundred years, there have been twenty-six different total or annular solar eclipses that were potentially visible from the North Island or South Island of Aotearoa/New Zealand (see Table 2.1), but there has been only one total solar eclipse since 1769 and the birth of scientific astronomy in Aotearoa/New Zealand. This eclipse occurred in 1885 (An earlier eclipse 1936).

Of all astronomical spectacles, impressive naked eye comets and total solar eclipses did more to popularize astronomy worldwide during the nineteenth century than any other type of astronomical object or event (e.g. see Cottam et al. 2011a, b, 2012; Cottam and Orchiston 2014). This chapter examines the role that the 1885 total solar eclipse played in popularizing astronomy in New Zealand by examining newspaper accounts of the spectacle that were published at the time. But first, let us provide a context within which to view this eclipse by examining the development of coronal science during the nineteenth century and the scientific observations that were made of the 1885 eclipse in New Zealand.[1]

16.2 Coronal and Chromospheric Science Prior to the 1885 Eclipse

By 1885 rapid advances had been made in solar science thanks to a succession of earlier well-observed total solar eclipses (e.g. see Cottam and Orchiston 2014; Pearson 2009; Ranyard 1879). From a research standpoint, undoubtedly the most important nineteenth century total solar eclipse occurred on 18 August 1868, and this was visible from Aden, India (Launay 2012; Orchiston et al. 2006), Thailand (Orchiston and Soonthornthum 2016) and the Dutch East Indies. It was at this time that the British solar specialist, Joseph Norman Lockyer (1836–1920; Fig. 16.1; Meadows 1972) renamed the 'sierra' the 'chromosphere' (Lockyer 1874). This was the first eclipse that was investigated spectroscopically (Clerke 1893), and Lieutenant John Herschel (the son of Sir John Herschel) from the Trigonometrical Survey of India, the noted French pioneer of astrophysics Pierre Jules César Janssen (1824–1907; Fig. 16.2; Launay 2012), Madras Observatory Director Norman Robert Pogson (1829–1891; Fig. 16.3), Paris Observatory astronomer Georges-Antoine-Pons Rayet (1839–1906), and the Indian-based British astronomer and surveyor Major James Francis Tennant (1829–1915) all recorded a number of emission lines which showed that the corona and chromosphere were composed mainly of hydrogen and iron. Pogson also noted a line close to the D3

[1]A companion paper (Orchiston and Rowe 2016) examines in detail the astronomers who observed this eclipse, their instruments and their observations; discusses the place of the 1885 eclipse in the overall pattern of professional and amateur astronomy in New Zealand between 1874 and 1885; and examines the contribution that this particular eclipse made to our international understanding of solar physics.

Fig. 16.1 Norman Lockyer in 1909 (https://en.wikipedia.org)

line of sodium that he associated with an unidentified element, and Lockyer and Janssen subsequently named this 'helium' (see Kragh 2009; Nath 2013). A conspicuous feature of the chromosphere on this occasion was a enormous prominence dubbed 'The Great Horn' (Fig. 16.4), which spectroscopic observations revealed to be composed principally of hydrogen (Orchiston et al. 2006).

The following year offered another observable total solar eclipse, on 7 August 1869, when further spectroscopic observations of the corona were a priority. On this occasion 11 emission lines were detected, six of which were new ones. One of these, dubbed the 'green emission line', was assigned to the new element 'coronium', and only much later would it be identified as a forbidden line of iron (Golub and Pasachoff 1999).

The eclipse of 1871, which also was visible from India, was the first in which a realistic representation of the corona as seen visually was recorded photographically, and details of coronal form were apparent on the photographs (e.g. see Fig. 16.5).

Subsequent eclipses offered astronomers further opportunities to investigate the nature and composition of prominences and the solar corona. As Pearson (2009: 14) has noted:

Fig. 16.2 Oil painting of Jules Janssen by his friend, the well-known French artist Jean-Jacques Henner (https://en.wikipedia.org)

> Photography, of the 1878 corona, gained over drawing in earnest in its role of making permanent coronal records. Large aperture short focal length portrait lenses were used to record significantly more detail within the inner and outer corona. General coronal form was photographed to one lunar diameter. The photographic process was still not of sufficient sensitivity to record the longer-fainter coronal streamers and rays … [but] At the 1883 eclipse, the extensions of the streamers were photographed beyond that visible to the naked eye or telescopic observer …

Further high quality coronal photographs were taken during the 1882 and 1883 eclipses, so by 1885 the principal research topics of international interest were:

1. The form and structural detail of the inner and outer corona, and the relationship between coronal features and photospheric phenomena such as sunspots, faculae and flares;
2. The precise gaseous composition of, and particle-gas distribution within, the inner and outer corona; and
3. The precise location and the identity of the green coronal line.

If observations of the New Zealand eclipse were to play a major role in furthering coronal science these were the sorts of issues that had to be addressed.

Fig. 16.3 Norman Pogson (www.dhinakarrajaram.blogspot.com)

Fig. 16.4 Drawing of 'The Great Horn' (after Tennant 1869)

16.3 Scientific Observations of the 1885 Eclipse

The eclipse occurred on 9 September 1885 local time, and the path of totality extended from the Antarctic continent, across the southern Pacific Ocean, over central New Zealand to a point just east of the southeastern coast of Australia (see Fig. 16.6). As the British astronomer and Superintendent of the *Nautical Almanac* John Russell Hind (1823–1895) stressed in his announcement of the eclipse in *Monthly Notices of the Royal Astronomical Society*, "The only land traversed by the

Fig. 16.5 A combination of images of the 1871 eclipse obtained by Tennant and Davis showing fine coronal detail (after Ranyard 1879: plates)

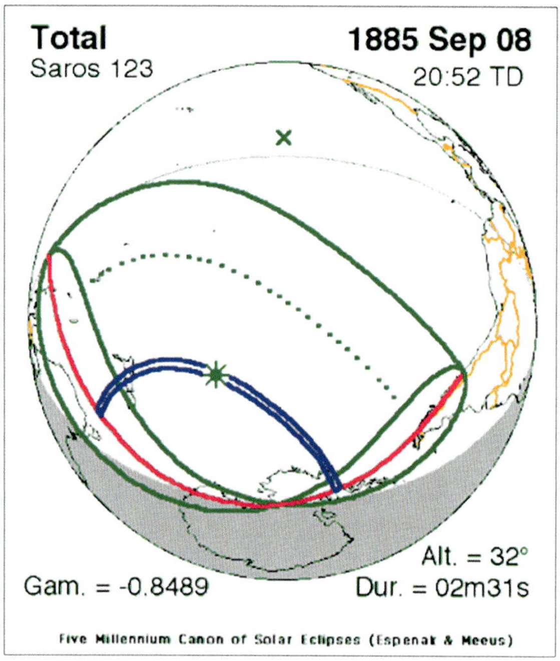

Fig. 16.6 The path of totality (double blue line) of the total solar eclipse of 8 September 1885

central line … will be the southern part of the north island and the northern part of the south island of New Zealand …" (Hind 1885). Thus, the path of totality passed through or very close to Castlepoint, Masterton, Otaki, Kapiti Island, Collingwood and West Wanganui Inlet, while the northern limit was near Wanganui, Waipawa and Waipukurau, and the southern limit near Foxhill and Blenheim (Harding 1885; *Manawatu Standard* 1885a). This region where a total eclipse would be visible—weather permitting—is illustrated in Fig. 16.7, and other North Island and South Island localities not marked there but mentioned in this chapter are shown in Fig. 16.8. Notable population centres within the 145-km (80-mile) wide strip of totality, but off the centre line, included Wanganui, Palmerston North, Feilding, Dannevirke, Wellington, Picton, Nelson and Stoke. As total solar eclipses go, totality was of comparatively short duration, lasting a maximum of just 2 minutes.

The Eclipse Committee of the Royal Society in London identified specific research objectives for this particular eclipse and these were circulated widely in New Zealand among those who planned to carry out scientific observations. There were four primary objectives (e.g. see Atkinson 1885: 212):

1. To conduct a spectroscopic analysis of the corona;
2. To measure the mean lateral extension of the corona;
3. To record the positions and extent of any coronal steamers; and
4. To ascertain whether the corona exhibited a line of "approximate symmetry".

Most of those situated along the path of totality were treated to uninterrupted views of the eclipse, and even experienced astronomers could not help but be moved by the spectacle:

Fig. 16.7 Map showing locations mentioned in this chapter that lay within the path of totality of the 9 September 1885 eclipse. Those locations at or very close to the centre line are shown in *black*, while locations at or very near the northern and southern limits are shown in *blue*. All other localities are in *red*

Fig. 16.8 New Zealand localities mentioned in this chapter, other than those marked in Fig. 16.7, are shown in *red*

> … a sight grander and more unique than the whole eclipse it is impossible to conceive. Even as the wind falls when the shades of evening close around, the very light breeze which had been blowing in the early morning gradually died away, and darkness increased. Birds ceased their twittering … except some paraquets, which were evidently much startled, and broke into the most noisy chattering as the sun disappeared, and flew away, it may be supposed, to their usual night haunts. Everything else became hushed; even the human voice had, or seemed to have, an unnatural sound. All nature seemed to bow its head, and stand in mute silence as the awful spectacle passed, and until the God of Day should again emerge from his temporary seclusion. The general appearance of things at the moment of totality, which was certainly not a period of complete darkness—for a soft and 'dim, religious light' was always present—was such as the observer can surely never forget. It was decidedly uncanny. The human face looked ghastly. The colours on mountain and field, on sea and sky, were weird, unearthly, and indescribable, such as one had never seen before. They had gradually deepened in hue as the eclipse proceeded, and just before totality the sky around the sun was of a dirty yellow, and quivering beams, of the colour of electric light, shot out from above and below the moon, giving it somewhat the appearance of a St. Andrew's cross with a circular centre.
>
> Generally speaking, during the sun's complete obscuration, the sky was of a mauve colour, except round about the luminary itself, where the intense brilliance of the silvery protuberances or the golden glory of the coronal rays diffused tints of dirty red and grey. The sea became black, the mountains across the bay iron-grey, while the sky above the latter assumed shades of dirty, ghastly yellow. A few patches of fleecy clouds hanging low over the sea took on the appearance of black cumulus heaps, and afterwards, on the emergence of the sun, donned garbs of varied colours. (Meeson et al. 1886: 377).

According to the accounts published in scientific journals and Government reports, at least twenty-two people obtained scientifically-useful photographs or prepared descriptions or sketches of the event, and these are discussed in Orchiston and Rowe (2016). Many observers were able to address objectives 2, 3 and 4 in the Royal Society's listing (above), and also supply information on the prominences that were present, but their drawings varied considerably (e.g. see Fig. 16.9), highlighting the shortcomings that Pang (2002) has alluded to that relate to this graphical form of representation of the corona. So the various accounts were contradictory, which in most cases limited their scientific usefulness.

Potentially more helpful were the many photographs reportedly taken, but none of these seems to have been published, and the few images that have survived do not show details of coronal structure, which was so sought after. An example is the image shown in Fig. 16.10, which was obtained by British-born James Raglan Akersten (1855–1928) from A.S. Atkinson's Observatory in Nelson.

So the visual and photographic records of the eclipse provided limited scientifically-useful information, and sadly, the one party that planned a spectroscopic assault on the eclipse was clouded out and therefore was unable to make any such observations (Hector 1885; McKerrow 1886). This somewhat disappointing scientific outcome, however, was countered to some degree by the amazing public spectacle and the role the eclipse played in fostering popular interest in astronomy.

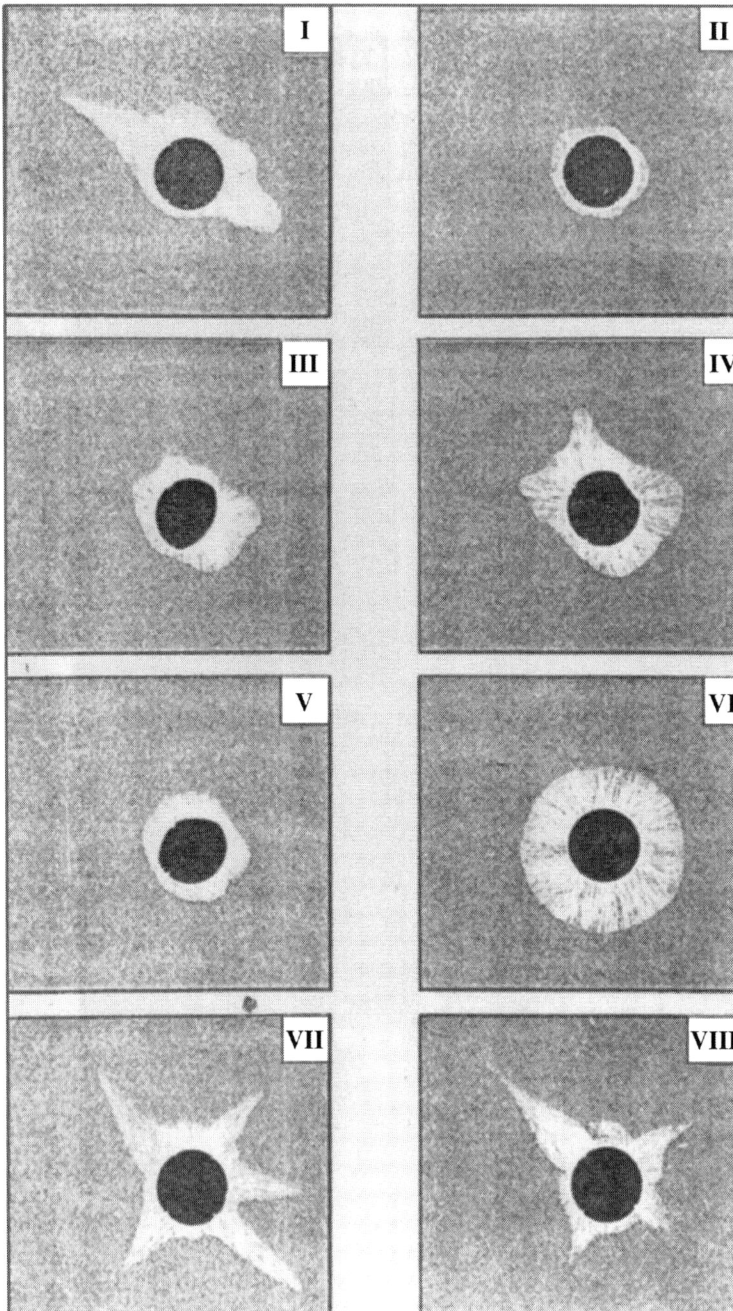

Fig. 16.9 Coronal drawings provided by different observers during the 1885 eclipse (after Meeson et al. 1886: Plate XV)

Fig. 16.10 A photograph of the 1885 solar eclipse taken by J.R. Akersten (*Courtesy* Archives of the Cawthron Institute, Nelson)

16.4 Popular Astronomy and the 1885 Eclipse

16.4.1 Astronomical Spectacles in the Years Leading up to the Eclipse

It is important to understand that by September 1885 astronomy already had a high public profile in New Zealand, even though a mere one hundred and sixteen years had elapsed since scientific astronomy was introduced by James Cook and Charles Green during Cook's First Voyage to the South Seas (see Orchiston 1998 and Chap. 4 in this book).

By 1885 there was a Colonial Observatory in Wellington (see Eiby 1977 and Chap. 9 in this book) where New Zealand's first professional astronomer (albeit a part-time one), Archdeacon Arthur Stock (Fig. 9.4) maintained a local time service. As we have seen, Stock had a passion for astronomy that went far beyond his official calling, and he maintained a private observatory where he carried out a wide range of astronomical observations (Orchiston 1985). He also was committed to popularizing astronomy (Orchiston 1986), and in 1874 took advantage of the transit of Venus to produce two little books (Stock 1874a, b), the first on astronomy ever written by a New Zealander.

Meanwhile, Christchurch had its own champion of astronomy, British-born Professor Alexander William Bickerton (Fig. 16.11; Burdon 1956), who arrived in 1874 as the first Professor of Chemistry at Canterbury University College. McIntosh (1970: 104) has described how Bickerton "… burst like a flaming meteor into New Zealand astronomy." The appearance of Nova Cygni in 1876 awakened

Fig. 16.11 A *Canterbury Times* photograph of Professor Alexander William Bickerton by Hemus Sarony (*Courtesy* Canterbury Museum, Bishop collection, ref 1923.53.500)

his dormant interest in astronomy, and two years later he published two research papers outlining his 'Partial Impact Theory' (Bickerton 1878a, b). He believe that the close approach of two stars could explain the formation of novae, other variable stars, double stars, star clusters and even comets and entire solar systems (Gilmore 1982)! Unfortunately, Bickerton did not provide a mathematical explication of his 'omnibus' theory, so it was not accepted by the world's professional astronomers, but at a local level this did not matter. Bickerton was a charismatic speaker and had no trouble winning over the public with his ever-popular public lectures and demonstrations; consequently, they were enamoured with astronomy.

In cities and towns across the country (see Fig. 16.8) amateur astronomers of note were beginning to make their presence felt (see Mackrell 1985; Orchiston 1998). Auckland had John Torrens Stevenson (Fig. 17.17) and Thomas Cheeseman (Fig. 14.30); in Thames there was John Grigg (Fig. 10.1); Palmerston North had Captain J.D. Hewitt; in Wellington there also was Thomas King (Fig. 9.8); Nelson had Arthur Samuel Atkinson (Fig. 14.31); Christchurch had James Townsend; and in Dunedin were Arthur Beverly (Fig. 14.34), Robert Gillies and Henry Skey (Fig. 14.35). Most of these men liked to promote astronomy through their local newspapers and by delivering public lectures, and some ran their observatories as de facto 'city observatories', offering a public information service in astronomy (and often meteorology) and public viewing nights. Through this network of notable amateur astronomers, knowledge of astronomy reached a much wider cross-section of the public than the few New Zealand professionals employed in astronomy and surveying alone were able to achieve.

Another which played a key part in promoting astronomy nationally was the famous British astronomer and author, Richard Anthony Proctor (1837–1888; Fig. 16.12), who toured New Zealand in 1880–1882 delivering popular public lectures (e.g. see Gibbs n.d.). One of those who attended a Proctor lecture in 1882 when he was just a schoolboy of 14 was Martin Maxwell Fleming Luckie (1868–1951) who would later play a key role in the founding and development of the Carter Observatory, and "… such was the impression made upon him that it continued throughout his life. Little did Proctor realize the seed he had sown." (Martin Maxwell … 1951: 23). Indeed, Proctor could be described as the 'Patrick Moore of the 1880s'. He was not only a gifted, entertaining public speaker, but he also penned a seemingly endless succession of books on different aspects of astronomy, from star charts, to popular works such as *Handbook of the Stars* (first edition 1866), *The Expanse of Heaven* (1873), *The Universe of Suns* (1874), *Our Place Among Infinities* (1875) and *Mysteries of Time and Space* (1883), as well as specialized tomes like *Saturn and its System* (1865), *The Sun* (1871) and *The Moon*

Fig. 16.12 Richard Proctor (https://en.wikipedia.org)

(1877). Through his books and lectures, Proctor reached a wide audience in New Zealand.

The early 1880s also offered a wealth of impressive naked eye comets, and these could not fail to attract the attention of even the most disinterested of individuals. These were: the Great Comets of 1880 (C/1880 C1; Fig. 16.13), 1881 (C/1881 K1; Fig. 16.14) and 1882 (C/1882 R1; Fig. 22.16), we well as two other notable naked eye comets, Schaeberle (C/1881 N1) and Wells (C/1882 F1) in 1881 and 1882 respectively.

In addition to this amazing cometary opulence there were two transits of Venus, in 1874 and 1882, both of which were visible from New Zealand. The 9 December 1874 event

> … promised to be the most important astronomical event in New Zealand since the European settlement of the country, and it attracted enormous public attention. Not only did the Government plan transit observations, but amateur astronomers throughout the nation readied themselves for this once-in-a-lifetime event which promised views of the entire transit and all four contacts … (Orchiston 2004: 261).

As we saw in Chap. 14, international parties from England, France, Germany and the USA also were based in the South Island of New Zealand or on surrounding islands to the east and to the south, but despite extensive preparations, most of New Zealand was clouded out on the vital day and only some of those in Auckland, the British at Burnham and the Americans at Queenstown (see Fig. 16.8 for localities) in the South Island and the German and American parties based on the Auckland Islands and Chatham Islands, respectively, saw the transit.

Fig. 16.13 The Great Comet of 1880, C/1880 C1 (Orchiston collection)

Fig. 16.14 The Great Comet of 1881, C/1881 K1 (Tebbutt) (*Courtesy* Williams College)

The focus then shifted to the all-important 1882 transit, which was the last chance for astronomers from around the world to use these transits to refine the value of the Astronomical Unit (see Dick et al. 1998). As we have already seen, even though only a part of the transit would be visible from New Zealand, on this occasion the weather was more cooperative and local observers in Auckland, Thames, New Plymouth, Martinborough, Wellington, Nelson, Christchurch, Dunedin and Clyde, plus transit teams from Britain and the United States stationed at Burnham and Auckland respectively, all saw the transit. So it proved to be a widely-viewed public spectacle and—hopefully—a portend for the much anticipated total solar eclipse of 1885.

16.4.2 The Media and the 1885 Eclipse

Given the majesty of the Great Comets of 1880, 1881 and 1882 and the 1882 transit of Venus, there also was a heightened public awareness of solar eclipses in the lead up to the 1885 event, and this is nowhere better reflected than in the following account of the partial solar eclipse seen throughout New Zealand on 11 November 1882. In Wellington,

> The eclipse of the sun, which occurred between 10 a.m. and noon to-day, attracted almost as many interested spectators in the streets as did the apparition of the comet in broad daylight some weeks ago. All along the streets amateur astronomers, *pro tem.*, might be seen staring fixedly up into the sky, some having their eyes protected with dark spectacles, some using pieces of coloured glass, others—the great majority—fragments of glass more or less efficiently smoked … As for the eclipse itself, it was of course a striking spectacle … (The solar eclipse 1882).

What is perhaps remarkable is that this high degree of interest was shown in an eclipse where only about half of the solar disk was obscured at mid-eclipse.

Six months later, on 7 May 1883, New Zealand once more was treated to a partial solar eclipse, and again there was widespread public interest in this event. However, it only was successfully observed in New Plymouth and between Castlepoint and Whangarei (Fig. 16.8), with Wellington and other towns and cities further south clouded out (The eclipse of the Sun 1883).

A little over three months later, New Zealand's leading scientist, Dr (later Sir) James Hector (Fig. 14.28), used the *Evening Post* newspaper to alert Wellingtonians to an article that had appeared in the 28 June 1883 issue of the British scientific journal *Nature*, containing information on the next major solar eclipse (Hector 1883). This would occur on 9 September 1885 and would be visible (weather permitting) across central New Zealand. The public therefore had more than two years in which to anticipate and prepare for this grand event, but it was only in mid-1885, less than three months before the eclipse that they were fed details of where and when the total eclipse would be seen. Thus, on 4 July 1885 the *Manawatu Standard* published the following account:

> In answer to a correspondent in last weeks *Otago Witness*, Mr Beverly supplies the following information with reference to the coming solar eclipse:— The solar eclipse next September will be of considerable magnitude all over New Zealand, but will be total in the vicinity of Cook Strait only. The central line passes near Seaford, Otaki, and Tainui; the northern limit near Wanganui, and Waipukurau, and the southern limit near Foxhill and Blenheim. Outside these limits the eclipse will be partial, and the proportion of the sun's disc unobscured will depend on the remoteness of the place from the limits of totality. (*Manawatu Standard* 1885a).

In Wellington, on 26 August 1885 Archdeacon Arthur Stock (Fig. 9.4) from the Colonial Observatory delivered a lecture on the approaching eclipse (*Manawatu Standard* 1885b), while in Napier T.D. Harding (1885) provided information about the eclipse in the *Hawkes Bay Herald*, in response to queries from readers. He explained the differences between solar and lunar eclipses, total and partial solar eclipses and then specifically related this information to the 9 September eclipse:

> … for an extent of about 80 miles N. and S. of a line passing through Castlepoint and Kapiti Island on the West Coast (say from the south of Waipawa as the northern limit) the eclipse of the 9th proximo will be visible in its totality, and it will be well worth the trouble of all who would know by experience what a total solar eclipse really is to visit some spot south of that place.

Harding noted that the nearest point to Napier on the path of totality that was accessible by rail was Tahoraite (see Fig. 16.15), and he suggested that the railways should run a special train from Napier (which they subsequently agreed to do). Finally, he noted that for those wishing to consult them, "At my office, Hastings-street, are exhibited three diagrams of this eclipse, as drawn by the Rev. J. Pearson, of Fleetwood, Lancashire, England, showing the phenomenon as it appears at Napier, Castlepoint, and Wellington respectively." (ibid.). Not to be outdone, three days later Dr. William Isaac Spencer (1885) provided readers of the *Hawkes Bay Herald* with further information on solar eclipses, and quoted distinguished overseas solar experts like Professor Joseph Norman Lockyer from the Royal College of Science in London and Professor Charles Augustus Young (1834–1908) from Dartmouth College in the USA. At the time Spencer (1831–1897) was the Mayor of Napier. Originally Spencer

> … was an army surgeon, attached to the 18th Royal Irish Regiment, and came to New Zealand with the troops that took part in the second Maori war. Subsequently, on account of his eminent qualities as a surgeon and physician, he was induced by the people of Napier to resign his commission, and to practise his profession in their town. Dr. Spencer was also a diligent scientist, who kept well abreast of the times in all matters of science, and who gave occasional lectures on scientific subjects. (Ex-Mayors 1908).

Wellington lay within the path of totality, and on 4 September 1885—just five days before the eclipse—Dr Hector (Fig. 14.28) delivered a lecture at the Petone Institute, which was reported in detail in the *Evening Post* Newspaper. In reflecting on Wellington's hilly terrain (see Fig. 16.16) and the fact that the eclipse would

Fig. 16.15 Tahoraite was a small relatively insignificant railway station on the main line between Napier and Wellington. This photograph was taken in 1912, long after the eclipse (https://en.wikipedia.org)

Fig. 16.16 A chromolithograph by William Potts (1859–1947) showing the city of Wellington in 1885 (*Courtesy* Alexander Turnbull Library, Reference Number: C-060-005)

occur between 7 and 8 in the morning with the Sun low in the eastern sky, Hector provided advice on the best places to observe from and what to expect:

> The best point to observe from is any rising ground or hill-top that has a steep slope to the N.E. The Flagstaff hill, the top of the Botanic Gardens or the hill above Wadestown, are all most favourable points for observation … at 7.35 of New Zealand meantime, the total phase will commence and last for about 1 minute 38 seconds. The sun at this time will have an altitude of 18 degrees and will bear about N.E. of the observer. The chief interest of the eclipse will lie in the observation of the scarlet prominences and of the silvery light of the corona or halo that surrounds the sun during the period of total darkness. The prominences may be expected to have great brilliancy, as for some weeks past the spots on the sun have shown that its surface is in a state of violent activity, and one of such unusual size as to be almost visible to the naked eye, will have reached such a position that it will coincide with the left hand edge of the sun at the time of the eclipse, and from that point unusually large flames should be looked for. (The solar eclipse 1885a).

In Nelson a Mr. T. Scott (1885) decided the best way to forewarn local residents of what to expect was to describe the spectacle provided by the 7 August 1869 total solar eclipse, which was widely viewed in the USA (see Cottam et al. (2011b) and Cottam and Orchiston (2014) for details).

All of those supplying newspaper information to the public were very aware that permanent eye damage could occur if the eclipse was not observed correctly, and they therefore provided advice on safe observing methods.

The Eclipse Committee of the Royal Society in London prepared guidelines for those who would like to try and make a scientifically-useful observation, and these were published by some newspapers:

> The following suggestions have accordingly been drawn up in the hope that they may be of service to those who are able to observe the total solar eclipse which is to be visible in New Zealand on the 9th of September.
>
> Drawings of the corona, as seen with the naked eye. Structure of the corona, as seen with the telescope.— Drawings of the corona have only seldom proved to be of great utility. If such drawings are attempted on the present occasion observers ought to pay attention to the general outline of the corona rather than to points of detail …
>
> Those who observe the eclipse with telescopes will not be able to determine well the outline and extent of the corona, but will be able to examine its structure. Any special structure observed should be described, and perhaps drawn … The observer should also try if he can see structure in the corona before and after totality …
>
> General observations.— Observers unpractised in accurate drawing will obtain more useful results by paying attention to certain features of the corona, than by attempting what can only be a very rough and inaccurate sketch of the corona. Definite answers to the following questions, for instance, would be of great value: —
>
> A: To what distance from the sun, estimated in solar diameters, can you trace the corona?
> B: Does it extend further in some directions than others, and what are the directions of greatest and least extent?
> C: Is there a line of approximate symmetry in the corona, and what is the direction of that line? (Dobson 1885).

Dr Hector pointed out that, ideally, those wishing to make such observations needed to use pieces of glass with different tints, and that the naked eye or field glasses were best employed rather than a telescope:

> For observing these flames [the prominences] it is necessary, in order to intensify their light, to use a fragment of rose-tinted glass. On the other hand, to observe the light of the corona to perfection, a very pale blue-tinted glass is necessary, so as to cut off the red light and intensify the pale silvery light of the corona. A telescope of high power is quite unsuitable to the observer, but a wide field opera glass will be useful. (The solar eclipse 1885a).

Meanwhile, an article in *the Nelson Evening Mail* reminded observers "… not to attempt to use a telescope or field glass, &c., unless properly protected by colored glass." (The total eclipse of the Sun 1885).

As we have seen, Napier lay north of the path of totality, and we learn from a report in the *Daily Telegraph* newspaper the day before the eclipse that

> The 3 p.m. train this afternoon left town crowded with those curious to see the solar eclipse to-morrow morning at Tahoraite, the nearest available place from here in the line of totality. Among the passengers we noticed fifty-two boys of the High School under the charge of the headmaster Mr Heath. (*Daily Telegraph* 1885a).

I am bound to say that school excursions were never so exciting in my day! Napier Boys' High School was founded in 1873 and at the time of the eclipse was located on the hills immediately north of the city known as 'Scinde Island'. Mr Neil Heath was the School's second headmaster, arriving in 1884 from

Auckland (www.nbhs.school.nz/) where he had been the Headmaster of the Auckland Girls' High School. From all accounts, this was his sole astronomical adventure.

Meanwhile, Napier's *Daily Telegraph* newspaper also offered a last-minute option for those who missed the train to Tahoraite but still wanted to see a total solar eclipse on the morrow:

> Scientifically disposed persons who do not suffer from *mal de mer* have a splendid opportunity afforded them of seeing the eclipse of the sun to-morrow morning in the s.s. Kiki, which vessel leaves here this evening, and will be off Castle Point to-morrow at sunrise, which will be almost in the exact line of totality. (*Daily Telegraph* 1885a).

Castle Rock, a distinctive headland near Castlepoint on the Wairarapa Coast, is shown in Fig. 16.17, and we can see from this newspaper article that 'eclipse cruises'—so popular these days—are not an entirely recent phenomenon. Unfortunately, there is no record as to how many people decided to chance their luck and sail on the *S.S. Kiki*, but perhaps later some wished that they had for from all accounts not all of those who decided to remain in Napier and view a partial eclipse were overjoyed:

> The solar eclipse was witnessed at Napier by a large number of persons. A number of these, having formed somewhat exaggerated notions of the phenomenon through reading or hearing of the more startling circumstances attending a *total* eclipse, professed to be, and doubtless were, disappointed at what they saw. The sight was one well worth seeing, however, especially shortly before, after, and at the period when the climax of the eclipse was seen. A kind of steel-tinted twilight obtained at this time, the sun being all but totally covered, and only showing a thin crescent equal to two per cent. of the whole disk. It seemed strange, indeed, with only such a small portion of the sun unobscured, that the darkness was not much greater. (Napier 1885).

Fig. 16.17 A view of Castle Rock near the town of Castlepoint (https://nzfrenzynorth.files.wordpress.com/2013/01/img_1581.jpg)

By a lucky coincidence, just after the eclipse ended those in Napier felt "… a slight but well-defined shock of earthquake …" (ibid.), so they alone among New Zealanders were privileged to witness a rare astronomical event and experience an all-too-familiar seismic one at pretty much the same time. Meanwhile, "Mr W.H. Neal, photographer, of Tennyson street, took a series of very successful views of the solar eclipse … When completed, these photographs will be extremely interesting, and should command a large sale." (*Daily Telegraph* 1885b). William Henry Neil ran a photography business in Napier from 1887 to 1896, and it is much to be regretted that his photographs and very few of the others that were taken during the eclipse can no longer be traced (see Orchiston and Rowe 2016).

For those towns that lay within the path of totality the weather was, on the whole, cooperative. For example, in Dannevirke a correspondent for the *Daily Telegraph* newspaper reported the sky was clear at the time and everyone saw a 90 second duration total solar eclipse. However, a cold wind dampened spirits, and an hour after the eclipse there was a light fall of snow. Nonetheless, "The hotels in Danevirke [*sic*] were crowded." (Eclipse of the Sun 1885).

Meanwhile, those who journeyed from Napier to Tahoraite by train were rewarded with views of the eclipse, and Mr Neil Heath (presumably the head of the school party) "… was able to secure a photograph at the moment of totality. Mr. S. Carnell has prepared from this plate several enlarged copies, and these are well worth inspection. The photograph shows the corona as an immense halo of irregular form." (*Hawkes Bay Herald* 1885b). We can identify 'Mr. S. Carnel' as Samuel Carnell (1832–1920; Fig. 16.18), yet another Napier professional

Fig. 16.18 Mr S. Carnell when he was the Mayor of Napier in 1904–1905 (after Napier Corporation 1908)

photographer. Carnell was born in England and learnt about photography before emigrating to Auckland, New Zealand, in 1862. After working in studios there and in Nelson he moved to Napier in 1869 and established his own studio. Later Carnell would enter politics and serve as the Napier Liberal Party member of Parliament; he also would enjoy two terms as the Mayor of Napier (Napier Corporation 1908).

Nearby at Woodville "… a magnificent view of the eclipse was obtained … The sky was perfectly clear, and the duration of totality one minute and forty-two seconds. The corona and red flame-like prominences were very brilliant and beautiful." (Eclipse of the Sun 1885).

Further south was Masterton, which was closer to the centre line of the path of totality and was where the scientist congregated prior to the eclipse with the intention of dispersing to a number of nearby observing sites:

> Dr Hector, Messrs McKerrow, Adams, Humphries, Stock, Travers, Beverly, and others are now in Masterton. The preparations for observing the sun's eclipse are now complete. Messrs McKerrow, Beverley, and Humphreys go to Otahamo. The two former take observations from the trig-station on the top, and the latter photographs about 400 feet lower down. Dr Hector and two other gentlemen proceed to Rangitumaki, a higher hill and more difficult of access, but nearer the centre line. Archdeacon Stock goes to Branespeth, Messrs Beethams' station, and Mr Travers will take photographs from Masterton. (The solar eclipse 1885b).

On the 9th there was a heavy southwest gale with rain during the night and the morning broke without any sign that this would change. Nonetheless,

> Mr McKerrow and party, who had camped at the foot of Otahuao, proceeded to the top and fixed their instruments amid driving snow and hail. Just before totality the sky cleared and all the phenomena were fairly visible. One photo was taken before totality, three during, and one after. The corona was distinctly visible for a full minute, an encircling ring of light radiating to a distance of about half the diameter of the sun. It was of a pale white color like the electric light and uniform width except at the sun's equator where it slightly protruded … (Eclipse of the Sun 1885).

Further details are provided later in the same newspaper article:

> Just at the time of the first contact the clouds parted, and the phenomenon known as Bailey's Beads was clearly observed. Mr Humphries, who was stationed 400 feet below the summit, secured some good photographs, using a five-inch lens. When totality set in the clouds again parted and red protuberances were very prominent. Mr Humphries obtained three good photographs at this stage. Jupiter and Mercury were plainly visible near the sun, as well as fixed stars here and there. Dr Hector, accompanied by Mr R. Gore and Mr Chapman, took observations from Dryertown on the centre line, having abandoned Rangituomai owing to the hill being covered with clouds. They had fair views at intervals. *It is expected that the observations will not prove of great scientific value*, but the photographs are considered splendid. (ibid.; my italics).

However, an article published in the *Poverty Bay Herald* has a slightly different account of Hector's exploits: "Dr Hector proceeded *to Dryerton, the centre line of the eclipse, on top of Rangitumau Hill*, and got fair observations." (The solar eclipse 1885c; my italics).

To the west of the Tararua Mountain Range in Feilding local residents were forewarned about the 9 September eclipse through a long article that was published in the *Feilding Star* the previous day. Most of the information was supplied by Dr Hector, who not only provided details of what to expect but also commented on the reaction the eclipse might elicit:

> In the lower animals a total eclipse of the sunlight of even a few seconds, excites fear and distress. In the savage it awakens awe, to the ignorant it lends force to their superstitions, but to the trained intelligence it is an occasion for the keenest exercise of the perceptive faculties … [totality] will last 1 min 58 sec. A short interval certainly, but yet one quite sufficient to display many phenomena of striking interest … If the morning is fine and clear the scene will be impressive beyond description. (The eclipse 1885a).

On 10 September Mr. Goodbehere (1885) reported favourable conditions for observation of the eclipse. As the Moon

> … slowly travelled across the sun's disc the varied crescent shaped phases of the sun, as seen through a small telescope, together with the gradual darkening and weird-like appearance of the landscape, were exceedingly beautiful and interesting. The grandest and most impressive part of the spectacle was, however, at the moment of totality, which was notified to observers by a feeling somewhat like an electric shock, when the beautiful white slivery light of the corona surrounding the moon (supposed to be the sun's atmosphere) suddenly appeared, together with the long streamers or rays of light called aigrettes, shining through the corona and extending to a considerable distance beyond it, which are always observed in total solar eclipses, but the nature of which are not at present thoroughly understood by astronomers. The end of totality also was most striking and impressive (perhaps to some observers more so than the commencement), when the first speck of light from the sun shot forth with electrifying brilliancy … during the totality … numerous stars were seen by different observers, as also were the planets Jupiter, Venus, and Mercury, which were close to the sun.

He also noted (ibid.) that "The only regret, if any, to the observers was the shortness of the duration of totality (about 90 seconds) and the spectacle would probably have been more impressive if it had taken place later in the day." This account was presumably written by the solicitor Samuel Goodbehere (1819–1899; Fig. 16.19), who regularly published weather reports in the *Feilding Star* and served one term as Mayor of Feilding and two terms as a Borough Councillor (Ex-Mayors 1897), rather than his son Edmund Goodbehere (1855–1938), who also lived in Feilding and worked as a land, estate, commission, insurance and general agent and a valuator (Professional. Commercial, and Industrial 1897).

Those in nearby Palmerston North also saw the entire eclipse:

> The solar eclipse this morning was watched with great interest … the sky being almost cloudless at the time, thus affording an excellent opportunity of watching the progress of the phenomenon. By 7.30 the eclipse was total, all Nature being darkened. In about two minutes thereafter the first bright gleam of Sol (we mean old Sol, of course) made its reappearance … The sky was particularly favourable to observation, and we learnt that Mr Shailer obtained some capital camera views. The Natives, of whom a large number watched the eclipse with much interest, and apparently a certain amount of superstitious apprehension, gave vent to their feelings by cheers when the first bright glimpse of the sun's rays reappeared. (Solar eclipse 1885).

Fig. 16.19 Samuel Goodbehere (*Courtesy* Feilding Public Library)

Wanganui also lay just within the band of totality and the *Wanganui Chronicle* reported that a total eclipse was witnessed:

> A large proportion of the adult population of Wanganui and a considerable number of children availed themselves of the opportunity yesterday morning of viewing the total solar eclipse from Cook's Gardens and other elevated positions, some with smoked pieces of glass and others with telescopes. The beginning of the eclipse could not be seen for a bank of clouds in the horizon, but at ten minutes to seven the sun shone through the bank and revealed about one-twentieth of his disc obscured by the moon … About 25½ minutes to eight an effect took place so sublime that it was almost unexpected. The whole aspect of heaven and earth underwent a change with regard to light and shade. Round the black sun was an irregular halo of coloured light, defining clearly and strongly the obscured orb. In some places this halo extended into longer gleams, especially in the upper side. The rapid diminution of light at the moment of total obscuration was sudden and startling, and a great many of the stars appeared, one in particular, close to the darkened sun. For about a minute only was the sun obscured, when suddenly as quick as lightning a ray of light shot from the left lower side of the moon like a very brilliant star, and at the same instant the halo disappeared. (The eclipse of the Sun 1885b).

A short interim account appeared in the other local newspaper, the *Wanganui Herald*, on the afternoon of eclipse day. Again we are told the Sun climbed above clouds along the eastern horizon and a total eclipse was viewed by all. According to this account totality last for about 1m 28s (The eclipse of the Sun 1885c), significantly longer than stated in the *Wanganui Chronicle* account.

Further south, those in the nation's capital, Wellington, who took Dr Hector's advice and climbed the hill above Wadestown or ventured to the upper Botanic Garden initially must have been disappointed as 9 September dawned cloudy, but then it began to clear and the eclipse finally came into view. Thus,

> By the time the total phase was reached the sun was sufficiently clear of clouds to give an uninterrupted view, and as totality was reached the scene was most impressive. As the darkness increased the western heavens became illuminated with a deep orange color shading off into the most delicate of yellows. A number of stars were plainly seen during the darkness. After about a minute and a half the sun again shone out and gradually increased. (The eclipse of the Sun 1885a).

Wellington's favourable view of the eclipse is also reported, in more detail, in the *Hawkes Bay Herald* newspaper. The Sun rose already partly eclipsed, then as totality approached

> The black disc of the moon became of a sudden clearly visible, with a very faint white light rounded off the sun. Quickly following this came the most impressive, if not the grandest, part of the phenomenon. Travelling quickly across the land and sea from the north-east came the dark shadow of totaltay [*sic*]. In ten or fifteen seconds the full light of day, which prevailed while only a fiftieth part of the sun's disc was visible, darkened to the obscurity of a moonless night, and no one could avoid experiencing a sensation of awe as the black mantle fell upon them, but awe changed to admiration at the beautiful flood of silver light which burst forth round the moon was the sun's last golden spark was extinguished. Not a cloud now disturbed the sight, which was seen in its full brilliancy. As the darkness approached the western heavens presented a singularly beautiful spectacle, the sky being a deep orange color, shading off into the most delicate tints of yellow towards the zenith. The rays of the corona could be traced to between two or three diameters distance. Next to the corona the red spots or flames were the most interesting objects of observation. One very large red prominence was conspicuous. Just to the right, and above the longest ray of the corona, two bright spots [small prominences] were also seen, almost diametrically opposite to the large one. During totality the stars shone out as at night. (Wellington 1885).

The duration of totality lasted from 7h 35m 4s local time to 7h 36m 46s, or 1 minute and 42 seconds (ibid.).

Across Cook Strait at the top of the South Island, a total solar eclipse was seen in Havelock, Nelson and adjacent towns, and in Collingwood, while Blenheim was right on the southern limit of the path of totality (see Fig. 16.7).

On 9 September, Blenheim's *Marlborough Express* reported that Havelock in the Marlborough Sounds experienced a fine morning and the eclipse was seen, but provided no details (The eclipse 1885b).

Nelson had a long association with scientific astronomy, starting with a series of public lectures on astronomy that the surveyor Charles Heaphy (1822–1881) gave at the Richmond Mechanics Institute in 1847 (Gibbs n.d.), so there was great interest in the eclipse. On 9 September 1885 a clear day dawned, and

> By six o'clock scores of people were to be seen climbing the hills around town, and the numbers increased as time drew on. Just before 7 o'clock the view from any of the hilltops was very impressive, but from that time was increasingly so. In the clear atmosphere every

feature in the town stood out distinctly, and the hills beyond with the snow capped ranges from Mount Arthur northwards, presented a picture scarcely to be rivalled. Upon the hills around the town were to be seen groups of watchers; upon the summit of the Zig zag Hill there were a great many congregated, and so there were upon the Port hills in various places, on the hills near "Long Look Out," and at the back of Mr Fell's establishment …

[From 7 am] … a feeling of uncannyness extended and affected all living nature; birds were no longer to be heard, and silence, broken only by the lowing of cows as the period of totality arrived, became oppressive, whilst all the strange surroundings imparted a feeling of awe that will not speedily be forgotten … [Immediately before totality] The light now caused all vegetation to assume unnatural colors, and the faces of observers became almost ghastly in appearance. In a second the glare from this small bright spot had ceased, and around the dark face of the moon was seen the corona, the light from which was silvery white. From this halo of glory extended rays of light similar in color, and at the instant when the sun's light was wholly obscured the planet Jupiter shone forth in all its brightness from just below the sun, while another planet was discernable within the corona, and overhead stars became visible. In the meantime strange shadows were creeping over the earth, and it seemed as though glass, faintly spotted and lined, was being passed before a strong lantern by an unsteady hand, whilst the quivering shadows made it appear as though the whole surface of the country was creeping. The view westerly, meanwhile, was extraordinary. At first a rose colored glow, deepening into a dull red, was visible over the sea, and then the atmosphere assumed an indigo tint, whilst as the shadow passed over the Mount Arthur range, under the strange light it seemed to alter its shape, and a dull leaden hue superseded the glistening of the snow. On looking back towards the sun, rays appeared to be projected beyond the corona, and it seemed as though a great commotion was going on there. All too quickly the period of totality passed away; a bright point of light, as from a diamond of wonderful brilliance, or from a particularly bright morning star, shot forth from the upper surface of the moon … [and] speedily extended as the moon's shadow passed downwards and to the right, and the majority of observers were soon hastening homewards, only disappointed that the period of totality had not been longer. (The eclipse of the Sun 1885a).

The extreme cold noted during totality was mentioned in this article: a thermometer that had registered 50° before the eclipse plunged to 30° immediately after totality. This newspaper article concluded by noting: "We in Nelson were particularly fortunate in being able to observe this wonderfully grand phenomenon under such favorable circumstances …" (ibid.). This long and detailed account appeared in the *Colonist*, while Nelson's other newspaper, the *Nelson Evening Mail*, provided a similar, though much shorter, report. Just before totality

… over the knots of spectators that were gathered here and there on the hilltops there came a hush as the moment approached when the wonderful sight they were all so anxious to witness was to be revealed to their eager gaze … [and after the event] Who among the thousands that witnessed the sublime scene can ever forget its unspeakable grandeur? … in the heavens was that glorious corona of light, a thing of such marvellous beauty as to be beyond the conception of those who have never witnessed the wondrous spectacle. (Total eclipse of the Sun 1885).

The *Marlborough Express* newspaper also reported that in Nelson some "Successful photos were taken." (The eclipse 1885b).

At nearby Wakefield "The total eclipse of the sun was witnessed here this morning by the whole populace, under the most favorable conditions possible, and developed the greatest astronomical enthusiasm." The only regret was the short period of totality, as Wakefield lay just within the path of totality, "However, the most was made of the privilege …" (Total eclipse of the Sun 1885).

At Motueka on the western side of Tasman Bay across from Nelson "The eclipse was beautifully visible here this morning … the atmosphere was perfectly clear. Totality lasted a little over a minute …" during which prominences were seen. "The sight was magnificent. Immediately before and after totality brilliant waves of light appeared to be flashing all over the ground presenting a most remarkable phenomenon." (ibid.).

Collingwood in Golden Bay, also was well within the path of totality and graced with clear skies, so

> A magnificent view of the eclipse was obtained here … the morning being beautifully fine and not a cloud visible … At the total phase the sight was grand beyond expression … A better observation could not possibly have been obtained … It is a matter for regret that some of our astronomers did not establish an observatory in Collingwood. (ibid.).

Of all those living in the extreme northern part of the South Island the inhabitants of Blenheim were perhaps the most unlucky, for their town was on the very southern limit of the path of totality. But this did not seem to dampen the spirits of local residents:

> Blenheim was favored this morning with a magnificently clear atmosphere and cloudless sky … [and] the wondrous spectacle was observed by hundreds of people in this district … Many observers took up positions at an early hour on the Opawa Bridge. The cricket ground, the road to Grovetown, and other places in the North Ward, which being within the belt of totality were considered to be favorable spots. A better morning for observation could not have been imagined. (The eclipse 1885b).

This long article, published by the local *Marlborough Express* newspaper, mentioned that "… totality … lasted but a bare moment …" as Blenheim was on the very edge of the path of totality. Nonetheless, "The corona and sun's flames [prominences] were plainly visible, and formed a spectacle which no mechanical contrivance can imitate and no art can reproduce." (ibid.). The article then continues:

> The eclipse was certainly a wonderful phenomenon, and almost as interesting to the non-scientific observer as to the scientist who viewed it through the powerful lens of his observatory. There were many attendant phenomena which were singular and novel. Just before the moment of totality lines of rippling shadows moved across the ground from east to west in a manner we never saw before, and just after the totality the shadows seemed to ripple back again along the ground in exactly an opposite direction.
>
> Observations of the eclipse were taken in the Cricket Ground by Mr Dobson C.E., and ten instantaneous photographs were secured by Mr W.H. Macey, the two gentlemen acting in conjunction. Mr Dobson's observations were made by the telescope and theodolite, the powerful telescope belonging to Mr Cullen of Mahikipawa … Both observation and photographing were interfered with by the spectators who crowded round, although requested to stand a little further away. We have since seen some of Mr Macey's plates, after being developed, and they contain valuable records of the eclipse at different stages. It seems very doubtful if the eclipse was absolutely total from the cricket ground … [and] Blenheim could not have been within the belt of totality. Had the observations been taken further north they would have been more fortunate. (ibid.).

The "Mr Dobdon C.E." referred to in this quote was almost certainly the London-born surveyor and civil engineer Alfred Dobson (1824–1887) of Blenheim rather than his son, Ernest Douglas Dobson (1863–1938), who also lived in Blenheim and also was a civil engineer. Alfred Dobson emigrated to New Zealand in 1851 and was Chief Surveyor in Nelson before accepting the post of Chief Surveyor and Provincial Engineer of Marlborough (*Otago Daily Times* 1887). Meanwhile, "Mr W.H. Macey" was William Henry Macey (1850–1931; Fig. 16.20) who also was born in London, came to New Zealand in 1857 and learnt photography from William Collie of Blenheim. He went into partnership with Collie in 1870, and four years later purchased the business outright. Macey also owned a bookshop. Later Macey would serve on the Blenheim Borough Council and two terms as Mayor of Blenheim (Macey n.d.).

Elsewhere to the north and south of the path of totality the eclipse was only visible as a partial event, and it was seen clearly in Ashburton, Timaru and Waimate, where there was a "... splendid uninterrupted view ..." (The eclipse 1885b), and in Kaikoura, Greymouth, Kumara and Oamaru in the South Island, while clouds only allowed occasional observations in Christchurch, Dunedin and Invercargill and prevented any observations being made from Lyttelton (see Fig. 16.8 for South and North Island localities). In the North Island there were clear skies in Napier, Gisborne, Tauranga, and New Plymouth, while clouds in Auckland, Hamilton and Cambridge limited the observations that could be made (*Grey River Argus* 1885; The eclipse 1885c, d; The eclipse of the Sun 1885a, d; The solar eclipse 1885b, d).

Fig. 16.20 Mr. W.H. Macey (after William Henry Macey n.d.)

In the weeks following the eclipse several interesting newspaper articles appeared. For instance, on 15 September, less than a week after the eclipse, the *Hawkes Bay Herald* (1885a) reported on a meeting of the Hawkes Bay Philosophical Institute which was held the previous evening and drew a "… crowded attendance including several ladies." After addresses by the Chairman, the distinguished botanist the Reverend William Colenso, and the biologist and ethnologist Augustus Hamilton,

> Mr Goodall then read an interesting paper containing his notes on the recent solar eclipse as witnessed at Tahoraite. Similar addresses were delivered by Messrs Brydon, Harding, Tanner, and Macdonald. The remarks of the first four speakers were illustrated by diagrams prepared by themselves, the drawings giving a very lucid idea of the wonderful phenomena attendant upon a total solar eclipse of the sun. The remarks of each of the speakers were intensely interesting, and at the conclusion of the addresses the president expressed his grateful appreciation of the intellectual treat presented to the meeting. (ibid.).

Presumably, "Mr Goodall" was John Goodall (1839–1905), the Napier Harbour Board Engineer who designed the harbor at Napier (Bell et al. 1884). It is interesting that the Mayor, Dr W.I. Spencer, was at the meeting (ibid.), and although he had published information about the eclipse in the *Hawkes Bay Herald* on 4 September, he did not make a presentation on this occasion. Was he one of those who remained in Napier on the vital day and was less than impressed with the view of the partial eclipse?

One day later, on 16 September, the *Nelson Evening Mail* noted that a Wellington newspaper had published a report on Archdeacon Arthur Stock who had just returned to Wellington from the Wairarapa without observing the eclipse, thanks to unfavorable weather. Stock was based at the Beetham's famous Brancepeth Station, which in a few short years would become one of the largest sheep stations in New Zealand (see Winter and Hedley 2012). While the Editor of the *Evening Mail* professes to sympathize with Stock he says

> … we cannot help thinking that a certain amount of stupid obstinacy was displayed by those who selected stormy Wellington in preference to sunny Nelson as the spot whence to take their observations. From any part of Blind Bay [Tasman Bay] Archdeacon Stock could have obtained a magnificent view, and, what is more, he might have calculated upon doing so almost with certainty. It is very much to be regretted that he preferred Wairarapa. (*Nelson Evening Mail* 1885).

There is more than a touch of venom in these comments, which really were unwarranted. Compared to Wellington, the Wairarapa is renowned for its fine weather and clear skies, and we should note that Stock was one of very few astronomers who ventured there and came back 'empty-handed'. As we have seen, the great majority had favourable views of the eclipse.

16.5 Concluding Remarks

Thanks to clear skies, the 9 September 1885 total solar eclipse was a spectacular event that was widely observed across much of the band of totality, and especially by those residing in Nelson, Wellington and much of the Wairarapa. Naked eye observations were published in newspapers and in the *Transactions of the New Zealand Institute*, and many successful photographs were taken. While useful information on the form and extent of the corona was obtained it was the non-scientific element of this eclipse that should be emphasized. It was an amazing public spectacle and allowed many thousands of ordinary New Zealanders to witness an unforgettable and once-in-a-lifetime event.

From this perspective the New Zealand eclipse was a resounding success, and by following as it did 'hot on the heels' of the 1880, 1881 and 1882 Great Comets and the 1882 transit of Venus, it also served as a catalyst for the future development of astronomy in New Zealand, and the involvement of New Zealand amateur and professional astronomers in local and overseas solar eclipse expeditions during the twentieth century (e.g. see Bateson 1957, 1965; Michie 1938).

Acknowledgements I am grateful to the following for their assistance: the late Robert Campbell (Oamaru, NZ), Wayne Hammond (Chapin Library, Williams College, USA), the late Peter Hingley (former RAS Librarian, London), Lynne Huddlestone (Cawthron Institute, Nelson, NZ), Jan Meeus (Belgium), Joanna Szczepanski (Canterbury Museum, Christchurch, NZ), the late Jackie St George (Sydney) and staff at the Alexander Turnbull Library (Wellington, NZ). I also wish to thank the Alexander Turnbull Library, Canterbury Museum, Cawthron Institute, Feilding Public Library (NZ), and Williams College for kindly supplying Figs. 16.10, 16.11, 16.14, 16.16 and 16.19.

References

An earlier eclipse. *Southern Stars, 2*, 158–160 (1936).

Atkinson, A. S. (1885). On the total solar eclipse of September 9, 1885 (in a letter to J.N. Lockyer, F.R.S.). *Proceedings of the Royal Society of London, 39*, 211–212.

Bateson, F. M. (1957). Eclipse islands of the South Pacific. *Sky and Telescope, 16*, 573–575.

Bateson, F. M. (1965). Eclipse expedition on Manuae Atoll. *Sky and Telescope,* August, 78–79.

Bell, C. N., Scott, W. H., & Goodall, J. (1884). *Report on Mr John Goodall's Design for a Harbor at Napier*. Napier: Napier Harbour Board.

Best, E. (1955). *The Astronomical Knowledge of the Maori*. Wellington: Government Printer (Dominion Museum Monograph No. 3).

Bickerton, A. W. (1878a). On temporary and variable stars. *Transactions of the New Zealand Institute, 11*, 118–124.

Bickerton, A. W. (1878b). Partial impact: a possible explanation of the origin of the solar system, comets, and other phenomena of the universe. *Transactions of the New Zealand Institute, 11*, 125–132.

Burdon, R. M. (1956). *Scholar Errant. A Biography of Professor A.W. Bickerton*. Christchurch: Pegasus.

Clerke, A. (1893). *A Popular History of Astronomy During the Nineteenth Century*. London: Adam and Charles Black.

Cottam, S., & Orchiston, W. (2014). *Eclipses, Transits and Comets of the Nineteenth Century: How Americas' Perception of the Skies Changed*. New York: Springer.

Cottam, S., Orchiston, W., & Stephenson, R. (2011a). The 1874 transit of Venus and the popularization of astronomy in the U.S.A. as reflected in *The New York Times*. In W. Orchiston, T. Nakamura & R. Strom, R. (Eds.), *Highlighting the History of Astronomy in the Asia-Pacific Region* (pp. 225–241). New York: Springer.

Cottam, S., Orchiston, W., & Stephenson, R. (2012). The 1882 transit of Venus and the popularization of astronomy in the U.S.A. as reflected in *The New York Times*. *Journal of Astronomical History and Heritage, 15*, 183–199.

Cottam, S., Pearson, J., Orchiston, W., & Stephenson, R. (2011b). The total solar eclipses of 7 August 1869 and 29 July 1878 and the popularization of astronomy in the U.S.A. as reflected in *The New York Times*. In W. Orchiston, T. Nakamura, & R. Strom (Eds.), *Highlighting the History of Astronomy in the Asia-Pacific Region* (pp. 339–375). New York: Springer.

Daily Telegraph, September 8, 1885a (p. 3).

Daily Telegraph, September 9, 1885b (p. 2).

Dick, S. J., Orchiston, W., & Love, T. (1998). Simon Newcomb, William Harkness and the nineteenth century American transit of Venus expeditions. *Journal for the History of Astronomy, 29*, 221–255.

Dobson, A. (1885). The eclipse. How to observe it. *Marlborough Express*, September 7 (p. 3).

Eclipse of the Sun. *Daily Telegraph*, September 9, 1885 (p. 3).

Eiby, G. (1977). The New Zealand Government time-service: an informal history. *Southern Stars, 27*, 15–34.

Ex-Mayors. In *The Cyclopedia of New Zealand [Wellington Provincial District]* (p. 1222). Wellington: The Cyclopedia Company (1897).

Ex-Mayors. In *The Cyclopedia of New Zealand [Taranaki, Hawke's Bay & Wellington Provincial Districts]* (p. 312). Christchurch: The Cyclopedia Company (1908).

Gibbs, F. G. (n.d.) Notes on the history of astronomy in Nelson. MS, Nelson Provincial Museum (MSS 289).

Gilmore, G. (1982). Alexander William Bickerton: New Zealand's colourful astronomer. *Southern Stars, 29*, 87–108.

Golub, L., & Pasachoff, J. M. (1999). *The Solar Corona*. Cambridge: Cambridge University Press.

Grey River Argus, September 10, 1885 (p. 2).

Harding, T. D. (1885). The solar eclipse. *Hawkes Bay Herald*, September 1 (p. 3).

Hawkes Bay Herald, September 15, 1885a (p. 2).

Hawkes Bay Herald, October 10, 1885b (p. 2).

Hector, J. (1883). The next great total solar eclipse. *Evening Post*, August 22 (p. 3).

Hector, J. (1885). On the total solar eclipse of September 9, 1885 (in a letter to Professor Stokes, Sec. R.S.). *Proceedings of the Royal Society of London, 39*, 208–211.

Hind, J. R. (1885). Note on the track of the total phase on the solar eclipse of 1885, Sept. 8, in its passage across New Zealand. *Monthly Notices of the Royal Astronomical Society, 45*, 126–127.

Kragh, H. (2009). The solar element: a reconsideration of helium's early history. *Annals of Science, 66*, 158–175.

Launay, F. (2012). *The Astronomer Jules Janssen: A Globetrotter of Celestial Physics*. New York: Springer.

Lockyer, J. N. (1874). *Contributions to Solar Physics*. London: MacMillan and Co.

Mackrell, B. (1985). *Halley's Comet over New Zealand*. Auckland: Reid-Methuen.

McIntosh, R. A. (1970). Early New Zealand astronomy. *Southern Stars, 23*, 101–108.

McKerrow, J. (1886). *Surveys of New Zealand (Report for 1885–86)*. Wellington: Government Printer.

Macey, William Henry. In Early New Zealand Photographers and their Successors (n.d.).

Manawatu Standard, July 4, 1885a (p. 2).

Manawatu Standard, August 26, 1885b (p. 2).

Martin Maxwell Fleming Luckie O.B.E. [Obituary]. *Southern Stars, 15*(2), 21–24 (1951).
Meadows, A. J. (1972). *Science and Controversy. A Biography of Sir Norman Lockyer*. Cambridge (Mass.): MIT Press.
Meeson, J., Suter, Right Reverend Dr., Goodall, J., Atkinson, A. S., Hudson, Dr., & Coleman, A. (1886). The total eclipse of the Sun of the 9th September, 1885 … *Transactions and Proceedings of the New Zealand Institute, 18*, 375–394.
Michie, C. B. (1938). Report of the New Zealand total solar eclipse expedition to Canton Island, June 8, 1937. *Monthly Notices of the Royal Astronomical Society, 99*, 132–135.
Mr. Goodbehere, 1885. The solar eclipse. *Feilding Star*, September 10, 1885 (p. 3).
Napier. *Hawkes Bay Herald*, September 10, 1885 (p. 3).
Napier Corporation. In *The Cyclopedia of New Zealand [Taranaki, Hawke's Bay & Wellington Provincial Districts]* (1908, p. 306). Christchurch: The Cyclopedia Company.
Nath, B. B. (2013). *The Story of Helium and the Birth of Astrophysics*. New York: Springer.
Nelson Evening Mail. September 16, 1885 (p. 2).
Orchiston, W. (1985). A survey of manuscript material in the Mitchell Library (Sydney) relating to the history of astronomy in New Zealand. *Southern Stars, 31*, 225–238.
Orchiston, W. (1986). Taking Stock of early astronomy books. *Southern Stars, 31*, 326–327.
Orchiston, W. (1998). *Nautical Astronomy in New Zealand. The Voyages of James Cook*. Wellington: Carter Observatory.
Orchiston, W. (2000). A Polynesian astronomical perspective: the Maori of New Zealand. In H. Selin & X. Sun (Eds.), *Astronomy Across Cultures: The History of Non-Western Astronomy* (pp. 161–196). Dordrecht: Kluwer.
Orchiston, W. (2004). The nineteenth century transits of Venus: an Australian and New Zealand overview. *Journal of Astronomical Data, 10*, 219–308.
Orchiston, W., Chen, K.-Y., Lee, E.-H., & Ahn, Y.-S. (2006). British observations of the 1868 total solar eclipse from Guntoor, India. In K.-Y. Chen, W. Orchiston, B. Soonthornthum, & R. Strom (Eds.), *Proceedings of the Fifth International Conference on Oriental Astronomy* (pp. 23–34). Chiang Mai: Chiang Mai University.
Orchiston, W., & Rowe, G. (2016). New Zealand astronomy and the total solar eclipse of 1885. In T. Nakamura, & W. Orchiston (Eds.), *The Emergence of Astrophysics in Asia: Opening a New Window on the Universe*. New York: Springer.
Orchiston, W., & Soonthornthum, B. (2016). French observations of the 18 August 1868 total solar eclipse from Wah-koa, Thailand. In S. Nakamura & W. Orchiston (Eds.), *The Emergence of Astrophysics in Asia: Opening a New Window on the Universe*. New York: Springer.
Otago Daily Times. September 7, 1887.
Pang, A. S.-K. (2002). *Empire and the Sun. Victorian Solar Eclipse Expeditions*. Stanford: Standford University Press.
Pearson, J. (2009). *The Role of the 40 Foot Schaeberle Camera in the Lick Observatory Investigations of the Solar Corona*. Unpublished Ph.D. Thesis, James Cook University.
Professional, Commercial, and Industrial. In *The Cyclopedia of New Zealand [Wellington Provincial District]* (1897, pp. 1233–1234). Wellington: The Cyclopedia Company.
Ranyard, A. C. (1879). Observations made during total solar eclipses. *Memoirs of the Royal Astronomical Society, 41*.
Scott, T. (1885). A total eclipse of the Sun. *Nelson Evening Mail*, September 8 (p. 2).
Solar eclipse. *Manawatu Standard*, September 9, 1885 (p. 2).
Spencer, W. I. (1885). Total eclipse of the sun. *Hawkes Bay Herald*, September 4 (p. 3).
Stock, Archdeacon [Arthur] (1874a). *Astronomy for Beginners Who Live in the Southern Hemisphere*. Wellington: Bowden.
Stock, Archdeacon [Arthur] (1874b). *December 9, 1874. The Transit of Venus, and How to Observe It*. Wellington: Bowden.
Tennant, J. F. (1869). Report on the total eclipse of the sun, August 17–18, 1868 … *Memoirs of the Royal Astronomical Society, 37*, 1–55.
The eclipse. *Feilding Star*, September 8, 1885a (p. 3).
The eclipse. *Marlborough Express*, September 9, 1885b (p. 2).

The eclipse. *Bay of Plenty Times*, September 10, 1885c (p. 2).
The eclipse. *Timaru Herald*, September 10, 1885d (p. 3).
The eclipse of the Sun. *Marlborough Express*, May 7, 1883 (p. 2).
The eclipse of the Sun. *Colonist*, September 10, 1885a (p. 1).
The eclipse of the Sun. *Wanganui Chronicle*, September 10, 1885b (p. 2).
The eclipse of the Sun. *Wanganui Herald*, September 9, 1885c (p. 2).
The eclipse of the Sun. *Oamaru Mail*, September 9, 1885d (p. 3).
The solar eclipse. *Evening Post*, November 11, 1882.
The solar eclipse. *Evening Post*, September 7, 1885a (p. 3).
The solar eclipse. *Poverty Bay Herald*, September 9, 1885b (p. 2).
The solar eclipse. *Poverty Bay Herald*, September 10, 1885c (p. 2).
The solar eclipse. *Waikato Times*, September 10, 1885d (p. 2).
The total eclipse of the Sun. *Colonist*, September 8, 1885 (p. 3).
Total eclipse of the Sun. *Nelson Evening Mail*, September 9, 1885 (p. 2).
Wellington. *Hawkes Bay Herald*, September 10, 1885 (p. 3).
Winter, G., & Hedley, A. (2012). *In the Boar's Path: Brancepeth A Journey to the Heart of a Pastoral Kingdom*. Masterton, Hedley's Bookshop.
Zirker, J. B. (1984). *Total Eclipses of the Sun*. New York: Van Nostrand Reinhold.

Chapter 17
John Grigg, and the Genesis of Cometary Astronomy in New Zealand

Abstract John Grigg was New Zealand's leading amateur astronomer during the first decade of the twentieth century, and independently discovered four comets, three of which now bear his name. In addition, he pioneered astronomical photography in New Zealand, and also applied this technology to comets. Grigg is best remembered internationally for his discovery of Comet 26P/Grigg-Skjellerup, which has one of the shortest periods of any known comet. In this chapter Grigg's cometary work is discussed, along with other New Zealand astronomers who also discovered comets.

17.1 Introduction

The second half of the nineteenth century was an important period for New Zealand astronomy, with the establishment of the first professional observatory and the emergence of a small number of notable amateur astronomers (McIntosh 1970; Mackrell 1985; Orchiston 1998), spurred on largely by an impressive succession of naked eye comets, two transits of Mercury, two transits of Venus, and a total solar eclipse, all potentially visible from New Zealand between 1874 and 1887.

By the turn of the century arguably New Zealand's premier amateur astronomer was John Grigg (Orchiston 1998, 2001b; see also Chaps. 10 and 22 in this book) who lived in the thriving Coramandel Peninsula mining town of Thames (see Fig. 17.1 for localities mentioned in the text). In addition to operating a successful furnishing and music store, Grigg found time to observe known comets and search successfully for new comets.

Biographical details of Grigg (Fig. 17.2) and information on his telescopes and two observatories (e.g. see Fig. 17.3) have already been provided in Chap. 10 and need not detain us further here. Instead, in this chapter we review Grigg's achievements in cometary astronomy before mentioning two other New Zealand astronomers who observed comets during the nineteenth century and the five other

W. Orchiston, *Exploring the History of New Zealand Astronomy*,
Astrophysics and Space Science Library 422,
DOI 10.1007/978-3-319-22566-1_17

Fig. 17.1 New Zealand localities mentioned in this chapter are shown in *red*

Fig. 17.2 Another photograph of John Grigg (Orchiston collection)

Fig. 17.3 A view showing part of Grigg's first observatory, in Pollen Street, and the rear of his business premises (after Grigg 1970)

New Zealand resident astronomers who subsequently added their names to the annals of astronomical history by discovering new comets.[1]

17.2 Comets

17.2.1 Introduction

Grigg's main interest was in comets, both the systematic observation of known ones and the search for and tracking of new ones. In both endeavours he was successful, but since his observing books have not survived information is limited on his observations of known comets. Data on his comet discoveries are preserved in contemporary journals and in letters he and others wrote to the noted Australian cometary astronomer John Tebbutt (see Orchiston 1985b, 2004a).

17.2.2 Discoveries

Although it is stated in the *Family History* (Grigg 1970) that Grigg began systematic comet searches in 1894, a letter he wrote to Tebbutt in 1906 reveals otherwise. After describing his observations of C/1886 X1 (Fabry) in May-June 1886, Grigg writes:

> My little observatory had been erected about a year and a half and I had not then decided what branch of astronomy I would take up. The pleasure I found in following a moving object and mapping its path, decided me to begin comet seeking … (Grigg 1906d).

And in August 1902 Grigg informed Tebbutt that "I have been searching regularly for comets for the past 15 years …" (Grigg 1902b). Grigg therefore initiated his comet-search program in 1886 or 1887.

According to his daughter, Grigg's usual procedure was to conduct morning searches, typically from 2.00 to 5.00 a.m. (McIntosh 1958: 23), using the 3.5-inch telescope, but in a 1902 letter to Tebbutt he mentions "… my monthly search of the evening sky …" (Grigg 1902b). In order to evaluate any 'suspect' object that came into view Grigg used a working list of nebulae (Grigg 1902d), which he had assembled from his own observations between declination −20° and −90° (Grigg 1902b) and from the very limited astronomical literature at his disposal. In a 1903 letter Grigg

[1]Although Hughes (1991), Mackrell (1985: 75–79), McIntosh (1958, 1970); Orchiston (1985b, 1998) and Rumsey (1974) all provide brief accounts of Grigg, the first detailed study of his contribution to cometary astronomy was published by Orchiston in 1993. This chapter is an up-dated version of that paper.

listed these as the *Nautical Almanac*, Loomis' 1500 catalogue, and Proctor's Atlas (Grigg 1903j).[2]

While Grigg recovered 2P/Encke in 1898 (see "Observations of Other Comets", below), his first discovery of a new comet occurred on 22 July 1902, during systematic comet-sweeping, when he detected what was later to become known as Comet 26P/Grigg-Skjellerup. Grigg described the comet at the time as "… like a faint nebula about 3 or 4 times the diameter of Jupiter."[3] Further observations were made on 23, 26 and 29 July and on 1 and 2 August (Grigg 1902d)[4] as the comet moved through Virgo, but by 2 August he noted that "… its increasing faintness seemed to have reached the limit of my optical means …" (Grigg 1902b).

On each of the above evenings Grigg attempted to determine the position of the newcomer by "… placing the comet centrally in the field, noting the time, reading the circles of his equatorial, and then doing the same for the nearest "Nautical Almanac" star available." (*Journal of the British Astronomical Association* 1907). Not unexpectedly, such a procedure produced very crude results, and Grigg (1903g) was sorely aware of this: "… I regret that with my best efforts I fail to obtain close positions." Nonetheless, he was able to use the values obtained for 23, 26 and 29 July to calculate the comet's orbital elements, and these were published in *Astronomische Nachrichten* (Grigg 1902d).

John Grigg was the only astronomer to observe this new comet. Because it was magnitude 9.5 (Hughes 1991: 2) at the time (close to the limiting magnitude of his telescope) and bad weather intervened, Grigg had to wait until 26 July before he had proof this was indeed a comet. Only then did he notify the Press Association, and write to one of Sydney's leading amateur astronomers, Walter Gale (1865–1945: Fig. 17.4; Wood 1981) and Melbourne Observatory Director, Pietro Baracchi (Fig. 17.5) of the discovery, but as we have seen the positions he was able to provide were imprecise.

By the time his letter arrived, Gale (1902) was unable to locate the comet, but he did comment on the abundance of nebulae in the general vicinity: Virgo was certainly one of the worst possible constellations in which to search for a faint new comet, position unknown! For his part, when Baracchi received his letter he was in a quandary. At that time Melbourne Observatory was the designated Australian and New Zealand centre for the dissemination of data on newly-discovered comets, but what should he do? His solution was to seek the advice of Australia's leading

---

[2]Loomis' 1500 catalogue is the "Catalogue of 1500 Stars" in Loomis (1861: 400–459). Proctor's "Atlas" was presumably one of the various editions produced of *A Star Atlas for the Library, The School, and The Observatory*, London, Longmans and Green (see Luther 1989: 198).

[3]This account appears on the bottom of an untitled sheet of paper in the 'Tebbutt Letters' in the Mitchell Library, Sydney, containing Grigg's orbital computations. Grigg sent this to Tebbutt with his letter of 12 August 1902 (Grigg 1902b).

[4]Hughes (1991) has problems reconciling the "… six very rough observations … made between 1902 July 22 and 1902 August 3 …" with Kronk's (1984: 255) claim that Grigg observed the comet on 14 occasions. Grigg clarifies this in his 12 August 1902 letter to Tebbutt: "I recorded 14 obsn. of the comet (also comparison stars) on 6 nights till Aug 3 …" (Grigg 1902b).

Fig. 17.4 Walter Gale (*Courtesy* British Astronomical Association Archives)

Fig. 17.5 Pietro Baraccho (*Courtesy* Museum Victoria)

astronomer and cometary expert, John Tebbutt (Fig. 17.6), who maintained a well-equipped private observatory at Windsor, near Sydney (Orchiston 2001a, 2004b):

Fig. 17.6 John Tebbutt (after Tebbutt 1908: Frontispiece)

> I feel hardly justified in sending the announcement of the Discovery of this Comet to Kiel owing to the vague character of the information given by the discoverer. I wish to have your advice as to what course I should take in order to be fair to all concerned. (Baracchi 1902a).

Tebbutt felt that Dr Heinrich Carl Freidrich Kreutz (1854–1907; Fig. 17.7) of Kiel should be notified, but when Baracchi received a further letter from Grigg, including right ascension and declination positions for 3 August 1902, and he too could not locate the comet (see *Monthly Notices of the Royal Astronomical Society*, 242, 1903), he decided instead to simply write Kreutz (Baracchi 1902b). By this

Fig. 17.7 Dr H.C.F. Kreutz (https://en.wikipedia.org)

time the comet was fading rapidly, and there is no evidence that the larger instruments in the observatories at Adelaide, Perth, Sydney, Wanganui or Windsor were involved in the search. Often the Australian professional observatories left it to Tebbutt to follow up on newly-discovered comets, but on this occasion "… there were not sufficient data for finding so faint an object, and [as] I was closely engaged in minor planet work, I did not make a search." (Tebbutt 1903c: 11).

Grigg's comet thus was lost to science, that is until a fortuitous discovery made from South Africa by the Australian cometary astronomer, John Francis (Frank) Skjellerup (Fig. 17.8), on 17 May 1922 (*Monthly Notices of the Royal Astronomical Society,* 288, 1923). This new comet was followed for four months, and subsequently Dr Gerald Merton (1893–1983; Dewhirst 1996) noted that its orbital elements were almost identical to those of the 1902 comet. In this way periodic comet 26P/Grigg-Skjellerup came into being (Merton 1927). At just 5.1 years, this has one of the shortest periods of any known periodic comet (see Marsden 1989), and it is now known to be associated with the Pi Puppids meteor shower.

Grigg continued his search program, and discovered his second comet, C/1903 H1 (Grigg) on 17 April 1903 in Eridanus (Grigg 1903e; Vsekhsvyatskii 1964).[5] He observed the comet on a further 29 nights, through to and including 26 May (Grigg 1903c, d, j), noting on 24 May that it was "… extremely faint though still large …" in his small telescope (Grigg 1903i).

This time Grigg made certain that others would have the chance to observe the comet by cabling details of his discovery to Baracchi on 19 April and following up with letters to Baracchi, Tebbutt and Gale on the very same day (Grigg 1903g). Tebbutt found the comet on 25 April, and was able to obtain a long series of precise micrometric positions, through to 28 May (Tebbutt 1903b, 1904: 9; cf. *Monthly Notices of the Royal Astronomical Society* 1904). The Sydney mathematical astronomer and cometary expert, Charles James Merfield (Fig. 18.9) and Keil's Dr Kreutz calculated the comet's orbital elements (Tebbutt 1903a) using Tebbutt's observations of 27 and 30 April and 2 May (see Baracchi 1903) and the elements proved to be very similar, but Grigg noticed that there was a considerable difference between these and his own elements, computed from his observations of 22, 25 and 28 April (Grigg 1903f, 1903h; Merfield 1903). To Merfield (ibid.) the reason was obvious: "I fancy his positions are only rough ones, his method of observing, as explained in his letter, would lead to very large errors in his results …" Grigg (1903d) recognized this, admitting that "… the majority of the observations have

[5]In Orchiston (1985b: 233) I incorrectly stated that Grigg discovered a second comet in 1902, bringing his tally to five, not four. A letter dated 2 May in the 1902 volume of *Letters to J. Tebbutt* in the Mitchell Library (Sydney) documents the discovery on 17 April of a new comet, and its subsequent observation (Grigg 1902a). I originally interpreted this as an independent discovery by Grigg of C/1902 G1 (Brooks) just two days after the initial discovery, but a subsequent re-analysis revealed that the letter in question dates from 1903 and relates to C/1903 H1 (Grigg). For some reason this letter was misbound in the 1902 volume of *Letters to John Tebbutt*. The *Family History* (Grigg 1970), which contains a far from accurate account of Grigg's cometary discoveries, also states (on p. 28) that he found a second comet in 1902.

Fig. 17.8 Frank Skjellerup and his wife photographed in their garden in Cape Town about the time he re-discovered Grigg's comet (Orchiston collection)

been repeated once or twice on both comet and star, but the method, instruments, and observer, are evidently imperfect." The Sydney amateur astronomer, T.W. Close, also calculated orbital elements, using positions provided by Gale, Grigg and Tebbutt, and on plotting out all of the various observations found Grigg's to be the least reliable (Wright 1903).

On 2 May 1903 Baracchi (1903) had cabled the announcement of the discovery to Kiel, giving positions for 17 April (after Grigg) and 27 and 28 April (both from Tebbutt). Kreutz was then able to distribute an ephemeris of the comet to non-Australian and non-New Zealand observatories. As a result, Comet Grigg 1903 III was observed through to 28 May (*Monthly Notices of the Royal Astronomical Society* 1904).

After the comet became too faint to observe Grigg sent Edward Walter Maunder (1851–1928) of the British Astronomical Association a full set of his observations, including drawings of the comet relative to different stars in the eyepiece field (Grigg 1903j), but this material was never published.

Buoyed on by his success, Grigg continued his regular comet-searching, and on 13 May 1904 wrote Tebbutt reporting a new discovery only to withdraw this claim the following day (Grigg 1904a, b): his 'discovery' was a false alarm!

Further successes only came in 1906, but on both occasions Grigg was destined for disappointment. On 10 February 1906 he discovered what he thought was his third comet (Grigg 1906b), only to learn that this was in fact Comet/1905 X1 which the French astronomer Michel Giacobini from Nice had first detected on 8 December 1905 (Marsden and Williams 1999). Given the long interval between the original discovery and Grigg's detection, we can hardly claim this as an independent discovery for Grigg. Were it not for the problems that Australia and New Zealand sometimes encountered when it came to the international communication of new cometary discoveries, Antipodean comet observers would already have been aware of this comet long before Grigg made his 'discovery'.

Then on 19 March, a little over one month later, Grigg (1906a) discovered a new comet in Cetus and issued a report to the local newspaper on the 20th. The following day he forwarded details of his discovery (including two positions) to Tebbutt, noting "Possibly this is the comet reported from Australia as having been found in "Sculptor" but of which no positions were given." (Grigg 1906c). A further position was obtained on 27 March (Grigg 1906d), following which Grigg found that his suspicions were correct and that one of Melbourne's leading amateur astronomers, David Ross (Fig. 17.9), had a legitimate claim to the prior discovery of this comet (see Reports of the Observing Sections … 1906). The *Family History* (Grigg 1970) does not mention this independent discovery.

However, through Ross' inability to obtain precise positions and Baracchi's indifference, the comet was almost awarded to Grigg by default. Sydney amateur astronomer, Hugh Wright (1906), explains these circumstances:

> He [Ross] said that he discovered the comet on Feb. 14, but only rough positions were obtained, & Baracchi would not accept rough positions, nor would he search. It was only after much delay & worrying that B. said he would give Ross 3 minutes with the telescope

Fig. 17.9 David Ross (Orchiston collection)

> to pick up the comet. He failed, the following night was cloudy, & on the following night Ross showed Baracchi the object. A position was taken, & later announced; but none too soon as Grigg of N.Z. independently detected it the following night. So officialism nearly robbed a discoverer of his credit!

Note that Vsekhsvyatskii (1964: 366) incorrectly lists 18 March as the discovery date.

Grigg made his final independent comet discovery on 8 April 1907 when he detected a faint rapidly-moving "... 15′ diameter ... Nebulous mass varying in distribution of brightness, but no distinct nucleus or tail." (Kreutz 1907; cf. *Astronomische Nachrichten* 1907; Grigg 1907a). Six days later this comet was independently discovered by the American astronomer, John Edward Mellish (Fig. 17.10), and it is now known as C/1907 G1 (Grigg-Mellish). However, once it was reported by Mellish, an image of the comet was found on a photographic plate taken by Edward Emerson Barnard (1857–1923; Fig. 17.11; Sheehan, 1995) of Yerkes Observatory (while searching for C/1907 E1 (Giacobini) one night before Mellish's independent discovery (*Monthly Notices of the Royal Astronomical Society* 1908).

The same evening he detected the comet, Grigg sent a report to the local newspaper office, and provided further details the next evening (Grigg 1907a). He also notified his close friend, Joseph Thomas Ward (Fig. 11.4) in Wanganui. Grigg's report to the Thames newspapers reached other New Zealand newspapers through the Press Association, and Grigg assumed (erroneously, as it turned out) that following usual practice they would also be communicated to the leading

Fig. 17.10 J.E. Mellish in about 1910 (https://en.wikipedia.org)

Fig. 17.11 E.E. Barnard (https://en.wikipedia.org)

Australian newspapers. As 'a matter of form', therefore, on 10 April 1907 he sent Melbourne Observatory the following simple cable: "Baracchi Melbourne, comet northeast alpha columbae Grigg." The next day Grigg sent both Pietro Baracchi and Sydney Observatory Director Henry Alfred Lenehan (1843–1908; Fig. 17.12; Wood, 1958) details of his observations to date, but what he did not do was contact John Tebbutt. Subsequently, Tebbutt commented on this in a letter published in *The Observatory* (Tebbutt 1907), which prompted Grigg to point out that

> … had I been a wealthy enthusiast, I might have sent full particulars by cable to several of my correspondents in the Commonwealth – and – possibly have been rewarded as I was the previous year by finding, after the lapse of six months, that the honour of the discovery had been awarded to an Australian gentleman for one days precedence. (Grigg 1907c; his underlining).

Not only was cometary astronomy an expensive hobby if one lacked limitless means, but Grigg was obviously still riled by the unfortunate 'Ross Comet' episode of 1906. Meanwhile, once the new comet was too faint for further observation Grigg forwarded copies of all of his records to Merfield, who by now was working at Sydney Observatory (see Orchiston 2015).

Grigg reported observations that he made on 8, 9, 10 and 11 April, and also observations by two other New Zealand amateur astronomers, J.T. Ward on 10, 11 and 12 April, and a single observation by Wellington's George Vernon Hudson (Fig. 17.13) on 11 April. At the time, Hudson provided this description of the comet: "It is diffuse, with no defined tail or nucleus. Its substance is extremely attenuated, the faintest stars shining through the mist of it." (see *Journal of the British Astronomical Association* 1907). Bad weather prevented Grigg from observing the comet on the 12th, and by the time clear skies returned the comet was

Fig. 17.12 H.A. Lenehan (adapted from Russell 1892)

too far north for observations from New Zealand (Grigg 1907c). Numerous observations of C/1907 G1 (Grigg-Mellish) were then made from the Northern Hemisphere, the last one on 14 May 1907 (ibid.).

Given the exceedingly vague position contained in the cable, Baracchi felt quite unable to publicise the discovery (if, indeed, that was what this cryptic communication indicated!), and it was only after Merfield had computed the orbital elements (from the positions sent him by Grigg) that Baracchi was able to cable these to Kiel (Barrachi 1907; Merfield 1907a). Grigg (1907b) was far from impressed: "… I feel very much disappointed with Mr Baracchi …" but Tebbutt (1907) was equally peeved with Grigg:

> … I cannot understand … why the discoverer himself, who is a correspondent of mine and who knows that nearly all the comet observations in Australia for many years past have devolved upon me, ignores the existence of my Observatory.

In this instance, Grigg chose not to write to Tebbutt as he understood that Tebbutt had already officially retired from observational astronomy (see Tebbutt 1908).

Baracchi's behaviour on this occasion proved to be the last in a series of incidents involving Melbourne Observatory which led to a successful petition by the

Fig. 17.13 G.V. Hudson in 1907 (https://en.wikipedia.org)

New South Wales Branch of the British Astronomical Association for the transfer of the Australasian Astronomical Information Centre to Sydney Observatory (Merfield 1907b, c; see also Orchiston 1999b). Baracchi finally was forced to pay for that air of indifference which Grigg first witnessed when he discovered his 1902 comet, but this is not to say that Grigg was without fault. In a letter to Merfield, Barrachi (1907) writes:

> Your remarks about amateurs who can't determine good positions of the objects they discover are indeed appropriate. I go much further in my appreciation of their services to astronomy. Grigg seems to be a beauty of the same type as D. Ross.[6]

Precisely when Grigg retired from systematic comet-searching is not known, though it would have been sometime between 1908 and 1912, and more likely in 1911 or 1912. In this latter year Gale discovered a comet, and on 31 October Ross (1912) wrote: "It is to be regretted that Mr. J. Grigg, F.R.A.S., of Thames Observatory, New Zealand, has retired from this work, which he so zealously supported." Grigg was 74 years of age at the time.

In the course of his observational 'career' as a cometary astronomer Grigg independently discovered four different comets, three of which now bear his name. The Astronomical Society of the Pacific awarded him Donohoe Medals for his 1902

[6]In Orchiston and Brewer (1990: 177) this quote is incorrectly attributed to Merfield.

Fig. 17.14 Photographs of the obverse (left) and reverse (right) of the Donohoe Medals that Grigg received from the Astronomical Society of the Pacific for his discoveries of Comet P26/Grigg-Skjellerup (1902) and Comet C/1903 H1 respectively (after Grigg 1970)

and 1903 discoveries (Crommelin 1921; Grigg 1970), and these are shown here in Fig. 17.14.

17.2.3 Observations of Other Comets

According to the *Family History*, between 1886 and 1914 Grigg observed every comet visible in his telescope from New Zealand, and in addition to plotting their paths investigated their orbits. In the absence of any observing books we cannot test the first part of this claim, but if true, then Grigg would have observed about 35 different comets,[7] in addition to those he independently discovered. In that case, it is extremely unlikely that he computed orbital elements for all of these, otherwise he would surely have followed international convention and published at least some of them in *Astronomische Nachrichten*, *Journal of the British Astronomical Association* or *Monthly Notices of the Royal Astronomical Society*.

Whatever Grigg's actual tally, we only have documentary evidence that he observed fourteen different comets (including his own four independent discoveries) and these are listed below in Table 17.1.

Grigg's first known telescopic cometary observations were of C/1885 X1 (Fabry) soon after he acquired the Wray telescope and installed this in his new observatory (Mr Grigg's observatory 1885), and came after what must have been impressive naked eye views of the 'Great Comets' of 1880, 1881 and 1882. We know that the thrill of tracking Comet Fabry on 33 different nights between 5 May and 25 June 1886 captivated Grigg (1906d), and converted him to regular cometary astronomy. Despite obtaining a long sequence of positional observations, he did not

[7]This figure was arrived at by consulting Marsden's *Catalogue of Cometary Orbits* (1989), and after allowing for their superior instrumental means also noting comets that actually were observed by David Ross (see Orchiston and Brewer 1990; Tebbutt 1908) during this period.

Table 17.1 Comets known to have been observed by John Grigg 1886–1910

Year	Name of Comet	Reference
1886	C/1885 X1 (Fabry)	Grigg (1906d)
1895	2P/Encke (1895)	Grigg (1970: 27)
1898	2P/Encke (1898: Grigg-Tebbutt)	Vsekhsvyatskii (1964)
1899	C/1899 E1 (Swift)	*Thames Advertiser* (1899)
1901	C/1901 G1 (Great Comet)	Grigg (1970: 27)
1902	26P/Grigg-Skjellerup (1902)	This book, this chapter
	C/1902 R1 (Perrine)	Grigg (1903a, b)
1903	C/1903 A1 (Giacobini)	Grigg (1903c)
	C/1903 H1 (Grigg)	This book, this chapter
1904	2P/Encke (1905)	Grigg (1904c)
1906	C/1906 F1 (Ross)	This book, this chapter
1907	C/1907 G1 (Grigg-Mellish)	This book, this chapter
	C/1907 L2 (Daniel)	*Thames Star* (1907b)
1910	1P/Halley (1910)	Grigg (1970: 27)

communicate these to any astronomical journal. At this time he probably was not aware that right ascensions and declinations obtained by reading the circles were imprecise, and could not be used for the calculation of reliable orbital elements.

Among all the different comets that he observed, 2P/Encke had special significance to Grigg, for it is the only comet known to have been observed by him on at least three different returns. According to the *Family History* (Grigg 1970: 27), late in 1894 when the comet was in northern skies "… a search ephemeris was prepared by John for its southern path, by which it was found in February 1895 …" Observations were carried out on both 24 and 25 February.[8] Backlund (1894) had published an 1894–1895 ephemeris in *Astronomische Nachrichten* the previous year, but Grigg did not have access to this journal. His observations turned out to be the last ones made of the comet on this return (Marsden 1989: 38), but were not published anywhere.

The 1898 return of 2P/Encke was an important event for Grigg, as he and Tebbutt were credited with the recovery (Vsekhsvyatskii 1964: 339). Using his own ephemeris, Grigg detected the comet on 7 June in Gemini, just three degrees above the horizon. He observed it on the following three nights, and used three of the four positions obtained to calculate the orbital elements which he sent to Gale (Gale 1898; *Observatory* 1898). Tebbutt was plagued by cloudy nights, and his independent recovery of the comet came four nights after Grigg's. By 25 June Tebbutt "… was surprised to find that, notwithstanding a beautifully clear sky, the object

[8]Tebbutt observed the comet in December, but because it was "… in northern declination, and on the confines of the evening twilight, was not observed with the micrometer." (Tebbutt 1895: 13), and he was prevented from making post-perihelion observations in 1895 March because of cloudy weather (Tebbutt 1896: 15).

was invisible. On the following evening it was seen as an extremely faint patch of light …" (Tebbutt 1898, 1899: 13–14). Tebbutt was using a 20.3-cm (8-in) Grubb refractor (see Orchiston 2001a) and by this time the comet was well beyond the light-grasp of Grigg's much smaller telescope.

Finally, when Grigg observed 2P/Encke in 1904 he made just one observation (Grigg 1904c), and like most of his observations of other comets this was not published.

Of all of the comets listed in Table 17.1, apart from 1P/Halley the one that created the greatest public interest was undoubtedly the Great Comet of 1901 (C/1901 G1). On 4 May the *Thames Star* reported that Grigg "… was besieged in his observatory, but with his usual courtesy he allowed many to inspect the aerial stranger through his telescope." (The comet 1901).

17.2.4 Comets and Astronomical Photography

Grigg was a pioneering New Zealand astronomical photographer, and obtained successful photographs of sunspots, a solar eclipse (in 1895), the Moon, and comets (Grigg 1902c, 1905b). He is known to have photographed two different comets: on 12 and 16 May 1901 he obtained photographs of the Great Comet of 1901 (C/1901 G1)

Fig. 17.15 A photograph by Grigg of Comet 1P/Halley in 1910 (Orchiston collection)

and one of his exposures is reproduced in Fig. 22.7, then on 13 May 1910 he photographed Comet 1P/Halley (see Fig. 17.15).

Further information on Grigg's involvement in astronomical photography is presented in Orchiston (1995) and in Chap. 22 in this book.

17.3 Discussion

From the number of letters—mainly about comets—that he wrote to Andrew Claude de la Cherois Crommelin (1865–1939) and Edward Maunder in England, Walter Gale, C.J. Merfield and John Tebbutt (e.g. see Table 17.2) in Australia, and others notable overseas astronomers, Grigg obviously relished the astronomical contacts he maintained with overseas colleagues, which partially compensated for the geographical and intellectual isolation of Thames. This would also have been one of the reasons he joined the British Astronomical Association in 1897, just seven years after its foundation. The British cometary expert, Crommelin from the Royal Greenwich Observatory notes that once he was a member of the Association Grigg worked "… with great zeal in the comet section." (Crommelin 1921). In 1906 he was elected a Fellow of the Royal Astronomical Society, after successfully soliciting Tebbutt's support (Grigg 1905a, 1906b), and membership of these two respected societies back 'home' (as it was perceived in those days) must have enhanced Grigg's international astronomical visibility and respectability.

John Grigg could rightly be called the 'founding father' of New Zealand cometary astronomy, but he was not the first astronomer in this country to discover a comet. This honour goes to Archdeacon Arthur Stock (Fig. 9.4), the Astronomical Observer at the Colonial Observatory in Wellington, who independently detected the Great Comet of 1881 just one day after Tebbutt (Stock 1881, 1882). As a nascent discipline, astrophysics was undergoing major development at this time (e.g. see Langford 1984; Meadows 1984), and as a result Comet C/1881 K1 (Tebbutt) proved to be one of the most significant comets of the nineteenth century (Orchiston 1999a).

Another New Zealander who was active in cometary astronomy at the same time as Stock was Auckland's John Torrens Stevenson (Fig. 17.16), and a short paper on his observations of the Great Comet of 1882 (C/1882 R1) was published in *Monthly Notices* (Stevenson 1882).

Table 17.2 Letters from John Grigg to John Tebbutt in the Mitchell Library, Sydney

Year	Number
1902	4
1903	7
1904	5
1905	2
1906	3
1907	3
Total	24

Fig. 17.16 J.T. Stevenson (Orchiston collection)

Comets discovered by New Zealanders *and formally credited to them* are shown in Table 17.3. Note that only six different observers are involved. Apart from Thames' John Grigg, Murray Arthur Geddes (Fig. 17.17; Thomsen 1945; Williams 2007) made his discovery from Otakura, Southland, using a 12.7-cm (5-in) Cooke telescope (see Geddes 1932). Geddes is known mainly for his research on aurorae, and as a prolific meteor observer. He was appointed founding Director of the Carter Observatory, only to die during WWII.

My dear friend the late Dr Albert Francis Arthur Lofley Jones (Fig. 17.18) made his two discoveries from Timaru and Nelson respectively, and holds the record for the longest number of years between two successive discoveries of any known comet observer. He also was the oldest astronomer to discover a comet when he detected C/2000 W1 (Rod Austin, personal communication 2014; John Drummond, personal communication 2014). Whilst comets certainly were one of Albert's interests, his undoubted forté was the visual observation of variable stars, and he achieved international renown for his effort in this area. Thus, he received many medals and other awards for his work, and an honorary doctorate from Victoria University of Wellington. By the time he died in 2013 Albert had reported well in excess of 500,000 individual magnitude estimates, a record which is unsurpassed in variable star history. His remarkable achievements are discussed by Austin (1994), Orchiston (1990), Sharma (2013), amongst others.

British-born Mike Clark (b. 1942; Fig. 17.19) worked as a professional astronomer at Mt John Observatory from 1971 to 1996 (the last sixteen of these as

Table 17.3 Comets discovered by New Zealanders

Comet	Reference
Credited Discoveries	
1902 O1 26P/Grigg-Skjellerup	Orchiston (1993), this book, this chapter
C/1903 H1 (Grigg)	Orchiston (1993), this book, this chapter
C/1907 G1 (Grigg-Mellish)	Orchiston (1993), this book, this chapter
C/1932 M2 (Geddes)	Marsden (1989), Vsekhsvyatskii (1964)
C/1946 P1 (Jones)	Marsden (1989), Vsekhsvyatskii (1964)
1973 L1 71P/Clark	Marsden (1989)
C/1982 M1 (Austin)	Marsden (1989)
C/1984 N1 (Austin)	Marsden (1989)
C/1989 X1 (Austin)	Austin (1990)
C/2000 W1 (Utsunomiya-Jones)	Gilmore and Jones (2001)
P/2007 Q2 (Gilmore)	Gilmore, 2007
Independent Discoveries	
C/1881 K1 (Tebbutt) Stock	Stock (1881, 1882)
C/1906 F1 (Ross) Grigg	This book, this chapter
C/1914 S1 (Campbell) Westland	Orchiston (1983), this book, Chap. 18
C/1923 T1 (Dubiago-Bernard) Morshead	Morshead (1945: 5–6)
C/1927 X1 (Skjellerup-Maristany) Rhind	Glover (1928)
C/1941 B2 (de Kock-Paraskevopoulos) McIntosh	McIntosh (1941), this book, Chap. 19
C/1947 S1 (Bester) Bateson	Bateson (1989: 151)
C/1966 P2 (Barbon) Thomas	Rod Austin, personal communication, 11 December 1991

Fig. 17.17 Murray Geddes (www.rasnz.org.nz/InfoForm/MurrayGeddes.hym)

Fig. 17.18 Albert Jones photographed on 15 June 2012 at age 91 (*Photograph* Wayne Orchiston)

Fig. 17.19 Michael Clark photographed on 16 December 2014 (*Courtesy* Michael Clark)

Superintendent), and discovered his comet in 1973 in the course of a variable star photographic patrol with the 10-cm Bamberg Camera (for details see Kronk and Meyer 2010: 379). Periodic comet 71P/Clark has "… a period of almost exactly 5.5 years. It is normally quite faint, so every 11 years it comes around at about 11th mag." (Clark 2014).[9]

[9]Rodney Austin, whose three comet discoveries are discussed in the next paragraph, notes that

> At that time I was working at Mt John myself, and was off duty that evening - comethunting (!) only about 100 metres away from the building where Mike was busy taking plates for the Bamburg Observatory patrol. He found the comet on inspecting his plates the following day. He started looking VERY closely at his plates after just missing the discovery of the SN in NGC 5253 the previous year. (Shanklin 2014).

Fig. 17.20 Rod Austin and his 15.2-cm f/8 refractor in 1982, a few days after discovering his first comet (*Courtesy* Rod Austin)

As Table 17.3 indicates, Rodney Richard Dacre (Rod) Austin (b. 1945) of New Plymouth discovered three comets during the 1980s while working as a photo-lithographer for Taranaki Newspapers Limited (after leaving a position at the Mt John Observatory). All of these comets were detected as a result of systematic searches, the first two with a portable 15.2-cm (6-in) refractor that was used in two different optical configurations (see Fig. 17.20), and the third with a 20.3-cm (8-in) Meade Schmidt-Newtonian telescope with a customized mounting especially designed for comet-seeking (Rod Austin, personal communication 2014).

Finally, the most recent discovery by a New Zealand astronomer also was made at Mt John Observatory, where Alan Charles Gilmore (b. 1944; Fig. 17.21) was carrying out research mainly on comets and minor planets:

> An apparently asteroidal object, found by Alan Gilmore on CCD images taken with the Mount John University Observatory 1-m reflector on August 22.58, has been shown to have a coma and tail after posting on the NEOCP. The comet was near perihelion at 1.8 AU and has an elliptical orbit with period around 13 years. (Shanklin 2014).

This comet emerged in the course of a Near Earth Asteroid program (Gilmore 2007).

Of the 'Independent Discoveries' in Table 17.3, official inactivity at the Carter Observatory (Wellington) very much reminiscent of Baracchi's day deprived Alan Thomas of the discovery of what should rightfully be known as comet C/1966 P2 (Thomas-Barbon), while current naming conventions, if in vogue from 1914, also would have resulted in C/1914 S1 (Campbell-Westland-Lunt)—see Chap. 18 in this book—and C/1927 X1 (Skjellerup-Rhind). Meanwhile, Archdeacon Arthur Stock should also be given credit for his co-discovery of the Great Comet of 1881 (i.e. C/1881 K1 (Tebbutt-Stock). Unfortunately, the late Dr Frank Bateson never got round to publishing further details of his independent discovery of Comet C/1947 S1 (Bester) which was picked up by the South African astronomer, Michiel John (Mike) Bester (1917–2005; Cooper 2005) from the Boyden Observatory on 28 September 1947 (Marsden and Williams 1999), but since Bateson first detected the comet in the summer of 1947–1948, months after its initial discovery, perhaps it is charitable to regard this as an 'independent discovery'. The same comment applies equally to the detection by Francis Joseph Morshead (1884–1967) of Comet C/1923 T1 (Dubiago-Bernard) long after the initial discovery.

Fig. 17.21 A recent photograph of Pam Kilmartin and Alan Gilmore at Lake Tekapo (*Courtesy* Alan Gilmore)

Fig. 17.22 Bill Bradfield (right) and his comet-seeker, with Paul Curnow from the Astronomical Society of South Australia in about 2004 (www.sydneyobservatory.com.au/2014/farewell-to-famous-australian-comet-hunter-bill-bradfield/)

In light of the listings in Table 17.3, New Zealand can boast a proud record of cometary discovery for a small country with so minute an astronomically-minded population, all the more so if the eighteen independent discoveries made in Australia by expatriate New Zealander, the late William Ashley (Bill) Bradfield (1927–2014; Fig. 17.22), are allowed for.

17.4 Concluding Remarks

John Grigg gained international recognition during the first decade of the twentieth century through his comet discoveries, and is now remembered for 26P/Grigg-Skjellerup, C/1903 H1 (Grigg) and C/1907 G1 (Grigg-Mellish). He also independently discovered C/1906 F1 (Ross) and independently recovered 2P/Encke in 1898.

Given his cometary record, and his involvement in astrophotography and the popularization of astronomy, Grigg was arguably New Zealand's leading amateur astronomer during the first decade of the twentieth century. But, like his Australian contemporaries Alfred Barrett Biggs (Orchiston 1985a)[10] and David Ross

[10]Both Grigg and Biggs were committed to the popularization of astronomy, and each constructed technical equipment to assist their astronomical pursuits, including a telephone. Biggs' main

(Orchiston and Brewer 1990)[11] he belonged to that second echelon—those active, highly-skilled amateurs who made a significant local impact, but rarely were given an opportunity to join the staff of a professional observatory. Despite this, New Zealand cometary astronomy largely owes its genesis to John Grigg, and New Zealand astronomy in general would have been much the poorer without him.

Acknowledgements I am grateful to the following for their assistance: Mrs Beverley Angus (Wellington, NZ), Rod Austin (New Plymouth, NZ), the late Graham Blow (formerly Carter Observatory, Wellington, NZ), the late Robert Campbell (Oamaru, NZ), the late George Eiby (Wellington, NZ), Nick Frost and Hazel McGee (British Astronomical Association, London), Richard Gillespie and Lorenzo Iozzi (Museum Victoria, Melbourne), Pam Kilmartin and Alan Gilmore (Mount John Observatory, NZ), the late Peter Hingley (former RAS Librarian, London), Dr Nick Lomb (formerly Sydney Observatory) and John Seymour (Palmerson North, NZ). I also wish to thank Rod Austin and John Drummond (Patutahi, NZ) for reading and commenting on the first draft of this chapter, and Rod Austin, the British Astronomical Association, Michael Clark, Alan Gilmore, and Museum Victoria for providing Figs. 17.4, 17.5, 17.19, 17.20 and 17.21.

References

The following abbreviation is used:

TL = In Letters to John Tebbutt (1860–1915), comprising 46 volumes of bound letters and documents in the Mitchell Library, Sydney (A3682-3727).

Astronomische Nachrichten, 175, 175–176 (1907).
Austin, R. (1990). The discovery of Comet Austin 1989c_1. *Astronomy Now, 4*(6), 19–21.
Austin, R. (1994). Albert Jones—the quiet achiever. *Southern Stars, 36*, 36–42.
Backlund, O. (1894). Ephemeride des Encke'schen Comenten. *Astronomische Nachrichten, 136*, 379–384.
Baracchi, P. (1902a). Letter to J. Tebbutt, dated August 6, in TL.
Baracchi, P. (1902b). Letter to J. Tebbutt, dated August 15, in TL.
Baracchi, P. (1903). Letter to J. Tebbutt, dated May 22, in TL.
Barrachi, P. (1907). Letter to C.J. Merfield, dated May 6, in TL.
Bateson, F. M. (1989). *Paradise Beckons*. Waikanae: Heritage Press.
Clark, M. (2014). One of Wadhurt's 'sons' came good in astronomy. *The Wadhurst Astronomical Society* (www.wadhurstastro.co.uk/news.htm).
Cooper, T. (2005). Michiel John Bester (1917–2005). *Monthly Notices of the Astronomical Society of Southern Africa, 64*, 118–119.
Crommelin, A. C. D. (1921). Obituary: John Grigg. *Monthly Notices of the Royal Astronomical Society, 81*, 258–259.
Dewhirst, D. W. (1996). Gerald Merton (1893–1983). *Quarterly Journal of the Royal Astronomical Society, 37*, 847–849.

(Footnote 10 continued)

research interest was double stars. Beyond astronomy, both men conducted choirs over many years, and composed music.

[11]Both Grigg and Ross observed known comets and searched successfully for new ones. Both men were pioneers in astronomical photography, and were committed to the popularization of astronomy. But unlike Grigg, Ross made his own (reflecting) telescopes.

Gale, W. (1898). Letter to J. Tebbutt, dated June 20, in TL.

Gale, W. (1902). Letter to J. Tebbutt, dated August 5, in TL.

Geddes, M. (1932). Letter to J.C. Begg, dated July 3. Copy in Orchiston collection.

Gilmore, A. C. (2007). Comet Gilmore P/2007 Q2. *Southern Stars, 46*(3), 5–6.

Gilmore, A. C., & Jones, A. F. (2001). Comet Utsunmiya-Jones C/2000 W1. *Southern Stars, 40* (1), 9–11.

Gingerich, O. (Ed.), (1984). *The General History of Astronomy. Volume 4: Astrophysics and Twentieth-Century Astronomy to 1950: Part A.* Cambridge: Cambridge University Press.

Glover, P. W. (1928). The orbit of Comet 1927 *k.* (Skjellerup). *Transactions of the Royal Society of New Zealand, 59*, 429–442.

Grigg, J. (1902a). Letter to J. Tebbutt, dated May 17, in TL.

Grigg, J. (1902b). Letter to J. Tebbutt, dated August 12, in TL.

Grigg, J. (1902c). Photography with a small telescope. *Journal of the British Astronomical Association, 12*, 125–126.

Grigg, J. (1902d). Über die Entdeckung eines neuen Cometen 1902 c. *Astronomische Nachrichten, 159*, 389–392.

Grigg, J. (1903a). Comet 1902 *b. Journal of the British Astronomical Association, 13*, 169–170.

Grigg, J. (1903b). Comet 1902 b (Perrine). *Journal of the British Astronomical Association, 13*, 284–285.

Grigg, J. (1903c). Comet 1903 *b. Journal of the British Astronomical Association, 14*, 77–78.

Grigg, J. (1903d). Comet 1903 *b. Journal of the British Astronomical Association, 14*, 78–79.

Grigg, J. (1903e). Discovery of Grigg's Comet, 1903 *b. Journal of the British Astronomical Association, 13*, 320.

Grigg, J. (1903f). Letter to J. Tebbutt, dated May 2, in TL.

Grigg, J. (1903g). Letter to J. Tebbutt, dated May 6, in TL.

Grigg, J. (1903h). Letter to J. Tebbutt, dated May 22, in TL.

Grigg, J., 1903i. Letter to J. Tebbutt, dated May 25, in TL.

Grigg, J. (1903j). Letter to J. Tebbutt, dated June 13, in TL.

Grigg, J. (1904a). Letter to J. Tebbutt, dated May 13, in TL.

Grigg, J. (1904b). Letter to J. Tebbutt, dated May 14, in TL.

Grigg, J. (1904c). Letter to J. Tebbutt, dated December 17, in TL.

Grigg, J., 1905a. Letter to J. Tebbutt, dated December 15, in TL.

Grigg, J., 1905b. Letter to J. Tebbutt, dated December 16, in TL.

Grigg, J. (1906a). Comet 1906 c. *Journal of the British Astronomical Association, 16*, 319.

Grigg, J. (1906b). Letter to J. Tebbutt, dated February 2, in TL.

Grigg, J. (1906c). Letter to J. Tebbutt, dated March 21, in TL.

Grigg, J. (1906d). Letter to J. Tebbutt, dated April 20, in TL.

Grigg, J. (1907a). Discovery of comet 1907b. *Journal of the British Astronomical Association, 17*, 364.

Grigg, J. (1907b). Letter to J. Tebbutt, dated May 22, in TL.

Grigg, J. (1907c). Letter to John Tebbutt, dated October 10, in TL.

Grigg, M. (Ed.), (1970). *John Grigg, F.R.A.S. of Thames, New Zealand. Together with some Genealogical Memorials of his Family and Glimpses of Pioneer Life in New Zealand.* Unpublished typescript (Orchiston collection).

Hughes, D. W. (1991). J. Grigg, J.F. Skjellerup and their comet. *Vistas in Astronomy, 34*, 1–10.

Journal of the British Astronomical Association, *18*, 55–56 (1907).

Kreutz, H. (1907). Zur Entdeckung des Kometen 1907b. *Astronomische Nachrichten, 175*, 175–176.

Kronk, G. W. (1984). *Comets, A Descriptive Catalog*. Hillside: Enslow.

Kronk, G. W., & Meyer, M. (2010). *Cometography. Volume 5, 1960–1982. A Catalog of Comets.* Cambridge: Cambridge University Press.

Langford, J. (1984). The impact of photography on astronomy. In Gingerich, 16–39.

Loomis, E. (1861). *An Introduction to Practical Astronomy* (pp. 400–459). New York: Harper.

Luther, P. (1989). *Bibliography of Astronomers. Books and Pamphlets in English By and About Astronomers. Volume 1. The Spirit of the Nineteenth Century*. Richmond: Willmann-Bell.
McIntosh, R. A. (1941). The history of Comet de Kock 1941a. *Southern Stars, 7*, 75–77.
McIntosh, R. A. (1958). The astronomical history of the Auckland Province. Part IV: John Grigg, F.R.A.S. *Amateur Astronomer [Journal of the Auckland Astronomical Society], 18*, 20–23.
McIntosh, R. A. (1970). Early New Zealand astronomy. *Southern Stars, 23*, 101–108.
Mackrell, B. (1985). *Halley's Comet Over New Zealand*. Auckland: Reed-Methuen.
Marsden, B. (1989). *Catalogue of Cometary Orbits*. Cambridge, MA: Minor Planet Centre.
Marsden, B. G., & Williams, G. V. (1999). *Catalogue of Cometary Orbits 1999* (13th ed). Cambridge, MA: Minor Planet Center.
Meadows, J. (1984). The origin of astrophysics. In Gingerich, 3–15.
Merfield, C. J. (1903). Letter to J. Tebbutt, dated May 17, in TL.
Merfield, C. J. (1907a). Letter to J. Tebbutt, dated April 29, in TL.
Merfield, C. J. (1907b). Letter to J. Tebbutt, dated May 22, in TL.
Merfield, C. J. (1907c). Letter to J. Tebbutt, dated May 30, in TL.
Merton, G. (1927). The periodic comet Grigg (1902 II): Skjellerup (1922 I) (1902–1927). *Memoirs of the Royal Astronomical Society, 64*, 47–104.
Monthly Notices of the Royal Astronomical Society, 63, 242 (1903).
Monthly Notices of the Royal Astronomical Society, 64, 345 (1904).
Monthly Notices of the Royal Astronomical Society, 68, 291 (1908).
Monthly Notices of the Royal Astronomical Society, 83, 288 (1923).
Morshead, F. J. (1945). New Plymouth Astronomical Society. A review of twenty-five years of astronomical activity—August, 1920, to June, 1945. *Southern Stars, 11*(6), 3–7.
Mr Grigg's observatory. *Thames Advertiser*, March 20, 1885 (p. 3).
Observatory, 21, 319 (1898).
Orchiston, W. (1983). C.J. Westland and Comet 1914IV: a forgotten episode in New Zealand cometary astronomy. *Southern Stars, 30*, 339–345.
Orchiston, W. (1985a). The contribution of Alfred Barrett Biggs to Australian and Tasmanian astronomy. *Records of the Queen Victoria Museum, 89*, 1–28.
Orchiston, W. (1985b). A survey of manuscript material in the Mitchell Library (Sydney) relating to the history of astronomy in New Zealand. *Southern Stars, 31*, 225–238.
Orchiston, W. (1990). Albert Jones: An interview. *Astronomy Now, 4*(9), 16–17.
Orchiston, W., & Brewer, A. (1990). David Ross and the development of amateur astronomy in Victoria. *Journal of the British Astronomical Association, 100*, 173–181.
Orchiston, W. (1993). John Grigg, and the genesis of cometary astronomy in New Zealand. *Journal of the British Astronomical Association, 103*, 67–76.
Orchiston, W. (1995). John Grigg, and the development of astrophotography in New Zealand. *Australian Journal of Astronomy, 6*, 1–14.
Orchiston, W. (1998). *Nautical Astronomy in New Zealand. The Voyages of James Cook*. Wellington: Carter Observatory.
Orchiston, W. (1999a). C/1881 K1: a forgotten "Great Comet" of the nineteenth century. *Irish Astronomical Journal, 26*, 33–44.
Orchiston, W. (1999b). Comets and communication: amateur-professional tension in Australian astronomy. *Publications of the Astronomical Society of Australia, 16*, 212–221.
Orchiston, W. (2001a). 'Sentinel of the Southern Heavens': the Windsor Observatory of John Tebbutt. *Journal of the Antique Telescope Society, 21*, 11–23.
Orchiston, W. (2001b). The Thames observatories of John Grigg. *Southern Stars, 40*(3), 14–22.
Orchiston, W. (2004a). Highlighting the history of nineteenth century Australian astronomy: the Tebbutt Collection in the Mitchell Library, Sydney. *Journal of Astronomical Data, 10*, 77–103.
Orchiston, W. (2004b). John Tebbutt and observational astronomy at Windsor Observatory. *Journal of the British Astronomical Association, 114*, 141–154.
Orchiston, W. (2015). The amateur-turned-professional (ATP) syndrome: two Australian case studies. In W. Orchiston, D. Green, & R. Strom (Eds.), *New Insights from Recent Studies in*

Historical Astronomy: Following in the Footsteps of F. Richard Stephenson (pp. 259–350). New York: Springer.

Reports of the observing sections—The Comet Section. *Journal of the British Astronomical Association, 16*, 271–272 (1906).

Ross, D. (1912). Correspondence. *Journal of the British Astronomical Association, 23*, 101–102.

Rumsey, N. (1974). Astronomy in New Zealand. *Southern Stars, 25*, 171–207.

Russell, H. C. (1892). *Observations of the Transit of Venus, 9 December 1874; Made at Stations in New South Wales*. Sydney: Government Printer.

Scholefield, G. H. (Ed.), (1940). *Dictionary of New Zealand Biography. Two Volumes*. Wellington: Department of Internal Affairs.

Shanklin, J. (2014). BAA Comet Section: Comets discovered in 2007. www.ast.cam.uk/~jds/coms07.htm#07Q2.

Sharma, A. A. (2013). Albert Jones (1920–2013). *Sky & Telescope, 128*(5).

Sheehan, W. (1995). *The Immortal Fire Within: The Life and Work of Edward Emerson Barnard*. Cambridge: Cambridge University Press.

Stevenson, J. T. (1882). Observations of the Great Comet (b) 1882, made at Auckland, New Zealand. *Monthly Notices of the Royal Astronomical Society, 43*, 54–55.

Stock, A. (1881). Letter to J. Tebbutt, dated June 7, in TL.

Stock, A. (1882). Letter to J. Tebbutt, dated March 28, in TL.

Tebbutt, J. (1895). *Report of Mr. Tebbutt's Observatory … For the Year 1894*. Sydney, Printed for the author.

Tebbutt, J. (1896). *Report of Mr. Tebbutt's Observatory … For the Year 1895*. Sydney, Printed for the author.

Tebbutt, J. (1898). Observations of minor planets and Encke's Comet at Windsor, N.S. Wales. *Astronomische Nachrichten, 147*, 314–318.

Tebbutt, J. (1899). *Report of Mr. Tebbutt's Observatory … For the Year 1898*. Sydney, printed for the author.

Tebbutt, J. (1903a). Comet 1903 *b*. *Journal of the British Astronomical Association, 14*, 79–80.

Tebbutt, J. (1903b). Planet and comet observations at Windsor, N.S. Wales. *Astronomische Nachrichten, 163*, 295–298.

Tebbutt, J. (1903c). *Report of Mr. Tebbutt's Observatory … For the Year 1902*. Sydney, printed for the author.

Tebbutt, J. (1904). *Report of Mr. Tebbutt's Observatory … For the Year 1903*. Sydney, printed for the author.

Tebbutt, J. (1907). Comet Grigg, 1907, April. *Observatory, 30*, 285–286.

Tebbutt, J. (1908). *Astronomical Memoirs*. Sydney (printed for the author).

Thames Advertiser, March 31, 1899.

The comet. *Thames Star*, May 4, 1901 (p. 3).

Thomsen, I. L. (1945). Obituary notices: Murray Geddes. *Monthly Notices of the Royal Astronomical Society, 105*, 88–89.

Vsekhsvyatskii, S. K. (1964). *Physical Characteristics of Comets*. Jerusalem: Israel Program for Scientific Translations.

Williams, T. R. (2007). Murray Geddes. In Hockey et al. (Eds.), *Biographical Encyclopedia of Astronomers* (p. 411). New York: Springer.

Wood, H. W. (1958). *Sydney Observatory 1858 to 1958*. Sydney: Sydney Observatory.

Wood, H. W. (1981). Gale, Walter Frederick (1865–1945). In *Australian Dictionary of Biography. Volume 8*. Melbourne: Melbourne University Press.

Wright, H. (1903). Letter to J. Tebbutt, dated May 15, in TL.

Wright, H. (1906). Letter to J. Tebbutt, dated November 29, in TL.

Chapter 18
C.J. Westland and Comet C/1914 S1 (Campbell): A Forgotten Episode in New Zealand Cometary Astronomy

Abstract New Zealand astronomer C.J. Westland independently discovered a comet on 18 September 1914, just a few hours after its initial sighting by Campbell in South America. Now known as C/1914 S1 (Campbell), this comet is one of a select band of comets known to exhibit sunward-pointing anti-tails. Westland observed and reported on this tail configuration, and obtained a photographic record of it on two different nights. After his discovery of this comet, Westland went on to become a professional astronomer under Dr C.E. Adams for a short time, before working for nine years as the director of a geomagnetic observatory, and then returning to amateur astronomy. Westland's numerous publications show his wide-ranging astronomical interests, but he is best remembered for his work in cometary astronomy.

18.1 Introduction

At first glance, New Zealand does not appear to have a particularly impressive record when it comes to cometary discoveries. Only ten comets have been officially credited to New Zealanders, and these are listed in Table 17.3 in the previous chapter. However, the true picture is somewhat brighter if we add recoveries of periodic comets (by John Grigg and the late Albert Jones, for example), make allowance for discoveries by expatriate New Zealanders like the late Bill Bradfield of Adelaide with his 18 comets, and consider independent discoveries that were not officially credited. The late Dr Frank Bateson (1982) has referred to some of these latter in a note appropriately titled "Hard Luck Stories", and it is to Comet C/1914 S1 (Campbell)—which was omitted from his listing—that this chapter is directed. This comet also was independently discovered by Joseph Lunt in South Africa and by the New Zealander, C.J. Westland. But before we review Westland's

W. Orchiston, *Exploring the History of New Zealand Astronomy*,
Astrophysics and Space Science Library 422,
DOI 10.1007/978-3-319-22566-1_18

observations of this comet, let us learn a little about this little-known but remarkable New Zealand astronomer.

18.2 C.J. Westland: A Biographical Sketch[1]

Charles James Westland (Fig. 18.1) was born in India in 1875, "… the eldest son of Sir James Westland (1842–1903), for many years a member of the Council of the Governor-General of India." (*The Cyclopedia of New Zealand* 1903: 567). He went to England for his education, and attended Clifton College, an independent public boarding school located in Bristol. Founded in 1862, and with an impressive assemblage of stately buildings (see Fig. 18.2), Clifton was noted for its emphasis on science in the curriculum (Clifton College n.d.)—which may go some way towards explaining Westland's passion for astronomy and geomagnetism.

After completing his schooling Westland decided on a farming career, and migrated to New Zealand, arriving in 1891. He then served as a 'cadet' on the Lake Coleridge and Tipapa Stations, and in 1894 acquired a property near Cheviot in North Canterbury when this region was opened up for European settlement (see Fig. 18.3 for New Zealand localities mentioned in the text). His 'Glenmuick Estate' was 3500 acres, and just nine years later, it was reported that "… great improvements have been effected, a splendid residence with wool-shed, stables, and out-buildings having been erected. The stock consists of crossbred sheep in which English strains predominate." (*The Cyclopedia of New Zealand* 1903: 567). Among the "out-buildings" was an astronomical observatory.

From the start, Westland determined to combine farming and astronomy, and in early 1914 he joined forces with Frederick Giles Gibbs (1866–1953; Mann 1977) and Edward George Leonard (Len) Morley (1894–1973; Eiby 1973) of Nelson[2] and the noted British solar astronomer John Evershed (1864–1956; Stratton 1957) and his astronomer wife, Mary Ackworth Orr Evershed (1867–1949; Brück 1998), from the Kodaikanal Observatory in India (Fig. 18.4) in a sight-testing program near Nelson. Their objective was to decide on a suitable location for the proposed Cawthron Solar Observatory (Collier 1984; Eiby 1973: 102).

Soon after discovering Comet C/1914 S1, Westland was forced to abandon farming when he was appointed Assistant to Dr Charles Edward Adams, the Government Astronomer of New Zealand (from 1912 to 1936) at the Hector

---

[1]Some of the following biographical details were supplied to me by the late Alan S. Westland, the son of C.J. Westland.

[2]Gibbs was the Headmaster of Nelson Central School from 1894 to 1924, and it was while he was a student there that Morley acquired a keen interest in astronomy. In discussing this, Collier (1984) states that Morley "… was one of those men whose exceptional talents were generally concealed from all but those of close acquaintance." This would explain his participation in the solar observatory survey at so youthful an age and while still a law student at Victoria University in Wellington.

Fig. 18.1 C.J. Westland (after *The Cyclopedia of New Zealand* 1903: 567)

Fig. 18.2 An etching of Clifton College made by Edward J. Brown in 1898. All of the buildings shown here were already in existence when Westland attended the school in the 1880s (https://en.wikipedia.org)

Observatory in Wellington (Fig. 18.5), and when Adams (Fig. 9.9) spent ten months at the Lick Observatory in 1915 Westland was Acting Government Astronomer (Eiby 1977: 25). Figure 18.6 shows Westland setting up the transit telescope at the Hector Observatory. Subsequently, Westland became the Director of the Apia Observatory in Samoa, which was concerned with geomagnetism rather than astronomy, and after nine years in this post he returned to New Zealand, to

Fig. 18.3 New Zealand localities mentioned in this chapter are shown in *red*

Fig. 18.4 Staff of the Kodaikanal Observatory in about 1907. John Evershed is seated in the middle of the second row. Mary Evershed, in the white dress, is sitting beside him (astronomy.snjr.net/blog/?page_id = 66)

farming and to amateur astronomy. Therefore, like his well-known trans-Tasman contemporaries Robert Thorburn Ayton Innes and Charles J. Merfield, Westland was a prime example of an 'ATP', an amateur-turned professional (i.e. an amateur astronomer who gained international recognition and then joined the ranks of the professional astronomers; see Orchiston 2015). But, unlike Innes and Merfield, Westland outlived his vocation in professional science and so could also become a 'PTA', a professional-turned-amateur.

In addition to his cometary interests, including the computation of orbital elements, Westland was particularly interested in solar eclipses and the history of astronomy, and he published 17 short papers on these topics, and others, in *Southern Stars*, the journal of the Royal Astronomical Society of New Zealand.

In a 1946 survey of New Zealand astronomy, just four years before his death, C.J. Westland is well represented:

> … Christchurch cannot be passed by without reference to Mr. C.J. Westland. In this one man observatory many diverse astronomical computations are carried out, varying from ordinary everyday affairs to eclipses which are due to appear in the far future as well as those which occurred sometime B.C. Calculation is not sufficient, and when necessary out comes his telescope to verify his work, or check up on some comet which may have been reported. Time is even found to experiment on sunspot photographs, or to make detailed auroral observations. (Anonymous 1946).

Fig. 18.5 The Hector Observatory at the time Westland worked there (Orchiston collection)

Fig. 18.6 C.J. Westland and the Hector Observatory transit telescope (Orchiston collection)

Just three years after establishing the Glen Muick Estate Westland had married Miss Fisher, daughter of the late Captain Fisher, of Christchurch (*The Cyclopedia of New Zealand* 1903), and their son, Alan S. Westland (d. 1970), not only inherited his father's interest in astronomy, but also the Glen Muick Estate and the observatory when his father died in 1950.

18.3 Comet Campbell-Westland-Lunt

The discovery of a new comet on 18 September 1914 by Dr Joseph Lunt (1866–1940) of the Royal Observatory at the Cape of Good Hope in South Africa was announced in the first issue of the *Journal of the British Astronomical Association* for 1914 (see Notes 1914a), and was followed in the next issue (Notes 1914b) by information on independent discoveries by the amateur astronomer C.J. Westland in New Zealand and the Harvard College Observatory variable star expert Leon Campbell (1881–1951; Fig. 18.7; Mayall 1951; Williams and Saladyga 2011) from Harvard's 'southern station' at Arequipa in Peru. The note indicated that Westland discovered the comet several hours before Lunt, while Campbell's detection came two of three days later. Despite this date differential, the note concluded by mentioning that in America and Europe the comet already was being referred to as "Campbell's Comet".

The confusion over actual discovery dates and times was resolved once and for all with the publication of a 'Discovery Table' in which the actual discovery times by the three observers were expressed in terms of G.M.A.T. (Notes 1914c). This revealed that Campbell's discovery in fact preceded Westland's by a mere seven hours, with Lunt a little over eight and a half hours later again (see Table 18.1).

The third 1914 issue of the *Journal of the British Astronomical Association* contains a paper by Westland on his discovery. He reports that shortly after 2300 h

Fig. 18.7 Leon Campbell (1881–1951) at Harvard College Observatory (*Courtesy* Smithsonian Archives, SIA Acc. 90–105)

Table 18.1 Discovery table for Comet C/1914 S1 (after Notes 1914c)

Observer	Discovery date (GMAT)
Campbell	1914 September 17.7
Westland	1914 September 17.99
Lunt	1914 September 18.35

New Zealand time on 18 September when turning his telescope to R Doradus he noticed a naked eye comet between α Doradus and ε Reticuli. An 18-minute exposure photograph taken that evening revealed a conspicuous tail which, by the following night, had grown to between 6° and 7° in length.

Throughout September and October 1914 this comet was a conspicuous object, reaching a visual magnitude of 3.8 on 21 September (Merfield 1914). It was widely observed in the Southern Hemisphere until February 1915, when it became too faint for observation (Marsden 1989). Up until 5 November 1914, Lunt (1914) recorded fifteen observations of it, while Westland's tally from 18 September to 7 October was thirteen (see Table 18.2). Westland (1914) also photographed the comet on six occasions during this period using a home-made astrocamera (see Fig. 18.8), and noted an interesting change in the nature of its tail. On 24 September he was

Table 18.2 Observations of Comet C/1914 S1 reported by Westland (1914)

Date	Observed	Photographed	Comments
September			
18	x	x	Considerable tail
19	x	x	Tail 6–7°
20	x	x	
21	x		
22	x		
23	x		
24	x	x	Two tails visible, 180° separation
25	x		
26	x		
27	x		
28	x		
29	x	x	Two tails visible, 130° separation
30			
October			
1			
2			
3			
4			
5			
6			
7	x	x	Only one short tail

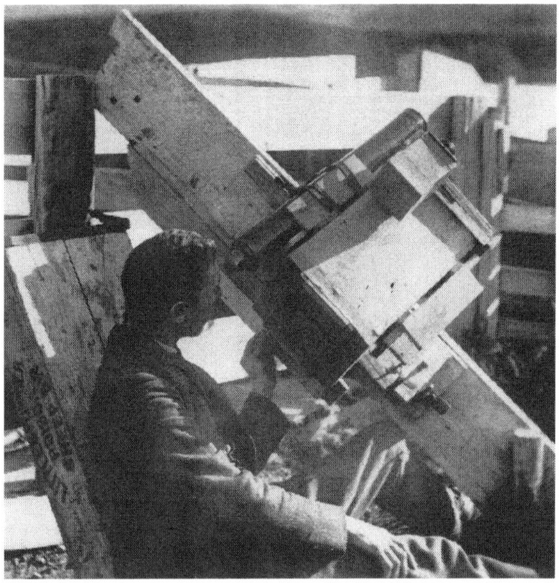

Fig. 18.8 C.J. Westland using his home-made astrocamera (*Courtesy* Alexander Turnbull Library, Wellington, Ref.1/2-151291-F)

surprised to see two tails, pointing in opposite directions. On the 29th, the angle between these tails had reduced to 130°, and by 7 October only one tail again was visible, a little over 1° in length.

The Melbourne-based amateur astronomer Davis Ross (Fig. 17.9), well-known for his own cometary discoveries (see Orchiston and Brewer 1990), also photographed Comet C/1914 S1 (Campbell) (Notes 1914b), while fourteen observations were made by Melbourne Observatory staff between 21 September and 21 October. Melbourne Observatory mathematical astronomy expert, C.J. Merfield (Fig. 18.9) used those recorded between 21 and 30 September to calculate the orbital elements of the comet.

18.4 Discussion

18.4.1 Naming Conventions

Conventions pertaining to the naming of newly-discovered comets have undergone considerable change over the last century or so. Prior to the formation of the International Astronomical Union in 1920 it was tradition to assign a single name only to each newly-discovered comet, but this practice appears to have been broken in the 1920s when Schwassmann and Wachmann were searching together and discovered comets (Brian Marsden, personal communication, April 1983). In this context the 'tyranny of distance' sometimes played its cruel hand as news of comet discoveries by Antipodean astronomers took too long to reach international

Fig. 18.9 Charles Merfield (*Courtesy* Royal Astronomical Society)

astronomical centres in Britain, Europe or North America, and the comets were unknowingly assigned to later independent discoverers (see Orchiston 1997). Once the name of a comet was adopted it was generally impossible to change it.

The present rule is to assign a maximum of three names to any new comet, with the names generally appearing in independent discovery order. Sometimes the discovery dates are widely separately, as in the case of Comet C/1970 K1 (White-Ortiz-Bolelli), but more often, as with Comet C/1914 S1 (Campbell), there were merely hours separating the initial sightings. Had contemporary rules of nomenclature applied in 1914 then Comet C/1914 S1 (Campbell) would be known officially to the world as C/1914 S1 (Campbell-Westland-Lunt), and New Zealand would have an additional recognized comet to its credit! However, Westland was not the only New Zealand astronomer to miss out in this way, as Table 18.3 indicates.

18.4.2 Anti-Tails

Comet C/1914 S1 (Campbell) was one of a comparatively select band of comets that exhibit 'anti-tails', which apparently point towards the Sun. Arguably the most famous of these was Comet C/1956 R1 (Arend-Roland), which was a remarkable spectacle in 1957 and exhibited an anti-tail that eventually reached 12° in length (Finson and Probstein 1968; Hendrie 1996; Sekanina 1968). Comet C/1973 E1

(Kohoutek) was another bright comet that also possessed a notable anti-tail (Hendrie 2000; Whipple 1976)[3].

We now believe that the apparent sunward orientation of the anti-tail is merely an illusion caused by the actual positions of the comet and the Earth in their respective orbits relative to the Sun. As the 'viewing angle' changes so too does the apparent orientation of the two tails, and this is in keeping with Westland's observations of C/1914 S1 (Campbell). Richter (1963: 101–103) presents a detailed examination of this phenomenon for Comet C/1956 R1 (Arend-Roland), and page 102 includes a diagram which illustrates clearly the illusion of a sunward-pointing tail.

18.4.3 Nucleus Splitting and C/1914 S1 (Campbell)

Boehnhardt (2004: 307) points out that

> Outbursts in the visual lightcurve of comets can indicate splitting events. There are prominent cases that demonstrate the temporal relationship between nucleus splitting and activity outbursts with amplitudes of 3 mag or more … [while] Smaller lightcurve peaks and nucleus break-ups are associated with Comets C/1899 E1 (Swift), C/1914 S1 (Campbell), C/1943 X1 (Whipple-Fedtke), C/1969 T1 (Tago-Sato-Kosaka), and C/1975 V1 (West) (see Sekanina 1982) …

Comet C/1914 S1 is therefore one of a comparatively small number of comets that is known to have experienced nucleus splitting, although Westland apparently did not notice the existence of two adjacent nuclei in the course of his visual or photographic observations.

18.4.4 Westland's Pioneering Efforts in Astronomical Photography

As we have seen, Westland photographed Comet C/1914 S1 (Campbell) on six different nights between 18 September and 7 October, inclusive (Table 18.3). These images were taken with a home-made astrocamera featuring a Ross portrait lens (Berry 1947a: 52), which he also used to photograph other comets, including 1P/Halley in 1910 (see examples in Mackrell 1985). The late Douglas Charles Berry (1918–2002), one-time Director of the Comet Section of the Royal Astronomical Society of New Zealand, identifies Westland as a pioneering New Zealand astronomical photographer, and he joins the likes of Thames' John Grigg and Meeanee's Reverend Dr David Kennedy, both of whom also photographed comets (see Orchiston 1995, 2002; and Chaps. 10 and 13 in this book).

[3]Another comet discovered by a New Zealander which also exhibited a conspicuous anti-tail was C/1984 N1 (Austin). Rod Austin (personal communication, 2014) noted that "… at one point the anti-tail was stronger than the main tail."

Table 18.3 Comets that were independently discovered by New Zealand astronomers but not named after them

Coment	Independent discoverer	Location	Reference
C/1881 K1 (Tebbutt)	Arthur Stock	Wellington	Stock (1881, 1882)
C/1906 F1 (Ross)	John Grigg	Thames	Orchiston (1993); this book, Chap. 17
C/1914 S1 (Campbell)	C.J. Westland	Cheviot	Orchiston (1983); this book, this chapter
C/1923 T1 (Dubiago-Bernard)	F.J. Morshead	New Plymouth	Morshead (1945: 5–6)
C/1927 X1 (Skjellerup-Maristany)	Rhind	New Plymouth	Glover (1928)
C/1941 B2 (de Kock-Paraskevopoulos)	McIntosh	Auckland	McIntosh (1941)
C/1947 S1 (Bester)	Bateson	Cook Islands	Bateson (1989: 151)
C/1966 P2 (Barbon)	Thomas	Mt. John	Rod Austin, personal communication
		Observatory	11 December 1991

18.5 Concluding Remarks

Charles James Westland is a remarkable New Zealand astronomer who managed to migrate successfully from amateur astronomer to professional astronomer, then into professional geomagnetic researcher, and finally back to amateur astronomer. While he pursued many astronomical topics and targets, I believe he should be best remembered for his contribution to cometary astronomy—including the independent discovery and tracking of Comet C/1914 S1 (Campbell), which, had circumstances been different, would now be known as C/1914 S1 (Campbell-Westland-Lunt). The late Douglas C. Berry reminds us that Westland was also a pioneering New Zealand astronomical photographer, and regards him as "… a father …" of New Zealand astronomy (Berry 1947b: 6). Meanwhile, that famous New Zealand son, Dr Leslie John Comrie (1893–1950; Massey 1952) who revolutionized astronomical computing, once referred to Westland as the Crommelin of New Zealand (Obituary 1950). This is high praise indeed, and merely serves to consolidate Westland's claim for a place in the annals of New Zealand astronomical history.

Acknowledgements I am grateful to the following for their assistance: Rod Austin (New Plymouth, NZ), Dr Grant Christie (Auckland, NZ), Jenni Chrisstoffels (Alexander Turnbull Library, Wellington, NZ), the late Peter Hingley (former RAS Librarian, London), Mary Markey (Smithsonian Institution Archives, Washington), the late Dr Brian Marsden (Smithsonian Astrophysical Observatory, USA) and the late Alan S. Westland (Cheviot, NZ). I also wish to thank John Drummond (Patutahi, NZ) and Rod Austin for reading and commenting on the first draft of this chapter, and the Alexander Turnbull Library, the Royal Astronomical Society (London) and the Smithsonian Institution Archives (Washington) for kindly supplying Figs. 18.7, 18.8 and 18.9.

References

Anonymous. (1946). Astronomical activity in New Zealand. *Southern Stars, 12*(2), 17–21.

Bateson, F. M. (1982). Hard luck stories. *Royal Astronomical Society of New Zealand Newsletter*, 2–3.

Bateson, F. M. (1989). *Paradise Beckons*. Waikanae: Heritage Press.

Berry, D. C. (1947a). Comet Section. *Southern Stars, 13*(3), 51–52.

Berry, D. C. (1947b). Comet Section report. *Southern Stars, 13*(1), 5–6.

Boehnhardt, H. (2004). Split comets. In M. C. Festou, H. U. Keller, & H. A. Weaver (Eds.), *Comets II* (pp. 301–316). Tucson: University of Arizona Press.

Brück, M. T. (1998). Mary Ackworth Evershed née Orr (1867–1949), solar physicist and Dante scholar. *Journal of Astronomical History and Heritage, 1*, 45–59.

Clifton College. n.d. (http://en.wikipedia.org/wiki/Clifton_College).

Collier, C. (1984). Edward George Leonard Morley—1894 to 1973. Nelson astronomer. *Journal of the Nelson and Marlborough Historical Societies, 1*(4), 19.

Eiby, G. A. (1973). Obituary. Edward George Leonard Morley, F.R.A.S. (1894–1973). *Southern Stars, 25*, 101–103.

Eiby, G. A. (1977). The New Zealand Government time service: an informal history. *Southern Stars, 27*, 15–34.

Finson, M., & Probstein, R. (1968). A theory of dust comets. II. Results for Comet Arend-Roland. *Astrophysical Journal, 154*, 353–380.

Glover, P. W. (1928). The orbit of Comet 1927 k. (Skjellerup). *Transactions of the Royal Society of New Zealand, 59*, 429–442.

Hendrie, M. J. (1996). The two bright comets of 1957. *Journal of the British Astronomical Association, 106*, 315–330.

Hendrie, M. J. (2000). Comet Kohoutek 1973f. *Journal of the British Astronomical Association, 110*, 9–19.

Lunt, J. (1914). Comet 1914e. *Journal of the British Astronomical Association, 25*, 140–142.

McIntosh, R. A. (1941). The history of Comet de Kock 1941a. *Southern Stars, 7*, 75–77.

Mackrell, B. (1985). *Halley's Comet Over New Zealand*. Wellington: Reed-Methuen.

Mann, S. (1977). *F.G. Gibbs, His Influence on the Social History of Nelson, 1890–1960*. Stoke: Nelson Historical Society.

Marsden, B. (1989). *Catalogue of Cometary Orbits*. Cambridge (Mass): Minor Planet Centre.

Massey, H. S. W. (1952). Leslie John Comrie 1893–1950. *Obituary Notices of Fellows of the Royal Society, 8*(21), 96–107.

Mayall, M. W. (1951). Leon Campbell 1881–1951. *Popular Astronomy, 59*, 356–358.

Merfield, C. J. (1914). Comet 1914e. *Journal of the British Astronomical Association, 25*, 142–143.

Morshead, F. J. (1945). New Plymouth Astronomical Society. A review of twenty-five years of astronomical activity—August, 1920, to June, 1945. *Southern Stars, 11*(6), 3–7.

Notes. *Journal of the British Astronomical Association*, 25, 53 (1914a).

Notes. *Journal of the British Astronomical Association*, 25, 106–107 (1914b).

Notes. *Journal of the British Astronomical Association*, 25, 146 (1914c).

Obituary. Charles James Westland. *Southern Stars*, 14, 129 (1950).

Orchiston, W. (1983). C.J. Westland and comet 1914IV: a forgotten episode in New Zealand cometary astronomy. *Southern Stars, 30*, 339–345.

Orchiston, W., & Brewer, A. (1990). David Ross and the development of amateur astronomy in Victoria. *Journal of the British Astronomical Association, 100*, 173–181.

Orchiston, W. (1993). John Grigg, and the genesis of cometary astronomy in New Zealand. *Journal of the British Astronomical Association, 103*, 67–76.

Orchiston, W. (1995). John Grigg, and the development of astrophotography in New Zealand. *Australian Journal of Astronomy, 6*, 1–14.

Orchiston, W. (1997). The “tyranny of distance” and Antipodean cometary astronomy. *Australian Journal of Astronomy, 7*, 115–126.
Orchiston, W. (2002). From Crossley to Carter: the life and times of an historic Cooke refractor. *Journal of the Antique Telescope Society, 24*, 9–24.
Orchiston, W. (2015). The amateur-turned-professional (ATP) syndrome: two Australian case studies. In W. Orchiston., D. Green & R. Strom (Eds.), *New Insights From Recent Studies in Historical Astronomy: Following in the Footsteps of F. Richard Stephenson* (pp. 259–350). New York: Springer.
Richter, B. N. (1963). *The Nature of Comets*. London: Methuen.
Sekanina, Z. (1968). A dynamic investigation of Comet Arend-Roland 1957 III. *Bulletin of the Astronomical Institute of Czechoslovakia, 19*, 343–350.
Sekanina, Z. (1982). The problem of split comets in review. In L. L. Wilkening (Ed.), *Comets* (pp. 251–287). Tuscon: University of Arizona Press.
Stock, A. (1881). Letter to J. Tebbutt, dated June 7, in TL.
Stock, A. (1882). Letter to J. Tebbutt, dated March 28, in TL.
Stratton, F. J. M. (1957). John Evershed. 1864–1956. *Biographical Memoirs of Fellows of the Royal Society, 3*, 40–51.
The Cyclopedia of New Zealand. (1903). Christchurch: The Cyclopedia Company Limited.
Westland, C. J. (1914). Comet 1914e. *Journal of the British Astronomical Association, 25*, 139–140.
Whipple, F. (1976). Comet Kohoutek in retrospect. *Proceedings of the American Philosophical Society, 120*, 1–6.
Williams, T. R., & Saladyga, M. (2011). *Advancing Variable Star Astronomy: The Centennial History of the American Association of Variable Star Observers*. Cambridge: Cambridge University Press.

Chapter 19
A Catholic Approach to Astronomy: The Remarkable Record of Ronald A. McIntosh

Abstract Ronald A. McIntosh was an Auckland amateur astronomer who made a significant contribution to meteor astronomy from the late-1920s through to the mid-1940s and became a leading authority on Southern Hemisphere meteor streams. But he also pursued telescopic astronomy, with emphasis on the planet Jupiter, and he independently discovered a comet in 1941. When his observing days ended McIntosh then focused his efforts on astronomical education and the development of astronomical societies.

19.1 Introduction

One of New Zealand's most accomplished twentieth century astronomers was R.A. McIntosh (Fig. 19.1), who lived in Auckland (see Fig. 19.2 for New Zealand localities mentioned in this chapter).

Brief general accounts of McIntosh already have been published by Bateson (1977), Orchiston and Wolf (1994)[1] and Orchiston (2000), while Luciuk and Orchiston (2016) provide a detailed examination of his meteor work. In this regard, Sect. 19.3 draws heavily on this last publication.

[1]After he left the Carter Observatory, during my directorship of that institution, Graham Wolf (1994) published a review paper on R.A. McIntosh in *WGN*, the journal of the International Meteor Organisation. This sole-authored paper, purportedly the results of his research, actually was based entirely on a Carter Observatory Information Sheet titled *Ronald McIntosh: A Remarkable New Zealand Astronomer* that I researched and wrote (Orchiston and Wolf 1994), a paper with precisely the same title and authors that I wrote and presented at that 1994 Annual Conference of the Royal Astronomical Society of New Zealand, and manuscript notes which I had prepared on McIntosh based upon an extensive collection of manuscript material and other related records that Jackie St. George acquired from McIntosh's descendants and then passed on to me. It is interesting that none of this is mentioned in Wolf's paper (nor does my name appear anywhere, even in the acknowledgements). It is a shame that the editor of WGN was unaware of these circumstances when he decided to publish Wolf's paper, and this case surely highlights the value of refereeing papers before they are accepted for publication.

W. Orchiston, *Exploring the History of New Zealand Astronomy*,
Astrophysics and Space Science Library 422,
DOI 10.1007/978-3-319-22566-1_19

Fig. 19.1 Ronald McIntosh (Orchiston collection)

19.2 Ronald Alexander McIntosh: A Biographical Sketch

Ronald Alexander McIntosh was born in Auckland to William John Alexander and Lucy Jane McIntosh. His early schooling was at St. Mary's Convent and the Marist Brothers School in Auckland. In 1922, McIntosh studied for the Professional Accountants' Examination and achieved passes in Economics and Mercantile Law; the following year he completed Company Law and Bankruptcy Law, and this ended his formal education. His astronomical and computational knowledge was entirely self-taught (Bateson 1977: 82).

During his working life McIntosh experimented with several occupations. His first job was as an office boy, then in 1926 he joined the *New Zealand Herald* newspaper in Auckland as a proofreader, and he eventually rose to the position of Cable Sub-Editor. McIntosh joined the New Zealand Army in June 1941 in the 1st Battalion, Auckland Regiment, and was posted to the Intelligence Section. A year later he went to the Intelligence Section of the Auckland Fortress Headquarters. He reached the rank of Sergeant, was discharged in 1944 and returned to the *New Zealand Herald*. In 1946, he resigned from the *Herald* to become editor of the *Young New Zealander*. After it ceased publication in 1948, he worked with *White's Flying Magazine* for two years. He then joined the Auckland Public Relations Office as Manager of Printing, Publicity and Information, a position he held until 1957. He then rejoined the *New Zealand Herald* for two years as Senior Sub-Editor,

Fig. 19.2 New Zealand localities mentioned in this chapter are shown in *red*

continuing part-time for eight more years in this position and for four additional years as their astronomical correspondent.

McIntosh's interest in astronomy began in 1910, at age six, when his parents showed him Comet 1P/Halley. This was a particularly dramatic apparition (e.g. see Mackrell 1985), and made a lasting impression on the young boy. By age 13, McIntosh was observing the Moon and comparing his observations with existing maps. In addition, "… he observed sunspots, Nova Aquilae 1918, the planets … a lunar eclipse and peculiar cloud formations." He started observing meteors in 1919 (Bateson 1977: 83). Although self-taught, McIntosh was able to build an international reputation as an amateur astronomer, and this was recognized in 1959 when he joined the Auckland War Memorial Museum Planetarium as their Lecturer-Demonstrator. He retired in 1972 (Bateson 1977: 82–83).

In 1930 Ronald McIntosh married Harriet Catherine Munro in Whitianga, and they had two children (Orchiston 2000). He died on 17 May 1977 after a long illness (Bateson 1977: 86).

19.3 Meteor Research

19.3.1 Introduction

In February 1928 the Meteor Section of the New Zealand Astronomical Society (later the Royal Astronomical Society of New Zealand) was formed. The foundation members were Ronald Alexander McIntosh of Auckland, Frank Maine Bateson (Fig. 19.3) of Wellington and Ivan Leslie Thomsen (Fig. 9.17; Eiby 1970) of Dannevirke, with McIntosh serving as Director. It is telling that all three ultimately would become professional astronomers. The first objective of the Section was:

> … the fixing of the many unknown centres of radiation in the southern hemisphere, the linking up of observed radiants to reveal the presence of unsuspected meteor streams of long duration and a careful study of the behavior of the principal showers. (McIntosh 1946b: 10–11).

Over its sixteen year period of activity, nine different observers at various times contributed to the Section. Apart from McIntosh, Bateson and Thomsen, others who were prominent were Maxwell (Max) Butterton (Fig. 19.4) of Wellington, Murray Geddes (Fig. 19.5) of Otekura, Southland, and Albert Jones (Fig. 19.6) of Timaru. Geddes also went on to become a professional astronomer.

From the start it was decided to concentrate on naked eye observing rather than photography. Certainly photography was available at this time and could be used to determine meteor paths more accurately than naked eye observations allowed, but it was quite expensive for amateur observers and had relatively poor sensitivity at the time so could not record faint meteors that were seen with the naked eye. Photography, therefore, really was only useful for recording fireballs and bolides.

Fig. 19.3 Dr Frank Bateson (http://taurangahistorical.blogspot.com/2013_10_01_archive.html)

Between 1925 and 1945 members of the Meteor Section recorded 15,627 meteors, about half of which were contributed by McIntosh. By pooling all observations and analyzing them, McIntosh was able to produce regular Section Reports, as well as a succession of research papers which were published mainly in overseas astronomical journals. Favoured publication outlets were the *Journal of the Royal Astronomical Society of Canada*[2] and *Monthly Notices of the Royal Astronomical Society*.

McIntosh's reports on Southern Hemisphere meteor showers concentrated on their activity, colors, magnitudes, heights and radiant locations. His findings did not include data on meteor shower zenith hourly rates and population indices, which were not in use until well after he became inactive in meteor astronomy.

[2]This particular journal was chosen because at this time the Canadian astronomer Peter McKenzie Millman (1906–1990) from the David Dunlop Observatory and later the Dominion Observatory was very active and had a large number of amateur astronomers observing meteors under his direction. Immediately after WWII, Donald William Robert McKinley (b. 1912) from the National Research Council began carrying out radar observations of meteors, in a radar-visual-photographic collaboration with Millman. Consequently, meteor astronomy had a high research profile in Canada, and the *Journal of the Royal Astronomical Society of Canada* regularly included papers on meteors (see Jarrell 2009).

Fig. 19.4 Max Butterton (left) and Bruce Stonehouse (right), two keen young Wellington amateur astronomers (Orchiston collection)

The Meteor Section's work was carried out entirely visually with the naked eye, which allowed the detection of meteors as faint as magnitude 6, but had some uncertainties with regard to radiant positions and velocity determinations. Although telescopic observations were attempted, they produced no significant results.

It is clear from descriptions of the procedures that the Section used in recording their observations that accuracy was highly emphasized. Meteor Section observers utilized procedures developed in 1917 by the Meteor Committee of the American Astronomical Society to accurately determine the positions of the radiants of different meteor showers:

> 1a. A radiant shall be determined by not less than four meteors whose projected paths all intersect within a circle of 2° diameter, and are all observed within a period of at most four hours on one night, by one observer.
> 1b. Or by three meteors on one night and at least two on the next night, seen during the same approximate hours of G.M.T., and all five intersecting as above.
> 1c. Or by one stationary meteor …
> 3b. Under no circumstances shall a meteor be used to determine a radiant-point whose projected path passes more than 3.5° from the adopted point, and it is recommended that 2.5° be generally adopted as the usual limit.
> 4. Three meteors which fulfill (1a) shall be considered enough to give a confirmatory radiant for one determined on the same date of a previous year - i.e., where L, the meteoric apex, differs by less than 2° (Olivier 1925: 90–91).

Fig. 19.5 Murray Geddes during WWII (Orchiston collection)

Fig. 19.6 Albert Jones in 1987, long after he had abandoned meteor observing in order to concentrate on variable stars and comets (Orchiston collection)

During this time, the effect of sporadic meteor pollution on determining low activity meteor shower radiants was not well understood. Professor Martin Beech has pointed out that:

> The modern day meteor astronomer now knows that probably only 10 to 30 per cent of the observed meteors actually belong to well-defined meteoroid streams. What this means is that the vast majority of observed meteors cannot in fact be traced to common radiant points. This principal while clear to modern astronomers was not, however, known to Denning or his contemporaries. They believed, in contrast, that all meteors could be traced to a radiant point, and that each radiant point could probably be associated with the orbit of a comet (Gyssens 1998: 86).

McIntosh's assessment of his meteor observations was expressed as follows:

> … the little we know about the meteor radiants and rates in the southern hemisphere comes from the labours of our own Meteor Section, which, handicapped by small personnel, has not been able in the few years of its existence to gain a complete picture of meteor distribution in this unexplored hemisphere. Even in the behavior of the principal annual showers there are gaps in our knowledge waiting to be filled. (McIntosh 1941b: 10).

The New Zealand climate was less than perfect for the observers. There were many rainy periods in Auckland, averaging 185 rain days annually with 100 days of at least 2.5 mm of rain. There was sunshine only 50 % of the time annually and only 42 % of the time in June and July. Average conditions in Wellington were quite similar (McLintock 1966). This kind of weather must have been frustrating for the Meteor Section, and would have made it difficult to verify radiants for minor showers on a subsequent night or year basis. In addition to weather, the Meteor Section had to take lunar conditions into consideration, which further limited observational opportunities.

The following sections present summaries of McIntosh's most significant meteor publications. They illustrate the broad range of his interests, from meteor radiant determination, observations of existing and new showers, meteor seasonal heights, meteor velocities, and meteor path lengths, to fireballs and bolides.

19.3.2 The Stationary Radiant Controversy

In "The Aquarid Meteors" McIntosh (1931) responded to Denning's (1915) claim of stationary Aquarid radiants. William Frederick Denning (1848–1931), one of Britain's leading meteor authorities, was an avid amateur astronomer who spent most of his life in Bristol and seems to have earned an income from scientific journalism (see Beech 1998). Although interested also in solar and planetary astronomy and furnished with a 25.4-cm (10-in) With-Browning reflector, from about 1870 Denning concentrated on naked eye observations of meteors, publishing his first meteor paper in *Monthly Notices of the Royal Astronomical Society* in 1872.

For the last forty years of his life Denning championed the existence of stationary radiants, contrary to theory and to the views of most astronomers at the time. McIntosh (1929b: 157–160) had earlier published his Meteor Section's observations on the movement of the Aquarid radiant, but Denning insisted that it was stationary. McIntosh responded that no Aquarid meteors had been observed by his group between 1 August and 15 December, so no stationary radiant existed.

He pointed out that Denning had only seen 140 Aquarid meteors in 52 years, between 10 August and 27 September, only about 3 Aquarids per 48-day period yearly. He correctly criticized Denning's observational methodology. Ronald McIntosh was not intimidated by criticism from a famous observer like Denning, as long as he had the supporting observational evidence.

19.3.3 The Eta Aquarid Radiant

The Eta Aquarid meteor shower is the most active in the Southern Hemisphere, so it received a great deal of Meteor Section attention. McIntosh (1935a) published an important paper accurately determining its radiant based on observations made from 1929 to 1933. He utilized an improved method he developed in an earlier paper to create an ephemeris of the motion of the Eta Aquarid radiant. Each radiant's celestial latitude and the longitudinal difference between the radiant and the apex of the Earth's motion was averaged to produce the right ascension and declination of the radiant:

> The principal advantage of this method is that the whole of the available data is used to determine the mean radiant, whereas under the earlier method the radiants observed on each day were treated separately, the mean daily positions being used to show the motion of the radiant-point. (McIntosh 1934: 584).

McIntosh derived 31 radiants from the five years of observations. The radiants were weighted based on the number of observed meteors per radiant. This improved method produced radiant position accuracy in the range of about ±0.5° The mean ecliptic longitude of the Aquarid radiant minus the longitude of the meteoric apex was 25.3° and its ecliptic latitude was +8.3°. The 4 May 1935 radiant position was determined to be at 335.7° right ascension and −1.3° declination. The meteor activity graph of the Section's observations in Fig. 19.7 shows a peak activity of 10 Eta Aquarid meteors per hour around 4 or 5 May. Jenniskens (2006: 705) puts the Eta Aquarid 6 May maximum at 28 meteors/hour. The German meteor and variable star astronomer, Dr Cuno Hoffmeister (1892–1968; Zejda 2014) from the Sonneberg Observatory, observed the Eta Aquarids in South Africa and determined a maximum activity of 36 meteors per hour (Lovell 1954: 264–265). McIntosh's lower maximum rate may have been due to a combination of non-zenith Eta Aquarid radiants and non-ideal sky conditions at the Section's observing sites.

19.3.4 The Search for New Southern Showers

The most important and by far the most referenced paper by McIntosh (1935b) was "An Index to Southern Meteor Showers". The paper listed 320 radiant positions determined from eight years of meteor observations by the Meteor Section.

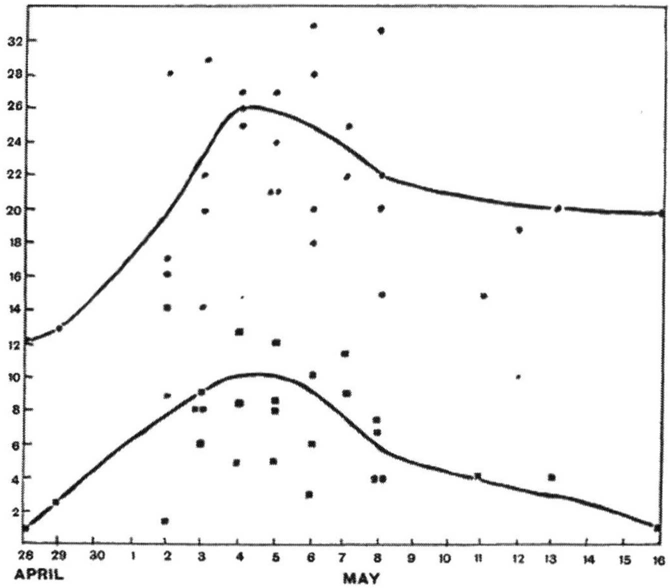

Fig. 19.7 Total meteor (upper curve) and Aquarid meteor (lower curve) hourly rates (after McIntosh 1935a: 601)

Also utilized were works by other sources that contained southern declination radiants. These sources were "Denning's General Catalogue of Meteor Radiants", the publications of the American Meteor Society, the *Journal* and *Memoirs* of the British Astronomical Association and Russian observations (McIntosh 1935b: 710). McIntosh's Meteor Section had strict procedures for the determination of shower radiants:

> The policy of the section has been to determine the radiants from meteors observed within a few hours on one night, not less than four paths intersecting within a circle 2° in diameter, or three meteors on one night and two on the following night, intersecting as described above, or one stationary meteor, being required to form a radiant. (McIntosh 1930: 462).

This policy follows guidelines which the American Astronomical Society's Meteor Committee under C.P. Olivier published in 1917.

McIntosh noted that four radiants appeared to be stationary, but with additional observations, they "… will be resolved into a number of minor streams all showing the motion required by theory." (McIntosh 1935b: 711). McIntosh's paper had the most comprehensive list of southern radiants until the use of radar after WWII. However, the paper listed many chance radiants of sporadic meteors rather than shower meteors. Kronk (1988: xxi) felt that the list was not sufficiently selective and "… many [over 52 %] of the radiants were based on only one or two nights of observations—making the probability of the inclusion of sporadic radiants quite high." During that era, many astronomers believed that all meteors were shower related. McIntosh is credited for the discovery of the following nine minor showers in Kronk's (1988) comprehensive book, *Meteor Showers*:

> Gamma Normids – "Ronald A. McIntosh (Auckland, New Zealand) discovered this meteor shower on March 10.1, 1929." (Kronk 1988: 38).
> Librids – "The primary sources for data supporting this steam's existence came from Ronald A. McIntosh's 1935 paper …" (Kronk 1988: 65).
> Northern May Ophiuchids – "During 1935 this shower experienced the beginning of recognition as an annual shower when Ronald A. McIntosh listed it in his paper 'An Index of Southern Meteor Showers'." (Kronk 1988: 79).
> Southern May Ophiuchids – "The first apparent recognition of this stream as a potential annual producer of meteors was in 1935, when Ronald A. McIntosh's paper …" (Kronk 1988: 81).
> Ophiuchids – "The first formal recognition that this shower was a producer of annual activity was R.A. McIntosh's …" (Kronk 1988: 103).
> Omega Scorpiids – "McIntosh's radiant list … It was the first time activity had been officially recognized from this stream …" (Kronk 1988: 111).
> June Scutids – "The discovery of this meteor shower should be credited to R.A. McIntosh, since it was listed in his classic work …" (Kronk 1988: 114).
> Tau Capricornids – "The meteor shower was discovered by Ronald A. McIntosh and was listed in his classic paper …" (Kronk 1988: 140).
> Alpha Puppids – "The discovery of this stream should be attributed to Ronald A. McIntosh … who listed this stream in his 1935 paper …" (Kronk 1988: 270).

Another example of how McIntosh's observations are referenced in a modern publication can be found in Table 7 of Jenniskens' *Meteor Showers and their Parent Comets* (2006: 691–746), in which McIntosh's paper is cited 19 times within the list of 240 meteor showers. Kronk (1988) makes reference to McIntosh 32 times in *Meteor Showers*. Although McIntosh's "An Index to Southern Meteor Showers" is dated, it still provides much useful information.[3]

19.3.5 Fireball Tracking and Reporting

Ronald McIntosh (1932: 194) took great pains reporting fireballs seen in New Zealand, defined them as a "… body of exceptional brilliance." He utilized observations from both experienced and inexperienced observers in order to calculate fireball trajectories McIntosh 1938), and reported on their heights, path lengths, locations, velocities, sounds, radiants and flight durations. To accomplish this, he had to carefully decide on the accuracy of the dozens of reports he received. He often created a map showing where the fireball traversed the country.

[3]McIntosh's 1935 paper also provided the inspiration for later New Zealand meteor observers, such as Morse (1981) of Gisborne (and later Wellington) and Morgan (1979) of Renwick. McIntosh's paper also provided me with the basic list of radiants I needed when I began serious meteor observing in 1959. First I looked at the Phoenicid meteor shower (see Orchiston 1963), but one of my objectives was to investigate some of McIntosh's minor radiants, and in the process I even discovered a new one to add to his list (see Orchiston 1964).

His observers occasionally supplied information on colors, sounds, brightness and trains. McIntosh's reports on fireballs illustrated his great interest in these spectacular objects, and his willingness to create a meaningful analysis from dozens of contradictory reports received from the public. In order to encourage the rapid reporting of fireball observations locally, in 1934 agencies of the New Zealand Astronomical Society's Meteor Section were established in Auckland, Hamilton, New Plymouth, Wanganui, Napier, Eketahuna, Wellington, Nelson, Christchurch, Dunedin and Otekura (see Fig. 19.2 for localities). These agencies also were responsible for promptly forwarding McIntosh press clippings relating to recently-reported fireballs (McIntosh n.d.: 1934).

In "Determination of the Real Paths of Fireballs," McIntosh (1938) detailed procedures used in gathering data and in carrying out the calculations. He was very organized in soliciting fireball observations from untrained observers, utilizing newspaper articles and radio interviews to ask for:

> (1) exact location of observer; (2) date and time of appearance of meteor; (3) the position of its path in the sky with reference to the stars, by compass bearings, or in relation to terrestrial objects; (4) particulars of any noises which accompanied or followed the meteor; (5) description of the brightness of the object and its trail, and (6) the length of time the meteor (and trail) remained visible. (McIntosh 1938: 3).

McIntosh then examined information sent to him and selected data that appeared most detailed and reliable. In Fig. 19.8 he illustrates the potential locations of observers relative to a fireball, and Table 19.1 describes the observer's view of the fireball, based upon where he or she is located in Fig. 19.8. These perceptions

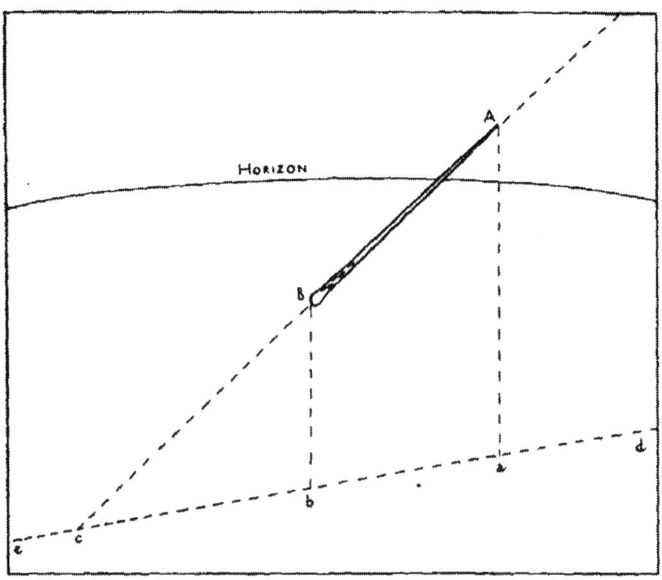

Fig. 19.8 Observer location versus fireball path (McIntosh 1938: 8)

Table 19.1 Meteor observer perceptions (after McIntosh 1938: 8)

Position in Fig. 19.9	'Explanation'
At a	Begins at zenith; falls vertically
At b	Rises vertically; ends at zenith
At c	Appears stationary
On da	Falls vertically
On ab	Rises and falls vertically; crosses zenith
On bc	Rises vertically
On ce	Falls vertically

helped McIntosh position an observer relative to the fireball, and he concluded his paper with an extensive explanation of the spherical geometry formulae that he used in order to compute the path of the fireball.

19.3.6 The Heights of Meteors

While McIntosh (1940a) was carrying out his study of the seasonal variation in the heights of meteors, he discovered that the Estonian-born Irish astronomer Ernst Julius Öpik (1893–1985) and his American colleague Harlow Shapley (1885–1972) had published a similar study based upon Öpik's Arizona Expedition observations (see Öpik and Shapley 1937). McIntosh's paper also included published non-New Zealand sources of meteor heights, and utilized a total of 2856 meteors. Denning's 950 meteors were a prime source and he made use of three additional British sources with 794 meteors. Finally, he employed a total of 1112 meteors from other countries. His analysis was divided into three groups: I. Denning; II. Denning plus additional British; III. All data. McIntosh (1940a: 511) described meteor heights as follows: "… the height of appearance (Ha), the height at mid-point (Hm), and the height at disappearance (Hd)."

The variation in sporadic meteor heights calculated with arithmetic means is shown in Fig. 19.9 (with heights in miles). There was general agreement in all three groups and for the appearance (Ha), mid-path (Hm) and disappearance (Hd) heights. Early in the year (February) heights were at their lowest and six months later (August) were at their highest. Meteor path lengths were longer in the winter. Note, also, that height variability was minimal in the Ha group and increased as heights decreased in the Hd group. The main reason for seasonal meteor height variations likely was due to seasonal changes in the atmospheric density of upper meteoroid paths. Greater atmospheric densities would create meteor appearances at higher altitudes.

The seasonal meteor height variation trends from the Öpik/Shapley study were similar to those found by McIntosh. The main difference was in the meteor path lengths. The meteor path lengths (Ha-Hd) determined by McIntosh were

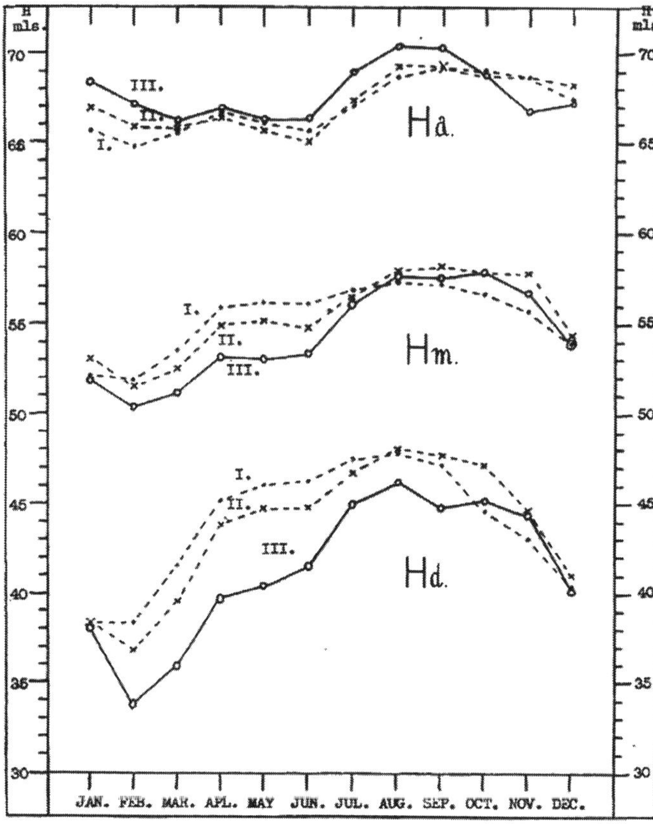

Fig. 19.9 Seasonal sporadic variation (height in miles) (after McIntosh 1940a: 515)

approximately 24 miles (35.6 km) while Öpik's were about nine miles (14.5 km) in length. On the basis of +3 visual magnitude meteors, the British-born American astronomer Gerald Stanley Hawkins (1928–2003) lists path lengths of 10–11 km (Hawkins 1964: 76). On the other hand, fellow-British astronomer John Guy Porter (1900–1981) from the Royal Greenwich Observatory derived sporadic meteor path lengths of about 34 km (see Lovell 1954: 194). It is not clear why McIntosh's and Porter's meteor path lengths differ so markedly from those of Öpik and Hawkins.

19.3.7 The Accuracy of Meteor Shower Radiants

In 1928 the Soviet astronomer V.A. Maltzev published a paper titled "Concerning the Fictitious Radiants of Meteoric Streams" where he concluded that more meteors than were typically utilized were needed to confirm a shower radiant point.

Later, Öpik described the complex statistical process his Arizona Expedition used to determine 223 radiants from 22,000 meteor observations, or about 100 meteor observations per radiant, and he claimed that "The majority of radiants published by various observers are probably only accidental configurations." (Öpik 1934: 2). The Maltzev and Öpik papers addressed meteor observer inaccuracies which then resulted in erroneous radiant determinations.

McIntosh (1940b) felt that Öpik's Arizona Expedition stated error rate of ±8.4° for differences in angle of two simultaneous trails of the same meteor was much greater than that experienced by his Meteor Section observers, so he carried out a statistical study based on his own results. His analysis of a random group of six radiants indicated a probable error of only ±1.7° (McIntosh 1940b: 376), substantially better than Öpik's Arizona Expedition experience. As a result, McIntosh concluded that his radiant procedures were adequate and that complex statistical tests were unnecessary for organizations whose error rates were as small as those of his Meteor Section:

> The present investigation has demonstrated to me that a statistical analysis can never be the sole criterion of reality of radiant deductions, for there are factors in the deduction of radiants which mathematical formulae cannot cope with. Briefly put, they are experience on the part of the investigator and the recurrence of meteor showers. (McIntosh 1940b: 379).

McIntosh felt that experienced meteor observers understood

> … that a certain meteor could not have emanated from a specific radiant, although apparently directed from it, while the laws of perspective act as a rough indication of radiation, the length of visible path being in direct proportion to the meteor's elongation from the radiant. Neither of these factors are [*sic*] taken into account in Öpik's statistical analysis, which is based solely on direction of flight. (ibid.).

However, it is now clear that additional meteor observations were in fact necessary for better discrimination between random sporadic and low-activity shower radiants. McIntosh sometimes used fewer than half a dozen observed meteors for a radiant determination. In his paper, McIntosh did not address Öpik's contention that only 15–26 % of meteors belong to genuine radiants.

19.3.8 The Velocities of Meteors and Fireballs

In "The Velocities of Meteors," McIntosh (1943) researched findings by other meteor observers on the accuracy of velocity determinations and the relationship of meteor appearance heights and their velocities. Some of the studies McIntosh considered were by Maltzev (using the data contained in the Von Niessl-Hoffmeister catalogue); by the German astronomer, meteorologist and geologist Alfred Wegener (1880–1930), for meteors brighter than Jupiter; and by Öpik.

Using Denning's data, the British astronomer and former Royal Astronomical Society President Henry Crozier Keating Plummer (1875–1946) determined the meteor height of appearance and velocity correlation coefficient to be 0.5 (Plummer 1941). Height determinations from Denning's observations and from meteors recorded by British Astronomical Association members were found by McIntosh to be similar. Figure 19.10 shows McIntosh's analysis of the appearance and disappearance of fireballs and meteors versus their velocities. Meteor and fireball heights were positively correlated with velocity for Solar System meteoroids, and negatively correlated for hyperbolic meteoroids. The graph confirms that fireballs appear higher and disappear lower than ordinary meteors, probably due to their greater meteoroid mass. Denning's hyperbolic meteor velocities are likely due to observational errors. Meteor path lengths in Fig. 19.10 averaged about 28 km.

However, fireball path lengths were more than 50 % greater than meteors. In Maltzev's analysis of fireballs from the Von Niessl-Hoffmeister catalogue, path lengths were positively correlated with velocity, with lengths ranging from 20 km at 10 km/s to 100 km at 110 km/s (Lovell 1954: 147). This differs markedly from the fireball trends in Fig. 19.10.

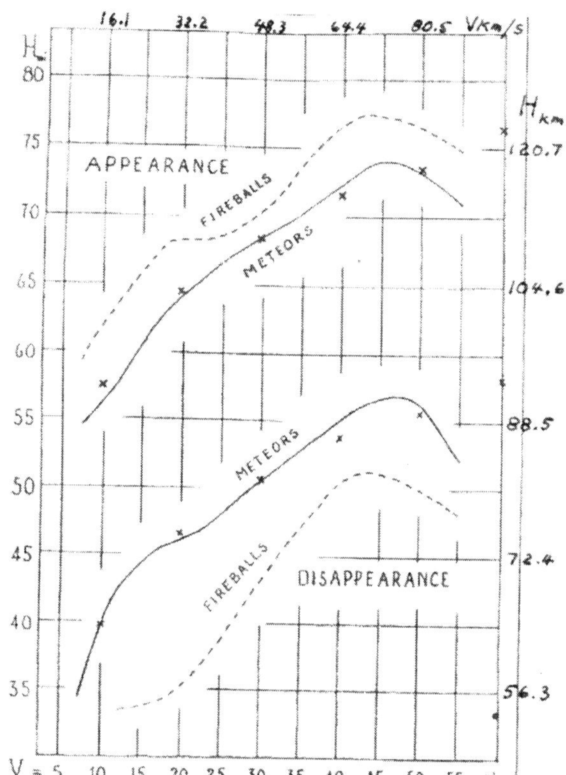

Fig. 19.10 Analysis of meteor/fireball divergences (adapted from McIntosh 1943: 314)

19.3.9 Meteor Path Lengths

McIntosh (1944) investigated the relationship between meteor path length and the path's angle of inclination as it enters the atmosphere. To accomplish this, he utilized meteor heights compiled by W.F. Denning. Since inclinations were not published, he calculated them using the following equation:

$$\sin i = (H_a - H_d)/P \tag{1}$$

where i is the meteor path angle of inclination, H_a is the height of meteor appearance, H_d is the height of meteor disappearance and P is the meteor path length.

Path lengths determined from Denning's meteor heights are shown in Fig. 19.11. The sporadic meteor path lengths (dotted lines) are longer in winter than in summer for the 809 meteors. However, mean heights of appearance for sporadic meteors are higher in summer than in winter. The inclination to path length trends of the two

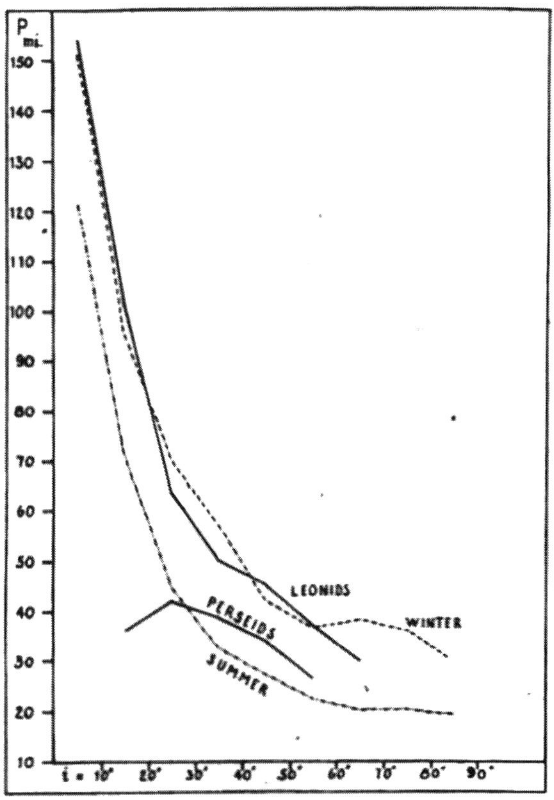

Fig. 19.11 Sporadic and meteor shower inclination versus path lengths (after McIntosh 1944: 197)

showers is similar to that of sporadic meteors. There were 41 Leonid meteors in the study and 126 Perseids. McIntosh (1944: 200) concluded that

> The path lengths of meteors are shown to be dependent upon the inclination of their paths in the atmosphere, the length of path varying inversely with the altitude of the radiant above the horizon.

McIntosh defended his use of Denning's data for the study from criticisms by Porter (1943: 134), who claimed that Denning's lists were "… not all computed by a rigid method, but each path was liable to adjustment by the computer exercising an independent and non-methodical judgement." McIntosh (1944: 199) used Denning's data-set because it "… was the largest and most homogeneous contribution to the study of real heights at the time and … without it the various correlations and statistical inferences could not have been drawn." The viability of Denning's data was also rationalized because "… the mean difference between observed and theoretical velocities for 165 meteors amounted to only −0.3 miles per second [−0.48 km/s]." (ibid.).

19.3.10 The Sequel

After WW II, radar detection of meteors became common worldwide, and in 1952 Clifford Darfield Ellyett (d. 2006) began research at the newly-formed Rolleston field station to the south of Christchurch. By the time he, Dr Colin Keay and other associated staff and graduate students moved *en masse* to the University of Newcastle in Australia in 1964 Ellyett's dynamic group at the University of Canterbury had investigated various aspects of southern hemisphere meteors and meteor streams (see Keay 1965) and acquired an international reputation. Fortunately there was only a brief hiatus after this mass exodus before Professor Jack Baggaley arrived from England in order to continue this line of research. The story of New Zealand's early exploits in meteor radar astronomy deserve to be properly documented, and would make an ideal Ph.D. thesis topic for an interested student with a passion for meteor astronomy *and* history.

19.4 Binocular and Telescopic Astronomy

19.4.1 Introduction

McIntosh was not only interested in naked eye observations of meteors. He also was keen to carry out binocular and telescopic observations of different celestial objects and, as we saw in Chap. 12, on 30 June 1927 he ordered a 35.6-cm (14-in) equatorially-mounted Newtonian Reflector from the estate of Joseph Thomas Ward (see Fig. 11.4)[4] of Wanganui, "… the 'founding father' of amateur telescope-making in New Zealand …" (Orchiston 2002: 13). Initially McIntosh mounted the telescope outdoors (Fig. 12.4), but after moving houses in 1931 he erected a roll-off roof observatory for it in the backyard of his new property and this is shown in Fig. 19.12. Meanwhile, Fig. 19.13 is a close-up of McIntosh with the telescope.

19.4.2 The Moon

In 1917, when he was just 13 years of age, McIntosh made his first observations of the Moon using binoculars, and on 27 October 1920 he observed a total lunar eclipse, which "… formed the subject of my first published article, which appeared in English Mechanic for January 7, 1921." (McIntosh n.d.: 1920). However, it was only after he acquired the 35.6-cm telescope in 1927 that he commenced serious lunar observing. His account of observations made in 1927 includes the following:

> The Moon was observed on a number of occasions and drawings of the various features were made. A study was begun of the crater Eratosthenes under vertical illumination, the object being to determine the variation alleged to take place, possibly due to lunar vegetation. Drawings were made on five nights. The bright areas on the floors of Plato and Archimedes were mapped.
>
> It appears that the local atmosphere is not sufficiently steady to enable such minute changes to be detected here. (McIntosh n.d.: 1927).

This last sentence explains why he never published his Eratosthenes study, but despite the poor seeing over the next thirty years he went on to study various craters and other lunar features. His first published lunar paper was a 2-page account of the anomalous crater Linné, which appeared in 1936 in *Southern Stars*, the journal of the New Zealand Astronomical Society (McIntosh 1936c).

McIntosh's focus on the Moon may to us appear surprising, but in 1953 he stressed that

> Three things normally deter amateur astronomers from selecting the moon as an object of study—they believe that the beautiful photographs taken with the world's largest telescopes reveal more than they could with their own modest instruments; they believe their own telescopes are too small for serious study and profitable results and, finally, they feel that the lunar detail is too difficult for them to sketch.
>
> Actually such beliefs are quite erroneous …

[4]Joseph Thomas Ward also was instrumental in founding the Wanganui Astronomical Society and securing a 24.1-cm (9.5-in) Cooke refractor for its observatory. This telescope was for many years the largest operational refracting telescope in New Zealand (see Orchiston 2001; Chap. 11 in this book).

Fig. 19.12 A view of R.A. McIntosh, the 35.6-cm reflector and his new roll-off roof observatory (Orchiston collection)

Fig. 19.13 A close up of R.A. McIntosh and his 35.6-cm (14-in) reflector (Orchiston collection)

> It is the constant returning to the same scene, observing it under every angle of illumination and libration, which enables the selenographer to build up a reasonably complete picture of the object he is studying. Do not imagine that your sketches will show the same details at every lunation. (McIntosh 1953: 227–228).

In this same paper McIntosh (1953: 228) mentions that he has "… been sketching the lunar crater of Posidonius on every favourable occasion for more than twenty years and no two sketches are exactly alike." Another lunar crater which he subjected to a long-term detailed study was Aristarchus. This 10-yr project began in 1950, and culminated in the publication of a 9-page paper in the *Journal of the British Astronomical Association* (McIntosh 1961a).

19.4.3 Jupiter

Jupiter was one of McIntosh's favourite Solar System targets, and features in his 'Observatory Report' for 1927, the year in which he acquired the large reflecting telescope: "Jupiter was regularly observed, drawings of its surface features being made on 12 occasions. Satellite phenomena were also recorded." (McIntosh n.d.: 1927). More Jovian drawings followed in 1928 and 1929 (McIntosh n.d.). Initially McIntosh was interested in the appearance of the different belts (for reference purposes see Fig. 19.14), and over many years he made a large number of full-disc drawings (e.g. see Fig. 19.15).

However, from the mid-1930s McIntosh paid particular attention to:

1. The changing colours of the Equatorial Zone and the Great Red Spot; and
2. Eruptions in the Equatorial Zone.

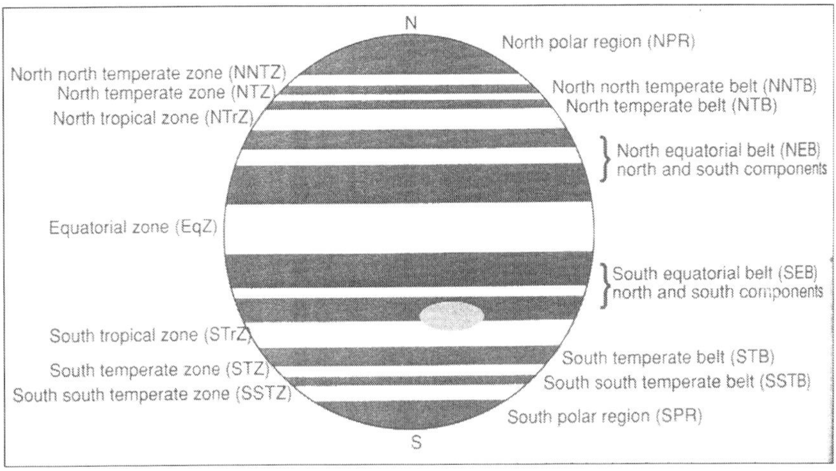

Fig. 19.14 The system of belts and zones on Jupiter

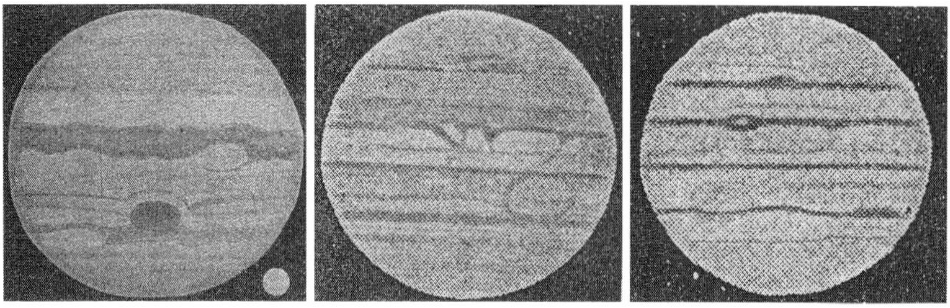

Fig. 19.15 Three of McIntosh's full-disk drawings of Jupiter that he published in his newspaper columns. The size of the Earth is indicated by the small circle in the left hand panel (after McIntosh 1957c, 1959d, 1960 respectively)

In his Observatory Report for 1936 he wrote that "Jupiter was well observed throughout the year, no less than 27 disc drawings being secured. The Red Spot was found to be much redder than I have ever seen it before." (McIntosh n.d.: 1936). Much later, in McIntosh (1961b), he also wrote about the "Recent behaviour of the Red Spot on Jupiter."

McIntosh (1936a) also was interested in whether there was any periodicity on the changing colour of Jupiter's Equatorial Zone, and in 1936, after nine years of observations, he forecast a period of 7.35 years. At this time he also reported that the Equatorial Zone was distinctly yellow (McIntosh 1936b), although other observers saw it as off-white (Rogers 1995: 149–160). In 1959 he published a further paper on Jupiter and its belts and zones (McIntosh 1959a).

In 1949 and 1962 McIntosh noticed outbursts in the South Equatorial Belt, and he published papers on these (see McIntosh 1950; 1965). According to an article that he wrote under the *nom-de-plume* 'Aries' for his newspaper column 'The Newest in Science', the 1949 outbreak was first noticed on 19 July when he

> … saw a small bright spot appear in the [South Equatorial] belt. Ten days later it had expanded to fill the whole width of the belt (approximately 10 degrees on Jupiter) and began to move along the belt in the opposite direction to the rotation of the clouds.
>
> As the spot moved away from its original position another formed at that place and began to follow it, until the 20th spot appeared four months later. The spots, which then extended about three-quarters of the distance around the planet, gave the appearance of pearls upon a string.
>
> There were many other interesting features of the eruption. But the most significant one … was the fact that the new spots always formed in the same longitude. It was as if some upset below the clouds were forcing huge white bubbles, each more than 6000 miles in diameter, to the top of the atmosphere. (Aries 1954).

McIntosh (1965) also mentioned further outbreaks that occurred in the 1950s.

At the time McIntosh was observing Jupiter the true nature of the various belts and zones (see Fig. 19.14), let alone the outbursts, was not understood, but we now know

The zones and belts are zonal jet streams moving with velocities up to 400 miles/hr. Wind direction alternates between adjacent zones and belts. The light colored zones are regions of upward moving convective currents. The darker belts are made of downward sinking material. The two are therefore always found next to each other. The boundaries of the zones and belts (called bands) display complex turbulence and vortex phenomenon.

Note: Upward moving gases in Jupiter's atmosphere bring white clouds of ammonia/water ice from lower layers. Downward moving gases sink and allow us to view the lower, darker layers. (Jupiter the giant planet, n.d.).

We should note that McIntosh was not the first New Zealand astronomer to publish internationally about Jupiter. His predecessor, also from Auckland, was a "Miss Hirst", and a paper about her observations of the planet using a 21.6-cm (8.5-in) Browning-With reflector appeared in *Monthly Notices of the Royal Astronomical Society* in 1875 (see Lambert 1875). She was observing at a time when the changing colours of the Jovian belts attracted worldwide attention (see Hockey 1992, 1999).

19.4.4 Comets

After his juvenile glimpse at 1P/Halley in 1910, McIntosh's first serious observations of a comet were made in June–July 1927 when he observed periodic comet 7P/Pons-Winnecke (McIntosh n.d.: 1927). In his "Record of Early Astronomical Work" McIntosh (n.d.) indicates that from that date up to and including 1936 (when the detailed listing of his annual observations ceased) McIntosh observed comet C/1927 X1 (Skjellerup-Maristany) on five nights between 5 and 10 December 1927 (McIntosh, n.d.: 1927), followed by C/1931 O1 (Nagata). According to Marsden and Williams (1996), this comet was visible from 18 July 1931 until 9 April 1932, but McIntosh does not indicate when he made his observations, or how many, only that on the basis of his observations "My first cometary ephemeris was calculated, showing the motion of Comet Nagata in the southern sky. This was communicated to the Dominion Observatory." (McIntosh n.d.: 1931).

The next comet that McIntosh had the very great pleasure of observing was C/1932 M2, which was discovered by his friend and fellow meteor-observer, Murray Geddes. McIntosh (n.d.: 1932) observed this comet on thirteen different nights between 30 June and 15 July, on each occasion noting its appearance and magnitude, and its position relative to nearby stars.

Exactly five and a half months later, on 1 January 1933, McIntosh made the first of two observations only of C/1932 Y1 (Dodwell-Forbes), the other observation being on 29 January. Four and a half years later he observed Comet C/1936 K1 (Peltier) on eighteen nights between 7 and 28 August 1936, making fifteen positional determinations and taking one photograph.

Of all of the afore-mentioned comets, C/1927 X1 (Skjellerup-Maristany)[5] is of great interest to us given McIntosh's observations of it. By December 1927 this was a conspicuous object in southern skies, and the Director of the Comet Section of the British Astronomical Association, Royal Greenwich Observatory's Andrew Crommelin, reported McIntosh's observations in the *Journal*, and included two drawings he had submitted (Crommelin (1928a). These are reproduced here in Fig. 19.16. Meanwhile, in his report on C/1927 X1 (Skjellerup-Maristany) published in *Monthly Notices of the Royal Astronomical Society* Crommelin (1928b: 298) reveals that

> Mr. R.A. McIntosh of Ponsonby, Auckland, New Zealand, observed it on the evening of December 5, but I gather from his letter that he had heard of its previous discovery elsewhere. *He obtained some fairly accurate positions …*
>
> A drawing, made by Professor G. van Biesbroeck at Yerkes before sunrise on December 21 … shows the nucleus placed eccentrically in the tail, and a bright fan on the sunward side of the nucleus, produced by the intersection of two paraboloid envelopes. The latter feature is also shown in Mr. McIntosh's drawing of December 6; this shows also a dark, narrow stripe down the middle of the tail; the whole appearance closely resembles that of Coggia's comet of 1874 … (Crommelin 1928b; my italics).

Glover (1928) also discusses McIntosh's observations in his paper on the orbit of this comet.

McIntosh was to continue his association with comets discovered by his trans-Tasman colleague Frank Skjellerup when he independently discovered Comet C/1941 B2 (de Kock-Paraskevopoulos) on 27 January 1941. However, investigations soon revealed that this comet had previously been discovered by the Cape

[5]The naming of this comet is unfortunate and totally unwarranted. While Skjellerup has a rightful claim Edmundo Maristany from the La Plata Observatory in Argentina does not. Crommelin (1928a) maintains:

> Mr. Maristany's name should not be retained in the official title of this comet. His discovery came too late for this; when he sent his telegram the comet was already known (even to 'the man in the street') in New Zealand, Australia and South Africa; the mere fact that he sent a telegram to Europe does not entitle him to be singled out from the very numerous previous discoverers …

The New Zealand astronomer, Percy William Glover ($\sim$1877–1947) who computed and published the orbital elements of the comet, confirms this. He lists Skjellerup of Melbourne, a Mr Rhind of New Plymouth and McIntosh of Auckland as the first three persons to report accurate positions to professional observatories, Maristany being the fourth (Glover 1928: 432). So, by rights, this comet should have been named C/1927 X1 (Skjellerup-Rhind), and it should be included in New Zealand's tally of independent cometary discoveries (see Chap. 17 in this book).

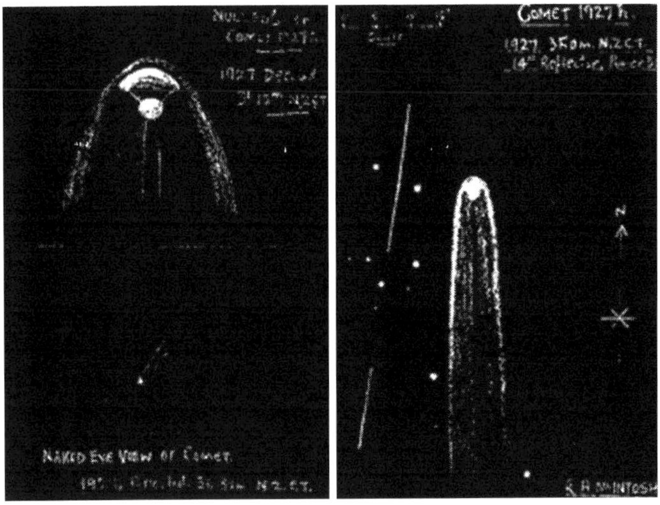

Fig. 19.16 Drawings of Comet C/1927 X1 (Skjellerup-Maristany) made by R.A. McIntosh (after Crommelin 1928a, b)

Town amateur astronomer Reginald Purdon (Rex) de Kock (1902–1980; Glass 1986; Williams 1986.)[6] on 15 January and a Mr. Barnes and Frank Skjellerup (see Fig. 17.8) of Melbourne, Australia, on 21 January. The irony is that Dr John Stefanos Paraskevopoulos (1889–1961; Fig. 19.17) Director of the Boyden Observatory near Bloemfontein, South Africa, was actually the fourth person to independently discover the comet, yet through the vagaries of war-time international telecommunications his name ended up being assigned to the comet even though Skjellerup had the rightful claim. From an historical viewpoint the naming of new comets is sometimes an area of concern—especially where national aspirations are involved—and Comet C/1941 B2 is merely one of a number of comets independently discovered by Australian amateur astronomers who, for one reason or another, were not officially credited with their discoveries (see Orchiston 1997). An image of Comet C/1941 B2 taken at the Yerkes Observatory, USA, is shown in Fig. 19.18.

Back in New Zealand, McIntosh continued to observe the comet, which was known locally as 'Comet McIntosh' or 'Skjellerup-McIntosh' (*Southern Stars* 1941), and on 8 February calculated its orbital elements, which illustrates his computational skills (McIntosh 1941a).

[6]As a result of this discovery de Kock was invited to join the Royal Observatory at the Cape of Good Hope (usually referred to simply as the Cape Observatory) as a computer, and he remained there for the rest of his working life. So he is yet another Antipodean example of an enthusiastic amateur who became a professional astronomer—the so-called 'amateur-professional syndrome' (see Orchiston 2015).

Fig. 19.17 John Paraskevopoulos (*Courtesy* University of Chicago Photographic Archive, apf6-04190, Special Collections Research Center, University of Chicago Library)

19.4.5 Other Observations

In 1918—when he only had access to binoculars—McIntosh began observing sunspots on every possible occasion, and throughout his long observing career—but with a number of notable gaps—he continued to observe sunspots and to make full disk drawings of the Sun. Surprisingly, given his record of publication, he never wrote research papers on any of his solar observations, not even of the 1936 annular eclipse, which

> … was well observed at Auckland, under almost perfect conditions. A party of six observers carried out a very full program of observations at my observatory. Nineteenth photographs were secured, including three showing Bailey's Beads. One also revealed the moon's limb projected against the solar atmosphere. (McIntosh n.d.: 1936).

Instead, accounts of his solar observations were occasionally included in his astronomical column that appeared in the *New Zealand Herald* newspaper.

In 1920 McIntosh (n.d.: 1920) also commenced "A special study of lunar and solar haloes … my first original research." (ibid.). These observations continued throughout 1921, and in 1922 McIntosh (n.d.:1922) recorded that

Fig. 19.18 A photograph of Comet C/1941 B2 (de Kock-Paraskevopoulos) taken with the 61-cm (24-in) reflector at the Yerkes Observatory (*Courtesy* University of Chicago Photographic Archive, apf6-02105, Special Collections Research Center, University of Chicago Library)

> The research was concluded on lunar and solar haloes. This showed haloes to be visible chiefly in light cirrus clouds or mackerel sky. Interesting notes on the colours and dimensions of bows recorded. It was found that lunar rainbows were a very rare occurrence.

These conclusions were never published, but he did publish a short account of paraseleane, or mock moons, seen in New Zealand (McIntosh 1929a).

McIntosh carried out his first variable star observations in 1924, "… showing that useful work can be accomplished with binoculars." He was proud to record that in November 1924 these observations were published by the New Zealand Astronomical Society (McIntosh n.d.: 1924) which would have been the first time that many New Zealand astronomers encountered the name of this 20-year old amateur astronomer from Auckland. The following year he religiously followed Nova Pictoris from June to September, recording the suite of magnitude estimates listed in Table 19.2, from which "… an excellent light-curve [was] determined."

Table 19.2 McIntosh's visual magnitude estimates of Nova Pictoris 1925

Date (1925)	m_V
25 May	2.5
8 June	1.2
27 June	3.4
28 June–6 July	3.4
9 July	2.8
10 July-18 July	2.8
19 July	1.8
27 July	1.7
4 August	3.2
9 August	1.7
23 August	2.8
22 September	3.5
9 October	3.6
27 October	3.8
6 November	4.4

Subsequently this was published in the 'Monthly Notes' of the New Zealand Astronomical Society (McIntosh n.d.: 1925), reinforcing the view that this keen young Aucklander was someone worth watching.

After this other observing priorities prevailed—and especially once he acquired the 35.6-cm (14-in) telescope in 1927 and built a roll-off roof observatory for it—so it is a little surprising to see him return to variable stars briefly in 1934: "Some naked eye observations of the variables l Carinae and Beta Doradus for the Variable Star Section were made in co-operation with Lembang Observatory." (McIntosh n. d.: 1934). This is rather strange since Lembang Observatory near Bandung, Java, in what was then the Dutch East Indies (now present-day Indonesia), specialised in photographic observations of binary stars. I can only think that Frank Bateson, Director of the NZAS's Variable Star Section, had a special program running involving l Car and Beta Dor, suggested by his Lembang colleagues, and that he called on old friends for support. As we have seen, Bateson was a very active member of McIntosh's NZAS Meteor Section, so McIntosh was only too happy to oblige.

In addition to sunspots, solar and lunar haloes and variable stars, McIntosh (n.d.) carried out occasional observations of Mercury, Venus, Mars and Saturn; aurorae; nebulae; double stars; star clusters and the Zodiacal Light. He also observed a transit of Mercury. So despite his penchant for meteors and passion for Jupiter, he displayed a catholic approach to observational astronomy.

19.5 Other Astronomical Interests and Activities

19.5.1 The History of New Zealand Astronomy

In additional to observational astronomy, McIntosh was interested in the history of New Zealand astronomy, both Maori and non-Maori.

In 1946 he wrote a paper on "Maori meteor lore", and described how the appearance of *Rongomai* (a bright meteor or fireball) foretold the outcome of current or imminent hostilities (McIntosh 1946a). Then between 1957 and 1959 (inclusive) he published an invaluable series of short papers on the "Astronomical History of the Auckland Province" in *Amateur Astronomer,* the journal of the Auckland Astronomical Society. These dealt with Maori astronomy (McIntosh 1957a); Cook's visits to New Zealand (McIntosh 1957b); Auckland's role in the observation the two nineteenth century transits of Venus, and early local developments in astronomical photography (McIntosh 1958a); the Thames astronomer John Grigg, who early in the twentieth century arguably was the New Zealand astronomer with the highest international profile (McIntosh 1958b) thanks largely to his succession of cometary discoveries—as discussed elsewhere in this book; Thomas Cheeseman, other leading nineteenth century Auckland astronomers, and formation of the city's first astronomical society (McIntosh 1959b); and Les Comrie (Fig. 19.19), Auckland's first astronomer to achieve fame overseas as a professional astronomer (McIntosh 1959c). What made this review of Auckland region astronomy so valuable was the fact that, as a journalist, McIntosh was keenly aware of the wealth of astronomical information buried in the pages of old local newspapers, and he made a point of mining these when researching his papers.

An excellent project for the Auckland Astronomical Society given its up-coming centennial, or those at Star Dome, would be to publish a monograph that not only faithfully reprints all six McIntosh papers but includes additional short chapters on later key developments in the history of astronomy in Auckland. For example, these later chapters could discuss the important contribution that Leigh and Piha played in solar and non-solar radio astronomy between 1945 and 1948 inclusive (see the last two chapters in this book); the founding of Auckland Observatory and its development; the history of the Auckland Astronomical Society; pioneering research in variable star photometry by Grant Christie, the late Brian Marino, Stan Walker and the late Roger Freeth; more recent developments in photometry by enthusiasts like Jenny McCormick (and others); amateur telescope-making in Auckland—perhaps with individual short chapters on Graham Loftus and Harry Williams; Lionel Warner's outstanding efforts in popularizing astronomy; the birth and development of astronomy at the University of Auckland; the launch of radio astronomy at the Auckland University of Technology; and of course McIntosh's own role in promoting astronomy throughout the region (see Sect. 19.5.2). I am sure that Auckland astronomers can immediately think of additional chapters that would demand a place in such a monograph.

Fig. 19.19 Dr Leslie Comrie (after Massey 1952)

We should note that McIntosh (1946b) did not restrict his historical writing to the Auckland region: as Director, he also wrote a paper on the history of the Meteor Section of the Royal Astronomical Society of New Zealand, and in 1970 he wrote an important review paper titled "Early New Zealand astronomy" (McIntosh 1970). Apart from an obituary that he penned the following year (McIntosh 1971), this was to be his last known astronomical publication, and it appeared when he was sixty-six years of age. With the benefit of hindsight, one can safely say that McIntosh himself is now seen as an important part of New Zealand's astronomical history!

19.5.2 Popular Astronomy

As well as his research interests, McIntosh had a strong commitment to promote astronomy among the general public (see St. George 1987), and he did this mainly in three different ways.

From 1927 he began an astronomy column titled initially "The New Zealand Skies Month by Month" and later "Glory of the Stars", and finally "The Sky This Week", which was published in the *New Zealand Herald* newspaper for more than 45 years. As well as containing descriptions of the night sky, his columns included information on important up-coming astronomical events (such as eclipses or stellar and planetary occultations by the Moon), and descriptions of objects (e.g. comets, individual variable stars, sunspots) or events (e.g. auroral displays, eclipses, meteor showers, planetary conjunctions, transits of Mercury) once they had occurred or been witnessed. Over the years, he also wrote about all of the planets, about exoplanets, gaseous nebulae, lunar craters, Maori astronomy, meteorites, the Russian space program, the Star of Bethlehem and the Auckland Astronomical

Society. Note that this is merely a selection, designed to show the wide coverage of topics in his column. But McIntosh was interested also in other areas of science, so from 1949 to 1955 he wrote a weekly column titled "Latest in Science", which covered a wide range of topics, although astronomy articles tended to predominate. Through these two columns and various other articles that he wrote, McIntosh was able to bring the glory of astronomy to a wide audience.

In 1959 McIntosh was appointed the Lecturer-Demonstrator at the Auckland Planetarium (Fig. 19.20) which was housed in the Auckland Institute and Museum, and he continued in this role until his retirement in 1972 at the age of 68. In the speech he delivered at this farewell on 5 October 1972 McIntosh reflected on his first day in the job:

> I must particularly mention my first boss in the Museum – Sir Gilbert Archey. Opening day was an abnormal one. I gave six lectures during the day, four more in the evening and – the bitterest blow of all – a member of the Council who had missed all the lectures insisted on a solo audience.
>
> Somehow I staggered through the eleventh demonstration and emerged in front of Sir Gilbert. He took one glance at me and herded me into his office, where he pumped two stiff whiskeys into me.
>
> I don't know how I got home. I suspect my wife had to carry me. (McIntosh 1972).

Fortunately, not all days were to be like this, and over the years McIntosh used the planetarium to popularize astronomy. By the time he retired he had delivered about 310,000 lectures (ibid.), and as Bateson (1977: 85) has pointed out,

> He spent many hours preparing each lecture to make certain that it was correct and presented in an interesting manner. His journalistic training and wide general knowledge were great assets in his conduct of public sessions.

Fig. 19.20 R.A. McIntosh giving a lecture-demonstration in the Auckland Planetarium in 1959, soon after the opening of this popular facility (Auckland's new planetarium … 1959)

The third way in which McIntosh enjoyed bringing the excitement of astronomy to a wide audience was by conducting 'Public Nights' at the Auckland Observatory (see Fig. 19.21). Through these popular sessions, and his planetarium programs McIntosh was able to profoundly influence the astronomical perceptions, attitudes and interest of a great many people, from members of the general public to amateur astronomers, school children and special interest groups. He also was able to bring the magic of the New Zealand sky on a dark moonless night to visitors from the Northern Hemisphere who wisely decided to include the Museum and a planetarium show in their tour itinerary.

Apparently McIntosh also somehow found time to conduct Adult Education classes on astronomy. The foregoing account merely outlines Ronald McIntosh's long and very important role in promoting astronomy in New Zealand, and it is clear that this whole topic warrants a separate detailed study.

19.5.3 Founding of the Auckland Observatory

From the early 1950s Ronald McIntosh began lobbying for the establishment of a public observatory in Auckland, pointing out in his newspaper columns the ways in which the Carter Observatory in Wellington, the Wanganui Observatory and the New Plymouth Astronomical Society's observatory served to foster popular interest in astronomy in these cities, and provide facilities where astronomers could carry out serious observations. As New Zealand's largest city Auckland was missing out … Fig. 19.22 shows a sketch of a suggested public observatory for Auckland that McIntosh published in 1953 in his column 'The Newest in Science' (Aries 1953).

Further designs followed, and by the time McIntosh's dream looked like becoming a reality a very different-looking Auckland Observatory was presented to the public (see Fig. 19.23). The opening finally took place in 1967 (see McIntosh 1967) after a long hard slog, as Bateson (1977: 85) recounts:

> [McIntosh] … commenced to formulate plans as early as 1953, and battled to have his ideas accepted, even having to overcome opposition from within the ranks of the Auckland Astronomical Society in the mid-1950's. He was the first Secretary of the Auckland Observatory and Planetarium Trust Board, whose first members were carefully selected by himself. The final selection of a suitable site and the erection of the Observatory constituted a fitting memorial to one who was the true founder of astronomy in Auckland.

While there were others who also were actively involved in these developments, it is true that McIntosh played a key role, and hopefully this will be fully documented in the up-coming centennial publication(s) of the Auckland Astronomical Society.

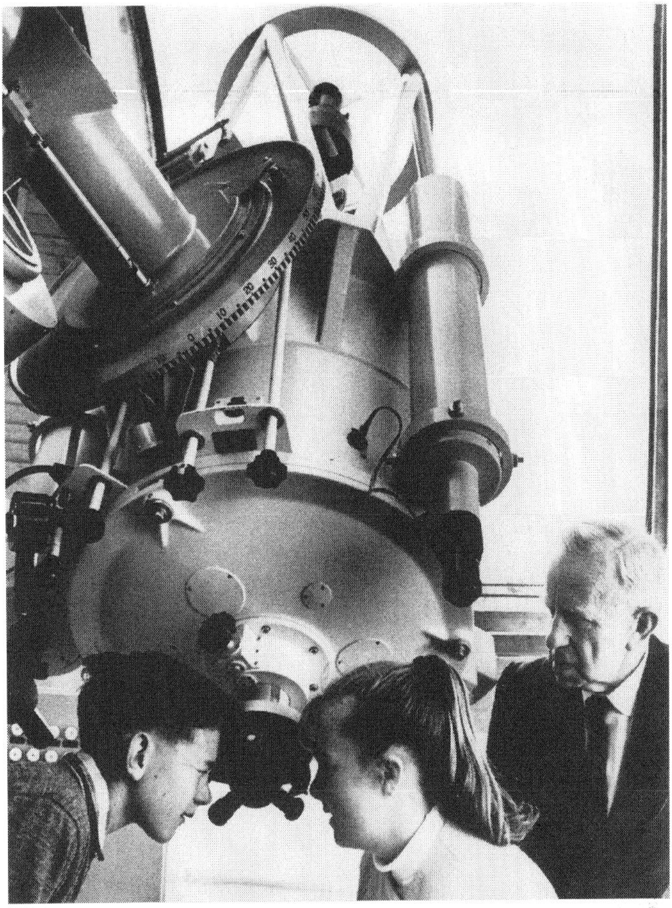

Fig. 19.21 R.A. McIntosh and two visitors with the Auckland Observatory's 50.8-cm (20-in) Zeiss reflector in 1978 (Orchiston collection)

Fig. 19.22 A design for a public observatory in Auckland suggested in 1953 (after Aries 1953)

Fig. 19.23 The design of the Auckland Observatory displayed in the *New Zealand Herald* on 24 January 1959 (after Observatory design, 1959)

19.5.4 Involvement in Astronomical Societies

In a submission to Otago University regarding the Mechaelis Memorial Award, McIntosh (1968: 3) lists society memberships and awards, and the former are shown here in Table 19.3.

Given that he lived and worked in Auckland most of his life it is only natural that McIntosh's main societal involvement was with the Auckland Astronomical Society, which as Table 19.3 indicates, he joined in 1928, when he was 23 or 24 years of age and was already starting to become known locally and nationally for his meteor work. In this context, it is interesting that in his manuscript "Record of Early Astronomical Work" McIntosh closes his account of the year 1926—just prior to purchasing the 35.6-cm (14-in) telescope—with the following self-evaluation:

> This closes the earlier years of my interest in astronomy, when my knowledge had gradually grown from zero to a very fair state and although I had not done much observing on account of my various roaming [e.g. for a little while he lived in Hamilton] and the fact that I spent all of this time boarding still I had gained some practical experience. *At the same time I had begun writing articles on meteors and had attained come reputation as an observer.* (McIntosh, n.d.: 1926; my italics).

In his Observatory Report for 1928 McIntosh tells us that "My first lecture was delivered this year, a general lecture on comets entitled "Comet Skjellerup, 1927k" being delivered on September 12 to the Auckland Astronomical Society at the University." (McIntosh n.d.: 1928). This was the same year in which he joined the Society, and the topic was particularly apposite given his independent discovery of Comet C/1927 X1 (Skjellerup-Maristany) the previous year and all the publicity that this generated at the time. His next lecture to the Society, on 11 December 1930, was on his favourite research theme, "Meteors", and was well attended (McIntosh n.d.). By the time he gave his next lecture to the Society, on 9 November 1932, McIntosh was well known and respected throughout New Zealand as a serious observer who got results and published them, and his topic was particularly appropriate: "Research for Amateur Astronomers". As might be expected, this drew

Table 19.3 R.A. McIntosh's society involvement

Society	Year joined	President	Remarks
American Meteor Society	1919		
Auckland Astronomical Society	1928	1940–1948 1952–1954 1956–1957	Elected Life Member in 1961
British Astronomical Association	1927		
International Astronomical Union	1936		Member of Commission 22
New Zealand Astronomical Society (later the Royal Astronomical Society of New Zealand)	1921	1942–1943	Vice-President 1943–1948 Elected Fellow of the RASNZ 1965 Meteor Section Director 1928–1955
Royal Astronomical Society	1929		Elected a Fellow
Royal Astronomical Society of Canada	1931		
Societé Astronomique de France	1928		

a good audience (ibid.). His next lecture, on 11 July 1934, was on yet another topic dear to his heart, meteorites and impact cratering, the title being "Our Stone-Pelted Planet", then the following year he was able to expound on yet another topic when on 11 December 1935 he lectured to Society on "Exploring the Atmosphere" (ibid.).

Over the years, Ronald McIntosh made a major contribution to the growth and development of the Auckland Astronomical Society, as Table 19.3 foreshadows, and I look forward to reading the account of this that undoubtedly will be included in the Society's Anniversary history.

19.6 Concluding Remarks

Ronald Alexander McIntosh made an important contribution to Auckland and New Zealand astronomy over a period of more than four decades, starting in the late 1920s, and his meteor work was of international importance and plugged the observational gap that existed in the Southern Hemisphere. The formation of the Meteor Section of the New Zealand Astronomical Society (later the Royal Astronomical Society of New Zealand) was a testament to his leadership skills, and he was justly proud of the research that he and the dedicated members of the Section were able to accomplish and their record of publication in local and international journals. However, these very same publications attracted criticism from some in the Society, to which McIntosh (1941c) replied:

> To the writer's view the amateur's task consists of observing, recording and publishing. If the last of those is not done, the time spent on the former two is wasted. It is all very well for a New Zealand observer to watch for meteors, to discover that they come from new radiants, and to store the information in his mind. If he does not publish it the knowledge will die with him and no one will know about the new radiants.

But as we have seen, not all McIntosh's observational efforts were directed towards meteors. Once he acquired a 35.6-cm (14-in) reflector in 1927 he also paid attention to the Moon and Jupiter, making a great many drawings and ultimately publishing a number of papers based on these observations. His research interests also extended to the history of astronomy, and even now his papers on the history of astronomy in the Auckland region make interesting reading.

The other area in which McIntosh made a major contribution was in popularizing astronomy, mainly through his astronomical columns and reports in various newspapers, but particularly the *New Zealand Herald*; his role as Planetarium Lecturer-Demonstrator at the Auckland Institute and Museum; and through public nights at the Auckland Observatory (which he played a key role in founding). We cannot do full justice here to his exploits in astronomy education and popular astronomy, which deserve to be fully documented by others with a more intimate knowledge of the way in which Auckland astronomy evolved and developed. Only then will the full story of Ronald Alexander McIntosh be told.

Acknowledgements I am grateful to Michael Luciuk (USA) and the late Jackie St. A (Sydney) for their assistance. I also wish to thank John Drummond (Patutahi, NZ) for reading and commenting on the first draft of this chapter, and the University of Chicago Library for kindly supplying Figs. 19.17 and 19.18.

References

Aries (i.e. McIntosh, R. A.). (1953). Observatory for Auckland. In 'The Newest in Science', newspaper article dated April 28 (copy in Orchiston collection).

Aries (i.e. McIntosh, R. A.). (1954). Does Jupiter erupt? In 'The Newest in Science', newspaper article dated June 29 (copy in Orchiston collection).

Auckland's new planetarium shows all … the stars at your command. *The Weekly News*, November 4, 1959.

Bateson, F. M. (1977). Ronald Alexander McIntosh (1904–77). *Southern Stars, 27*, 82–87.

Beech, M. (1998). W.F. Denning—the doyen of amateur astronomers. *WGN, the Journal of the IMO, 26*, 19–24.

Crommelin, A. C. D. (1928a). Notes. *Journal of the British Astronomical Association, 38*, 124–126.

Crommelin, A. C. D. (1928b). Reports on comets in 1927. *Monthly Notices of the Royal Astronomical Society, 88*, 293–299.

Denning, W. F. (1872). Observations of luminous meteors. *Monthly Notices of the Royal Astronomical Society, 33*, 93–95.

Denning, W. F. (1915). Peculiarities of meteoric radiants. *The Observer, 38*, 334–335.

Eiby, G. (1970). Obituaries. Ivan Leslie Thomsen. *Southern Stars, 23*, 113–116.

Glover, P. W. (1928). The orbit of the Comet 1927k. (Skjellerup). *Transactions of the Royal Society of New Zealand, 59*, 429–442.

Gyssens, M. (Ed.), (1998). *WGN, The Journal of the IMO*, 85–92.

Hawkins, G. S. (1964). *Meteors, Comets and Meteorites*. New York: McGraw-Hill.

Hockey, T. (1992). Seeing red: observations of colour in Jupiter's Equatorial Zone on the eve of the modern discovery of the Great Red Spot. *Journal for the History of Astronomy, 23*, 93–105.

Hockey, T. (1999). *Galileo's Planet: Observing Jupiter before Photography*. Bristol: Institute of Physics.

Jarrell, R. A. (2009). Canadian meteor science: the first phase, 1933–1990. *Journal of Astronomical History and Heritage, 12*, 224–234.

Jenniskens, P. (2006). *Meteor Showers and their Parent Comets*. Cambridge: Cambridge University Press.

Jupiter the giant planet (n.d.). (http://abyss.uoregon.edu/~js/ast121/lectures/lec19.html).

Keay, C. S. L. (1965). Meteor research in New Zealand. *Southern Stars, 21*, 9–19.

Kronk, G. W. (1988). *Meteor Showers*. Hillside: Enslow Publishers Inc.

Lambert, S. J. (1875). Remarks on drawings of Jupiter made by Miss Hirst, at Auckland, New Zealand. *Monthly Notices of the Royal Astronomical Society, 35*, 401–403.

Lovell, A. C. B. (1954). *Meteor Astronomy*. Oxford: Oxford University Press.

Luciuk, M., & Orchiston, W. (2016). Ronald A. McIntosh: pioneer Southern Hemisphere meteor observer. *Journal of Astronomical History and Heritage, 19* (in press).

McIntosh, R. A. (n.d.). Record of early astronomical work. Manuscript (Copy in Orchiston collection).

McIntosh, R. A. (1929a). Paraselenae observed in New Zealand. *Popular Astronomy, 38*, 250–251.

McIntosh, R. A. (1929b). The meteor swarm of Halley's Comet. *Monthly Notices of the Royal Astronomical Society, 90*, 157–160.

McIntosh, R. A. (1930). Report of the section for the observations of Meteors, 1927–1928. *Transactions of the New Zealand Institute, 60*, 448–464.

McIntosh, R. A. (1931). The Aquarid meteors. *The Observatory, 54*, 78–80.

McIntosh, R. A. (1932). Meteors. *Journal of the Royal Astronomical Society of Canada, 26*, 193–210.

McIntosh, R. A. (1934). Ephemeris of the radiant-point of the Delta Aquarid meteor stream. *Monthly Notices of the Royal Astronomical Society, 94*, 583–587.

McIntosh, R. A. (1935a). Ephemeris of the radiant-point of the Eta Aquarid meteor stream. *Monthly Notices of the Royal Astronomical Society, 95*, 601–604.

McIntosh, R. A. (1935b). An index to Southern meteor showers. *Monthly Notices of the Royal Astronomical Society, 95*, 709–718.

McIntosh, R. A. (1936a). Colour changes on Jupiter. *Southern Stars, 2*, 34.

McIntosh, R. A. (1936b). Colour variations on Jupiter. *Southern Stars, 2*, 60–61.

McIntosh, R. A. (1936c). Linné. *Southern Stars, 2*, 106–107.

McIntosh, R. A. (1938). Determination of the real paths of fireballs. *Journal of the Royal Astronomical Society of Canada, 32*, 1–24.

McIntosh, R. A. (1940a). Seasonal variation in the height of meteors. *Journal of the Royal Astronomical Society, 100*, 510–528.

McIntosh, R. A. (1940b). An investigation of the accuracy of meteor radiants. *Journal of the Royal Astronomical Society of Canada, 34*, 378–382.

McIntosh, R. A. (1941a). The history of Comet de Kock 1941a. *Southern Stars, 7*, 75–76.

McIntosh, R. A. (1941b). The observation of meteors. *Southern Stars, 7*, 9–14.

McIntosh, R. A. (1941c). Meteor Section. *Southern Stars, 7*, 45–46.

McIntosh, R. A. (1943). The velocities of meteors. *Journal of the Royal Astronomical Society of Canada, 37*, 305–315.

McIntosh, R. A. (1944). The luminous paths of meteors. *Journal of the Royal Astronomical Society of Canada, 38*, 193–200.

McIntosh, R. A. (1946a). Maori meteor lore. *Southern Stars, 12*, 7–9.

McIntosh, R. A. (1946b). A history of the Meteor Section. *Southern Stars, 12*, 10–15.
McIntosh, R. A. (1950). Disturbance on Jupiter's South Equatorial region. *Journal of the British Astronomical Association, 60*, 247–250.
McIntosh, R. A. (1953). Your piece of the Moon. *Journal of the Royal Astronomical Society of Canada, 47*, 227–228.
McIntosh, R. A. (1957a). Astronomical history of the Auckland Province. Part I: The Maoris. *Amateur Astronomer* (Journal of the Auckland Astronomical Society), *17*(2), 20–22.
McIntosh, R. A. (1957b). Astronomical history of the Auckland Province. Part II: Eighteenth century. *Amateur Astronomer* (Journal of the Auckland Astronomical Society), *17*(3), 40–42.
McIntosh, R. A. (1957c). Jupiter's thunderbolts heard 500 million miles away … In 'The Science of Space', November 5 (copy in Orchiston collection).
McIntosh, R. A. (1958a). Astronomical history of the Auckland Province. Part III: Nineteenth century. *Amateur Astronomer* (Journal of the Auckland Astronomical Society), *18*(1), 3–6.
McIntosh, R. A. (1958b). Astronomical history of the Auckland Province. Part IV: John Grigg, F.R.A.S. *Amateur Astronomer* (Journal of the Auckland Astronomical Society), *18*(2), 20–23.
McIntosh, R. A. (1959a). Amateur observations of Jupiter. *Southern Stars, 18*, 5–8.
McIntosh, R. A. (1959b). Astronomical history of the Auckland Province. Part V: First astronomical society. *Amateur Astronomer* (Journal of the Auckland Astronomical Society), *18* (3), 46–48.
McIntosh, R. A. (1959c). Astronomical history of the Auckland Province. Part VI: Early twentieth century. *Amateur Astronomer* (Journal of the Auckland Astronomical Society), *18*(4), 60–65.
McIntosh, R. A. (1959d). Clouds of Jupiter in constant change. 'The Sky This Week'. *New Zealand Herald*, May 18.
McIntosh, R. A. (1960). Clouds of Jupiter pose problems. 'The Sky This Week'. *New Zealand Herald*, June 13.
McIntosh, R. A. (1961a). Features of the lunar crater Aristarchus. *Journal of the British Astronomical Association, 71*, 380–388.
McIntosh, R. A. (1961b). Recent behaviour of the red spot on Jupiter. *Southern Stars, 19*, 71–79.
McIntosh, R. A. (1965). The 1962 eruption on Jupiter. *Southern Stars, 21*, 20–26.
McIntosh, R. A. (1967). Auckland's new observatory. *Southern Stars, 22*, 93–96.
McIntosh, R. A. (1968). Letter to University of Otago re: Mechaelis Memorial Award. Manuscript. Copy in the Orchiston collection.
McIntosh, R. A. (1970). Early New Zealand astronomy. *Southern Stars, 23*, 101–108.
McIntosh, R. A. (1971). Obituary. Allan Bryce. *Southern Stars, 24*, 19.
McIntosh, R. A. (1972). My Oration on my retirement from the Museum 1972 October 5. Copy in Orchiston collection.
McLintock, A. H. (Ed.) (1966). *An Encyclopaedia of New Zealand.* Auckland City (http://www.teara.govt.nz/1966/A/AucklandCity/Climate/en).
Mackrell, B. (1985). *Halley's Comet Over New Zealand.* Auckland: Reed Methuen.
Maltzev, V. A. (1928). Concerning the fictitious radiants of meteoric streams. *Astronomische Nachrichten, 234*, 244–247.
Marsden, B. G., & Williams, G. V. (1996). *Catalogue of Cometary Orbits 1996.* Cambridge (Mass.): Minor Planet Center, Smithsonian Astrophysical Observatory.
Massey, H. S. (1952). Leslie John Comrie, 1893–1950. *Obituary Notices of Fellows of the Royal Society, 8*, 96–107.
Morgan, J. E. (1979). Southern meteor radiants. *Southern Stars, 28*, 39–49.
Morse, K. I. (1981). *Meteor Observers' Handbook.* Wellington: Meteor Section, Royal Astronomical Society of New Zealand.
Observatory design. *New Zealand Herald*, January 24, 1959.
Olivier, C. P. (1925). *Meteors.* Baltimore: Williams and Wilkins Company.
Öpik, E. (1934). Results of the Arizona expedition for the study of meteors, I. Statistical analysis of group radiants. *Harvard College Observatory Circular, 388*, 1–38.

Öpik, E. J., & Shapley, H. (1937). Results of the Arizona expedition for the study of meteors, V. On the distribution of heliocentric velocities of meteors. *Harvard College Observatory Circular, 391*, 1–52.

Orchiston, D. W. (1963). The return of the 1956 Phoenicid meteor shower. *Southern Stars, 20*, 23–25.

Orchiston, D. W. (1964). A minor meteor radiant in Libra. *Southern Stars, 20*, 162–163.

Orchiston, W., & Wolf, G. (1994). *Ronald McIntosh: A Remarkable New Zealand Astronomer* (pp. 4). Wellington: Carter Observatory (Information Sheet No. 4).

Orchiston, W. (1997). The "tyranny of distance" and Antipodean cometary astronomy. *Australian Journal of Astronomy, 7*, 115–126.

Orchiston, W. (2000). R.A. McIntosh. In C. Orange, et al. (Eds.), *The Dictionary of New Zealand Biography. Volume 5. 1941–1955*. Auckland: Auckland University Press and Department of Internal Affairs.

Orchiston, W. (2001). The English Equatorial Mounting and the history of the Fletcher Telescope. *Journal of Astronomical History and Heritage, 4*, 29–42.

Orchiston, W. (2002). Joseph Ward: pioneer New Zealand telescope-maker. *Southern Stars, 41*(2), 13–21.

Orchiston, W. (2015). The amateur-turned-professional (ATP) syndrome: two Australian case studies. In W. Orchiston, D. Green, & R. Strom (Eds.), *New Insights From Recent Studies in Historical Astronomy: Following in the Footsteps of F. Richard Stephenson* (pp. 259–350). New York: Springer.

Plummer, H. C. (1941). On the heights and velocities of meteors. *Monthly Notices of the Royal Astronomical Society, 101*, 76–83.

Porter, J. G. (1943). An analysis of British meteor data. *Monthly Notices of the Royal Astronomical Society, 103*, 134–153.

Rogers, J. S. (1995). *The Giant Planet Jupiter*. Cambridge: Cambridge University Press.

Southern Stars, *7*, 20–22, 25–28 (1941).

St. George, J. (1987). *History of Auckland Astronomy*. Manuscript (copy in Orchiston collection).

Wolf, G. W. (1994). Ronald Alexander McIntosh—not just a Southern meteor pioneer. *Proceedings IMC Belogradchik, 1994*, 78–85.

Zejda, M. (2014). Hoffmeister, Cuno. In T. Hockey et al. (Eds.), *Biographical Encyclopedia of Astronomers* (2nd ed, pp. 991–993). New York: Springer.

Part VI
Other Notable Astronomers and Their Activities

Chapter 20
Great Comets, and Wellington's Earliest European Astronomers

Abstract During the 1840s, soon after the initial European settlement of Port Nicholson, Wellington gained its first scientific astronomers, and these included the surveyors R. Sheppard and W.M. Smith. The early settlers also included J.H. Marriott, who had trained as a scientific instrument-maker in England, but there is little evidence that he practised this trade once he reached New Zealand.

20.1 Introduction

Long before the arrival of the first Europeans in Port Nicholson, Maori astronomers practised their craft in the Wellington area. However, their system of celestial beliefs was intricately interwoven with religion and mythology (see Best 1922; Orchiston 2000a; Chap. 2 in this book), and as such differed markedly from the nautical and positional astronomy typical of European observatories at the end of the eighteenth century.

During the 1860s a professional astronomical tradition began to emerge in Wellington, largely through initiatives undertaken by the Provincial Government, and these developments are well-known to historians of science (e.g. see Eiby 1977, 1978; Orchiston 1996, 1998c, 2001a). What is not so widely known is that astronomers were in fact among the very first European settlers of the Wellington region (see Orchiston 2001b).

In the 1840s, surveyors and the captains and officers of sailing ships learnt and regularly used astronomy in the course of their occupations, but few of these developed a passion for astronomy and were driven to carry out astronomical observations for the mere pleasure of it or through scientific curiosity. Those who did, and were committed astronomers, often published their observations at this time in *Monthly Notices of the Royal Astronomical Society* and/or in local newspapers.

During the nineteenth century, three different types of astronomical events or objects created enormous public interest worldwide, and they were transits of

W. Orchiston, *Exploring the History of New Zealand Astronomy*,
Astrophysics and Space Science Library 422,
DOI 10.1007/978-3-319-22566-1_20

Venus (Cottam et al. 2011a, 2012), total solar eclipses (Cottam et al. 2011b) and impressive naked eye comets (Cottam and Orchiston 2014; Olson and Pasachoff 1998). In the case of Australia, for instance, and comet C/1861 K1 (the Great Comet 1861, discovered by Australia's John Tebbutt), I have shown how Australian astronomers, and even those with a passing interest in astronomy, made effective use of the Sydney newspapers to publicize their observations or debate their views on cometary astronomy (see Orchiston 1998a, b).

There were no transits of Venus or total solar eclipses visible from New Zealand in the 1840s, when Wellington received its initial influx of European settlers, but two impressive naked eye comets were visible in Wellington skies in 1843 and 1844–1845. In this chapter we will use these two comets to try and identify Wellington's first European astronomers. As such, this chapter builds on Orchiston (2001b).[1]

20.2 The Great Comets of 1843 and 1844–1845

During the 1840s, those in the southern hemisphere were treated to two unforgettable astronomical spectacles: the Great Comets of 1843 (C/1843 D1) and 1844–1845 (C/1844 Y1).

The famous British astronomer Sir John Herschel (1792–1871) described the appearance of the 1843 comet:

> In more southern latitudes, however, not only the tail was seen, as a magnificent train of light extending 50° or 60° in length; but the head and nucleus appeared with extraordinary splendour, exciting in every country where it was seen the greatest astonishment and admiration. (Herschel 1851: 366; cf. Kronk 2003: 129–137).

Sekanina and Chodas (2008: 1415) regard C/1843 D1 as "... the first spectacular member of the Kreutz system of sungrazing comets discovered in the course of the past two centuries ..." and conclude that it has a period of between 600 and 800 years. They believe it is "... possibly the most massive, fragment of the celebrated sungrazer X/1106 C1 ..." (Sekanina and Chodas 2007). They identify two different Kreutz system subgroups (Sekanina and Chodas 2004) that can be traced back to an earlier 'ancestral sungrazer'. The most prominent members of Subgroup I are comets C/1843 D1 and C/1880 C1, and of Subgroup II are comets C/1882 R1, C/1963 R1 and C/1965 S1.

In comparison, the Great Comet of 1844–1845 may not have the research importance of the earlier Great Comet, but it was still an impressive object.

[1]Note that an earlier version of this paper in which archival data from Cambridge were unfortunately omitted, was published in *Southern Stars* (Orchiston 2000b). After I had completed the first draft of this chapter of the book, Brown (2014) published a very useful little paper—also in *Southern Stars*—that provided some new information on early European astronomy in Wellington and newly-sourced photographs of 'key' individuals, which I have been able to use here.

According to Harvard College Observatory's George Phillips Bond (1825–1865; Turner 2014), C/1844 Y1

> … was first seen at the Cape of Good Hope, on the 18th of December, 1844. On the day following, it was seen by Captain Wilmot, R.A., and by Mr. Maclear of the Cape Observatory. In the Monthly Notices of the Astronomical Society, it is usually designated as "Wilmot's" comet. It continued to be visible to the naked eye from the date of its discovery until the end of January, and was followed with the telescope up to the 12th of March. (Bond 1850: 97; cf. Kronk 2003: 145–149).

Both comets drew the attention of many New Zealanders to the wonders of astronomy, and their appearance at the very time when Wellington gained its first European astronomers was particularly fortuitous.

20.3 Observations of the Two Great Comets from Wellington

20.3.1 Comet C/1843 D1

When Comet C/1843 D1 first appeared Wellington boasted two different newspapers, and each described the welcome visitor to Wellington skies. On 7 March *The New Zealand Colonist and Port Nicholson Advertiser* (1843c), reported:

> The inhabitants of Wellington were surprised on Saturday and Sunday evenings last, by the appearance of a splendid comet in the south-west of Port Nicholson. As it set at an early hour in the evening we had no opportunity of seeing it entire.
> In a few evenings, we have no doubt, it will be seen to much more advantage, when we shall be enabled to give a fuller account of the "Tail of the Comet."

The following day it was the turn of the *New Zealand Gazette and Wellington Spectator* (1843):

> On Saturday and Sunday evenings, a comet appeared in the south-west quarter of the heavens, and was seen to great advantage by a large number of the inhabitants. The Maoris hailed it as an evil omen, and commenced howling very pathetically. It will most likely again be seen in the course of a few evenings.

The comet continued to impress over the following weeks, and to provide readers with background information on comets, the *New Zealand Colonist and Port Nicholson Advertiser*, published a 4-part series on "The Cometary System" (1843b) between 14 March and 24 March, with data extracted mainly from the *Magazine of Popular Science.*

Then on 19 April, when the comet was well past its prime, a detailed account of it was published by W.M. Smith (1843). An identical report by him appeared on both newspapers, and I have chosen to reproduce the version in the *New Zealand Gazette and Wellington Spectator*:

This phenomenon which has been for some time visible, has now nearly if not completely vanished. Its appearance has not been foretold in the Ephemeris for the present year, but it was first seen from Wellington, I believe, on the night of the 4th March, travelling away from the Sun in a North Easterly direction. The following observations were taken by a Sexant [*sic*] the best instrument at my command, and will enable some of your numerous readers, to trace its route on the Celestial Globe, or on a Planisphere which may easily be constructed on a Mercator's projection from the right ascensions and declensions [*sic*] of the principal fixed stars, given in the Nautical Almanack [*sic*].

On Wednesday, 8th March.—The Comet became visible about three-quarters of an hour after the Sun had set. The Angular distance from Rigel a large star at the foot of Orion, to the Nucleus was 60° 30′, bearing 17° south of west, or nearly W. by S. ½ S., from Sirius to the Nucleus 81° 48′. The length of the tail about 35° taking a direction from the Nulcleus [*sic*] a little to the south of Sirius.

On Friday, 24th March.—The distance from Sirius to the Nucleus was 48° 27′. The length of the tail about 36°, Rigel and Nucleus 24° 43′, a Eridani or Achonar [*sic*] and Nucleus 55° 50′, a Argu [*sic*] or Canopus 56° 5′.

On Thursday 30th March.—The distance from Sirius to the Nucleus was 49°, the length of the tail still about 36°, from Rigel to the Nucleus 17° 18′, from Achonar [*sic*] to the Nucleus 58° 8′, Canopus to Nucleus 53° 41′.

The tail of the present Comet, appears to be much greater than that of any other Comet, which has been seen from the earth within the present century … it appears that the length of the tail of a Comet, may appear to differ very much in length, when seen from different parts of the earth; and I am told that the Comet's tail as seen from Wanganui, which is distant in a straight line from Port Nicholson about 80 miles, measured 45°, while from hence it only subtended an angle of 36° …

This 'Letter to the Editor' was written on 17 April 1843 by "W.M. Smith, Capt., H.P. Royal Artillery." We can identify him as William Mein Smith (1798–1869), the former Surveyor-General of the New Zealand Company from 1840 to 1843. Subsequently, he sometimes advertised in the Wellington newspapers as a "Surveyor, Draughtsman, and Land Agent" (e.g. see Smith 1844).

Then, just two days later, *New Zealand Colonist and Port Nicholson Advertiser* catered further to the assumed cometary interest of its readers by publishing a long account of comet C/1843 D1 taken directly from the *Sydney Morning Herald* (The Comet 1843a).

After reporting on the comet in the local newspapers we might expect Captain Smith to have submitted an account for possible publication in the *Monthly Notices of the Royal Astronomical Society*, but this never happened, even though it was foreshadowed in an anonymous letter to the Astronomer Royal dated 11 March 1843 that is now the Airy Papers in Cambridge and reports Wellington observations of the comet:

Since the 2$^{\mathrm{nd.}}$ of this month we have been awe-struck with wonder at a most magnificent Comet which appears close to us. Capt. Smith *who told me he would send you some account of it*, found it by his instruments (imperfect for such a fine operation) to be included in angles of 36°, which is enormous …

I have seen some celebrated Comets, but this one is "Father of the Comets", as it is called here. The head is a brilliant ball of fire, and the length of its Tail, and its velocity especially, are quite incomprehensible … I am very glad to think I have seen such a "Comet Monstre". (Anonymous 1843a; my italics).

Between the two paragraphs in the original letter is a drawing showing the comet in relation to nearby stars, and the appearance of its tail, which is marked as 36° in length.

A second anonymous letter in the Airy Papers, written on 17 March 1843 and also from Wellington, contains the following brief description of the comet:

On Thursday, the 2[d] of March, a large Comet was seen from Nelson, and on the 4th it became distinctly visible at this place [Wellington]. It appeared to be of immense size, and … it measures, from the Nucleus to the end of the Tail, an arc of 36° ….

The Comet is now increasing its distance and its altitude: its tail is about 20° from, and pointed to, the shoulders of the constellation Orion. (Anonymous 1843b; his underlining).

These two letters would appear to have been forwarded to London through diplomatic channels, but neither was published in *Monthly Notices of the Royal Astronomical Society*.

Apart from the anonymous sketch of the comet mentioned above, no New Zealand drawings of Comet C/1843 D1 are known to exist, but Mary Morton Allport has provided an excellent rendition of it from Hobart, the capital of the island of Tasmania, to the south of the Australian mainland (see Fig. 20.1). Note that Hobart has a similar latitude to Wellington (cf. −42.9° vs. −41.3°), so those in Wellington would have had a similar view of the comet at this time.

20.3.2 Comet C/1844 Y1

The Great Comet of 1844 (C/1844 Y1) also attracted public notice in Wellington, and the following account if it, also by William Mein Smith, was published in the *New Zealand Spectator and Cook's Strait Guardian* on 18 January 1845:

In my short notice of the Comet given in the *Spectator* of Saturday last, I said that it was approaching the constellation Pavo. The weather has been unfavourable for observations, and I had mistaken A Gruis for A Pavenis. The following observations may perhaps prove interesting to some of your readers; they are the angular distances taken with a sextant from Sirius, (the Dog Star) A. Centauri, B. Centauri, and A. Gruis, taken as soon after 9 o'clock at night, as the state of the weather would permit.

From these observations, it would appear that the Comet is moving in a north-westerly direction,—towards the north at the rate of nearly a degree and three-quarters daily, and towards the west at the rate of nearly one degree. It cannot, I think, be visible to our friends in England at present; but, weather permitting, they should see it in about a fortnight hence. I will not pretend to say whose, or what Comet it is, for as its appearance is not predicted in the Astronomical Ephemeris, either for 1844 or 1845, it must, I presume, be an unexpected visitor …

December 17, 1844.
From Sirius to Nucleus 118° 33′ 15″
"A. Centauri to Nucleus 53 22 45

Fig. 20.1 A drawing of Comet C/1843 D1 made by Mary Morton Allport (1806–1895) in March 1843 from Hobart, Tasmania (https://en.wikipedia.org)

The night cloudy, the Comet only seen at short intervals.

December 30, 1844
From Sirius to Nucleus 113° 30′ 15″
"A. Centauri to Nucleus 55 47 45
"A. Gruis to Nucleus 13 7 15
The length of the tail about 9°, its extreme breadth about 2°. Fine night.

January 1, 1845
From Sirius to Nucleus 110° 0′ 0″
"A. Centauri to Nucleus 57 40 0
"A. Gruis to Nucleus 9 50 15
The length of the tail about 10°, the night cloudy, the comet only seen at short intervals.

January 6, 1845
From Sirius to Nucleus 101° 14′ 45″
"A. Centauri to Nucleus 66 32 15
"A. Gruis to Nucleus 4 13 0
The length of the tail about 10°. Fine night.

The very same 18 January 1845 issue of the *New Zealand Spectator and Cook's Strait Guardian* also contained a letter about this comet by Robert Sheppard from the New Zealand Company's Survey Office in Wellington. This supplies readers with the following information:

> It [the comet] was first seen on December 22, 1844 travelling through the constellation Sagittarius, near the northern part of Corona Australis, At that time it was very faint. It attained its greatest brilliance on January 1, 1845, when the nucleus was equal in brightness to a star of 2½ magnitude.
>
> As the beard [tail] becomes gradually invisible at a distance from the nucleus, it is almost impossible to define its exact limits. It may, however, be stated that its length (January 1) was observed to be not less than 8°, nor more than 10°, and its width at the extremity about 1° 30′. It extends from the nucleus nearly in the direction in which the Comet is moving; it is brightest along the centre line.
>
> The following table is sufficiently near the truth, to enable any person to mark its track on a celestial globe or map of the stars; the distances and observations given below will assist our English Astronomers (to whom the Comet is not visible) in calculating its elements.
>
> The following observations were taken between 9 and 10 P.M.–
>
> January 3.– The Nucleus in a direct line with Epsilon and Alpha in the constellation Grus. Observed distance from Alpha 4° 31′ 35″.
>
> January 4.– Observed distance from Alpha (in Grus) 3° 16′ 40″, from Beta (in Grus) 7° 46′ 53″, and from Fomalhaut 18° 15′ 40″.
>
> January 7.– From Fomalhaut 14° 51′ 40″, and from Beta (in Grus) 3° 3′ 13″.
>
> January 8.– From Archenar 29° 18′ 40″. The rising clouds prevented any further observation.
>
> January 9.– Observed distance from Fomalhaut 14° 28′ 27″; Archenar 27° 54′ 57″; Alpha (in Grus) 10° 2′ 47″; from Theta (in Grus) 00° 57′ 13″.
>
> January 10.– From Fomalhaut 13° 48′ 20″; Alpha (in Grus) 12° 7′ 40″; Archenar 26° 38′ 10″.
>
> January 12.– From Fomalhaut 14° 37′ 35″; Alpha (in Grus) 16° 17′ 00″; Archenar 24° 27′ 2″; Iota (in Phoenix) 00° 33′ 40″; and from Alpha (in Phoenix) 9° 13′ 42″.
>
> The beard of the Comet is (January 12) about 7° in length, and the nucleus equal in brilliance to a star of the 5th or 6th magnitude.
>
> The direction in which the Comet is now moving is a little inclined to the north, but, as it is at the same time receding from the earth, and will in a few days become invisible, it will entirely escape the observation of Astronomers in England. while it will be visible a little above the horizon in France. (Sheppard 1845a).

Following Sheppard's signature, his address and the date of the letter (15 January 1845) is the list of cometary positions that is shown below in Table 20.1.

Subsequently, this same 'R. Sheppard' submitted the following short report to the Royal Astronomical Society, and this was published in *Monthly Notices*:

> The comet was first seen on December 22, 1844. On December 26, the length of the tail was from 3° to 4°. On January 1, the tail was fan-shaped and no less than 10° in length, the

Table 20.1 Positions of Comet C/1844 Y1 (after Sheppard 1845a)

Date 1845	Right Ascension	Declination (S)
1 January	317° 30′	44° 00′
3 January	324° 40′	44° 45′
4 January	327° 40′	45° 00′
6 January	334° 10′	45° 00′
7 January	337° 30′	44° 55′
9 January	343° 00′	44° 20′
10 January	346° 00′	44° 00′
12 January	351° 50′	43° 00′

> width at the extremity being about 1° 30′; the nucleus was equal in brilliancy to a star of the 2.3 magnitude. Some observations of distances from neighbouring stars are given. (Sheppard 1845b).[2]

The following information was kindly supplied by Susan Djabri, a desendent of Robert Sheppard's brother Frederick, who is mentioned below (but see, also, Papers … n.d.). Robert Sheppard (1810–1896; Fig. 20.2), was born in Brighton, England, and "… may have been trained as a surveyor and engineer by Thomas Hughes, a trusted assistant of Thomas Telford, who is known to have imparted the basic principles of surveying to his younger brother, Frederick Augustus Sheppard …" (Susan Djabri, personal communication, 2014). He then worked in London and Horsham as an engineer and surveyor, and on 14 September 1841 married Eliza Jemima Lucas in Horsham. Less than three weeks later, on 2 October 1841, Robert and Eliza, and young Frederick (who had just completed his training as a surveyor) sailed for New Zealand on the *Brougham*, because

> Robert had been given a 3 year contract as an assistant land surveyor with the New Zealand Company … His companions on the ship included S.C. Brees, the principal surveyor, five other assistant land surveyors, and ten cadet surveyors. The "Brougham" arrived in Port Nicholson [Wellington Harbour] on 9 February 1842. In May 1842 Robert took charge of surveying and laying out sections in the Manawatu district. Then in the next two years he was surveying in the vicinity of Wellington including the Porirua Road, and Karori Stream areas, until January 1845. (ibid.).

It was while he was surveying in the vicinity of Wellington that Sheppard observed Comet C/1844 Y1. As a surveyor he was trained in observational astronomy, but the absence of any other reports by him in *Monthly Notices of the Royal Astronomical Society* would suggest that astronomy was not a serious hobby (although we do know that in his later years, at very least, he owned a small

[2]This was not the only account from New Zealand to appear in *Monthly Notices of the Royal Astronomical Society* at this time: Sheppard's short report was joined by one of equally brevity penned by John Collingwood Haile (1845; cf. The comet 1845) of Auckland, who is listed incorrectly in *Monthly Notices* as C.W. Haile.

Fig. 20.2 Robert Sheppard, about the time he went to New Zealand (*Courtesy* Susan Djabri)

refracting telescope, which is included in a family photograph). In February 1845, just one month after reporting on the comet, Robert and Eliza Sheppard ended their association with New Zealand and returned to Horsham. Robert died in East Redford, Nottingham, at the age of 86.

20.4 Discussion

20.4.1 Who Was Wellington's First Resident European Astronomer

As the observer and chronicler of *both* Great Comets of the 1840s we are justified in identifying William Mein Smith (Fig. 20.3) as Wellington's first resident European astronomer, and after observing the comets he had the added distinction of remaining in Wellington or the Wairarapa for the rest of his life and becoming a *bona fide* New Zealander.

So who was William Mein Smith? He was born in Cape Town, South Africa, in 1799 to a British Navy purser, and "Like so many more of those who figured prominently in the early days of New Zealand colonization, his origins were military." (Jones 1966). After attending school in Devon (England) he joined the army as a 'gentleman cadet' at the age of 14, eventually rising to the rank of Captain.

Fig. 20.3 William Mein Smith, Wellington's first European astronomer (Gibraltar-intro.blogspot.com/2012/05/appendix-2.html)

After serving in Ireland and Canada, on Gibraltar and in England he was employed in 1839 by the New Zealand Company as their Surveyor General. On 5 January 1840 Smith and his team of three other surveyors arrived in Port Nicholson, and one of his first tasks was to lay out the towns of Wellington and Petone. Later that year he surveyed land near Port Nicholson and in December 1841 made a reconnaissance of the Manawatu. In September 1842

> … Smith was directed to map the harbours of the South Island's east coast, and he explored as far as Bluff, Stewart Island and the Chatham Islands. Unluckily, in November his cutter, *Brothers*, sank in Akaroa Harbour, with his sketches, charts and instruments … but his report, written from memory, was of little use to the company in deciding sites for future settlement. (Smith 1990).

It was in this same year, 1842, that Smith formally retired from the British Army, and in 1843 he sold his commission (Jones 1966).

In early 1845, soon after saying farewell to Comet C/1844 Y1, Smith and his family moved from Wellington to near Martinborough in the Wairarapa, where he stayed for the remainder of his life. Initially he took up farming in partnership with Samuel Revans (1808–1888) on Huangarua Station. This comprised 22,000 acres of freehold land and 30,000 acres of leasehold land (ibid.), and Smith and Revans built this into a successful enterprise. But surveying tugged at Smith's heart-strings, and he soon returned to his favoured profession. He carried out contract work for the New Zealand Company in 1849 and 1850, and served as the Government

District Surveyor for Wairarapa from 1853 to 1857. In addition, "For years [he was] Wairarapa's sole resident magistrate." (Smith 1990). He also dabbled in politics: "... in 1858 he was elected to represent the Wairarapa in the Wellington Provincial Council, a post which he held until his retirement in 1865." (Jones 1966). In early January 1869 he died at Woodside, near Greytown, where he and Revans had established a sawmill in 1865.

Apart from serving as Wellington's first European astronomer, William Mein Smith should be

> ... remembered chiefly for his exact sketches and watercolours of early Wellington, the Hutt Valley and Wairarapa. But it is as a surveyor, and a teacher of young surveyors such as Kettle, that his central contribution to frontier society was made. Arguably the best theoretician in the colony, he tackled new problems scientifically, being an early exponent of triangulation. His career is controversial, however, because he planned the city of Wellington, which is notable for its poor design. Wellington's topography made nonsense of the company's scheme. (Smith 1990).

20.4.2 The Curious Case of James Henry Marriott: Was He Really New Zealand's First Professional Telescope-Maker?

James Henry Marriott (1799/1800–1886; Fig. 20.4; Downes 1990) was born in London and, following his father's calling, trained as an optician and scientific instrument-maker. On 27 July 1842 Marriott emigrated to New Zealand on the *Thomas Sparks*, arriving in Wellington on 31 January 1843 (Scholefield 1939: 28) after an eventful 6-month voyage (see Neale 1982). The late Eiby (1978: 123) has promoted Marriott as Wellington's—and indeed New Zealand's—first professional maker of astronomical telescopes, but what evidence is there for this? Despite Eiby's claim (ibid.) that during the 1840s Marriott advertised as a "... maker of astronomical telescopes." I could not find any such advertisements by him in the *New Zealand Colonist and Port Nicholson Advertiser*, the *New Zealand Gazette and Wellington Spectator* or the *New Zealand Spectator and Cook's Strait Guardian.* Instead all three newspapers are peppered with advertisements and reports relating to his theatrical activities, including his Dancing Academy, which is consistent with Downes' claim (1990) that Marriott was known as the 'founding father' of New Zealand theatre. Pendreigh Brown (2014) has also researched Marriott, and he found the same thing: "Like you I couldn't find the reference to astronomical telescopes implied by George Eiby." (personal communication, 2014).

Yet, in a "List of all persons qualified to serve as Jurors for the District of Port Nicholson ..." (1847) Marriott's occupation is given as "optician", and "Various rolls in the 1840s referred to him as either an optician or an engraver; and he had a shop on Lambton Quay until 1885." (Pendreigh Brown, personal communication, 2014).

Fig. 20.4 James Henry Marriott (*Courtesy* Warwick Marriott)

So what telescopes did he make? A survey of New Zealand museum collections failed to reveal any marine or astronomical telescopes made by him,[3] and the only instrument that has come to light is a telescope now owned by the U.S. collector, Carl Romick and acquired from a Hawaiian antique dealer in 2004 (see Orchiston and Romick 2015). This is a brass single-draw 'spyglass' (marine telescope) with a 44-mm achromatic objective, and a four-element eyepiece comprising a wide-angle field lens, an erecting lens and two eyepiece lenses. In its closed state this telescope is 51 cm long, and when the draw tube is fully extended—as in Fig. 20.5—it measures 82 cm, and magnifies 12×. The body of the telescope is covered in

[3]It is significant that when the late George Eiby searched for Marriott telescopes in the Wellington region some years ago he also was unable to locate any (Laureen Sadlier, personal communication, 4 February 2005).

Fig. 20.5 Romick's Marriott telescope on a tripod mounting (*Photograph* Carl Romick)

Fig. 20.6 The inscription on Carl Romick's 'Marriott' telescope (*Photograph* Carl Romick)

brown leather. On the draw tube near the eyepiece is the following inscription: "J.H. Marriott Maker Wellington N.Z. 1844" (Fig. 20.6). For astronomical observing, Romick has the telescope mounted on a tripod (Fig. 20.5).

In an interesting paper titled "Telescopes for land and sea", Deborah Jean Warner (1998) from the National Museum of American History in Washington explores small non-astronomical telescopes made, mainly in London, during the eighteenth and early nineteenth centuries. This was a time in England when "... an industrial revolution and expanding empire went hand in hand, and where numerous soldiers, sailors and civilians were able and eager to purchase the[se] latest high-tech consumer goods." (Warner 1998: 33). As the title of her papers foreshadows, Warner describes and discusses the various types of portable telescopes designed for terrestrial and maritime use, by day or night, and shows that the tubes of single-draw telescopes like the one made by Marriott were first made of metal towards the end of the eighteenth century (Warner 1998: 43).[4] Warner (1998) identifies the following types of land and sea telescopes: perspective glasses, pocket telescopes and day and/or night telescopes. Marriott's 'spyglass' (as Romick likes to call it) falls comfortably into this last category.

So we know that while resident in New Zealand J.H. Marriott made at least one land and sea telescope, in 1844, while a newspaper article dating to 1931 documents a second telescope, also made in this same year (unless Carl Romick's telescope and the following 'relic' are in fact one and the same):

> An interesting souvenir of the days when watermen used to ply between the shore and the sailing ships which came to Wellington is now in the possession of the Wellington Harbour Board. It is a telescope in excellent condition and of very creditable workmanship, made, according to the inscription, by T.H. Marriott,[5] of Wellington, in 1844. The telescope was presented to Mr. A.G. Barnett, secretary of the board, by Mr. J. Thompson, aged 86, who is the last surviving member of the company of watermen who were such an interesting feature of Wellington's early life. (Last of the Watermen 1931).

A literature and newspaper search revealed one further Marriott telescope, which was sent from Wellington among a consignment destined for the 1865 New Zealand Exhibition in Dunedin (List of Articles ... 1864). There is no description of this instrument, so we do not know if it also was a small land and sea telescope or a larger astronomical telescope. What we do know, however, is that Marriott frequently advertised as a "Telescope Manufacturer" in the *Wellington Independent* newspaper during the 1860s and early 1873 (for details see Orchiston and Romick 2015), and "... in the Wellington Almanack of 1872, p. 57; and advertised "telescopes made and repaired" in the Wellington Almanack of 1862, p. 37, and 1865, p. 21." (Pendreigh Brown, personal communication, 2014).

We also know that on 26 August 1852 Marriott gave a well-received lecture, "On the Telescope" at the Wellington Athenaeum and Mechanics' Institute in Wellington:

> On Thursday evening Mr. Marriott delivered a very interesting lecture on the telescope, during which he gave a brief history of the instrument, and explained in a clear and practical manner the various operations connected with its manufacture, both as regards the nice operations of cutting the lenses, and the construction of the metal tubes, which latter

---

[4]Prior to this, pasteboard and wood commonly were used.

[5]Upon referring to Fig. 20.6 it is easy to see how the 'J.' in Marriott's name could easily be mistaken for a 'T.'

> was more fully illustrated by a neat little model, constructed by Mr. Marriott for the occasion, of the machine used for this purpose; numerous diagrams were also exhibited to show the interior construction of the different kinds of telescopes. The lecture was well attended, and the audience, by their frequent applause, gave sufficient evidence of being well pleased with the manner in which the subject was treated. (Athenaeum and Mechanics Institute 1852).

It is clear from this account, and a similar one that appeared at the same time in the *Wellington Independent* newspaper (1852), that Marriott was still *au fait* with manufacturing of refracting telescopes at this time, but what is equally notable is the absence of any completed Marriott telescopes on display during the lecture. Here was *the* perfect opportunity for him to promote his telescopes before a captive audience, and the fact that he did not do so suggests to me that he had little—if any—personal involvement in telescope making at this time. This reinforces the view that after moving to New Zealand Marriott made very few telescopes, which is consistent with the observation in his obituary that

> By occupation he was an optician and mathematical instrument maker, but as might be expected at the time he arrived [in Wellington], he found very little to do in his own line …

Partly this was because the few people in and near Wellington interested in purchasing telescopes were already adequately catered for by three different firms that imported telescopes in 1843 (for details see Orchiston and Romick 2015).

During the 43 years that Marriott lived in New Zealand, Wellingtonians were treated to major naked eye comets in 1843, 1844, 1858, 1861, 1865, 1874, 1880, 1881 (two of them) and 1882 (again two of them—see Orchiston 1998a: 107). There also were transits of Venus in 1874 and 1882 (Orchiston 2004; Chap. 14 in this book), and the year before Marriott died a total solar eclipse was visible from central New Zealand (Orchiston and Rowe 2016; Chap. 16 in this book). Collectively, these astronomical spectacles generated enormous public interest in astronomy, which should have translated into an ever-increasing demand for telescopes. Marriott was in the ideal position to respond to this demand, and the fact that he did not do so, despite advertising extensively, would suggest that this anticipated demand never materialized, which is perhaps just as well for it is clear that Marriott's heart lay with the theatre and not with astronomy. While he did indeed manufacture two or three telescopes—a handful at most—it is better that we salute him as the founding father of New Zealand theatre rather than as the nation's first professional telescope-maker.

20.4.3 Astronomy in the Classroom: The Case of the Grace Academy

Thanks entirely to the recent research paper, "More on astronomy in Wellington in the 1840s" by Pendreigh Brown (2014), I became aware of yet another very early

European resident of Wellington with an interest in—or at very least a rudimentary knowledge of—astronomy, in the form of a Scottish school teacher named Charles Grace (ca. 1794–1857).

Brown (2014: 7) recounts that in May 1840, just two months after arriving in Wellington, Grace began advertising his intention of establishing a school, where the curriculum would include "... natural philosophy, astronomy, chemistry, anatomy, physiology, and natural history ..." Later in the year, Grace announced the opening of his "Wellington Academical Institution" on 1 December 1840, but

> Unfortunately, his optimistic Academy did not flourish. Soon after gold was discovered in Victoria, Australia in 1851, the family departed to [*sic*] the Port Philip Colony. Charles Grace was the first male teacher in the Port Nicholson settlement, and *yes, there was tuition in astronomy*. (ibid; my italics).

It would seem that Charles Grace was an anachronism: while he should be lauded as the first to introduce astronomy into *a* New Zealand school syllabus, it would take nearly a century and a half for astronomy to assume its rightful place as a regular and accepted part of *the* New Zealand school curriculum at both primary and secondary levels (see Leather et al. 1998).

20.4.4 The (Almost) Invisible Joseph Hurley

One of Marriott's contemporaries was Joseph Hurley, who has been described by descendants as "... a draughtsman, engineer and astronomer." (Bremner 1981: 3). An Englishman by birth, Hurley emigrated to New Zealand with his wife and daughter, reaching Wellington in February 1842. The following year, he and three associates opened a sawmill in what is now the Wellington suburb of Ngaio, and he eventually acquired significant land holdings and owned three sawmills before experiencing financial ruin. Later, during the 1860s, he set up an engineering shop in Wellington. Although Hurley had a sea chest "... full of indecipherable writings on astronomy, in longhand and criss-crossed with corrections and references ..." (Bremner 1981: 8), there is no record that he owned a telescope, built an observatory or carried out any serious observational astronomy. According to Bremner (ibid.), his sea chest "... was appreciated by neither publisher nor Observatory, and was destroyed in the 1920s."

20.5 Concluding Remarks

After eight hundred years of Polynesian-oriented astronomy, a new style of astronomy reached New Zealand in 1769 with the arrived of the *Endeavour*, and during the 1770s nautical astronomy held sway as Europeans explored the coasts of this strange new southern land. Despite Cook's preoccupation with neighbouring

Queen Charlotte Sound (see Orchiston 1998c), Port Nicholson was not party to these exploits, and it only witnessed the arrival of scientific astronomy during the 1840s.

The earliest documented European astronomers in Wellington were W.M. Smith and R. Sheppard, both of whom were trained as surveyors, and we are justified in recognizing William Mein Smith as Wellington's first resident European astronomer. A contemporary of Smith and Sheppard was J.H. Marriott who arrived in Wellington with a background in scientific instrument-making, but with little demand for telescopes he largely turned his back on this calling and devoted himself to the theatre. Another with astronomical knowledge was a Scottish school-master named Grace, who from December 1840 pioneered the teaching of astronomy at his short-lived Wellington Academical Institution. During the 1840s there must have been others in Wellington with a fascination for astronomy—people like J. Hurley—but the need to survive in this hash new land left little time for them to indulge such interests. This state of affairs only changed in the 1860s with the emergence of notable astronomers like Stephen Carkeek, Henry Jackson and Archdeacon Arthur Stock, and the founding of New Zealand's first Government observatory (see Chap. 9 in this book).

Acknowledgements I am grateful to the following for their assistance: Pendreigh Brown (Wellington, NZ), Susan Djabri (Horsham, England), the late Tony Dodson (Wellington, NZ), Judy Siers (Wellington, NZ), Laureen Sadlier (Museum of Wellington, NZ) and staff at the Alexander Turnbull Library. I also am grateful to the Donovan Astronomical Trust (Sydney) for helping fund my archival research in Cambridge. Finally, I wish to thank Pendreigh Brown and Susan Djabri for reading and commenting on the first draft of this chapter, and Susan Dabri, Warwick Marriott (Wellington, NZ) and Carl Romick (Richmond, California, USA), for kindly supplying Figs. 20.2, 20.4, 20.5 and 20.6.

References

Anonymous (1843a). Letter dated March 11, in Airy Papers, RGO Archives, Cambridge University Library (RGO 6/92).

Anonymous (1843b). Letter dated March 17, in Airy Papers, RGO Archives, Cambridge University Library (RGO 6/92).

Athenaeum and Mechanics' Institute. *New Zealand Spectator and Cook's Strait Guardian*, August 28, 1852 (p. 2).

Best, E. (1922). *The Astronomical Knowledge of the Maori*. Wellington: Dominion Museum.

Bond, G. P. (1850). On the Great Comet of 1844–45. *Astronomical Journal, 1*, 97–103.

Bremner, J. (1981). Onslow settlers of the 1840s: Joseph Hurley of the Kaiwarra sawmill. *The Onslow Historian, 11*(2), 3–8.

Brown, P. (2014). More on astronomy in Wellington in the 1840s. *Southern Stars, 53*(3), 6–8.

Cottam, S., & Orchiston, W. (2014). *Eclipses, Transits, and Comets in the Nineteenth Century: How America's Perception of the Skies Changed*. New York: Springer.

Cottam, S., Orchiston, W., & Stephenson, R. (2011a). The 1874 transit of Venus, and the popularisation of astronomy in the U.S.A. as reflected in *The New York Times*. In W. Orchiston, et al. (pp. 225–241).

Cottam, S., Orchiston, W., & Stephenson, F. R. (2012). The 1882 transit of Venus and the popularization of astronomy in the USA as reflected in *The New York Times*. *Journal of Astronomical History and Heritage, 15*, 183–199.
Cottam, S., Pearson, J., Orchiston, W., & Stephenson, R. (2011b). The total solar eclipses of 7 August 1869 and 29 July 1878 and the popularisation of astronomy in the U.S.A. as reflected in *The New York Times*. In W. Orchiston, et al. (pp. 339–375).
Downes, P. (1990). Marriott James Henry 1799/1800?–1886. In W. H. Oliver (Ed.), *The Dictionary of New Zealand Biography. Volume One. 1769–1869* (p. 271). Wellington: Allen and Unwin New Zealand Ltd. and Department of Internal Affairs.
Eiby, G. (1977). The New Zealand Government time-service: an informal history. *Southern Stars, 27*, 15–34.
Eiby, G. (1978). History, astronomy and New Zealand. *Southern Stars, 27*, 118–124.
Haile, C. W. (1845). Observations made at Auckland, New Zealand [Great Comet of 1844–5 (Wilmot's)]. *Monthly Notices of the Royal Astronomical Society, 7*, 10.
Herschel, Sir J. F. W. (1851). *Outlines of Astronomy* (4th ed). London: Brown, Green, and Longmans.
Jones, R. (1966). Smith, William Mein. In A. H. McLintock (Ed.), 1966. *An Encyclopaedia of New Zealand* (see http://www.TeAra.govt.nz/en/1966/smith-william-mein).
Kronk, G. W. (2003). *Cometography: A Catalogue of Comets. Volume 2: 1800–1899*. Cambridge: Cambridge University Press.
Last of the Watermen. *Evening Post*, December 19, 1931 (p. 8).
Leather, K., Andrews, F., Hall, R., & Orchiston, W. (1998). Coping with a new curriculum: the evolving schools program at the Carter Observatory, New Zealand. In L. Gouguenheim, D. McNally & J. Percy (Eds.), *New Trends in Astronomy Teaching* (pp. 267–272). Cambridge: Cambridge University Press.
List of all persons qualified to serve as Jurors for the District of Port Nicholson for the year, 1847. *New Zealand Spectator and Cook's Strait Guardian*, February 10, 1847 (p. 1).
List of Articles Forwarded from Wellington to the New Zealand Exhibition, Dunedin. *New Zealand Spectator and Cook's Strait Guardian*, December 24, 1864 (p. 3).
Neale, J. E. (1982). *Pioneer Passengers: To Nelson by Sailing Ship March 1842–June 1843*. Nelson: Anchor Press.
New Zealand Gazette and Wellington Spectator, March 8, 1843 (p. 2).
Olson, R. J. M., & Pasachoff, J. M. (1998). *Fire in the Sky: Comets and Meteors, the Decisive Centuries, in British Art and Science*. Cambridge: Cambridge University Press.
Orchiston, W. (1996). Observatories in the Wellington Botanic Garden. *New Zealand Historic Places*, March 9–11.
Orchiston, W. (1998a). Illuminating incidents in Antipodean astronomy: John Tebbutt and the Great Comet of 1861. *Irish Astronomical Journal, 25*, 167–178.
Orchiston, W. (1998b). Mission impossible: William Scott and the first Sydney Observatory directorship. *Journal of Astronomical History and Heritage, 1*, 21–43.
Orchiston, W. (1998c). *Nautical Astronomy in New Zealand: The Voyages of James Cook*. Wellington: Carter Observatory.
Orchiston, W. (2000a). A Polynesian astronomical perspective: the Maori of New Zealand. In H. Selin & X. Sun (Eds.), *Astronomy Across Cultures: The History of Non-Western Astronomy* (pp. 161–196). Dordrecht: Kluwer.
Orchiston, W. (2000b). Wellington astronomers of the 1840s. *Southern Stars, 39*(2), 55–57.
Orchiston, W. (2001a). Stephen Carkeek and New Zealand's oldest surviving observatory. *Journal of the Antique Telescope Society, 20*, 19–24.
Orchiston, W. (2001b). Wellington astronomers of the 1840s. *Southern Stars, 40*(4), 5–7.
Orchiston, W. (2004). The nineteenth century transits of Venus: an Australian and New Zealand overview. *Journal of Astronomical Data, 10*, 219–308.
Orchiston, W., Nakamura, T., & Strom, R. (Eds.), (2011). *Highlighting the History of Astronomy in the Asia-Pacific Region*. New York: Springer.

Orchiston, W., & Romick, C. (2015). James Henry Marriott: New Zealand's first professional telescope-maker. *Journal of the Antique Telescope Society* (in press).

Orchiston, W., & Rowe, G. (2016). New Zealand astronomy and the total solar eclipse of 1885. In T. Nakamura & W. Orchiston (Eds.), *The Emergence of Astrophysics in Asia: Opening a New Window on the Universe*. New York: Springer.

Papers relating to Frederick and Robert Sheppard. National Library of New Zealand Te Puna Mātauranga o Aotearoa (MS-Papers-10715).

Scholefield, G. H. (1939). Edgar Wallace and Wellington—four generations—James Henry Marriott of Wellington. *The New Zealand Railways Magazine, 74*, 28–31.

Sekanina, Z., & Chodas, P. W. (2004). Fragmentation hierarchy of bright sungrazing comets and the birth and orbital evolution of the Kreutz System. I. Two-superfragment model. *Astrophysical Journal, 607*, 620–639.

Sekanina, Z., & Chodas, P. W. (2007). Fragmentation hierarchy of bright sungrazing comets and the birth and orbital evolution of the Kreutz System. II. The case of cascading fragmentation. *Astrophysical Journal, 663*, 657–676.

Sekanina, Z., & Chodas, P. W. (2008). A new orbit determination for the bright sungrazing comet of 1843. *Astrophysical Journal, 687*, 1415–1422.

Sheppard, R. (1845a). [Letter to the Editor]. *New Zealand Spectator and Cook's Strait Guardian*, January 18 (p. 3).

Sheppard, R. (1845b). Observations made at Wellington, New Zealand [Great Comet of 1844–5 (Wilmot's)]. *Monthly Notices of the Royal Astronomical Society, 7*, 11.

Smith, W. M. (1843). The comet. *New Zealand Gazette and Wellington Spectator*, April 19 (p. 2).

Smith, W. M. (1844). W.M. SMITH, LATE, Surveyor, Draughtsman, and Land Agent. Tinakori-road. In *New Zealand Gazette and Wellington Spectator*, August 21 (p. 1).

Smith, W. M. (1845). To the editor. *New Zealand Spectator and Cook's Strait Guardian*, January 18 (p. 3).

Smith, P. M. (1990). Smith, William Mein. In C. Orange (Ed.), *The Dictionary of New Zealand Biography. Volume One. 1769–1869* (pp. 399–400). Wellington: Allen and Unwin New Zealand Ltd. and Department of Internal Affairs.

The comet. *Daily Southern Cross*, February 8, 1845 (p. 3).

The comet. *New Zealand Colonist and Port Nicholson Advertiser*, April 21, 1843 (p. 2).

The cometary system. *New Zealand Colonist and Port Nicholson Advertiser*, March 14, 1843 (p. 3); March 17, 1843 (p. 4); March 21, 1843 (p. 5); March 24, 1843 (pp. 3–4).

The New Zealand Colonist and Port Nicholson Advertiser, March 7, 1843 (p. 2).

Turner, S. (2014). Bond, George Phillips. In T. Hockey, et al. (Eds.), *Biographical Encyclopedia of Astronomers* (2nd ed, pp. 264–266). New York: Springer.

Warner, D. J. (1998). Telescopes for land and sea. *Rittenhouse, 12*, 33–54.

Wellington Independent, August 28, 1852 (p. 3).

Chapter 21
Henry Severn: Thames' Talented Transitory Astronomer

Abstract Thames' first amateur astronomer was an Englishman named Henry Severn, who lived in the thriving gold-mining town for six years during the 1870s. At that time Severn owned the largest reflecting telescope in New Zealand. A keen observational astronomer, the most important project that he made a commitment to during his brief residence in Thames was the 1874 transit of Venus. It would seem that Severn's astronomical efforts partially served to rekindle John Grigg's dormant interest in astronomy, thereby furnishing New Zealand with its first cometary astronomer of international repute.

21.1 Introduction

During the second half of the nineteenth century, astronomy blossomed in New Zealand. The central Government funded the Colonial Observatory in Wellington (Eiby 1977), the 1874 and 1882 transits brought teams of overseas astronomers to our shores (Orchiston et al. 2000; Orchiston 2004), Christchurch's Professor Bickerton was busy promoting his novel 'partial impact theory' and popularising astronomy (Gilmore 1982), the 1885 total solar eclipse attracted astronomers from across the country to the Wellington-Nelson region (Orchiston and Rowe 2016) and the first fledgling astronomical societies were formed. In addition, amateur astronomers began to appear across the nation (see Orchiston 1998).

Thames was a thriving gold-mining centre at this time and could boast two amateur astronomers of note, Henry Severn and John Grigg. While Grigg's name is well-known to New Zealand astronomers (see Orchiston 1993, 1995, 1996, 2001b and Chaps. 10, 17 and 22 in this book), few will have heard of Severn. This short chapter documents his astronomical activities during an all-too-brief sojourn in Thames and builds substantially on what was published in Orchiston (2001a).

W. Orchiston, *Exploring the History of New Zealand Astronomy*,
Astrophysics and Space Science Library 422,
DOI 10.1007/978-3-319-22566-1_21

21.2 Henry Severn: A Biographical Sketch

Henry Augustus Severn (1833–1883; Fig. 21.1) was an Englishman who was born in Rome. His father was Joseph Severn, the artist friend who accompanied the dying Keats to Rome (see Brown 2009). Although Henry inherited some talent as a painter, printmaker and photographer, just like his brothers and a sister (Kerr 1992: 715), he "… was the practical son, interested in electricity, mechanics and all things scientific." (Martin 2002).

In 1854, though barely out of his teens, Henry Severn emigrated to Australia to accept a job as a clerk at the newly-opened Sydney Mint. There, he soon learnt the techniques of assaying, a skill which was then much in demand in Australia and New Zealand, awash as they were with gold strike after strike. While in Sydney, Severn married Francis Allan (also from England) and they started a family, eventually ending up with three boys and three girls. He also found time to indulge his other passion, science—particularly astronomy—and became an active member of the newly-formed Philosophical Society (later Royal Society) of New South Wales.

In 1864 Severn, with his young family (by now there were four children), moved to Melbourne and a position as assayer at the Union Bank of Australia. There two further children were born and in 1871 the family was on the move again, this time to Thames in New Zealand (ibid.), where he worked as an assayer, initially at the Caledonian Mine, and from about 1875 with the Bank of New Zealand (Isdale 1999). When he arrived, Severn and his wife found Thames to be a thriving centre, and with about 12,000 inhabitants it was the largest town in the North Island.

Fig. 21.1 Henry Severn, photographed by William Stanley Jevons in 1857 (*Courtesy* John Rylands Library, University of Manchester)

At this time, the Caledonian Mine was the town's premier mine and was at the peak of its production, and the gold won from local mines that year sold for just over £1,000,000, a prodigious sum in those days (Isdale 1967). Thereafter, production slowly decreased and with it the town's population, but by 1874 it still stood at 11,000 (ibid.). While employed by the Caledonian Mine one of Severn's tasks was to reduce the level of theft by the workers, and in order to achieve this he "… lined up the men of a battery on boards near the furnaces used for retorting the gold-mercury amalgam, till the heat brought heavy silver globules dropping noisily onto the boards." (Isdale 1967: 34).

In early 1877 the Severns decided to return to England and Henry was farewelled on 7 February and "… left Thames the wiser for his presence." (*Thames Advertiser* 1877a). *En route* they stopped off in Sydney, and on six successive nights, between 25 October and 1 November he delivered a series of six popular 2-hr 'Experimental Science' lectures at the Guild Hall in Castlereagh Street. His topics were

> October 25. "The Earth and its Satellite" &c.
> October 26. "The Sun and Modern Solar Discoveries" &c.
> October 27. "Galvanism, Electro-Magnetism" &c.
> October 30. "Terrestrial Magnetism, Polarity" &c.
> October 31. "Light and Spectral Analysis" &c.
> November 1. "Spectrum Analysis (continued), Electric Lamp, Fusion of Metals" &c. (*Sydney Morning Herald* 1877a).

Patrons were fore-warned that

> In order to make these lectures thoroughly interesting to the public, the Lecturer has provided himself with the best possible apparatus, including Browning's scientific lantern, mechanical diagrams, photographs, induction coil, radiometer, automatic electric lamp, spectrum analysis apparatus, oxy-hydrogen microscope, 1000 slides and views, kaleidoscope, &c.; and no pains will be spared to make these scientific lectures at least equal to any yet given in the colony. (ibid.).

If we can believe the reports published after each lecture, most were well attended. But more than this, the reports tell us much about Severn, and his talent as a lecturer. For instance, on 26 November (his second lecture),

> There was a very numerous audience … Mr Severn delivered an interesting and instructive address … The illustrations and experiments were exceedingly beautiful, and performed with skill, while the explanations given with such rendered them still more attractive. (*Sydney Morning Herald* 1877b).

His next lecture, on "Galvanism and Electro-magnetism",

> … was certainly the most interesting lecture he has yet given … His experiments were numerous and exceedingly beautiful … there can scarcely be any of his audience who, if they pay ordinary attention, can come away without a better understanding of the subjects discussed during the evening … one lecture leads on to another by an easy gradation, and Mr. Severn, while taking care to impress facts on his audience, is never prolix or wearisome. (*Sydney Morning Herald* 1877c).

In the report on Severn's next lecture we are informed: "It is noteworthy that Mr. Severn's course of lectures is being attended by two hundred pupil teachers of the schools under the Board of education …" (*Sydney Morning Herald* 1877d). Yet even when he did not draw a maximal audience and the topic was not so popular (like his first one on spectral analysis), Severn was able to perform admirably:

> The audience was not a numerous one – about half the hall and half the gallery being filled – but it was a highly intelligent one, and was intensely interested throughout the evening in the intellectual entertainment afforded it. As a lecture there is nothing remarkable or even attractive about Mr. Severn's discourse, but he possesses that eloquence which makes itself intelligible, and his remarks prepare the way for a series of beautiful experiments which it is not too much to say enrapture those who witness them … (*Sydney Morning Herald* 1877e).

Clearly Henry Severn was a talented lecturer who not only captivated his audiences but also entertained them.

In November 1877 Severn must have left Sydney happy … and financially rewarded (as attendance to his lectures was 2 shillings and 6 pence for a seat in the hall and 1 shilling for a seat in the gallery). So he and his family headed back to England, and whilst they were *en route* the *Thames Advertiser* tells its readers that

> Mr Henry A. Severn, whose lectures on experimental science have made quite a sensation in New Zealand and Australia, is on his way to England, where he will commence a course of lectures on light and the spectral analysis; electro-magnetism and its relation to the telegraph, telephone, &c.; terrestrial magnetism, the principle of the mariner's compass, and kindred subjects. Mr Severn is said to be unrivalled in his experiments, which are conducted with apparatus of an unusually extensive and superior kind. (*Thames Advertiser* 1878a, b).

Meanwhile, on 19 December 1878 the *Thames Star* newspaper up-dates it readers about Severn. He has just arrived in England, and arrangements have been made for him to lecture in some of main towns in England and Scotland. Furthermore,

> The reception he met with in all parts of the Colonies, and the extraordinary success of the lectures, cannot fail to be most gratifying to his relations and friends in England … It is no exaggeration to say that Mr Severn has accomplished during the past two years what has never been accomplished before. Three hundred and sixty lectures have been delivered by him in all parts of the Colonies, to about 300,000 people, including 33,000 teachers and scholars. He has moved about with six and eight cwt, and sometimes a ton of valuable apparatus. As there is a growing demand for more technical education in the Mother Country, Mr Severn's lectures are likely to prove very popular. We are glad to hear that it is his intention after a few years to return to New Zealand …

By the end of March 1879 Severn had delivered lectures in London and was engaged to visit Glasgow (Lectures on Modern Scientific Discovery 1879). However, soon after he was appointed manager of a gold-mining company in southern India and was on the move again. After this we hear nothing more of him until 4 July 1883, when the *Evening Post* newspaper in Wellington reports his death:

> News has been received in Wellington of the death of Mr. H.A. Severn, who, it will be remembered, delivered a course of scientific lectures in this city some years ago. The demise occurred in the Wynaad, Develah, India, on 17 April, at the early age of 48 … He leaves a wife and several grown-up children (one of whom is in the Bank of New Zealand at Auckland) …

Meanwhile, in listing his death, the *Thames Advertiser* (Deaths … 1883) also lists his society affiliations: he was a Member of the Institute of Civil Engineers, a Fellow of the Royal Geographical Society, a Fellow of the Royal Astronomical and a 'Mem. Soc. Tel. E.' (could this be a Member of the Society of Telegraphic Engineers?).

21.3 Astronomical Activities

21.3.1 Introduction

Of all the physical sciences, Severn had a passion for astronomy, and in the late 1850s, when he was living in Sydney he made a 27.9cm (11-in) f/11 Newtonian reflector with a speculum mirror (*Lake Wakatip Mail* 1874; Severn 1858, 1860) and read a paper on this, titled "The construction of specula for reflecting telescopes" at the 8 June 1859 meeting of the Philosophical Society of New South Wales (see Philosophical Society of New South Wales 1859: 231).

In a letter to John Tebbutt dated 1 April 1860 Severn indicated that he planned to embark on the construction of a larger, 61-cm (24-in) reflector, also with a speculum mirror (Severn 1860), but it is now apparent that this ambitious instrument was never completed (cf. Orchiston 2003) as it was the 11-in telescope that he took to New Zealand when he shifted to Thames in 1871. Severn (1860) also reports that in 1859 he owned a "very fine" 7.6-cm (3-in) Simms refractor.

After he moved to Melbourne in 1864 Severn was able to continue to pursue his astronomical interests. These included observing comets and the enigmatic variable star Eta Carinae (see Severn 1858, 1870). When he first began monitoring this star in 1855 or 1856, Eta Argus—as it was then referred to—was still one of the most prominent stars in the southern sky (Orchiston 2000), but it then began to fade in brightness and by the end of the 1860s was too faint to detect with the naked eye (see the light curve in Frew 2004).

21.3.2 The 1874 Transit of Venus and Other Astronomical Observations

When he moved to Thames in 1871 Severn brought his two telescopes with him, and at this time his reflector ranked as easily the largest telescope in New Zealand (followed by Henry Skey's 23.5-cm (9.25-in) Browning-With reflector in Dunedin). Normally an instrument of this size would have been housed in an observatory, but a drawing of it based on a photograph and reproduced in the *Illustrated London News* in 1875 (see Fig. 21.2) shows that it was located outdoors, in the open. This illustration also shows that Severn had fabricated an elaborate-looking altazimuth mounting for the telescope, and one that differed markedly from a German or an

Fig. 21.2 Henry Severn and his reflector. This drawing, based on a photograph, appeared in the 20 March 1875 issue of the *Illustrated London News* (https://en.wikipedia.org)

English equatorial mounting and thus would have made the tracking of celestial objects challenging to say the least. Yet presumably he was able to make it work. A subsequent newspaper report indicates that Mr Severn made the telescope himself, and that it was located at his residence in Willoughby Street, Thames.

The engraving, meanwhile, was "… no doubt copied from a photograph taken by Mr Bishop of Pollen street, altho' the likeness of Mr Severn is scarcely recognizable." (*Thames Advertiser* 1875b).

Whether Severn continued to monitor Eta Argus once he was settled in Thames is not known, but we do know that he planned to observe the 9 December 1874 transit of Venus, and was one of those selected by Major Henry Spencer Palmer, leader of the British expedition based at Burnham, to man 'official stations' throughout New Zealand (Palmer 1881). Severn was responsible for the Grahamstown (i.e. Thames) station (*Lake Wakatip Mail* 1874; Palmer 1881), even though Palmer had some reservations about his reflector and the lack of a transit telescope. In a letter to the Astronomer Royal, dated 22 November 1874, he wrote:

> I do not know what the exact value of good observations with such an instrument may be, for the immediate purpose in view; but I thought it best to include M[r.] Severn in my list of co-workers, and to send him time on December 9. He has two or three good chronometers. (Palmer 1874a).

Given the circumstances, to have been selected by Palmer as one of the collaborating astronomers was a great honour.

In order to successfully observe the transit, Severn needed not only a telescope but also an accurate local time service, and the precise longitude of his observatory had to be determined. Consequently, a great deal of preparation was necessary, and on 3 November Palmer wrote up the following information for Severn and other local astronomers:

> For the subsidiary stations for observing the Transit, all that will be necessary on Dec. 9 is that (in addition of course to adequate telescopic means) each station shall be furnished with (1) a clock or chronometer which can be trusted to go steadily for a few hours; and (2) means of exchanging reciprocal time-signals with the main station at Burnham, shortly before and shortly after the Transit: so that all the observations may be directly referred to the Burnham time.
>
> The determination of accurate local time at the secondary station (for differences of longitude from Burnham) is a separate matter, which may be left till the Transit is over. (Palmer 1874b; his underlining).

Because it was a major population centre, Palmer knew that there would be no problem gaining access to the telegraph in order to exchange time signals between Thames and Burnham. Meanwhile, he arranged for the Governor of New Zealand, Sir James Fergusson (1832–1907) to loan Severn "… a fine chronometer …", and he also had access to two others (Preparing for the transit of Venus 1875) (Fig. 21.3).

As it turned out, all these preparations proved fruitless because "… rain and a gloomy [cloudy] sky …" on 9 December (see Fig. 20.3, which Dr William Tobin was kind enough to direct me to) prevented Severn from observing the long-awaited transit (*Daily Southern Cross* 1874). Indeed, inclement weather blanketed New Zealand, and the only site that did provide a totally satisfactory view of the transit was Queenstown, where a U.S. observing party was stationed (Orchiston et al. 2000; and Chap. 15 in this book). For Severn, the vagaries of the weather dashed his one

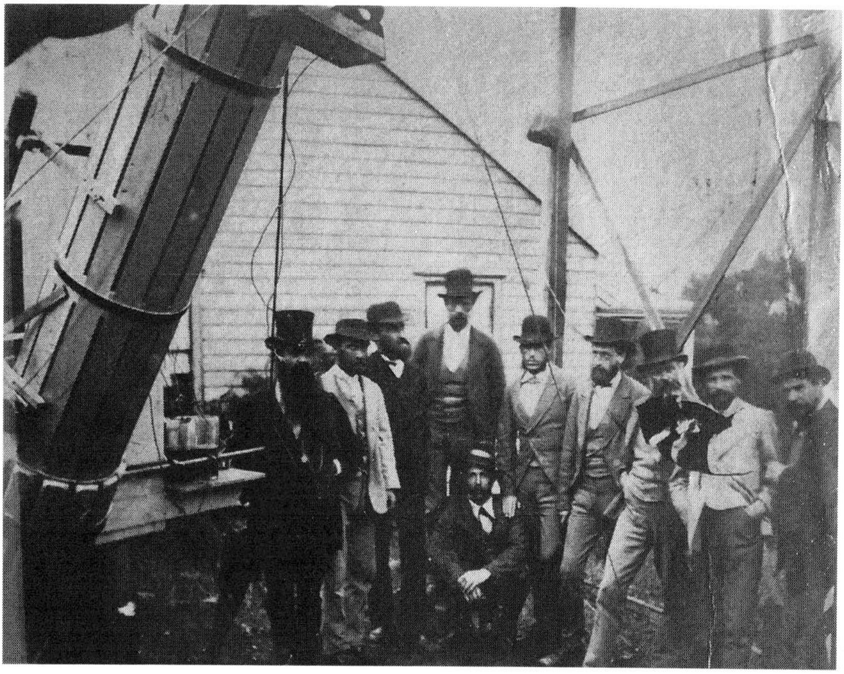

Fig. 21.3 A photograph titled "Transit of Venus 1874. Pouring rain" showing Severn (unidentified, but perhaps the man sitting on the ground), some of his colleagues and part of his telescope at or near the time of the transit (http://ketewestcoast.peoplesnetworknz.info/image_files/0000/0000/9287/Transit_of_Venus_1874__pouring_rain.jpg)

chance to participate in a major international collaboration *and* make a significant contribution to science. Fellow Thames resident, Vicesimus Lush (1874), could offer nothing but heart-felt sympathy: "Mr. Severne [*sic*] especially we feel for—for the last three months he has been preparing for this important event and his disappointment must be very great."

21.3.3 Educating the Public in Science and Astronomy

As we have seen, Severn's main avocational interest was science. Consequently, he was a strong supporter of the Mechanics' Institute in Thames, ultimately becoming its Vice-President (see *Thames Advertiser* 1875a, 1876b, c, 1877a).

In 1872 Severn is known to have given a series of lectures on science (Isdale 1999), and on 17, 24 June and 1 July 1875 he delivered a course of three "Experimental Lectures on Light and Spectrum Analysis" through the Academy of Music,

> … illustrated by Diagrams, Models, Oxy-hydrogen Light, Browning's Electric Lamp Lantern, and 50 cell Galvanic Battery, Prisms, &c., with a view to explain the laws and theories of Light, and the extraordinary modern discoveries made throughout the Universe by means of Spectral Analysis. (*Thames Advertiser* 1875).

His Thames lectures were so successful that he was called to Auckland "… and delivered a successful series of scientific lectures in connection with the opening of the Auckland Institute." (*Thames Advertiser*, 1876a).

When Severn was preparing to leave Thames he

> … gave a farewell lecture on "the Solar System," illustrated by some very fine mechanical diagrams. As usual the subject was treated in a plain, straight-forward manner. We were much pleased to see so good an attendance, and the deep interest taken in the matter was the best evidence that the lecture was interesting to the end. Mr Severn has quite lately imported much new apparatus from Mr John Browning, of London, all of the best kind, and the latest discoveries in astronomical matters were placed before the audience on the large screen by lime light Thus the solar system in motion, the theory of the tides, the form of the earth, &c. were all more clearly exhibited than we have seen them before … We understand Mr Severn begins a course of lectures at the Choral Hall, in Auckland, on Tuesday next, and then proceeds South, prior to his leaving for England in April. (*Thames Advertiser* 1877b).

In reporting Severn's death in April 1883 the *Evening Post* newspaper stated:

> He was an enthusiast in scientific pursuits, and the lectures which he delivered throughout New Zealand [including in Wellington] on astronomical and other subjects will not long be forgotten by those who had the pleasure of listening to them. (*Evening Post* 1883).

After the frustration of the 1874 transit of Venus we hear no further word on Severn's astronomical activities—if any—and in February 1877 he left for England, thus ending Thames' initial flirtation with scientific astronomy.

21.4 The Sequel

Despite this unsatisfactory outcome, Severn's focus on the 1874 transit did have a pleasant sequel in that it served to awaken John Grigg's dormant interest in astronomy. A Thames shop-owner and music teacher by profession, Grigg went on to achieve an international reputation for his cometary discoveries and he also pioneered astronomical photography in New Zealand. By the end of the nineteenth century, he was New Zealand's foremost amateur astronomer.

Acknowledgments I am grateful to the following for their assistance: the late Alistair Isdale (Thames, NZ), Megan Martin (Historic Houses Trust of New South Wales, Sydney), the late Jackie St. George (Sydney), staff at the Mitchell Library (Sydney) and National Archives (Wellington, New Zealand), and particularly Dr William Tobin (Vannes, France). I also am grateful to the Donovan Astronomical Trust (Sydney) for helping fund my archival research in Cambridge. Finally, I wish to thank Dr Grant Christie (Auckland, NZ) and Dr Tobin for reading and commenting on the first draft of this chapter, and the John Rylands Library at the University of Manchester for kindly supplying Fig. 21.1.

References

Brown, S. (2009). *Joseph Severn. A Life. The Rewards of Friendship*. Oxford: Oxford University Press.

Daily Southern Cross, December 10, 1874.

Deaths Severn. *Thames Advertiser*, July 2, 1883 (p. 2).

Eiby, G. A. (1977). The New Zealand Government time-service: an informal history. *Southern Stars, 27*, 15–34.

Evening Post, July 4, 1883 (p. 2).

Frew, D. J. (2004). The historical record of Eta Carinae. I: The visual light curve, 1595–2000. *Journal of Astronomical Data, 10*(6), 1–76.

Gilmore, G. (1982). Alexander William Bickerton: New Zealand's colourful astronomer. *Southern Stars, 29*, 87–108.

Isdale, A. (1967). *History of the River Thames New Zealand*, Thames, printed for the author.

Isdale, A. (1999). Henry A. Severn, MS notes.

Kerr, J. (Ed.). (1992). *The Dictionary of Australian Artists*. Oxford: Oxford University Press.

Lake Wakatip Mail, November 24, 1874.

Lecture. *Thames Advertiser*. June 16, 1875 (p. 2).

Lectures on modern scientific discovery. *Thames Star*, March 29, 1879 (p. 3).

Lush, V. (1874). *The Thames Journals of Vicesimus Lush*, copy in possession of A. Isdale.

Martin, M. (2002). Henry Severn. A Sydney Mint personality. *Insites, 31*, 1 (Newsletter of the Historic Houses Trust of New South Wales).

Orchiston, W. (1993). John Grigg, and the genesis of cometary astronomy in New Zealand. *Journal of the British Astronomical Association, 103*, 67–76.

Orchiston, W. (1995). John Grigg, and the development of astrophotography in New Zealand. *Australian Journal of Astronomy, 6*, 1–14.

Orchiston, W. (1996). John Grigg. In C. Orange, et al. (Eds.), *The Dictionary of New Zealand Biography*. (Vol. 3, pp. 188–189, 1901–1920). Auckland, Auckland University Press and Department of Internal Affairs.

Orchiston, W. (1998). *Nautical Astronomy in New Zealand. The Voyages of James Cook*. Wellington: Carter Observatory.

Orchiston, W. (2000). John Tebbutt of Windsor, New South Wales: a pioneer southern hemisphere variable star observer. *Irish Astronomical Journal, 27*, 47–54.

Orchiston, W., Love, T., & Dick, S. (2000). Refining the Astronomical Unit: Queenstown and the 1874 transit of Venus. *Journal of Astronomical History and Heritage, 3*, 23–44.

Orchiston, W. (2001a). Henry Severn: Thames' other nineteenth century astronomer. *Southern Stars, 40*(2), 8–10.

Orchiston, W. (2001b). The Thames observatories of John Grigg. *Southern Stars, 40*(3), 14–22.

Orchiston, W. (2003). Amateur telescope making in Australia. An historical perspective. In W. J. Cook (Ed.), *The Best of Amateur Telescope Making Journal* (Vol. 2, pp. 208–239). Richmond: Willmann-Bell.

Orchiston, W. (2004). The nineteenth century transits of Venus: an Australian and New Zealand overview. *Journal of Astronomical Data, 10*, 219–308.

Orchiston, W., & Rowe, G. (2016). New Zealand astronomy and the total solar eclipse of 1885. In T. Nakamura & W. Orchiston (Eds.), *The Emergence of Astrophysics in Asia: Opening a New Window on the Universe*. New York: Springer.

Palmer, H. S. (1874a). Letter to Sir G.B. Airy, dated 22 November, in Airy Papers, RGO Archives, Cambridge University Library (RGO 6/273).

Palmer, H. S. (1874b). Time signals. Transit of Venus, MS (Lands & Survey AAFV, Series 6865/1b, National Archives, Wellington).

Palmer, H. S. (1881). Transit of Venus, 1874. Part V. Expedition to New Zealand. In S. G. B. Airy (Ed.), *Account of Observations of the Transit of Venus, 1874*, December 8 (p. 481), London, Her Majesty's Stationery Office.

Philosophical Society of New South Wales. *The Sydney Magazine of Science and Art, 2*, 230–231 (1859).
Preparing for the transit of Venus. December 9, 1874. *Illustrated London News*, March 20, 1875.
Severn H. A. (1858). *Sydney Morning Herald*, October 26.
Severn, H. A. (1860). Letter to John Tebbutt, dated April 1, in Letters to John Tebbutt, Mitchell Library (Sydney).
Severn, H. A. (1870). Note respecting η Argûs. *Monthly Notices of the Royal Astronomical Society, 30*, 180.
Sydney Morning Herald, October 18, 1877a (p. 1).
Sydney Morning Herald, October 27, 1877b (p. 4).
Sydney Morning Herald, October 29, 1877c (p. 4).
Sydney Morning Herald, October 31, 1877d (p. 4).
Sydney Morning Herald, November 1, 1877e (p. 4).
Thames Advertiser, May 11, 1875a (p. 2).
Thames Advertiser, May 31, 1875b (p. 2).
Thames Advertiser, June 13, 1876a (p. 2).
Thames Advertiser, July 5, 1876b (p. 2).
Thames Advertiser, July 8, 1876c (p. 2).
Thames Advertiser, February 1, 1877a (p. 2).
Thames Advertiser, February 9, 1877b (p. 2).
Thames Advertiser, November 29, 1878a (p. 2).
Thames Star, December 19, 1878b (p. 1).

Chapter 22
John Grigg, and His Pioneering Astronomical Photography

Abstract John Grigg was New Zealand's leading amateur astronomer during the first decade of the twentieth century. While best known for his comet discoveries, one of his other claims to fame was as a local pioneer of astronomical photography. In this chapter we review his achievements in astronomical photography and discuss other New Zealand amateur astronomers who subsequently established reputations in this area of astronomy.

22.1 Introduction

Because of the relative neglect of astronomy by provincial governments during the nineteenth century, amateurs played an important part in the development New Zealand astronomy. The nation's leading amateur astronomer at the close of the nineteenth century arguably was John Grigg, who operated a successful furnishing and music store in Thames (for New Zealand localities mentioned in this chapter see Fig. 22.1). Grigg was active at a time when both professional and amateur astronomers worldwide were experimenting with the application of photography to astronomy (see Lankford 1984; Norman 1938; Turner 1912), and he was New Zealand's pioneer astronomer in this regard.

Brief general accounts of Grigg have already been published by Hughes (1991), McIntosh (1958, 1970), Mackrell (1985), Orchiston (1998: 107–109) and Rumsey (1974), and I have reviewed his Thames Observatory (Orchiston 2001) and his formative role in New Zealand cometary astronomy (Orchiston 1993), but until I published "John Grigg and the development of astrophotography in New Zealand" (Orchiston 1995) no-one had examined his achievements in astronomical photography.

Background information about John Grigg, and particularly his cometary and transit of Venus observations, is presented in Chaps. 10, 14 and 17 in this book. This chapter examines the equipment he used for astronomical photography, some of his results, and the subsequent development of astronomical photography by New Zealand amateur astronomers. As such, this paper draws heavily on Orchiston (1995).

W. Orchiston, *Exploring the History of New Zealand Astronomy*,
Astrophysics and Space Science Library 422,
DOI 10.1007/978-3-319-22566-1_22

Fig. 22.1 New Zealand localities mentioned in the text are shown in *red*

22.2 John Grigg: Astronomical Photographer

22.2.1 Introduction

Nowadays, 'astrophotography' is a challenge for many amateur astronomers, as ideally it calls for a telescope with an equatorial mounting, reliable circles and an accurate drive, a CCD camera and the ability to effectively use this and produce quality images (e.g. see Ratledge 2005; Stuart 2006). In Grigg's day there were, of course, no CCD or digital cameras, but the art of successfully exposing photographic glass plates and later developing them had to be learnt. Grigg's telescope also had to be modified: it came with an equatorial mounting and circles (Grigg 1902b), but as we have seen there was no drive, so Grigg set to work and constructed one himself. Financial constraints (e.g. see Grigg 1902c) meant that he also had to construct the cameras to use with this telescope.

By trial and error he built a crude prime-focus camera (ibid.), and this was attached to the eyepiece end of the telescope (see Fig. 22.2). This camera is no longer extant, and Grigg never published any technical details of it, but by scaling up from dimensions in Figs. 22.2 and 10.3 and 10.4 it would appear to have been designed for standard 4 × 5-inch glass plates (Frank Andrews, personal communication 1994). Grigg also constructed a simple astrocamera, which he mounted on a platform attached to the telescope tube and this is clearly shown in Fig. 10.3, a photograph that Grigg gave to his friend and fellow astronomer J.T. Ward from Wanganui. On the back of this photograph Grigg wrote the following short description: "Equatorial with home-made star camera attached." Again, the star camera has not survived and there are no known descriptions of it in published or manuscript records, but by measurements taken from Figs. 22.2 and 10.3 this camera would appear to have had a 1. 5-inch lens system with an f-ratio of between 4.5 and 6.3, and to have used standard glass quarter plates (3.25 × 4.25 inches). The few extant photographs we know of that were taken with this camera are 3.25 cm square (suggesting they were trimmed), cover a 22° field, and record stars down to a visual magnitude of 8.3 (based upon magnitudes shown in *The AAVSO Variable Star Atlas*—see Scovil 1980).

Having the requisite equipment was merely part of the story, for Grigg also had to learn how to use it effectively, and an 1896 newspaper article reveals that at first this was no easy task:

> … Mr Grigg … has for many years been grappling with the difficulties attendant upon the reproduction by means of photography of views of the heavens—difficulties that are even puzzling to a professor at a large observatory; but an amateur who has to make experiments as he goes along, and has to adapt his apparatus accordingly, is continually met with rebuffs requiring much perseverance to overcome and skill to enable him to formulate a basis to start afresh … [Suitable] apparatus is generally beyond the means of amateurs, and comparatively few trouble to further cope with the difficulty, but Mr Grigg, who is nothing if not persevering, set himself, as we have said, to bring photography to the aid of his researches. He had not only to read up all the details for the work, but to make his apparatus, and we are glad to see that he has so effectively accomplished his task as to secure beautiful views of the moon, sun, and stars. (Astronomical photographs 1896).

Fig. 22.2 Grigg's home-made prime focus camera (*Courtesy* Ward (Wanganui) Observatory Archives)

Since Grigg's observing books have not survived, we are forced to rely on newspaper reports and his published papers for details of his astronomical photographic work, so what follows below is no more than an outline of his achievements. We can, nonetheless, establish that Grigg was successful in photographing the Sun, the Moon, comets and stars, and these fields are dealt with individually below. He employed the prime focus camera shown in Fig. 22.2 to successfully photograph the Moon, a partial solar eclipse, sunspots and at least one comet, while photographs of star fields and comets were obtained with the 'star camera'.

When I researched my 1995 paper on Grigg's astronomical photography I did not have ready access to historic Thames newspapers, and on the basis of the data then available to me I concluded that "It would appear that Grigg confined his astrophotographic activities to two distinct periods: 1890–95 and 1901–1910." (Orchiston 1995: 7). This implied that most of the photographs were taken after his retirement as a full-time businessman so that he could devote more time to astronomy, and after he had relocated his observatory to Queen Street. We now know—

thanks to additional information included in newspaper articles, letters and reports—that Grigg continued his *involvement* with astronomical photography unabated from 1890 through to at least 1910, although his little paper in a 1902 issue of the *Journal of the British Astronomical Association* (Grigg 1902c) makes it clear that he did not actually take many new photographs from 1895 or 1896 until the arrival of the Great Comet of 1901. Thus, some if not all of his astronomical photographs that were on display in his shop in February 1896 (see the *Thames Advertiser* 1896) were probably among the last images he took prior to the arrival of the 1901 comet.

22.2.2 Solar Photography

The first indication that Grigg had access to a telescope that could be used for astronomical photography dates to December 1882, and the transit of Venus. While not strictly 'Solar Photography', this topic best fits here given there is no separate section on "Grigg's Planetary Astronomy" and it did involve photographing the Sun. As we saw in Chap. 14, in 1882 Grigg had yet to build an observatory, so he fitted out a temporary 'observing station' for the transit in his business premises, and a Mr Foy successfully photographed the transit using a 7.6-cm refractor that was on loan to Grigg (Transit of Venus 1882). There is no reference to the equipment used, but it is logical to assume it was some form of prime focus camera. Nor do we know which Foy took the photographs. From 1872, James Foy and his brother Joseph ran a successful photographic studio in Pollen Street, Thames (Early New Zealand Photographers and their Successors n.d), but other than this reference to the transit there is no evidence that either brother had a particular interest in astronomy, and there is no mention in the available literature that any of the surviving photographs that can be attributed to their studio includes astronomical subjects.

The first indication that John Grigg himself was engaged in astronomical photography dates to 12 December 1890 when he used the prime focus camera on his Wray telescope to obtain a series of photographs of a solar eclipse (Fig. 22.3) that was partial when viewed from New Zealand. More than a decade later he wrote:

> My first attempt [at astronomical photography] was with a solar eclipse. The lens being removed from the camera, the latter was fixed to the eye-tube of the telescope, and the sun's image focussed in it. The object-glass was then stopped down to about an inch and covered with a very light and loose cap, which could be removed and replaced rapidly without shaking the instrument. Half a dozen exposures were made, and good pictures obtained, which were afterwards mounted, together with a computed diagram, thus forming an interesting record. (Grigg 1902c: 125).

In this quotation, Grigg refers to an "interesting record", and this is reproduced here as Fig. 22.3.

This particular eclipse was a hybrid eclipse, where the path of totality was very narrow (a mere 24 km) and the duration of totality was only 28 seconds. However, a total solar eclipse was only experienced in the middle section of the totality path (the blue path in Fig. 22.4), with an annular eclipse visible at the two ends (the red

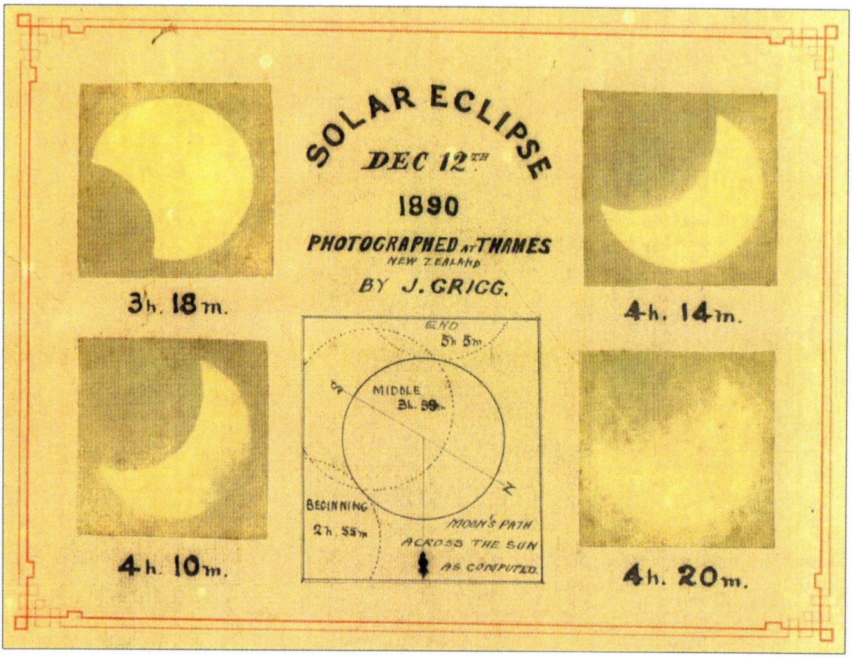

Fig. 22.3 Images of the partial solar eclipse of 12 December 1890 obtained by John Grigg (Orchiston collection)

Fig. 22.4 A map prepared for the 12 December 1890 solar eclipse showing the path of totality (the blue and pink line) and those areas where a partial eclipse would be visible, between the blue line and the upper green line (after Espanek and Meeus 2006)

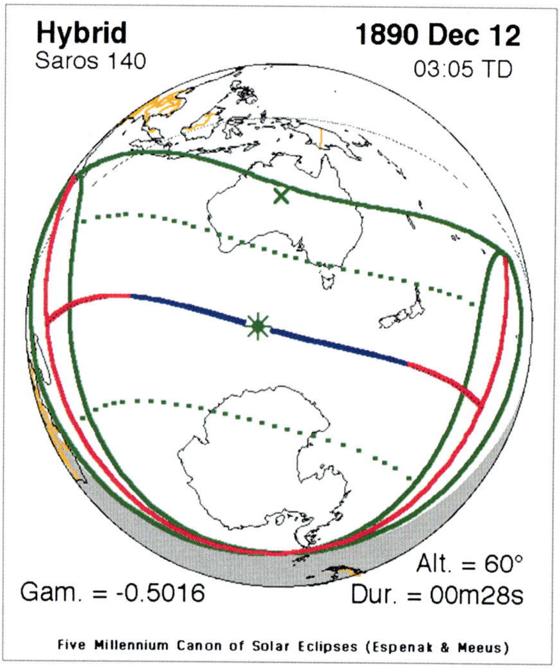

sections) of the path of totality. As Fig. 22.4 clearly shows, in the case of this particular eclipse, the path of totality did not cross any major (or even minor) land masses, but a partial eclipse was visible from throughout New Zealand and over the southern half of Australia (weather permitting, of course).

Another partial solar eclipse that drew Grigg's attention occurred on 2 February 1897, when the event already was in progress at sunrise in Thames. The *Thames Advertiser* reported that Grigg "… obtained eight photographs of different stages of the eclipse …" (Eclipse of the Sun 1897).

Apart from solar eclipses, Grigg also took 'regular' photographs of the Sun. For instance, two photographs he planned to submit to the April 1896 Inter-colonial Photographic Exhibition in Melbourne included images of "… the sun, showing the spots there …" (Astronomical photographs 1896). Then in 1905 he took advantage of the solar maximum to photograph sunspots (Grigg 1905), although none of these photographs can now be traced.

22.2.3 Lunar Photography

Grigg was not just interested in the Sun, as an interesting 1894 newspaper article reports:

> A picture of unusual interest is now on exhibition in the window of Mr A. Iles, the photographer. It is an enlargement of one of a series of photographs of the recent lunar eclipse taken by Mr Grigg at his observatory, and shows the central phase … the eclipse being only partial. On that portion of our satellite, which was clear of the earth's shadow, the various formation[s] on the surface are well shown, the skillful manipulation during the enlarging process bringing out details not easily discernible in the original negative. Where the penumbra or fringe of the shadow is shown, the surface markings become gradually more indistinct, and finally disappear under the shadow itself, a careful examination of which reveals the curvature of the earth's edge. Photographing the moon is a very difficult operation as special optical apparatus has to be prepared as well as a very accurate means of driving the telescope to follow the moon's motion during exposure, and we congratulate Mr Grigg on the success which has attended his efforts, extending over a long period of designing and experimenting. (The late eclipse of the Moon 1894).

This long but informative quotation indicates that by 1894 Grigg had already developed the ability to take successful astronomical photographs. We also see that in this instance he took a series of photographs, even though this was only a partial eclipse. Meanwhile, the photograph referred to in the quotation was on display in the rooms of Arthur James Iles (1870–1943), a well-known Lawrence photographer who "Arrived in Thames in the early 1890s, and specialised in Maori photography. He then worked at the Falk Studios in Sydney, before returning to NZ in 1901 and setting up business in Rotorua." (http://muse.aucklandmuseum.com/databases/librarycatalogue/41970.detail).

Inspired perhaps by these eclipse photographs, between March and October 1895 Grigg photographed the Moon on a number of occasions, securing images

Fig. 22.5 Three photographs of the Moon at different phases taken by Grigg in 1895 (Orchiston collection)

showing first quarter, full moon and last quarter, and copies of these images have survived (e.g. see Fig. 22.5).

Then, the following year Grigg submitted two astronomical photographs in the scientific class in the April 1896 Intercolonial Photographic Exhibition in Melbourne, which included

> … views of the moon in all its stages, the sun, showing the spots there, and of the stars—the Southern Cross, the Dog, Orion's Belt, and the Pleiades, etc. The pictures are all beautifully taken, those of the Lunar Queen being especially fine and undoubtedly a work of art. (Astronomical photographs 1896).

In 1902 Grigg took another series of lunar photographs (Grigg 1902c), similar to those obtained in 1895. Extant prints range from a 3-day old crescent through to a full moon.

Grigg was a popular public lecturer in Thames, but the first time we read about him using his own astronomical photographs in one of his lectures was when he went to Wanganui in August 1904 at the invitation of his friend J.T. Ward, a little over a year after the opening of the Wanganui Astronomical Society's Observatory (see Chap. 11 in this book). The following account, which appeared in the *Wanganui Chronicle* and was copied by the *Thames Star*, indicates that his lunar photographs were particularly appealing:

> A highly interesting and instructive lecture was given at the [Wanganui] Museum by Mr John Grigg … [who] proceeded to display some of the finest astronomical slides we have had an opportunity of seeing … Mr Grigg's photographs were all taken by himself … [and] The views displayed by the moon and comets were exceptionally fine. (Mr Grigg on "Astronomy" 1904).

22.2.4 Cometary Photography

Grigg is known to have photographed two different comets, both of which were 'Great Comets'.

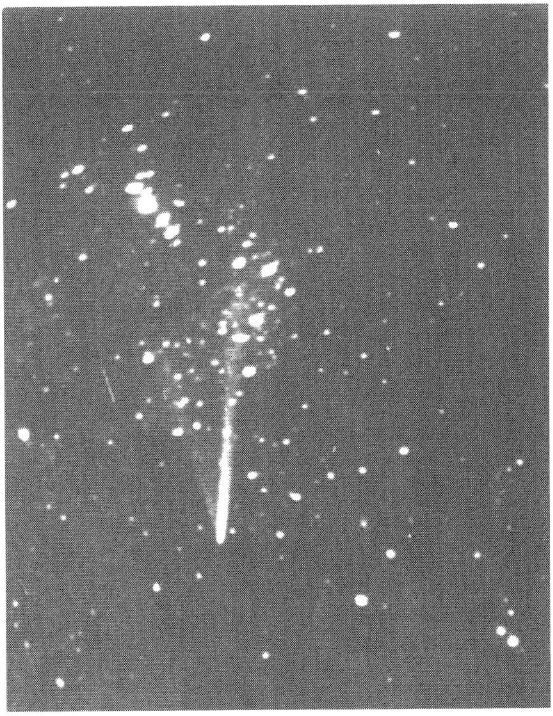

Fig. 22.6 Photograph by Grigg of Comet C/1901 G1 taken on 12 May 1901 (*Courtesy* Mitchell Library, Sydney)

The first of these was the Great Comet of 1901 (C/1901 G1),[1] which Grigg photographed on 12 and 16 May 1901, as it passed through Orion, and on 12 August 1902 Grigg (1902a) he sent copies to Australian astronomer John Tebbutt. Grigg's two photographs are bound in the 1902 volume of the *Letters to John Tebbutt* in the Mitchell Library, Sydney, immediately after Grigg's letter. One of these photographs is reproduced here as Fig. 22.6. Clearly this photograph was taken with the 'star camera' located on the platform attached to the Wray telescope.

Comet C/1901 G1 was a brilliant comet that first became visible in New Zealand and Australia in late April 1901 and was described at the time by Tebbutt: "… it is a striking and beautiful object. In addition to the principal or bright tail, a secondary much longer but fainter one made an angle with it toward the south of about 35 or 40 degrees." (Tebbutt 1908: 76). The clearly-distinguished dust and ion tails referred to by Tebbutt are evident in Fig. 22.6. Several days before this photograph was taken, two additional minor tails were discernible between these major tails (Vsekhsvyakstii 1964: 352). The comet faded rapidly between 12 and 16 May, and on the 15th split into two components that were separated by 1″ and differed by 1 magnitude in brightness (ibid.). This significant development is not shown in

[1]Cometary terminology used in this chapter is after Marsden (1989).

Grigg's 16 May photograph, which lacks the requisite definition. From all accounts, Comet C/1901 G1 was an impressive object, and it is not surprising that Chambers (1909: 158–159) includes it in his chapter on "Remarkable Comets".

In reference to this comet, following Grigg's death the *Thames Star* claimed that Grigg "… had the distinction of taking the only successful photo in the world of the 1901 comet. Copies were asked for and used as an official record by the R.A.S." (Obituary 1920). This is all very flattering to Grigg, but it simply is not true. Vsekhsvyakstii (1964) reveals that astronomers at the Cape Observatory in South Africa (e.g. see Gill 1901) and E.E. Barnard at the Yerkes Observatory in the USA also successfully photographed this comet, but because Grigg never published his photographs they were unknown to Vsekhsvyakstii (who also missed the photographs taken at the Perth Observatory in Australia—see Cooke 1901).

The other 'Great Comet' that Grigg is known to have photographed, also with the star camera, was 1P/Halley. According to the *Family History* (Grigg 1970: 27), Grigg took several successful photographs of the comet, all on 13 May 1910, and a copy of one of these is shown in Fig. 17.15. On the back of this photograph, Grigg wrote:

> With the naked eye, the Comet's Tail could be faintly traced considerably further than it appears in the picture, but the approaching dawn rendered it necessary to limit the exposure … the planet Venus … was shining brilliantly [just off the photograph, at the top right]. The star on the left is Gamma Pegasi (Algenib).

McIntosh (1958: 22, 1970: 103) errs in claiming that Grigg was the only New Zealand astronomer to successfully photograph 1P/Halley, for Cheviot's C.J. Westland and Brother J.J. Cullen and I. von Gottried at the Meeanee Observatory all succeeded in obtaining images of this comet (e.g. see Mackrell 1985: 55, 58, 110–111, 115, 120, 139, 143 and 197; Orchiston 1983; and Chaps. 13 and 18 in this book).

It would appear the Comet 1P/Halley was Grigg's 'swan-song', for there is no evidence that he continued to experiment with astronomical photography after 1910.

22.2.5 *Stellar Photography*

As we have seen, by 1896 Grigg had honed his photographic skills to the point where he could "… secure beautiful views of the moon, sun, and stars." (Astronomical photographs 1896). From this same long, informative newspaper article we also learn that Grigg planned to submit two astronomical photographs in the April 1896 Intercolonial Photographic Exhibition in Melbourne:

> Mr Grigg is entering in the scientific class two pictures, in which are reproduced views of the moon … the sun … [and] the stars—the Southern Cross, the Dog, Orion's Belt, and the Pleiades, etc. The pictures are all beautifully taken … (ibid.)

Now to quite a different topic. On 22 July 1903 Grigg wrote Tebbutt as follows:

> I had occasion to write Mr Russell [Sydney Observatory Director] some months ago about an apparent bright star which appeared on a reproduction of one of his photos, But was not shown on one of mine taken subsequently, asking him to compare his negative with the repron … (Grigg 1903).

Russell did not reply, so Grigg never did find out whether he had indeed picked up a known variable star, or perhaps even a nova.

22.3 Discussion

22.3.1 International Trends

Astronomical photography (Hughes 2012) was largely pioneered during the 1850s and 1860s, through the efforts of Warren De la Rue (1815–1889; Fig. 22.7; Hirshfield 2014) in England and both George Phillips Bond and Lewis Morris Rutherfurd (1816–1892; Fig. 22.8; Gould 1895) in the United States. As we saw in Chap. 14 in this book, photography was used as a primary research tool during the 1874 and 1882 transits of Venus—with somewhat disappointing results (e.g. see Lankford 1984)—but during the 1880s Andrew Ainslie Common (Fig. 22.9; Baum 2014) and Isaac Roberts (1829–1904; Fig. 22.10; Abbey 2007) in England, the Henry brothers, Pierre Paul (1848–1905; Fig. 22.11) and Prosper-Mathieu (1849–1903; Fig. 22.12; Barthalot 2014), at the Paris Observatory, and Sir David

Fig. 22.7 Warren de la Rue (https://en.wikipedia.org)

Fig. 22.8 Lewis Rutherfurd (https://en.wikipedia.org)

Fig. 22.9 Andrew Common (https://en.wikipedia.org)

Gill (1843–1914; Fig. 22.13; Forbes 1916) at the Royal Observatory, Cape of Good Hope, were able to demonstrate its potential for stellar and nebula astronomy (e.g. see Figs. 22.14 and 22.15). These and other notable developments led in the late

Fig. 22.10 Isaac Roberts (https://en.wikipedia.org)

1880s to the adoption of the International Astrographic Project (Turner 1912), and to the recognition of photography as a reliable—and, indeed, indispensable tool—of the professional astronomer. The 1890s also witnessed the emergence of serious amateur astronomical photographers in various countries, including New Zealand.

Fig. 22.11 Pierre Paul Henry (after Hughes 2012)

Fig. 22.12 Prosper-Mathieu Henry (after Hughes 2012)

Because of their comparatively tenuous nature, comets posed a special photographic challenge, and had to await the development of the dry gelatine plate and more accurate telescope drives. The first comet to be successfully photographed was C/1881 K1 (Tebbutt), which as the name infers, was discovered by Australia's

Fig. 22.13 Sir David Gill (https://en.wikipedia.org)

Fig. 22.14 A photograph of the Orion Nebula taken by Andrew Common in 1883 (https://en.wikipedia.org)

Fig. 22.15 The Great Comet of 1882 (C/1882 R1) (*Courtesy* South African Astronomical Observatory)

Fig. 22.16 Henry Draper (after *Popular Science Monthly* 1883)

John Tebbutt, and both Dr Henry Draper (1837–1882; Fig. 22.16; Barker 1888) and Jules Jannsen (Fig. 16.2) obtained good clear images of the head and tail (see Orchiston 1999). Meanwhile, in 1892 while still at the Lick Observatory, E.E. Barnard (Fig. 17.11) was responsible for the first photographic discovery of a comet (Clerke 1893: 447). So cometary photography was still in its infancy at the turn of the century when Grigg captured images of Comet C/1901 G1.

22.3.2 *The Development of Astronomical Photography in New Zealand*

In February 1896 the *Thames Advertiser* reported that to their knowledge, John Grigg was "… the only astronomer in New Zealand who has so far advanced [in astronomical photography] …" (Astronomical photographs 1896), and this is probably a fair assessment. Certainly there were other New Zealanders who succeeded in using photography in their astronomical endeavours at an earlier date, but these were generally one-off experiments. Thus, as we saw in Chap. 14, photography was successfully used in Auckland by Lambert, Martin, Pond and Redfern during the 1874 transit of Venus and by Lambert and Stuart during the 1882 transits, while a Mr Foy photographed the latter transit from Thames. Meanwhile, two weeks before the 1882 transit the afore-mentioned Auckland professional photographer, Josiah Martin, is reported to have exhibited a very successful photograph of the Sun at a meeting of the Auckland Institute (McIntosh 1959).

The 9 September 1885 total solar eclipse also tempted New Zealand professional photographers, and at least seven different individuals from Nelson, Wellington, Masterton and Taonui are known to have taken successful photographs of the event (e.g. see Orchiston and Rowe 2016, and Chap. 16 in this book). In Auckland the eclipse was partial (95 % total), and a Mr Cranwell "… climbed Mt Eden and secured a few photographs through gaps in a cloudy sky …" (McIntosh 1959: 47–48).

The foregoing summary indicates that a number of professional photographers and one or two astronomers successfully experimented with astronomical photography before Grigg did, but he was the first person to conduct on-going photographic work on behalf of astronomy. However, it would appear that his motivation throughout was the technological challenge involved, and to obtain interesting celestial images that could serve as visual props during his public lectures, rather than to systematically use astronomical photography for research purposes (see *Wanganui Herald* 1904).

If the purported photograph of the Great Comet of 1882 (C/1881 R1) reproduced by Mackrell (1985: 27) is not genuine—which on the evidence is likely—then Grigg would appear to have been the first New Zealand astronomer to have obtained a successful photograph of a comet, when he recorded the Great Comet of 1901 (C/1901 G1). In 1907 the Reverend Dr David Kennedy (Fig. 13.7) and his Meeanee Observatory assistants, J.J. Cullen and I. von Gottfried, successfully photographed Comet C/1907 L2 (Daniel)—see Fig. 13.9—with the 23-cm (9-in) Cooke refractor (Fig. 22.17) that would later go to the Wellington City Observatory and then the Carter Observatory and, as we have already noted, in 1910 Cullen and von Gottfied would use this same telescope to take many successful photographs of 1P/Halley (e.g. see Fig. 22.18 and other images in Mackrell 1985).

The only other New Zealander to successfully experiment with astronomical photography at this time was Cheviot-based C.J. Westland (see Fig. 22.19; and Chap. 18 in this book) who photographed Comet 1P/Halley and Comet C/1914 S1 (Campbell) using a simple home-made astrocamera (Fig. 18.8). Some of Westland's photographs of Comet 1P/Halley are reproduced by Mackrell (1985: 115, 130).

In the seventy-five years after Grigg's death, a number of other New Zealand amateur astronomers successfully experimented with astronomical photography. For instance, during the 1930s Murray Geddes (Fig. 22.20) and Douglas Berry (Fig. 22.21) carried out two-camera auroral photography, initially from Ermedale and Invercargill and later from South Hillend and Invercargill in the southern part of the South Island of New Zealand (Berry 1938), as part of an elaborate research program organized by Geddes as Director of the Auroral Section of the New Zealand Astronomical Society (see Geddes 1939). Soon after, Geddes would be appointed founding Director of the Carter Observatory in Wellington, only to die prematurely during WW II.

Fig. 22.17 The 23-cm (9-in) Cooke photovisual refractor at the Meeanee Observatory (Orchiston collection)

While Geddes and Berry were active in the far south of New Zealand, Charles William Bloomfield Michie (1890–1989; Fig. 22.22) from Kaitaia, in the far north of the North Island, began what was to become a lifelong commitment to solar eclipse photography (e.g. see Michie 1937, 1965; St. George 1991). Meanwhile,

Fig. 22.18 A Meeanee photograph of Comet 1P/Halley (after *Astronomical Photographs…* n.d.)

C.J. Westland continued photographing solar eclipses almost until his death in 1950 (see Thomsen 1937, 1949).

During the 1940s, Doug Berry moved to Dunedin and redirected his attention to comets. For a number of years he carried out photographic patrols in search of new comets using an astrocamera with a Ross 8.5-cm (3.3-in) portrait lens that gave a limiting photographic magnitude of around 11 (Berry 1946a, b). Although several new comets apparently were recorded, these turned out to be spurious images, and

Fig. 22.19 A sketch of C.J. Westland (Orchiston collection)

Fig. 22.20 Murray Geddes (Orchiston collection)

Fig. 22.21 Douglas C. Berry in 1986 (Orchiston collection)

Fig. 22.22 Charles Michie at the Auckland Observatory (Orchiston collection)

Carter Observatory Director, Ivan Thomsen (Fig. 13.15) was far from impressed (see Thomsen 1946a, b).

This situation changed markedly in 1947 when the New Zealander L.J. Comrie (Fig. 19.19) from the Nautical Almanac Office in London arranged for the British Astronomical Association to loan Berry its 15.2-cm (6-in) Charles Waller Astrograph (Berry 1947a), and he used this instrument (Fig. 22.23) to obtain successful images of Comet C/1947 F1 (Rondanina-Bester) and the 'Eclipse Comet of 1948, C/1948 V1 (Berry 1947b, 1949).

These very same comets were responsible for another South Island amateur astronomer, Robert Francis Joyce (1886–1961; Fig. 22.24; Howell 1967; Tunnicliffe 2009) of Kaiapoi, taking an active interest in astronomical photography. In early 1950 he replaced his original astrocamera in his Neptune Observatory with a new one featuring a 7.6-cm (3-in) Wray portrait lens (Joyce 1950), and by 1955, when he received the Murray Geddes Prize, he had "... accumulated a fine collection of photographs of star fields ... including notable plates of the Clouds of Magellan [see Fig. 22.25] of which he may well be proud." (Murray Geddes Prize Award—1955 1956). Some of Joyce's plates showed novae (e.g. Fig. 22.26) and comets.

One of Joyce's contemporaries and a close friend was the North Canterbury farmer, Alan S. Westland (Fig. 22.27), the son of Charles James Westland. By this time A.S. Westland had acquired the Charles Waller Astrograph and he used this to good effect photographing star fields.

The late 1960s and first half of the 1970s saw the focus of New Zealand amateur astronomical photography shift to the North Island, where Barrie Ward (Tirau) and Ronald G. Welsh (Auckland) were prominent. Both carried out systematic patrols,

Fig. 22.23 The Charles Waller Astrograph (Orchiston collection)

Fig. 22.24 R.F. Joyce and the 4.5-in (11.4-cm) Wray refractor at his Neptune Observatory in about 1958. The astrocamera is attached to the tube of the telescope, and part of it can be seen in this photograph (Orchiston collection)

and were rewarded with the discovery of a nova and new variable stars. A summary of their discoveries as at the end of 1972 is presented in a monograph commemorating the first fifty years of the Auckland Astronomical Society:

> No summary of observational research during the past five years is complete without reference to the patrol photography carried out by Ron Welsh and Barry Ward. Welsh initially discovered two important variable stars in 1969, a nova in Sagittarius and an unusual red variable in Crux. Ward has discovered several variable stars also, and using Ward's photographs Welsh has now found approximately twenty new stars. When the variable stars discovered in the sequence programme and in the red star programme are added, we find that about 50 new variables, mostly unclassified, have been found in Auckland. (Sale 1972).

The Ward-Welsh years were halcyon ones for New Zealand amateur astronomical photography. In October 1971 the formation of a 'Photographic Section' of the Royal Astronomical Society of New Zealand was seriously contemplated (see Allen 1971), and a year and a half later, on 12–13 May 1973 a 'Conference of Photographic Observers' was held at the Carter Observatory (see Bateson 1974).

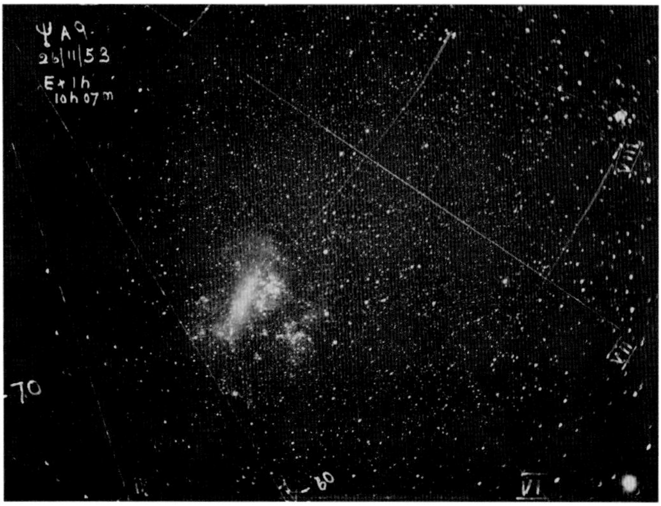

Fig. 22.25 One of Joyce's photographs of the Large Magellanic Cloud (Orchiston collection)

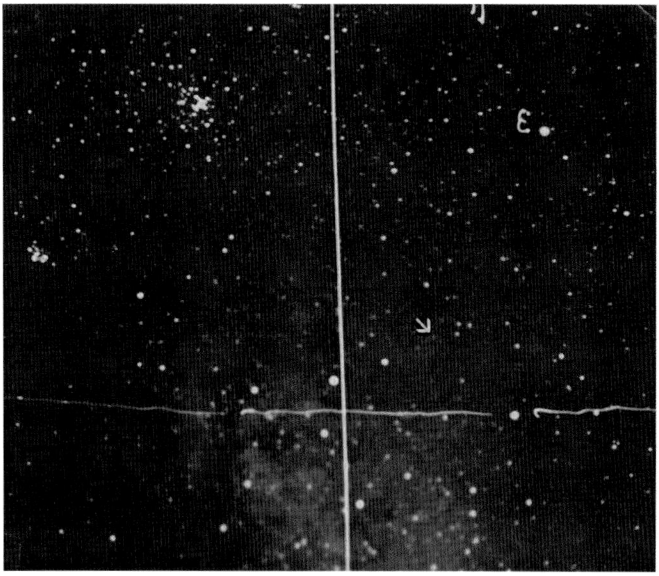

Fig. 22.26 Enlargement of a photograph taken by R.F. Joyce on 4 March 1952 showing Nova 1952 Haro (indicated by the arrow), which according to Taboada (1952) at that time was at about apparent visual magnitude 9.3 (Orchiston collection)

Fig. 22.27 Alan Westland in 1960, standing beside the observatory which at that time housed the Waller Astrograph (*Photograph* Wayne Orchiston)

In more recent years, the astronomical photographic research of Harry Williams of Auckland stands out. Using his 52.1-cm (20.5-in) Cassegrainian reflector (see Fig. 12.10), over a number of years during the early 1980s Williams studied the variable centre star of the planetary nebula NGC 2346 while Brian Marino (~ 1935–2002; Fig. 22.28) and Stan Walker (Fig. 22.29) carried out parallel

Fig. 22.28 Brian Marino in 1988 (Orchiston collection)

Fig. 22.29 Stan Walker (Orchiston collection)

photoelectric observations (Marino and Williams 1982, 1983, 1984; Marino et al. 1984).

Over the intervening years the very nature of amateur astronomical photography has changed markedly with the advent of CCD imaging and digital cameras, and from the successful results obtained of late by people like Maurice Collins of Palmerston North (e.g. see Collins 2009, 2010), John Drummond of Patutahi, near Gisborne (see Drummond 2003), Dr George Ionas of Palmerston North and Stuart Parker of Oxford (e.g. see Parker 2009, 2010), to name just a few, it is clear that New Zealand amateur astronomical photography is in safe hands.

22.4 Concluding Remarks

As arguably New Zealand's leading astronomer at the end of the nineteenth century, John Grigg is best known for discovering Comets 26P/Grigg-Skjellerup (1902), C/1903 H1 (Grigg) and C/1907 G1 (Grigg-Mellish), but he has an additional claim to fame as the nation's first serious astronomical photographer.

Using two home-made cameras, Grigg succeeded in taking photographs of the Sun, the Moon, comets and star fields. Although none of these was published at the time, many played an important educational role, for Grigg used them as popular visual aids during his numerous public lectures.

In New Zealand John Grigg was responsible for initiating astronomical photography as a distinct sub-discipline of amateur astronomy. Since his death in 1920, a number of other New Zealand amateur astronomers have used photography with distinction to obtain beautiful (aesthetically-appealing) celestial images or to carry out serious astronomical research projects.

Acknowledgements I am grateful to the following for their assistance: Bill Allen (Blenheim, NZ), Frank Andrews (Wellington, NZ), Mrs Beverley Angus (Wellington, NZ), Rod Austin (New Plymouth, NZ), Samantha Bennett (South African Astronomical Observatory, Cape town), the late Doug Berry (Auckland, NZ), Dr Grant Christie (Auckland, NZ), John Drummond (Patutahi, NZ), the late R.F. Joyce (Kaiapoi, NZ), Hazel McGee and Nick Frost (British Astronomical Association, London), Dr Nick Lomb (formerly Sydney Observatory), John Seymour (Palmerston North, NZ), the late Jackie St. George (Sydney) and the late Alan Westland (Cheviot, NZ). I also am grateful to the Donovan Trust for funding visits to the Mitchell Library and Sydney Observatory, during the original data-gathering phase of this project. Finally, I wish to thank Grant Christie, Rod Austin and John Drummond for reading and commenting on the first draft of this chapter, and the Mitchell Library (Sydney), the South African Astronomical Observatory and the Ward (Wanganui) Observatory—through John Seymour—for kindly supplying Figs. 22.2, 22.6 and 22.15.

References

The following abbreviation is used:

TL = In Letters to John Tebbutt (1860–1915), comprising 46 volumes of bound letters and documents in the Mitchell Library, Sydney (A3682-3727).

Abbey, L. B. (2007). Roberts, Isaac. In T. Hockey, et al. (Eds.), *Biographical Encyclopedia of Astronomers* (pp. 976–978). Berlin: Springer.

Allen, W. H. (1971). *Photographic Astronomy in New Zealand. A Report on the Possibility of Forming a Photographic Section* (pp. 7). Unpublished report, Carter Observatory, New Zealand.

Astronomical Photographs. *Thames Advertiser*, February 13, 1896 (p. 3).

Astronomical Photographs Taken at the Meeanee Observatory Napier, N.Z. (n.d.). Copy in Carter Observatory Library, New Zealand.

Barker, G. F. (1888). Memoir of Henry Draper, 1837–1882 (pp. 81–139). Washington: *Biographical Memoirs of the National Academy of Sciences*.

Barthalot, R. (2014). Henry, Paul Pierre and Prosper-Mathieu. In Hockey et al. (Eds.), *Biographical Encyclopedia of Astronomers* (2nd ed., pp. 483–485). Berlin: Springer.

Bateson, F. M. (1974). Research possibilities in photographic observing. *Southern Stars, 25*, 131–136.

Baum, R. (2014). Common, Andrew Ainsley. In T. Hockey, et al. (Eds.), *Biographical Encyclopedia of Astronomers* (2nd ed., pp. 447–449). Berlin: Springer.

Berry, D. C. (1938). Auroral observing. *Southern Stars, 4*, 47–48.

Berry, D. C. (1946a). Comet Section. *Southern Stars, 12*, 5–7.

Berry, D. C. (1946b). Astronomical photography for amateurs. *Southern Stars, 12*(82–85), 99–103.

Berry, D. C. (1947a). Comet Section report (for 1946). *Southern Stars, 13*, 5–6.

Berry, D. C. (1947b). Comet Section. *Southern Stars, 13*, 51–52.

Berry, D. C. (1949). The eclipse comet 1848 l. *Southern Stars, 14*, 74–78.

Chambers, G. F. (1909). *The Story of the Comets*. Oxford: Clarendon Press.

Clerke, A. (1893). *A Popular History of Astronomy During the Nineteenth Century*. London: Adam and Charles Black.

Collins, M. (2009). Lunar imaging techniques. *Southern Stars, 48*(2), 8–12.

Collins, M. (2010). Supersaturated colour images of the Moon. *Southern Stars, 49*(1), 3–4.

Cooke, W. E. (1901). Observations of Comet 1901 I observed at Perth Observatory, Western Australia. *Monthly Notices of the Royal Astronomical Society, 61*, 630–635.

Drummond, J. (2003). Possum Observatory. *Southern Stars, 42*(2), 3–5.
Early New Zealand Photographers and their Successors. Gibbs, William Brickell (n.d.). (canterburyphotography.blogspot.com/2013/02/gibbs-william-brickell.html).
Eclipse of the Sun. *Thames Advertiser*, February 3, 1897 (p. 3).
Espanek, F., & Meeus, J. (2006). *Five Millennium Cannon of Solar Eclipses: −1999 to +3000*. NASA Technical Publication (also available online).
Forbes, G. (Ed.). (1916). *Sir David Gill, Man and Astronomer*. London: John Murray.
Geddes, M. (1939). The photographic determination of the height and position of aurorae observed in New Zealand during 1937. *New Zealand Journal of Science and Technology, 20*, 389B–305B.
Gill, Sir David (1901). The Great Comet of 1901, as observed at the Royal Observatory, Cape of Good Hope. *Monthly Notices of the Royal Astronomical Society, 61*, 508–512.
Gould, B. A. (1895). Lewis Morris Rutherfurd, 1816–1892. *Biographical Memoirs of the National Academy of Sciences*, 415–441.
Grigg, J. (1902a). Letter to J. Tebbutt, dated August 12, in TL.
Grigg, J. (1902b). Letter to J. Tebbutt, dated December 3, in TL.
Grigg, J. (1902c). Photography with a small telescope. *Journal of the British Astronomical Association, 12*, 125–126.
Grigg, J. (1903). Letter to J. Tebbutt, dated July 22, in TL.
Grigg, J. (1905). Letter to J. Tebbutt, dated December 16, in TL.
Grigg, M. H. (Ed.), (1970). *John Grigg, F.R.A.S. of Thames, New Zealand. Together with some Genealogical Memorials of his Family and Glimpses of Pioneer Life in New Zealand*, unpublished typescript.
Hirshfield, A. W. (2014). De la Rue, Warren. In T. Hockey, et al. (Eds.), *Biographical Encyclopedia of Astronomers* (2nd ed., pp. 546–547). Berlin: Springer.
Howell, P. (1967). Robert Francis Joyce. An appreciation. *Southern Stars, 22*, 51–53.
Hughes, D. W. (1991). J. Grigg, J.F. Skjellerup and their comet. *Vistas in Astronomy, 34*, 1–10.
Hughes, S. (2012). *Catchers of the Light: The Astrophotographers' Family History*. Cyprus (published by the author).
Langkford, J. (1984). The impact of photography on astronomy. In O. Gingerich (Ed.), *The General History of Astronomy. Volume 4 Astrophysics and Twentieth-Century Astronomy to 1950: Part A* (pp. 16–39). Cambridge: Cambridge University Press.
McIntosh, R. A. (1958). The astronomical history of the Auckland Province. Part IV: John Grigg, F.R.A.S amateur astronomer. *Journal of the Auckland Astronomical Society, 18*, 20–23.
McIntosh, R. A. (1959). The astronomical history of the Auckland Province. Part V—first astronomical society. *Amateur Astronomer, 18*, 46–48.
McIntosh, R. A. (1970). Early New Zealand astronomy. *Southern Stars, 23*, 101–108.
Mackrell, B. (1985). *Halley's Comet Over New Zealand*. Auckland: Reed-Methuen.
Marino, B. F., & Williams, H. O. (1982). An interesting star in a planetary nebula. *Southern Stars, 29*, 266–270.
Marino. B. F., & Williams, H. O. (1983). Photographic observations of the central star in the planetary nebula NGC 2346. *Information Bulletin of Variable Stars, 2266*, 1–2.
Marino, B. F., & Williams, H. O. (1984). Photographic observations of NGC 2346 during 1983–84. *Information Bulletin of Variable Stars, 2583*, 1–3.
Marino, B. F., Williams, H. O., & Walker, W. S. G. (1984). Further observations of the central star in the planetary nebula NGC 2346. *Information Bulletin of Variable Stars, 2467*, 1–3.
Marsden, B. (1989). *Catalogue of Cometary Orbits*. Cambridge (Mass.): Minor Planet Centre.
Michie, C. B. (1937). Annular eclipse 1936, December 14. *Southern Stars, 3*, 40–43.
Michie, C. B. (1965). Total solar eclipse 1965 May 31. Observations at Tipene. *Southern Stars, 21*, 77–79.
Mr Grigg on "Astronomy". *Thames Star*, August 23, 1904 (p. 2).
Murray Geddes Prize Award—1955. Mr. R. Francis Joyce. *Southern Stars, 17*, 6 (1956).
Norman, D. (1938). The development of astronomical photography. *Osiris, 5*, 560–594.
Obituary. Mr. John Grigg. *Thames Star*, June 21, 1920 (p. 2).

Orchiston, W. (1983). C.J. Westland and Comet 1914IV: a forgotten episode in New Zealand cometary astronomy. *Southern Stars, 30*, 339–345.
Orchiston, W. (1993). John Grigg, and the genesis of cometary astronomy in New Zealand. *Journal of the British Astronomical Association, 103*, 67–76.
Orchiston, W. (1995). John Grigg, and the development of astrophotography in New Zealand. *Australian Journal of Astronomy, 6*, 1–14.
Orchiston, W. (1998). *Nautical Astronomy in New Zealand. The Voyages of James Cook.* Wellington: Carter Observatory.
Orchiston, W. (1999). C/1881 K1: a forgotten "great comet" of the nineteenth century. *Irish Astronomical Journal, 26*, 33–44.
Orchiston, W. (2001). The Thames observatories of John Grigg. *Southern Stars, 40*(3), 14–22.
Orchiston, W., & Rowe, G. (2016). New Zealand astronomy and the total solar eclipse of 1885. In T. Nakamura & W. Orchiston (Eds.), *The Emergence of Astrophysics in Asia: Opening a New Window on the Universe*. New York: Springer.
Parker, S. (2009). In search of dying stars. *Southern Stars, 48*(3), 3.
Parker, S. (2010). Journey to supernova 2010jb in Galaxy IC 1615. *Southern Stars, 49*(4), 7–9.
Popular Science Monthly, January 22, 1883.
Ratledge, D. (Ed.). (2005). *Digital Astrophotography: The State of the Art.* Berlin: Springer.
Rumsey, N. (1974). Astronomy in New Zealand. *Southern Stars, 25*, 171–207.
Sale, E. V. (1972). *The First 50 Years*. Auckland: Auckland Astronomical Society.
Scovill, C. E. (1980). *The AAVSO Variable Star Atlas*. Cambridge (Mass.): Sky Publishing.
St George, J., & L. (1991). Charles Michie. Catching Halley twice. *Astronomy Now, 5*(1), 51–53.
Stuart, A. (2006). *CCD Astrophotography: High-quality Imaging from the Suburbs*. Berlin: Springer.
Taboada, D. (1952). Novas Haro en Sagitario y Escorpión. *Boletin de los Observatorios de Tonantzintla y Tacubaya, 1*(5), 1–5.
Tebbutt, J. (1908). *Astronomical Memoirs*. Sydney (printed for the author).
Thames Advertiser, February 15, 1896 (p. 2).
The late eclipse of the Moon. *Thames Advertiser*, March 29, 1894 (p. 2).
Thomsen, I. L. (1937). Preliminary report on observations of the annular eclipse of the Sun, 1936 December 13.9 G.M.T. *Southern Stars, 3*, 17–21.
Thomsen, I. L. (1946a). The probable object photographed by Mr. D.C. Berry. *Southern Stars, 12*, 117–119.
Thomsen, I. L. (1946b). Visual versus photographic methods of comet searching. *Southern Stars, 12*, 119–122.
Thomsen, I. L. (1949). Partial solar eclipse, 1949 October 21d, U.T. *Southern Stars*, 14, 96.
Transit of Venus. *Thames Advertiser*, December 8, 1882 (p. 2).
Tunnicliffe, A. (2009). Frank Joyce: a man of many parts (http://www.astronomy2009.org.nz/r-f-joyce/).
Turner, H. H. (1912). *The Great Star Map. Being a General Account of the International Project Known as the Astrographic Chart*. London: Murray.
Vsekhsvyatskii, S. K. (1964). *Physical Characteristics of Comets*. Jerusalem: Israel Program for Scientific Translations.
Wanganui Herald, October 19, 1904.

Part VII
Opening a New Window on The Universe: Early New Zealand Radio Astronomy

Chapter 23
Dr Elizabeth Alexander and the Mysterious 'Norfolk Island Effect'

Abstract During March-April 1945, solar radio emission was detected at 200 MHz by operators of a Royal New Zealand Air Force radar unit located on Norfolk Island. Initially dubbed the 'Norfolk Island Effect', this anomalous radiation was investigated throughout 1945 by British-born Elizabeth Alexander, head of the Operational Research Section of the Radio Development Laboratory in New Zealand. Alexander prepared a number of reports on this work, and in early 1946 she published a short paper in the newly-launched journal, *Radio and Electronics*. A physicist and geologist by training, Elizabeth Alexander happened to be in the right place at the right time, and unwittingly became the first woman in the world to work in the field that would later become known as radio astronomy. Her research also led to further solar radio astronomy projects in New Zealand in the immediate post-war year, and in part was responsible for the launch of the radio astronomy program at the Division of Radiophysics, CSIR, in Sydney, Australia.

23.1 Introduction

Although radio astronomy had its origins in the 1930s through the pioneering efforts of Karl Jansky and Grote Reber (Kellermann and Sheets 1983), it only blossomed as an emerging scientific discipline after WWII. Part of the reason for this was the technological developments that occurred during the war, particularly those relating to radar (e.g. see Lovell 1977).

One of the wartime discoveries that provided an impetus for the post-war focus on radio astronomy was the independent detection of solar radio emission in Denmark (Schott 1947), the United States (Reber 1944; Southworth 1945), England (Hey 1946), Australia (Orchiston and Slee 2002) and New Zealand. This paper is about Elizabeth Alexander's investigation of solar radio emission in New Zealand during 1945, and builds on Orchiston (2005).[1]

[1]Previously, short accounts of Elizabeth Alexander's research were presented by Orchiston (1994), Orchiston and Slee (2002: 25–26) and Sullivan (1988: 316).

W. Orchiston, *Exploring the History of New Zealand Astronomy*,
Astrophysics and Space Science Library 422,
DOI 10.1007/978-3-319-22566-1_23

23.2 Elizabeth Alexander: A Biographical Sketch

Frances Elizabeth Somerville Alexander neé Caldwell (Fig. 23.1) was born on 13 December 1908 at Merton, Surrey, but spent her early life in India, where her father was the first Professor of Chemistry at Patna Science College and later was its Principal. At the end of WWI she returned to England and after completing her secondary schooling entered Newnham College, Cambridge, to study physics (Mary Harris, personal communication, 2014). Later she transferred to geology, graduating in 1931 with First Class Honours and the Harkness Prize. In 1934 she was awarded a Ph.D. by the same University, with a thesis on the main outcrop of the Aymestry Limestone (A.J.B. 1959).

In July of the following year Elizabeth married a young New Zealand physicist, Dr (later Sir) Norman Stanley Alexander (1907–1997), and in 1936 they moved to Singapore when he accepted the Chair in Physics at Raffles College; Elizabeth then began a study of tropical weathering. Their three children, William, Mary and Bernice, were born in Singapore, in 1937, 1939 and 1941 respectively. In 1940 and 1941 Elizabeth worked with the British Royal Navy (at the Singapore Naval Base) on radio direction-finding, with the rank of Captain in the Naval Intelligence Service (In Memoriam … 1959; Mary Harris, personal communication, 2005; Obituary 1959).

In early 1942 as the Japanese invasion of Malaya neared Singapore, Elizabeth managed to take her children to safety in New Zealand,[2] intending to return to Singapore with radar equipment from Sydney. But Singapore surrendered before her children were settled, and she was told her husband was dead, so she was forced to remain in New Zealand. By this time she was already well known in Australia *and* New Zealand for her work on radio direction-finding in Singapore, so in April 1942 she was appointed head of the Operational Research Section of the Radio Development Laboratory (Fig. 23.2)[3] in the nation's capital, Wellington (New Zealand localities mentioned in this chapter are shown in Fig. 23.3). At 33 years of age, she was considerably older than most of those who worked under her, one of whom later recalled: "She was much respected by us much younger folk, because of her experience of the wider world." (E.R. Collins, personal communication 1999). Elizabeth remained head of the Operations Research Section until the end of the

[2]This was very much a last-minute arrangement as the Japanese quickly advanced on Singapore, and Mary Harris (personal communication, 2015) remembers their plane being shot at.

[3]The Radio Development Laboratory was a Division of the Government's Department of Scientific and Industrial Research, and was set up in 1941 under Dr Oliver Owen Pulley (1906–1966) with an initial staff of ~100. Based in Wellington, it had branches in Auckland and Christchurch (see Fig. 23.2). When Pulley returned to Australia in 1942, Dr Charles Norman Machell Watson-Munro (1915–1991) became Director of the Laboratory (for details, see Galbreath 1998). Elizabeth Alexander was ideal for the position of head of the Operational Research Section of the Laboratory. Through her work on radar in Singapore she was already well known to Dr Ernest Marsden (Director of Scientific Developments, DSIR) and Pulley, and to Joe Pawsey and Fred White who played key roles in the development of radar in Australia. Meanwhile, she and Taffy Bowen were old friends from their days at Cambridge (Mary Harris, personal communication, 2014).

Fig. 23.1 Dr Elizabeth Alexander, 1908–1958 (*Courtesy* Mary Harris)

War, while Norman Alexander (who in fact had survived) was interned in the Changi Prisoner of War Camp in Singapore.

The Operational Research Section of the Radio Development Laboratory was involved in New Zealand's wartime radar development program (see Unwin 1992; *World War II* … 1948), and Elizabeth Alexander was responsible for the Section's own research radar unit on Mount Victoria, Wellington, where radar prototypes were operated and tested. She was particularly interested in studying propagational effects and developing fundamental theory so that radar performance could be predicted from meteorological data, and vice versa (Fig. 23.4). Consequently, she became a key player in an ambitious US-British-Australian-New Zealand radio-meteorological project to investigate 'anomalous propagation', radiation that appeared to originate from over the horizon. Towards the end of the War, this would evolve into the 'Canterbury Project'. Another—but very different—instance of 'anomalous propagation' that she researched was the 'Norfolk Island Effect', which unwitting took her briefly into a field that would later become known as solar radio astronomy.

At the end of the War, Norman Alexander joined Elizabeth and the children in New Zealand, but in March 1946 he flew back to Singapore to help reopen Raffles College. Elizabeth and the children remained in Wellington, and in July 1946 they sailed for England, where Norman joined them for a while. He was busy acquiring equipment for Raffles College, while Elizabeth wrote up some of her pre-war geological research. Then, in 1947, Norman and Elizabeth returned to Singapore, leaving the children with a relative in England—where they eventually went to boarding schools (Alexander, early 1990s; Mary Harris, personal communication, 2005).

Fig. 23.2 The nondescript office building that housed the Radio Development Laboratory (*Courtesy* Mary Harris)

The next few years saw the evolution of Raffles College into the University of Malaya, with Elizabeth acting as Registrar during the transition period. She then became a geological consultant to various bodies, and in 1949 was appointed Geologist to the Government of Singapore, "… with a main task of surveying the island's resources of granite and other useful stone, and in 1950 published a report which included the first reasonably complete geological map of Singapore Island." (A.J.B. 1959: 140).

In 1952, the Alexanders moved to Ibadan, Nigeria, when Norman accepted the Chair of Physics at the University College. Elizabeth was appointed a Lecturer in Soil Science in the Department of Agriculture, and again began researching tropical weathering. When the University founded a Department of Geology, in 1958, she was promoted as a Senior Lecturer and became Head of the Department (Alexander, early 1990s). After just three weeks in this new post she had a stroke, and died a little over one week later, on 15 October 1958; Frances Elizabeth Somerville Alexander was just two months short of her 50th birthday. She was

Fig. 23.3 New Zealand localities mentioned in this chapter are shown in *red*

Fig. 23.4 Elizabeth Alexander busy with her calculations (*Courtesy* Mary Harris)

fondly remembered for the "... warm welcome for those who dropped into the Alexander household, and there are many who are grateful for the hours they have spent there, either quietly listening to Beethoven, or else to conversation in which above all an air of sanity prevailed." (Obituary 1959: 5). Meanwhile, those who worked with her in New Zealand recalled her "... sound training, unremitting energy, a remarkably co-operative personality, and high intellect ..." (*World War II* ... 1948: 489). Currently, one of Elizabeth Alexander's daughters is preparing a biography of her illustrious mother, and this will be published later this year (Harris 2015).

23.3 Investigating the 'Norfolk Island Effect'

Between 27 March and 1 April 1945, a very striking increase in 'radio noise' was noted by the officer in charge of the Royal New Zealand Air Force (RNZAF) 200 MHz COL radar unit located on Norfolk Island.[4] This enhancement was shown

[4]At an approximate longitude of 168°E and latitude of 29°S, Norfolk Island is located about 1400 km east of Brisbane (Australia) and 750 km northwest of the northern-most tip of New Zealand's North Island. Although considerably closer to New Zealand, it is Australian territory. When the USA set up separate command areas in the Pacific in 1941 Norfolk Island fell within the New Zealand (South Pacific) command rather than the Australian (Southwest Pacific) command, and this explains why a Royal New Zealand Air Force radar station was established there. The radar station was located "... at a height of 1,000 feet near the north western corner of Norfolk Island ... [with] an unobstructed view all round ..." Alexander (1945d: 2).

Fig. 23.5 The Whangaroa radar station, showing the 200 MHz broad-side array and associated technical building (Orchiston collection)

to originate from outside the radar antenna, the turning gear and the receiver, and only occurred within half an hour of the rising or setting of the Sun. Furthermore,

> The maximum increase of noise was on the bearing of the sun and rotation of the aerial showed noise fluctuations corresponding fairly closely to the radiation diagram of the aerial. At its maximum the noise reached saturation on the azimuth of the sun and peaks of noise were also observed on azimuths corresponding to the first and second pair of side lobes. Switching off the Transmitter had no effect on the noise … (Alexander 1945d: 1).

As a normal course of action this phenomenon—dubbed the 'Norfolk Island Effect'—immediately was reported to Elizabeth Alexander, and she instantly recognised its significance and decided to investigate it by arranging for the Sun to be monitored within an hour of sunrise and sunset at five different RNZAF radar stations. These were located on Norfolk Island, and at North Cape, Whangaroa, Maunganui Bluff and Piha in the northern sector of the North Island of New Zealand (see Fig. 23.3 for the locations of these radar stations). A contemporary photograph of the Whangaroa radar station is reproduced below in Fig. 23.5. All five radar stations were instructed "… to record the increase in noise and the azimuth of maximum increase every few minutes and the time observations were taken … [together with] A general description of the weather at the time of taking the observations …" (Alexander 1945d: 1).

The monitoring took place between 10 and 23 April, although few stations were able to continue with this project past 18 April, and solar detections were recorded at all five stations. But, as Table 23.1 illustrates, even when all five stations were tracking the Sun, solar radio emission was never detected at more than three of

Table 23.1 Days when solar monitoring took place (dashes) and when solar radio emission was detected (crosses) at the different RNZAF radar stations

Radar Stations	Date (April 1945)														
	10	11	12	13	14	15	16	17	18	19	20	21	22	23	No. of Days
Norfolk Island			X	–	X	–	–	–	–	–	X	X	–	X	5
North Cape		X	–	X	–	–	–	–							2
Whangaroa			–	–	X	X	X	X	X						5
Maunganui Bluff		–	X	–	X	X	X	X							5
Piha	–	–	–	X	–	–	–	–	X	X					3
Stations with Detections	0	1	2	2	3	2	2	2	2	1	1	1	0	1	

these. It should be noted that the data in Table 23.1 for Piha are slightly misleading in that on four sunrise observing periods and one sunset observing period the radar antenna had to be used to track aircraft, so solar monitoring was not possible.

Part of the reason for the inconsistent results indicated in Table 23.1 lay with the design of the COL radar antennas, which could only rotate in azimuth and detect solar radio emission as the Sun rose or set through the antenna beam. The output, meanwhile, was displayed as 'grass' on a cathode ray tube, and it was a matter of making a subjective assessment as to whether the amplitude of this 'grass' exhibited a meaningful increase. Because the increase was often marginal, this was a problem. In order to increase sensitivity—and provide some means of quantification—the Commanding Officer at the Whangaroa radar station decided to install a microammeter between the receiver output and the diode limiter, and "Immediately a change was apparent and results [were] obtained." (Marsden 1945: 1).[5] Meter readings of 'normal noise' were taken either side of the Sun and as the radar antenna was slowly swept across the azimuth of the Sun, and up to five sweeps were made when solar noise was detected. Interestingly, "Where more than one complete sweep was taken the azimuth of the noise peaks could be seen to have drifted in the same direction as the changing azimuth of the sun." (Alexander 1945d: 3). Following this altered *modus operandi*, solar radio emission was detected at both sunrise and sunset on five successive days (see Table 23.1).

Elizabeth Alexander analysed the April observations made at the five radar stations, and concluded that "… at sunrise and sunset a detectable amount of noise over and above normal noise is received from a direction roughly that of the sun." (Alexander 1945d: 4). She stressed that while the observations were crude "… they do seem to indicate that more energy is sometimes radiated from the sun on 200 Mc/s than would be expected on black body theory." (ibid.); in other words, the emission was non-thermal. And while the levels of solar noise seen at the New Zealand radar stations were so small that they might have been missed in the past,

[5]The author of this report, E.D.L. Marsden, was known to Elizabeth Alexander, and was the only son of Dr Ernest Marsden, who is mentioned above in Note 3 (Mary Harris, personal communication 2014).

she noted that "At Norfolk Island, however, the increase was quite striking." (ibid.). Elizabeth Alexander then briefed Dr Marsden on the results of her investigation.

In a later much shorter report, Elizabeth Alexander (1945c) noted that on 26 March 1945 'jamming' at the azimuth of the Sun was also detected for 15–35 minutes before sunset at two non-New Zealand radar stations. One of these was in the Lingayen Gulf in the Philippines and operated at 106.8 MHz, and the other was an Australian 200 MHz station sited on Montalivet Island in the Darwin area. She concluded that "In both cases the jamming was almost certainly due to Solar Radiation. Probably other stations [in the region] were affected but, as is frequently the case, the phenomenon was not recorded." (ibid.).

On the basis of the initial New Zealand results, Elizabeth Alexander planned an elaborate solar monitoring program for the second half of 1945, which would involve the original five Air Force radar stations, "... and any Army and Navy stations that can take the observations ..." (Alexander 1945d: 5), and all were to be supplied with specially-designed vacuum tube voltmeters (to be inserted immediately after the second detectors in the radar receivers) and calibrated signal generators. The following observing procedure was adopted:

> ... switch off the transmitter, turn the aerial about 90° from the sun where the noise is normal, connect the vacuum tube voltmeter and record the meter reading. The signal generator is then connected in and the attenuator adjusted until the vacuum tube voltmeter reading is a little above the normal noise reading. These meter readings and the attenuator reading are recorded. The signal generator is disconnected and the set is ready for noise observations.

When an increase in noise voltage is observed, one of two procedures is adopted. Either the aerial is swept backwards and forwards across the sun's azimuth from normal noise through maximum to normal noise again and the meter reading, the azimuth and the time are recorded every two degrees. Or the aerial is swung just sufficiently to determine the azimuth of maximum noise and meter reading, and

> ... azimuth and time is [*sic*] recorded every one or two minutes. These observations are to be carried out over the sun rise and sunset period daily operational requirements permitting. Weather observations, with particular reference to amount and position of cloud cover, are to be made for each set of measurements. (Alexander 1945d: 6).

Unfortunately, two factors combined to prevent widespread adoption of this program: (1) an early, successful, outcome of the War was anticipated, so there was on-going reduction of staff at the various radar stations, but despite this (2) there was "... the necessity of keeping up operational watches [which] placed considerable obstacles in the way of a regular observation programme." (Millar 1946a: 1). Nonetheless, the five original RNZAF radar stations were able to carry out some solar monitoring. The Norfolk Island station began observations on 24 July, but the other radar units were not in a position to join the program until September, and all continued through into December 1945 (Alexander 1945c).

Sadly, this monitoring revealed just the one short-lived period when the Sun was particularly active at 200 MHz, centred on 5 October (Millar 1946a) when "... violent surges of noise were observed at irregular intervals. These surges were of

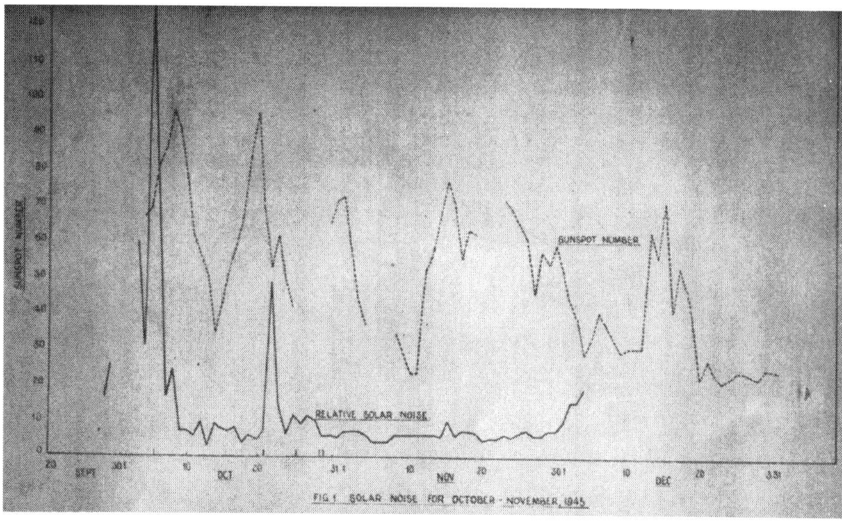

Fig. 23.6 Plots of sunspot numbers (the dotted curve), and 200 MHz solar noise recorded at the Norfolk Island and Piha radar stations, September–December 1945 (after Millar 1946b: 5)

momentary duration and sent the noise meter needle hard over." (Alexander 1945c). When this occurred, the officer in charge of the Piha radar station installed a simple Yagi aerial that could track the Sun, and although the gain was much less than that of the adjacent radar antenna, "… the noise could still be observed, its intensity remaining the same throughout the day whenever the aerial was directed towards the sun …" (Alexander 1946: 16). Alexander (1945c) also reported that "… The signals fluctuated rapidly but did not completely disappear until sunset." However, a plot of the solar noise observations made at this time (Fig. 23.6) reveals that solar radio emission also was detected on 21 October, and that both periods of solar activity correlated with enhancements in sunspot numbers. It is noteworthy that two later periods of higher than average sunspot numbers were not associated with 'solar noise'. Nonetheless, Elizabeth Alexander (ibid.) felt in a position to conclude:

> Such evidence as we have so far in New Zealand points to a direct correlation between sunspot number and solar noise … Though we have no absolute measure of the power received, there is strong evidence that during the periods of intense activity … Long wave Solar radiation is far removed from black body radiation.

The October activity reinforced her view that she was again dealing with non-thermal emission.

At the end of 1945, most of the RNZAF radar stations were closed down, and at the end of January 1946 the Radio Development Laboratory was disbanded and most of the staff returned to their previous positions or to interrupted university studies (Galbreath 1998; *World War II* … 1948). But before Elizabeth Alexander and her children sailed for England in July 1946 (Mary Harris, personal communication, 2002), she had one final radar-related task: to write up an account of her

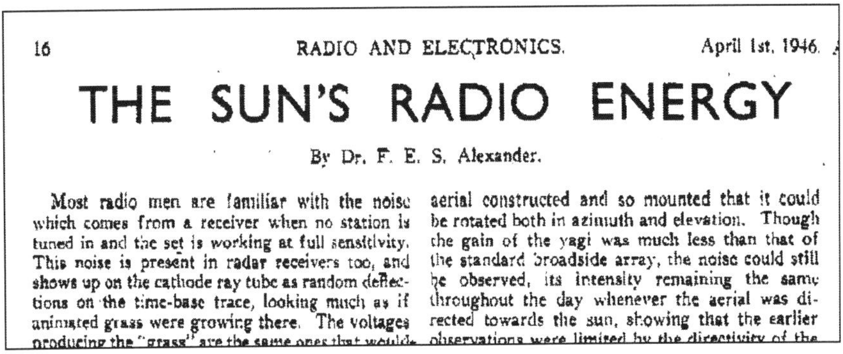

16 RADIO AND ELECTRONICS. April 1st, 1946.

THE SUN'S RADIO ENERGY

By Dr. F. E. S. Alexander.

Most radio men are familiar with the noise which comes from a receiver when no station is tuned in and the set is working at full sensitivity. This noise is present in radar receivers too, and shows up on the cathode ray tube as random deflections on the time-base trace, looking much as if animated grass were growing there. The voltages producing the "grass" are the same ones that would-

aerial constructed and so mounted that it could be rotated both in azimuth and elevation. Though the gain of the yagi was much less than that of the standard broadside array, the noise could still be observed, its intensity remaining the same throughout the day whenever the aerial was directed towards the sun, showing that the earlier observations were limited by the directivity of the

Fig. 23.7 Part of the title page of Elizabeth Alexander's 1946 paper in *Radio and Electronics*

1945 solar research project. This appeared as a 3-page research paper titled "The Sun's radio energy" (Alexander 1946; see Fig. 23.7),[6] and reported the March-April and October activity. The association between sunspots and solar noise—first put forward in her brief December 1945 report—was repeated, and it was noted that both March-April and October 1945 were periods of notable sunspot activity. She also mentioned the non-thermal nature of the 200 MHz emission, but thought that "To deduce solar temperatures of millions of degrees from this radiation, as has been suggested in some press reports, is absurd." (Alexander 1946: 20). However, later that same year, the Australian radio astronomers David Forbes Martyn (1906–1970) and Joseph Lade Pawsey (1908–1962) published papers in *Nature* in which evidence of a coronal temperature of 1 million degrees was presented (see Martyn 1946; Pawsey 1946).

In her paper, Alexander (1946) also discussed wartime observations of solar radio emission made by other researchers, in England, Australia and the USA; some of these projects were only undertaken after receipt of her reports on the 'Norfolk Island Effect'.

Elizabeth Alexander ended her 1946 paper by pointing to the need for further research, and suggesting a role for amateur astronomers and radio enthusiasts:

[6]Elizabeth Alexander's daughter, the London academic Mary Harris, has carried out a detailed investigation of her mother's research on the 'Norfolk Island Effect' and thinks

> … it is very likely that [in addition to the short paper published in *Radio and Electronics*] Elizabeth would have written up the work formally for DSIR, but that her report did not survive the writing of the Narrative and the destruction of the reports that became part of it. (Mary Harris, personal communication, 2014).

The 'Narrative' she refers to is the unpublished official history of New Zealand's war-time involvement in radar (see *World War II* … 1948), and many classified documents were destroyed during and following its preparation. I also am convinced that Elizabeth Alexander prepared a detail report on the 'Norfolk Island Effect' for the DSIR.

> What is required at the moment is more experimental evidence on the following aspects:
>
> 1. The way the received power varies with wavelength, over the whole band of the shortest microwaves up to the largest wavelengths (15-metres) which will penetrate the ionosphere (it almost looks at present as if the power might be greater at longer wavelengths).
> 2. The variation, at all wavelengths, with angle of elevation of the sun. This will permit evaluation of atmospheric absorption.
> 3. Exact times of onset, cessation, or change in character. This will permit correlation either with visible changes in the sun, or with other associated phenomena, magnetic storms, etc., and might lead to methods of predicting radio fadeouts.
> 4. Seasonal fluctuations, and variation with geographical position, particularly latitude.
>
> In observations of this type, amateurs can play an important part. Anyone who cares to build an ultra-shortwave receiver can be fairly sure of collecting useful information. The time is appropriate, since the sun is just entering a new phase of activity, and sunspots may be expected with increasing frequency over the next few years. (Alexander 1946: 20).

The tragedy is that Elizabeth Alexander's interesting little paper appeared in the inaugural issue of a New Zealand-based journal titled *Radio and Electronics*, which at the time had virtually no international visibility, and certain never came to the attention of radio astronomers—or indeed those researching the history of radio astronomy—until very recently. Nor did Elizabeth send out reprints (if indeed any were issued), or maintain a correspondence with colleagues she had met who were involved in the formative days of British and Australian radio astronomy. Mary Harris (personal communication, 2014) explains why:

> There were several reasons … a major one of which was time. Her contract ended on the last day of the war but was obviously extended to allow her to wind up her work properly … having written up her (assumed) report to DSIR and the more popular one for *Radio and Electronics*, there simply weren't enough hours in the day for an article for a relevant professional journal [prior to leaving New Zealand and joining her husband in England].

Mary Harris (ibid.) has also made the point that her mother was not a career physicist. When she arrived in New Zealand in 1942 "… she needed a job because she had 3 small children to rear, had no income and believed her husband to be dead." So her DSIR job was simply that—a job—even if it did just happen to take her on an interesting excursion into what would soon become radio astronomy. Then, when the war was over and she had completed her investigation of the 'Norfolk Island Effect', she simply "… wrote it up, signed off and left, rarely to speak of it again." (ibid.). As a result, her pioneering research was soon forgotten, only to be rescued from obscurity when Woody Sullivan (1988) came upon an archival reference to the 'Norfolk Island Effect' in the course of his research on the world-wide development of radio astronomy.

23.4 Discussion

23.4.1 *The Australian CSIR Division of Radiophysics Solar Research Program*

On 1 August 1945, Elizabeth Alexander penned a letter to J.L. Pawsey at the Radiophysics Laboratory in Sydney, and included as an attachment a copy of her R.D. 1/518 report on the 'Norfolk Island Effect', which "… describes our present and proposed investigations …" (Alexander 1945a). Pawsey passed the letter and report on to the Chief of the Division, John N. Briton (b. 1907), and they were also seen and initialled by Deputy-Chief, Dr Edward George ('Taffy') Bowen (1911–1991), Frank John Kerr (1918–2000), Lindsay Leslie McCready (1910–1976) and Ruby Violet Payne-Scott (1912–1981). The contents of Alexander's letter and report created considerable interest in Sydney, and were partly responsible for the launch of the Division's solar radio astronomy research program. After perusing this material, Taffy Bowen (as Acting Chief of the Division of Radiophysics) wrote to Dr (later Sir) Ernest Marsden (1889–1970) in New Zealand: "We were very interested to hear about the radar observations … and will attempt to repeat them here in Sydney." (Bowen 1945). However, this was not the only intelligence on solar radio emission to arrive in Sydney at this time, for Ruby Payne-Scott (1945) reveals

> … the almost instantaneous arrival of three reports in the [Radiophysics] laboratory, one recording severe noise interference in the direction of the sun on G.L. stations (55–85 Mc/s.) on 27th and 28th February 1942, coinciding with the passage of a large sunspot, another reporting … [the] Norfolk Island [Effect] … and a recent report by Reber in which he mentioned that, using his equipment for plotting "cosmic static" on 160 Mc/s., he obtained considerable radiation when the aerial was pointed at the sun.

Payne-Scott's report (ibid.) also indicates that it was these three reports that collectively inspired her boss, Joe Pawsey, to conduct the first Australian search for solar radio emission.[7]

These observations were made from 3 to 23 October 1945 using the 200 MHz COL radar antenna at the Royal Australian Air Force's radar station at Collaroy, a northern Sydney beach-side suburb, and immediately revealed the presence of solar radio emission (see Orchiston et al. 2006). Moreover, variations in mean intensity were found to correlate with changes in the total area of visible sunspots. When they came to write up these observations for *Nature*, Pawsey et al. (1946)

[7]In my initial paper about Elizabeth Alexander (Orchiston 2005) I described her as the first female to carry out a successful radio astronomy research program (in 1945) and to publish a paper on radio astronomy (Alexander 1946). However, her Australian counterpart, Ruby Payne-Scott, and the head of the Radiophysics radio astronomy group, Dr Joe Pawsey, made an earlier unsuccessful attempt to detect solar radio emission. There is therefore some debate as to which lady should be identified as the world's first female radio astronomer (e.g. see Goss and McGee 2009; Goss 2013), and although Woody Sullivan (2009) assigns priority to Elizabeth Alexander it is clear to me that both deserve full recognition for their pioneering efforts in 1945.

specifically mentioned Elizabeth Alexander's work, and one of the four references listed at the end of the paper was her 1945 report.

The paper was completed on 23 October 1945 and immediately submitted to *Nature*, but by the time it appeared in the 9 February 1946 issue it had been foreshadowed by two other contributions. The first of these was penned by Sir Edward Victor Appleton (1892–1965), dated 24 September 1945, and was published on 3 November. It discussed the evidence for non-thermal radio emission from the Sun (Appleton 1945). The other paper, dating from 17 October 1945, was by James Stanley Hey (1909–2000), and it appeared in the 12 January 1946 issue of *Nature* (Hey 1946). This important contribution reported on the 1942 detection of solar radio emission at radar stations in England, publication of which was only possible once the war was over and the research no longer carried a classified 'tag'.

When quizzed by the Australians about the publication order of the three papers, and the lengthy delay in the appearance of the Sydney paper, Appleton replied:

> I am sure your people were only anxious to acknowledge prior work … I have been much concerned with this solar noise myself and so know the history of it fully. Fortunately for us, my letter and Hey's letter to Nature preceded the Australian one, so no harm was done, as it turned out, so far as we are concerned; *though I feel rather sorry about the New Zealand people who were the next to conclude that it was really radio noise in their Norfolk Island experiments*. (Appleton 1946; his underlining; my italics).

Despite these protestations of concern, it is telling that Appleton neglected to mention Elizabeth Alexander's research in his paper, even though he was fully aware of the New Zealand work and the non-thermal nature of the emission by mid-1945. More than this, at some time prior to 1 August 1945 Appleton claims to have even gone to the trouble of replicating the New Zealand observations (see Alexander 1945a), yet he also fails to mention this additional evidence for non-thermal emission in his *Nature* paper. By this stage, Sir Edward was already beginning to gain a reputation for claiming credit for other people's work, and helping delay the publication of papers submitted to *Nature* by those he viewed as 'competitors' (e.g. see Bowen 1985; Kerr 1987). At the time, Taffy Bowen (1946) wrote Dr (later Sir) Frederick William George White (1905–1994) at CSIRO Head Office: "I am sorry that Appleton is making a song and dance about our letter to "Nature", but I suppose he is just expressing his well-known "ownership" of all radio and ionospheric work." (Bowen 1946).

After their initial observations at Collaroy, Pawsey and his Radiophysics colleagues decided to mount an even more ambitious solar monitoring program, extending from October 1945 through to March 1946. At first this involved the 200 MHz radar units at Collaroy and Dover Heights, but the limitations of sea interferometers soon prompted them to install 200 MHz steerable Yagi arrays at Collaroy, Dover Heights, the North Head radar station, and at Mount Stromlo Observatory. A few observations were also made at 75 and 3000 MHz. The results of this major investigation were published in the prestigious *Proceedings of the Royal Society* (McCready et al. 1947)—again after inordinate and inexplicable delays—and Elizabeth Alexander's work was once more referred to and referenced:

Fig. 23.8 Dr Bob Unwin (www.niwa.co.nz/atmopshere/facilities/lauder-atmospheric-research-station/lauder-photo-gallery)

the authors specifically state that "Observations on 200 Mc/s. *similar to those of the New Zealand stations* were begun by us towards the end of 1945 …" (McCready et al. 1947: 358, my italics). McCready et al. were not only able to confirm the New Zealand findings, but to make important new contributions to solar radio astronomy, thus placing Australia's foray into this emerging discipline on a firm observational and theoretical footing (for further details, see Orchiston et al. 2006).

23.4.2 The New Zealand Sequel

One of Elizabeth Alexander's principal preoccupations in 1944–1945 was radio-meteorology, which led to the 'Canterbury Project'. This joint British-New Zealand research project was approved in 1945, but the end of the War delayed its launch as funding had to be re-negotiated under civilian peace-time conditions (see Alexander 1945b). Given Elizabeth Alexander's relocation to England, the Canterbury Project came under the direction of one of her colleagues, Dr Robert S. Unwin (Fig. 23.8),[8] and in early 1947 an ex-WWII radar antenna and field trucks were set up at Wakanui Beach on the Canterbury coast, 85-km southwest of Christchurch (Fig. 23.9). In October 1947 the antenna was transferred to Ashburton Airport (Unwin 1947). Although this equipment was intended for the study of radio propagation across the Canterbury Plains under varying meteorological conditions,

[8]Back in 1942 Bob Unwin (later Dr Unwin) had been appointed as Elizabeth Alexander's assistant (Mary Harris, personal communication 2014).

Fig. 23.9 Installation of the 'Canterbury Project' 97.5 MHz radar antenna at Wakanui Beach near Ashburton. In the background are associated equipment trucks (Orchiston collection)

Dr Unwin arranged for his staff to observe the Sun at 97.5 MHz for an hour and a half after sunrise and before sunset from March to December 1947, and "… a large number of solar bursts of short duration were detected. On many occasions these occurred when sunspots and other visual signs of solar activity were in evidence." (Orchiston 1994: 68). The observations were forwarded to Ivan Leslie Thomsen (1910–1969) at the Carter Observatory in Wellington, and it is to be regretted that he never published the results of this interesting study.

On 23 November 1945 Elizabeth Alexander (1945b) warned Joe Pawsey that New Zealand's solar radio astronomical future looked bleak after the close-down of the COL radar stations at the end of the year: "I doubt that New Zealand will be able to put sufficient effort into building aerials adequate to investigate the phenomenon. It is a large scale job, if it is to be done properly, and the Canterbury Project is taking all available men and cash." This forecast proved remarkably accurate, and the part-time excursion into solar radio astronomy during the Canterbury Project was all that staff from the DSIR were able to attempt. Unlike in Australia, where many of the staff at the Radiophysics Laboratory were retained after the war and became the nucleus of the Sydney radio astronomy group (e.g. see Sullivan 1988), most of those in New Zealand's Radio Development Laboratory left the DSIR; and although Marsden succeeded in forming a Radar Section within the Dominion Physical Laboratory on 21 February 1946 (*World War II* … 1948), this began with a staff of just 13 (Atkinson 1976: 66) and would always be a small-scale operation with limited funding and research capability. Within post-War Government-funded science, there simply was no place for radio astronomy in New Zealand, and it was left to other institutions and individuals to progress this

Fig. 23.10 Alan Maxwell adjusting the 100 MHz twin Yagi antenna set up on the roof of the Biology Building at Auckland University College in 1947. The small hut below the antenna and mounting housed the receiver (*Courtesy* Alan Maxwell)

discipline. Those that were involved in solar radio astronomy prior to 1950 are discussed below.

While the Canterbury Project was in progress, a graduate student named Alan Maxwell (b. 1926) was engaged in a solar radio astronomy project for his M.Sc. degree at what is now the University of Auckland (see Maxwell 1948). His supervisor was Dr Karl S. Kreielsheimer, and he also found ready support from Professor Percy William Burbidge (1891–1984), both of whom were interested in astronomy, radiophysics and upper atmospheric physics. In mid-1947 Maxwell erected twin Yagis on the roof of the Biology Department (Fig. 23.10), and tracked the Sun at 100 MHz for the remainder of the year and into the second half of 1948. Maxwell (1948: 82) found that "In general, when solar noise was received there were sunspots on or near the sun's meridian." Of special interest was "… a period of solar activity between 1948 August 5–9, when there were numerous small-scale bursts of radio noise." (Orchiston 1994: 69), and "On at least two days an indication of a general solar noise background was noticed by pointing the array into and away from the sun … [Furthermore] A rough correlation of bursts with those observed in Canterbury [at Ashburton] has been established on several occasions." (Burbidge and Kreielsheimer 1947). Despite this being one of the first post-graduate theses on solar radio astronomy ever written anywhere in the world, Maxwell also failed to publish his work—it simply was not the custom at this time—and soon after completing his studies he moved to the dynamic astronomical environment of

Fig. 23.11 Ivan Thomsen (right), with two leading New Zealand amateur astronomers, Albert Jones (left) and E.L. Morley (after Eiby 1971: 19)

Jodrell Bank (at the University of Manchester) where he was quickly immersed in new research for a Ph.D. (Alan Maxwell, personal communication, 1993). Later he would go on to build an international reputation in solar radio astronomy while at Harvard College Observatory and develop the Radio Astronomy Station at Fort Davis, Texas (see Thompson 2010).

Carter Observatory in Wellington specialized in optical solar monitoring, and from the start its Director, Ivan Thomsen (Fig. 23.11), was vitally interested in the projects carried out by Elizabeth Alexander and Alan Maxwell and supplied both with relevant optical data. While his involvement in the solar radio monitoring associated with the Canterbury Project was disappointing (in that no results were published), he did in fact publish one paper pertaining to solar radio astronomy. Thomsen (Co-ordination … 1947) was particularly interested in the relationship between solar emission, sunspots and solar-terrestrial effects—such a short-wave radio fadeouts—and the 24 January 1948 issue of *Nature* features a paper where he compares Ryle and Vonberg's (1947) radio data for December 1946–April 1947 with sunspot records and finds "… a surprisingly general agreement." (Thomsen 1948: 134). Looking more closely, when the general level of radio emission for February–March 1947 was plotted against the position of photospheric features recorded at the Carter Observatory (Fig. 23.12), "… it was nearly always possible to ascribe some significant sunspot group to each of the maxima of the [radio] curve …" (Thomsen 1948: 134–135). Furthermore, the radio emission tended to coincide with the central meridian passage of the associated spot group, and "… in general, groups in the early stages of vigorous development, or showing activity by large

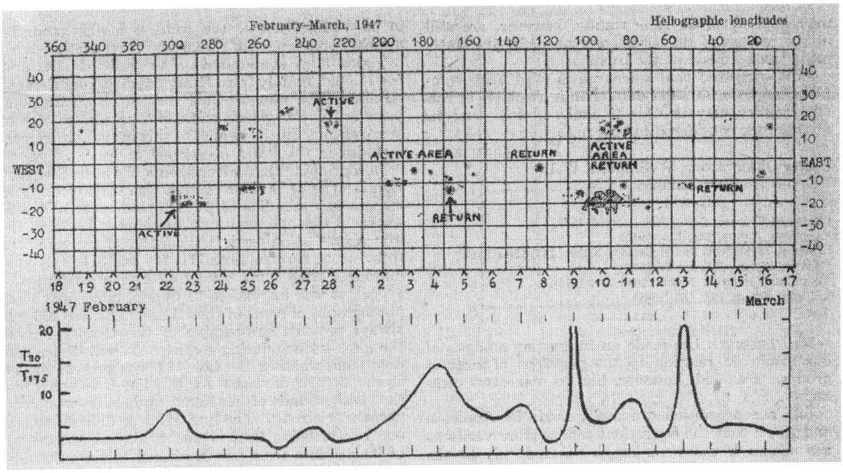

Fig. 23.12 Plot of solar radio emission in February–March 1947 and associated photospheric features (after Thomsen 1948: 135)

umbral movements and changes and accompanied by flares, give the greatest emission." (Thomsen 1948: 135).

Surprisingly, New Zealand can boast a fourth solar radio astronomy project dating to the immediate post-War years. Perhaps inspired by Elizabeth Alexander's 1946 'call to arms', Kaiapoi's Robert Francis Joyce (Fig. 23.13) constructed a corner reflector that could track the Sun, obtained an ex-WWII radar receiver, and from mid-1949 carried out solar monitoring at 515 MHz, sending his records to Ivan Thomsen at the Carter Observatory. By avocation, Joyce was a well-known amateur astronomer, whose Neptune Observatory boasted a much-used 11.4-cm (4.5-in) Wray refractor, and he was an ardent astronomical photographer (see Howell 1967; Murray Geddes Prize Award 1955; and the preceding chapter in this book). By vocation he ran a radio manufacturing and repair business, so he was in an ideal position to tap into the newly-emerging field of solar radio astronomy.

23.5 Concluding Remarks

New Zealand was merely one of several countries involved in the secret investigation of solar radio emission during WWII (see Sullivan 2009), and the person responsible for this Antipodean research was a British-born scientist, Dr Elizabeth Alexander. Rare among New Zealand-based scientists at the time, she was a woman, was married with a young family, held a senior post in the Department of Scientific and Industrial Research, and therefore moved in the upper echelons of New Zealand science (e.g. see Fig. 23.14). Being somewhat older than most of her colleagues, she had a wealth of research and international experience to fall back

Fig. 23.13 Kaiapoi amateur radio astronomer, R.F. Joyce (Orchiston collection)

Fig. 23.14 In September 1944 Dr Elizabeth Alexander poses with (left to right) Dr Ernest Marsden (Director of Scientific Developments, DSIR), Sir Cyril Newall (the Governor General of New Zealand), Dr Ian Stevenson (Director of the Radio Development Laboratory) and the Governor General's Aide de Camp (Orchiston collection)

upon when investigating the 'Norfolk Island Effect', and it is fortunate that her academic training as a geologist at Cambridge included units in physics and mathematics.

As it turned out, the 'Norfolk Island' project was but a momentary diversion in an eventful but all too short career dominated by geology and academia, but it was a significant diversion nonetheless, for by being in the right place at the right time she ended up being the first women to conduct research in the fledging new discipline of 'radio astronomy' (although it would take another five years of intensive research—mainly in England and Australia—before this term would begin to find common usage).

Elizabeth Alexander's solar work in 1945 was the catalyst that led to several other New Zealand research projects in solar radio astronomy in the immediate post-War years, yet unlike in neighbouring Australia, this discipline did not succeed in gaining a sustained foothold in New Zealand at this time. Had Elizabeth Alexander remained in New Zealand after the war and been able to continue her investigations in radio astronomy and radio-meteorology—instead of returning to geology and academia in Singapore—then the situation undoubtedly would have been very different.

Nonetheless, Elizabeth Alexander was a remarkable woman, and her place in the annals of radio astronomy deserves to be fully recognized and applauded.

Acknowledgements I am grateful to the following for their assistance: the late Gordon H. Burns (Wanganui, NZ), the late Professor E.R. Collins (Mahurangi East, NZ), Ross Galbreath (Tuakau, NZ), Professor Alan Maxwell (Harvard-Smithsonian Center for Astrophysics, USA), Dr Bruce Slee (CSIRO Astronomy and Space Sciences, Sydney, Australia), Professor Woody Sullivan (University of Washington, USA), the late Dr Bob Unwin (Wanaka, NZ), the late Dr R.M. Williams (Wellington, NZ), staff at the National Archives of New Zealand (Wellington), and especially Elizabeth Alexander's three children, Mary Harris (London), Bernice Jones (Edinburgh) and Bill Alexander (Leamington Spa). I also wish to thank Mary Harris for reading and commenting on the first draft of this chapter, and Mary Harris and Alan Maxwell for kindly supplying Figs. 23.1, 23.2, 23.4 and 23.10.

References

A.J.B. (1959). Frances Elizabeth Somerville Alexander (née Caldwell). *Proceedings of the Geological Society of London, 1572*, 140–141.

Alexander, F. E. S. (1945a). Letter to J.L. Pawsey, dated August 1. Copy in Sullivan collection.

Alexander, F. E. S. (1945b). Letter to J.L. Pawsey, dated November 23. Copy in Sullivan collection.

Alexander, F. E. S. (1945c). *Long Wave Solar Radiation*. DSIR, Radio Development Laboratory (Report).

Alexander, F. E. S. (1945d). *Report on the Investigation of the "Norfolk Island Effect"*. DSIR, Radio Development Laboratory Report dated 1 August (R.D. 1/518).

Alexander, F. E. S. (1946). The Sun's radio energy. *Radio and Electronics, 1*(1), 16–17, 20.

Alexander, Sir Norman, early 1990s, with a Footnote by Mary Harris dated 2002. Elizabeth's war work. Manuscript.

Appleton, E. V. (1945). Departure of long-wave solar radiation from black-body intensity. *Nature, 156*, 534–535.
Appleton, E. V. (1946). Letter to Dr F.W.G. White, dated March 4. Copy in Sullivan collection.
Atkinson, J. D. (1976). *DSIR's First Fifty Years*. Wellington: Department of Scientific and Industrial Research.
Bowen, E. G. (1945). Letter to Dr Ernest Marsden, dated August 27. Copy in Sullivan collection.
Bowen, E. G. (1946). Letter to Dr F.W.G. White, dated April 26. Copy in Sullivan collection.
Bowen, E. G. (1985). Letter to Woody Sullivan, dated August 7. In Sullivan collection.
Burbidge, P. W., & Kreielsheimer, K. S. (1947). *Report on Solar Noise Measurements Carried Out at the Physics Department, Auckland University College*. Unpublished report prepared for a DSIR meeting on November 6, 1947. Copy in National Archives, Wellington.
Co-ordination of Research on Auroral, Solar, Ionospheric, Geophysical and Radio Phenomena. Memorandum to Interested Parties. (1947). Report prepared for a DSIR meeting on November 6, 1947. Copy in National Archives, Wellington.
Eiby, G. (1971). Astronomy in New Zealand. *Sky and Telescope*, July, 18–20.
Galbreath, R. (1998). *DSIR. Making Science Work for New Zealand. Themes from the History of the Department of Scientific and Industrial Research*. Wellington: Victoria University of Wellington Press (in association with the Historical Branch, Department of Internal Affairs).
Goss, M. (2013). *Making Waves. The Story of Ruby Payne-Scott: Australian Pioneer Radio Astronomer*. Springer.
Goss, M., & McGee, R. (2009). *Under the Radar. The First Woman in Radio Astronomy. Ruby Payne-Scott*. Springer.
Harris, M. (2015). *Rocks, Radio and Radar: The Extraordinary Scientific, Social and Military History of Elizabeth Alexander*. London: Imperial College Press.
Hey, J. S. (1946). Solar radiations in the 4-6 metre radio wave-length band. *Nature, 157*, 47–48.
Howell, P. (1967). Robert Francis Joyce. An appreciation. *Southern Stars, 22*, 51–53.
In memoriam. Frances Elizabeth Somerville Alexander (née Caldwell), 1908–58. *Newnham College Roll Letter*, January, 36–37 (1959).
Kellermann, K., & Sheets, B. (Eds.), (1983). *Serendipitous Discoveries in Radio Astronomy*. Green Bank: National Radio Astronomy Observatory.
Kerr, F. (1987). Letter to Woody Sullivan, dated April 6. In Sullivan collection.
Lovell, B. (1977). The effects of defence science on the advance of astronomy. *Journal for the History of Astronomy, 8*, 151–173.
McCready, L. L., Pawsey, J. L., & Payne-Scott, R. (1947). Solar radiation at radio frequencies and its relation to sunspots. *Proceedings of the Royal Society, A, 190*, 357–375.
Marsden, E. D. L. (1945). *Report on Radio Sunset-Sunrise Observations Taken at No. 7 Radar Unit for the Period 14th until 18th April, 1945*. DSIR (Report).
Martyn, D. F. (1946). Temperature radiation from the quiet Sun in the radio spectrum. *Nature, 158*, 632–633.
Maxwell, A. (1948). *Enhanced Solar Radiation at 3 Metre Wavelengths*. Unpublished M.Sc. thesis, Physics Department, Auckland University College.
Millar, J. G. (1946a). *Long Wave Solar Research*. DSIR Report dated 15 January.
Millar, J. G. (1946b). *Observations of Noise From the Sun at Metre Wavelengths on Norfolk Island and in New Zealand*. DSIR, Radar Section, Dominion Physical Laboratory (Report 1/536).
Murray Geddes Prize Award. (1955). Mr. R. Francis Joyce. *Southern Stars, 17*, 6 (1956).
Obituary. (1959). Elizabeth Alexander. *Ibadan, 5*, 4–5.
Orchiston, W. (1994). Radio waves from the Sun: The New Zealand connection. In W. Orchiston, R. Dodd, & R. Hall (Eds.), *Astronomical Handbook for 1995* (pp. 65–69). Wellington: Carter Observatory.
Orchiston, W., & Slee, B. (2002). The Australasian discovery of solar radio emission. *Anglo-Australian Observatory Newsletter*, November, 25–27.
Orchiston, W. (2005). Dr Elizabeth Alexander: first female radio astronomer. In W. Orchiston (Ed.), *The New Astronomy: Opening the Electromagnetic Window and Expanding our View of Planet Earth* (pp. 71–92). Dordrecht: Springer.

Orchiston, W., Slee, B., & Burman, R. (2006). The genesis of solar radio astronomy in Australia. *Journal of Astronomical History and Heritage, 9*, 35–56.

Pawsey, J. L. (1946). Observation of million degree thermal radiation from the Sun at a wavelength of 1.5 metres. *Nature, 158*, 633–634.

Pawsey, J. L., Payne-Scott, R., & McCready, L. L. (1946). Radio-frequency energy from the Sun. *Nature, 157*, 158–159.

Payne-Scott, R. (1945). *Solar and Cosmic Radio Frequency Radiation. Survey of Knowledge Available and Measurements Taken at Radiophysics Laboratory to December 1st, 1945*. Sydney, CSIR, Radiophysics Division (Report SRP 501/27).

Reber, G. (1944). Cosmic static. *Astrophysical Journal, 100*, 279–287.

Ryle, M., & Vonberg, D. D. (1947). Relation between the intensity of solar radiation on 175 Mc./s. and 80 Mc./s. *Nature, 160*, 157–159.

Schott, E. (1947). 175 MHz emission of the Sun. *Physikalische Blätter, 3*, 159–160.

Southworth, G. C. (1945). Microwave radiation from the Sun. *Journal of the Franklin Institute, 239*, 285–297.

Sullivan, W. (1988). Early days of Australian radio astronomy. In R. Home (Ed.), *Australian Science in the Making* (pp. 308–344). Cambridge: Cambridge University Press.

Sullivan, W. (2009). *Cosmic Noise. A History of Early Radio Astronomy*. Cambridge: Cambridge University Press.

Thompson, A. R. (2010). The Harvard radio astronomy station at Fort Davis, Texas. *Journal of Astronomical History and Heritage, 13*, 17–27.

Thomsen, I. L. (1948). Solar radio emissions and sunspots. *Nature, 161*, 134–136.

Unwin, R. S. (1947). *Notes on Observations of Solar Noise at 3 Metres Wavelength Made at Canterbury Project, Ashburton*. DSIR Report dated 29 October (original in National Archives, Wellington).

Unwin, R. S. (1992). The development of radar in New Zealand in World War II. *The Radioscientist, 3*(1), 8–18.

World War II Narrative No. 3. Radar. Wellington: DSIR (1948). Original Manuscript in National Archives, Wellington (W3424, Box 16).

Chapter 24
John Bolton, Gordon Stanley, Bruce Slee and the Riddle of the 'Radio Stars'

Abstract New Zealand's role in unravelling the true nature of 'radio stars' during the 1940s is recounted, and more recent studies of galactic and extra-galactic radio emission undertaken in New Zealand are summarized.

24.1 Introduction

Radio astronomy is one of the newest astronomical disciplines, with a heritage that extends back little more than 80 years (see Reber 1983, 1984; Sullivan 1984b). Yet it was WWII and the development of radar which largely provided the impetus for the earliest major breakthroughs (Lovell 1977, 1983).

In his chronological analysis of radio astronomy, Verschuur recognises four different developmental phases. The first of these ended in 1950, and was characterised by:

> … an initial identity crisis as the new science tried to define itself. It was a time during which radio engineers made a series of startling discoveries, very few of which could be understood in terms of previous astronomical knowledge. Professional astronomers took relatively little notice of radio astronomy during this period. (Verschuur 1987).

Australia, Canada, England, France and the Soviet Union are the countries that generally are associated with the early post-war development of radio astronomy, with Australia and England leading the way (Sullivan 2009), and New Zealand generally does not rate a mention. However, New Zealand was directly involved in pioneering studies of solar radio emission (e.g. see Orchiston (2005a) and the previous chapter in this book), and also played a role in the identification of the first enigmatic 'radio stars'. This chapter, which builds extensively on two papers I published more than twenty years ago (Orchiston 1993, 1994), focuses on the New Zealand study of 'radio stars'.

W. Orchiston, *Exploring the History of New Zealand Astronomy*,
Astrophysics and Space Science Library 422,
DOI 10.1007/978-3-319-22566-1_24

24.2 Early Observations of 'Radio Stars': The New Zealand Connection

As Sullivan (1982) has indicated, the discovery of the first discrete radio source was an accident. In 1946, Hey et al. (1946b) were conducting a sky survey at 60 MHz and discovered a 'radio star' in Cygnus. Further observations disclosed "… short-period irregular fluctuations." (Hey et al. 1946a).

Although these source variations were later attributed to ionospheric scintillations, at the time they attracted the curiosity of radio astronomers from the Division of Radiophysics, C.S.I.R. (as it then was) in Australia, where research was being carried out on a number of different projects at scattered field stations in or near Sydney. These field stations and associated remote sites (Fig. 24.1) were a distinctive feature of

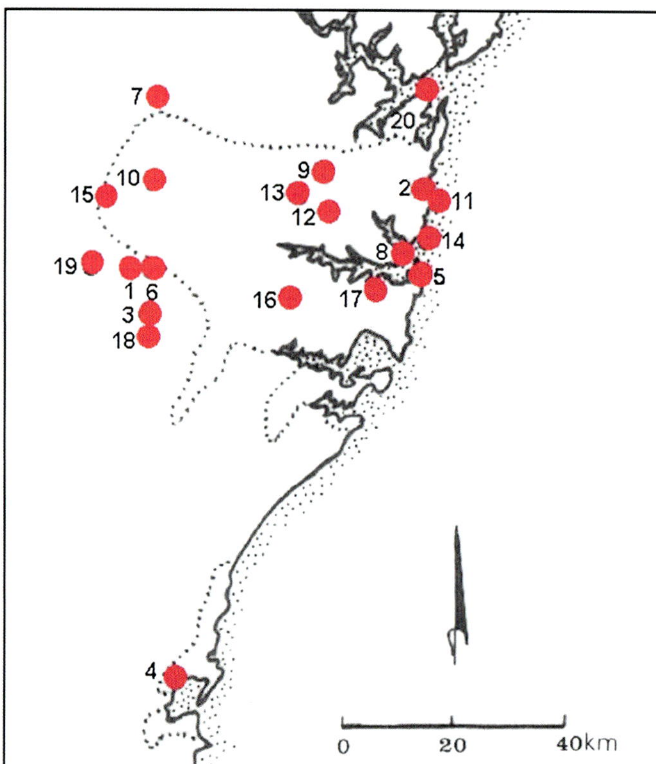

Fig. 24.1 Radio astronomy localities in the Sydney-Wollongong region. The dotted outlines show the current approximate boundaries of the Greater Sydney and Greater Wollongong regions. Key to localities: 1 Badgerys Creek, 2 Collaroy, 3 Cumberland Park, 4 Dapto, 5 Dover Heights, 6 Fleurs, 7 Freeman's Reach, 8 Georges Heights, 9 Hornsby Valley, 10 Llandilo, 11 Long Reef, 12 Marsfield (ATNF Headquarters), 13 Murraybank, 14 North Head, 15 Penrith, 16 Potts Hill, 17 Radiophysics Laboratory (Sydney University grounds), 18 Rossmore, 19 Wallacia, 20 West Head (*Map* Wayne Orchiston)

post-war radio astronomy in Sydney up to the mid-1960s, and largely were responsible for Australia's rapid emergence as one of the world's leading nations in the fledgling new field of radio astronomy. At these field stations, small groups of scientists and support staff built ever more sophisticated instrumentation designed to address specific research problems, maintained this equipment, carried out the observations, reduced the data and produced a succession of research papers that often featured a summary announcement in the leading British scientific journal *Nature*, followed soon afterwards by a much longer more detailed account in the *Australian Journal of Scientific Research* and its successor, the *Australian Journal of Physics*.

Some field stations were very focussed—for example, Dapto, Georges Heights and Penrith only dealt with solar radio astronomy; Badgerys Creek with discrete sources, and Murraybank with the H-line—while other field stations (Dover Heights, Fleurs, Hornsby Valley and Potts Hill) adopted more catholic approaches: at Dover Heights research was carried out on solar emission and discrete sources; at Hornsby Valley on lunar radar, solar emission and discrete sources; at Fleurs on Jovian decametric emission, discrete sources and solar emission; and at Potts Hill on Jovian decametric emission, solar emission, discrete sources and the H-line. In addition, solar and lunar research was carried out from the top of the Division's headquarters building (the Radiophysics Laboratory), which was located in the grounds of the University of Sydney (see Orchiston and Slee (2005) and Robertson (1992) for further details of these field stations).

From November 1961 I was one of those fortunate enough to work at the Fleurs field station, famous for its three distinctive cross-type radio telescopes (Orchiston and Slee 2002), before this facility was handed over to the University of Sydney and we at Radiophysics switched our allegiance to the newly-commissioned 64-m Parkes Radio Telescope (see Robertson 1992). These were the halcyon days of Australian radio astronomy, and even now, more than half a century later, I have very fond memories of the 1420 MHz Chris Cross solar radio telescope (see Orchiston 2004b; Orchiston and Mathewson 2009), the 85.5 MHz Mills Cross which initially focussed on discrete sources and galactic continuum emission (Orchiston and Slee 2005: 150–155) and the Shain Cross which also started its 'career' looking at galactic emission, but at the very low frequency of 19.7 MHz (see Orchiston et al. 2016). Elsewhere I have described how

> … those of us lucky enough to have lived through this era remember the field stations with genuine affection. There was a freedom not experienced by those back at the 'Lab' (as the Radiophysics Laboratory was known): the pervading sunshine, the clean fresh air, those incident-packed return trips from home to field station by Commonwealth car, and the sense that we were somehow making history. There were also snakes to contend with, wet days when antennas still had to be aligned and observations made, floods that had to be negotiated, and those times—fortunately they were few and far between—when vehicles became bogged and had to be rescued by a co-operative local farmer … Slide rules were the norm and computers but a future dream. Signal generators, not sources, provided calibrations, and results were displayed in real time on Esterline Angus and other all-too-familiar chart recorders. These were pioneering days! (Orchiston and Slee 2005: 122; cf. Christiansen 1984).

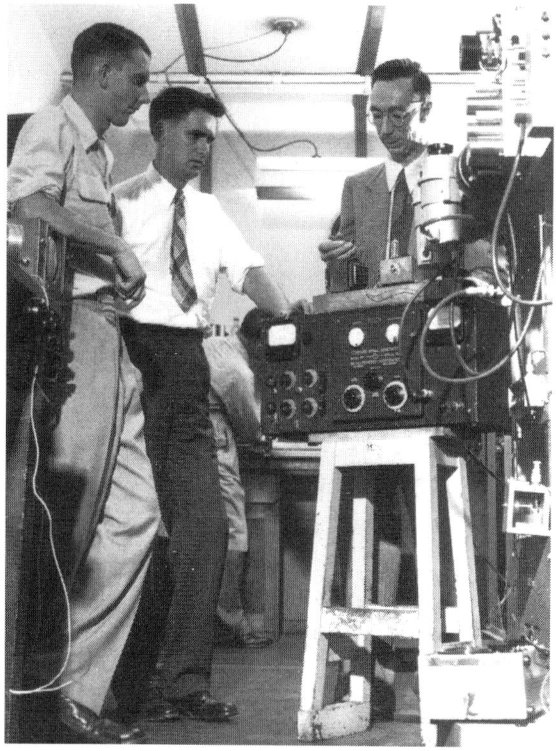

Fig. 24.2 John Bolton (left) and Gordon Stanley (centre), together with Dr Joe Pawsey, the head of the radio astronomy group within the Division of Radiophysics (*Courtesy* CSIRO RAIA)

Back in 1947 two of the Division's newer Research Scientists based at the Dover Heights field station, John Gatenby Bolton (1922–1993; Orchiston and Kellermann 2008; Robertson et al. 2014) and Gordon John Stanley (1921–2001; Fig. 24.2; Kellermann et al. 2005), decided to investigate Hey's variable radio source, which they soon detected and referred to as Cygnus-A (Bolton and Stanley 1948a, b). Later in the year they discovered a number of other discrete radio sources (Bolton 1948), and by the end of January 1948 the tally stood at five (Bolton 1982).

The radio telescopes used for these observations were 'cliff interferometers', which operated on the principle of a Lloyd's Mirror:

> Thc technique employed was to observe the region [of Cygnus A in this case] rising over the sea with the aerials situated on a high high cliff … Due to interference between the direct ray and the ray reflected from the sea, a lobe pattern is obtained, which gives rise to a succession of maxima and minima. An estimate of the size of the source can be made from the relative heights of maxima and minima, and an accurate position found from the times of occurrence of minima. (Bolton and Stanley 1948b).

Most of the cliff interferometer observations were carried out at 100 MHz, and indicated that sources less than 8 arcminutes in size were involved. These concentrated sources of radio emission, or 'radio stars', were an entirely unknown phenomenon to optical astronomers, and their explanation therefore posed a major challenge to the radio astronomers. The fundamental problem was to establish

Fig. 24.3 Bruce Slee just before he joined the CSIR Division of Radiophysics in 1946 (*Courtesy* Bruce Slee)

accurate source positions so that optical correlates of the radio sources could be sought, and it would only be possible to achieve this with the instrumentation at hand if rising *and* setting times of the sources were determined. Such observations could not be carried out in Sydney (where only the rising of the sources could be observed), but they were possible from near Auckland in Stanley's native New Zealand. The east coast and west coast of Northland not only offered ideal readily-accessible observing sites, with high cliffs, but Auckland also was home to a university with physics staff interested in radio astronomy. Bolton and Stanley therefore proposed an expedition to New Zealand, and Dr 'Taffy' Bowen, Chief of the Division, gave this his enthusiastic support. Meanwhile, the third member of the research team, Owen Bruce Slee (b. 1924; Fig. 24.3; Orchiston 2004a, 2005b), would remain in Sydney and conduct parallel observations at the Dover Heights field station. Partly this was in order to investigate the nature of the variable emission from Cygnus-A: parallel observations from these widely-separated trans-Tasman locations hopefully would indicate whether the variations were intrinsic to the source itself, or were imposed on the signal by the Earth's ionosphere or the interplanetary medium.

As a result, at the end of May 1948 an ex-Army-WWII gun-laying radar trailer (Fig. 24.4) containing a 100 MHz 4-Yagi array, a new receiver, recorders, chronometers and weather-recording equipment was shipped to Auckland. Bowen had arranged with New Zealand's Department of Scientific and Industrial Research for logistical support for the expedition, which resulted in the New Zealand Army supplying a truck that was used to haul the little mobile radio telescope to its first observing site, 'Pakiri Hill', a farm about 10 km from the township of Leigh and 70 km north of Auckland (for New Zealand localities mentioned in this chapter see Fig. 24.5). The site was at an altitude of about 300 m above mean sea level, and the

Fig. 24.4 The 100 MHz Yagi array and mobile field laboratory set up in Sydney prior to its departure for New Zealand (*Courtesy* CSIRO RAIA)

coastline ran roughly east-west so that Cygnus-A could be observed rising in the north-east, culminating at just 15°, and then setting in the north-west. Taurua-A also was visible from this location, but Centaurus-A and Virgo-A were too far south. The Yagi array had a horizontal beamwidth of 12° and a vertical beamwidth of 30°. In later years, Bolton (1982) was to reminisce:

> We spent nearly two months at Leigh, in periods of working 10 nights and then having four days' rest as tourists. Conditions were far from ideal; we had a long extension from an already overloaded power line and frequency variations caused variations in the recorder chart speed of at least 10%. The weather was sometimes appalling … Nevertheless we obtained about 30 nights' usable data on Cygnus-A and in mid-July five observations of Taurus-A …

A typical chart recording of Taurus-A is shown in Fig. 24.6, where it is compared with the 'discovery' chart record' obtained earlier at Dover Heights. The Pakiri Hill interference fringe amplitudes are greatly enhanced, because of the substantial increase in the height of the cliff-top at Pakiri Farm (300 m, vs. 79 m at Dover Heights). The fringe amplitude $P_{max} - P_{min}$ is given by:

$$P_{max} - P_{min} = 4P_o \tag{24.1}$$

and if

$$P = 2P_o[1 - \cos(4\pi h \sin\alpha/\lambda)] \tag{24.2}$$

then

$$P_{max} - P_{min} = 2\text{P}/[1 - \cos(4\pi h \sin\alpha/\lambda)] \tag{24.3}$$

where P is the power received by the antenna, P_o is the power received with the same antenna in space; h is the height of the antenna above mean sea level, α is the amplitude of the source, and λ is the wavelength (Bolton and Slee 1953: 421–422). As we can see, in sea interferometry the height of the cliff was critical. But more

Fig. 24.5 New Zealand localities mentioned in this chapter are shown in *red*

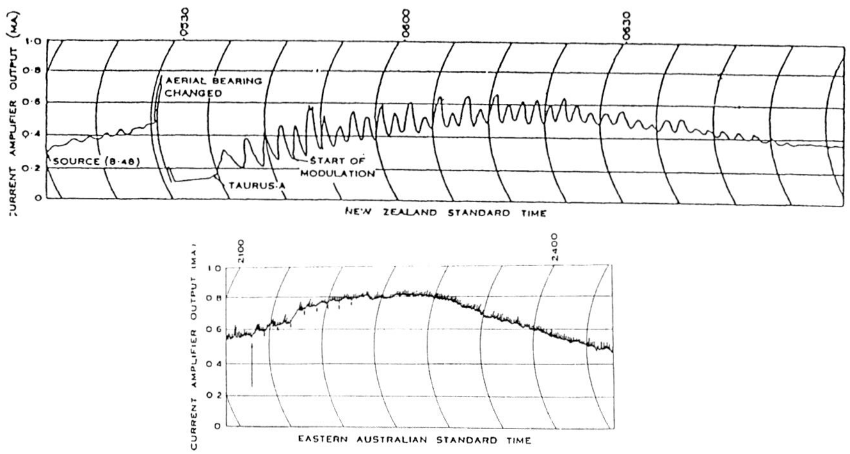

Fig. 24.6 Chart records, to the same scale, showing the Taurus-A interference fringes as seen at Pakiri Hill, New Zealand (top) and Dover Heights (bottom) (after Bolton and Stanley 1949: 140–141)

than this, the height of the cliff also impacted on the maximal size of the source responsible for the radio star. Thus, the upper limit to the size of the source, in terms of an 'equivalent radiating strip', W, is given by the following equation:

$$W = (\lambda/\pi h)\sqrt{(3R)} \tag{24.4}$$

where h is the height of the cliff and R is the ratio of the heights of the interference fringe maxima and minima above an extrapolated cosmic drift background level. But in practice, receiver instability and non-linear drift curves tended to place limits on the resolving power using this method.

At the end of July, Bolton and Stanley moved their mobile radio telescope to the deserted WW II radar station at Piha on the adjacent west coast of the Auckland Peninsula 30 km due west of the centre of Auckland (see Fig. 24.5). This also offered 300-m high cliffs, and an uninterrupted view of the western horizon. Conditions here were vastly superior to those encountered at Pakiri Hill: "The diesel plant for the radar provided a supply of electricity stable in both voltage and frequency, our receivers performed faultlessly and the weather was perfect." (Bolton 1982). Over the next two weeks successful observations were made of Cygnus-A, Centaurus-A, Taurus-A and Virgo-A as they set.

While these observations were underway in New Zealand, the third member of the 'radio sources' team, Bruce Slee, continued to observe these same sources from Dover Heights. He also succeeded in discovering one new source, Fornax-A.

After their return to Sydney, Bolton began reducing the New Zealand data. He set up a trestle table in the blockhouse at Dover Heights, and did all of the canculations and reductions there (Slee personal communication 2014). First he worked on Taurus-A, and after making the appropriate corrections for minor differences in latitude of the observing sites, curvature of the Earth's surface, and atmospheric refraction, he derived the following celestial coordinates for the radio source:

Right Ascension (1948) 5h 31m 20 ± 30s
Declination (1948) 22° 02 ± 8′

Lying within the 'position box' provided by the above co-ordinates was NGC 1952, the Crab Nebula, and Bolton and Stanley (1949) had no hesitation in associating Taurus-A with this object (Orchiston and Slee 2006). This identification showed that discrete radio sources could be related to known astronomical objects, in this case a remnant of a supernova that erupted in AD 1054, and was recorded by Arab, Chinese, Japanese and Korean astronomers (see Stephenson and Green 2002), but seems not to have been mentioned by European astronomers (Stephenson and Green 2003).

Four months later, Bolton et al. (1949) completed their pioneering paper "Positions of three discrete sources of galactic radio-frequency radiation" which was published in the leading English scientific journal, *Nature*. They found that "... all three sources correspond within limits of experimental error to positions of certain nebulous objects ...", and they were able to identify Centaurus-A with NGC 5128 (see Robertson et al. 2010), Virgo-A with NGC 4486 (M87), and confirm the Taurus-A Crab Nebula association. A much more detailed account of this work, which included the first radio spectra, was subsequently published in Australia (Stanley and Slee 1950). Despite this, these 'identifications' were regarded by many astronomers with suspicion, especially when "... other likely optical objects of the same classes turned out *not* to be radio sources." (Woody Sullivan personal communication 1992). This issue was finally resolved by Wilhelm Heinrich Walter Baade (1893–1960; Fig. 24.7; Osterbrock 2001) and Rudolph Minkowski (1895–1976; Fig. 24.8; Osterbrock 1983) with the publication in the *Astrophysical Journal* of their 1954 landmark paper titled "Identification of the radio sources in Cassiopeia, Cygnus A and Puppis A".

The other notable result deriving from the 1948 New Zealand field trip involved the enigmatic Cygnus-A source fluctuations first report by Hey et al. (1946a). Thanks to the parallel Australian and New Zealand observing sessions it was revealed that these puzzling source amplitude fluctuations were caused by terrestrial atmospheric scintillations and therefore were not intrinsic to the source itself (Stanley and Slee 1950). From all accounts, then, the New Zealand expedition of 1948 was judged to be a great success.

24.3 Discussion

24.3.1 From 'Radio Stars' to Discrete Sources

The 1948 expedition to New Zealand initiated a major breakthrough in astronomy, as the identification of discrete radio sources with known optical objects encouraged closer cooperation between optical and radio practitioners. It also clearly

Fig. 24.7 Walter Baade (*Courtesy* Observatories of the Carnegie Institution for Science Collection at the Huntington Library, San Marino, California)

Fig. 24.8 Rudolph Minkowski (left) and Bernie Mills (right) at the Fleurs field station in the mid-1960s (*Courtesy* CSIRO RAIA)

revealed that the term 'radio star' was a misnomer, as these sources were in no way connected with individual stars. In retrospect, Bolton (1982) also felt the Taurus-A work was a personal milestone: "The identification of the Crab Nebula was a turning point in my own career and for non-solar radio astronomy. Both gained respectability as far as the 'conventional' astronomers were concerned."

The New Zealand work therefore demonstrated that research into discrete radio sources was justified (see Bolton 1955), and led eventually to the various Cambridge and Sydney source catalogues, which, with their associated cosmological implications, were to generate considerable international controversy and even animosity (e.g. see Mills 1984; Smith 1984). Before this took place, however, there was the problem of Cygnus-A, which the New Zealand observations had not resolved.

24.3.2 *The Enigma of Cygnus-A*

In contrast to Centaurus-A, Taurus-A and Virgo-A, which were associated with obvious optical objects, the Cygnus-A source proved far more difficult to explain.

Table 24.1 shows the way in which British and Australian radio astronomers gradually pinned down the position of this source until Baade and Minkowski (1954) finally were able to publish an optical identification after subjecting the relevant region of sky to intensive scrutiny with the 5-m (200-inch) Palomar Telescope. The Cygnus-A source proved to be associated with a faint (15th magnitude) elliptical galaxy.

In fact Bernard Yarnton Mills (1920–2011; Fig. 24.8; Mills 2006)[1] and Adin B. Thomas from the Radiophysics Laboratory in Sydney made this very optical identification several year earlier, in 1949. After observing Cygnus-A from May to December 1949 with the 97 MHz Swept-lobe Interferometer at the Laboratory's Potts Hill field station (Wendt et al. 2011), Mills examined a photograph of the region that Minkowski had sent to Bolton and in the error box of their observations he noticed a faint extragalactic nebula. He felt that this could be the source of the emission and wrote to Minkowski suggesting this (Mills 1949). However, Minkowski (1949) advised Mills against claiming the identification, so when their paper on Cygnus-A finally appeared Mills and Thomas (1951) concluded that this faint galaxy was unlikely to be responsible for the emission. Ironically, the more precise position obtained in 1951 by Cambridge University's Francis (later Sir Francis) Graham Smith (b. 1923) confirmed that this galaxy was indeed responsible for the Cygnus-A emission.

[1]In a useful review of the history of New Zealand radio astronomy, Marilyn Head (2010: 9) proudly identifies Bernie Mills as a New Zealander. While Bernie was happy to employ a 'genuine' New Zealand radio astronomer (e.g. Bruce McAdam), he was by birth, domicile and sentiment entirely Australian, even if his mother was a New Zealander. He was born in Manly, a well-known seaside suburb of Sydney, and studied engineering at the University of Sydney, before joining the CSIR's Division of Radiophysics in 1948.

Table 24.1 Radio positions of Cygnus-A used in identifying its optical correlate (based on Baade and Minkowski 1954: 265)

R.A. (1950)	Dec. (1950)	References
19h 58m 47 ± 10s	+41° 47 ± 07′	Bolton and Stanley (1948a, b)
19h 58m 14 ± 60s	+40° 36 ± 10′	Stanley and Slee (1950)
19h 57m 46 ± 05s	+40° 30 ± 07′	Ryle et al. (1950)
19h 57m 37 ± 06s	+40° 34 ± 03′	Mills and Thomas (1951)
19h 57m 44 ± 2.5s	+40° 35 ± 1.5′	Mills (1952)
19h 57m 22 ± 25s	+40° 22 ± 16′	Hanbury Brown and Hazard (1951)
19h 57m 44.3 ± 1s	+40° 35 ± 01′	Smith (1951)

24.3.3 Media and Public Interest in the 'Radio Stars' Project

From the start, the New Zealand news media were excited by the idea that two young scientists from Australia should come to New Zealand to study the Universe. The publicity began when Dr Clabon Walter (Cla) Allen (1904–1987) stopped off in Auckland whilst *en route* to England a week or so before Bolton and Stanley were due to arrive. Although primarily an optical astronomer, he had collaborated with colleagues from the Division of Radiophysics on solar radio astronomy and published a paper (Allen 1947). He therefore was in a position to know about the forthcoming 'radio stars' field trip, and he sang its scientific praises in the pages of the *New Zealand Herald* ("Noises from the stars", 22 May 1948). He mentioned the Leigh observing site (but not Piha), and the advantages it offered over "… the Sydney headlands … where accuracy could not be attained."

Three days later, on 25 May, a further news item also titled "Noises from the stars" appeared in the same newspaper, with statements by Harley Weston Wood (1911–1984), the Director of Sydney Observatory, about the Cygnus-A radio source. In this article Arthur John Higgs (b. 1904) from the Radiophysics Laboratory was quoted as saying that the Bolton and Stanley work "… was the most advanced in the world in this field." Six days later, on 31 May, Stanley's arrival in New Zealand was briefly reported under the banner "Scientist to study space noises".

About two weeks after Bolton and Stanley arrived at Pakiri Hill they were visited by a reporter from the *New Zealand Herald*, who subsequently produced an interesting article which was published on 25 June. In it, he rightly speaks of this pair of 26-year olds as pioneers in radio astronomy, and identifies Bolton as an Englishman and Stanley as a New Zealander. He also reported on the response of the local population to these unusual intruders:

> The people of Leigh seem to have accepted the scientists with little comment. No one ever seems to have climbed the hill to visit the strange-looking trailer with its four antennae pointing horizontally out to sea. Professor P.W. Burbridge … and several members of the Department of Scientific and Industrial Research have been the only ones, apart from

> inquisitive newspaper men, to have broken the research workers' steady routine. This peace, of course, is what they want and what they have asked for, so that they can get on with their task. However, they proved willing enough to try to explain what they are doing.

In this context, Bolton proceeded to discuss their observations of the Cygnus-A source and, without knowing its origin, somewhat bravely postulated its distance as "… at least four [light] years, and possibly thousands of [light] years." We now know that a figure of 73×10^7 light years would have been closer to the mark. Scientists are often called upon to justify pure research, and Bolton, showing considerable political acumen for one so young, rose admirably to the occasion when the reporter raised this issue:

> Mr Bolton replied in terms of Michael Faraday's anaology of a new-born baby—who can say what it will become? Mr Bolton pointed out that the study of radio echoes from the ionosphere layer encircling the Earth led to the development of radar, and investigations of properties of uranium led to the atomic bomb. The idea so far was not to use these radiations but to find out something about them.

Through this lengthy article, the reporter offered an admirable insight into the exploits of the two radio astronomers. He concluded with the prophetic statement: "However, whether the public will ever fully understand what has been done and how it will affect scientific progress are … questions whose answers are hidden in the future." As it was to turn out, posterity would assign this field-trip and the ensuing research results a principal place in the annals of radio astronomy (see Robertson et al. 2014; Sullivan 2009).

Apart from newspaper reporters, during their Pakiri Hill sojourn Bolton and Stanley came to know Professor Percy Burbidge, Dr Karl Kreielsheimer and Alan Maxwell from the Physics Department at what is now the University of Auckland. Burbidge had a passion for, and considerable knowledge of, astronomy; Kreielsheimer specialised in upper atmospheric physics (A. Maxwell personal communication 1993); and Maxwell was their graduate student who was engaged in an M.Sc. on solar radio astronomy. This trio visited Pakiri Hill on at least one occasion (A. Maxwell personal communication 1992), and Bolton (personal communication 1992) also went to Auckland and gave a lecture at the University on 10 August. His lecture was written up by a local newspaper, in an article titled "New Zealand Lost Opportunity. Study of Radiations" (1948):

> New Zealand missed one of its greatest opportunities of leading in a scientific field by not following up observations of solar radiation obtained by radar operators in the Dominion in 1942 [in fact the observations were made in 1945—see the previous chapter]. This opinion was expressed by Mr J.G. Bolton the young scientist of the Australian Scientific and Industrial Research Council who has been studying cosmic noise at Leigh and Piha, in an address to a scientific gathering at the Auckland University College last night.
>
> Mr Bolton, who is among the world leaders in research into radiations reaching the earth from space, a field of study which has developed from observations of solar outbursts, described the work which had been done in this branch of science so far.

The reasons why Bolton and Stanley's observations were being conducted in New Zealand largely by-passed the local population, but this hardly effected Bolton and Stanley. An Auckland newspaper reporter who paid them a visit remarked on their "… enthusiasm for their abstruse studies in the Realm of pure science [which] derives largely from the knowledge that they are pioneers and perhaps leaders in one of the newest fields of science." (Cosmic 'noise' from region of the Milky Way 1948).

It is a little surprising that the local media did not make more of Stanley's New Zealand origins, following the appearance of this article. Obviously the ANZAC tradition was still prevalent, and issues of nationalism had yet to emerge. To all intents and purposes Stanley was an 'Australian' scientist—as was Bolton.

24.3.4 Later Developments in Non-solar Radio Astronomy in New Zealand

Despite this somewhat auspicious start, New Zealand does not have a notable reputation for galactic or extragalactic radio astronomy (see Gledhill and Liley 1984). Indeed, the first such work of note, following the Radiophysics expedition of 1948, was undertaken by an amateur, T.F. Mackrell, a science teacher based in the small North Island centre of Coromandel (see Fig. 24.5). In 1962 he constructed an antenna with a 250 square foot collecting area, and used this meridian transit instrument to carry out a 144 MHz survey of the southern sky from declination $-10°$ to $-50°$, but it is significant that he published his results in a geological rather than an astronomical journal (Mackrell 1963). It is also significant that this region of the sky had already been surveyed at much greater sensitivity by Bernie Mills, Bruce Slee and Eric R. Hill using the 85.5 MHz Mills Cross at the CSIRO Division of Radiophysics Fleurs field station near Sydney (see Mills et al. 1958, 1960). Subsequently, Mackrell (1967) used radio telescopes to record solar emission, while he was living in Christchurch and New Plymouth.

Between 1965 and 1978 Brian Egan, a Senior Lecturer in Electrical Engineering at the University of Auckland, coordinated a long-term radio telescope project involving a number of postgraduate students. Although Centaurus-A, Cygnus-A and Sagittarius-A all were detected at various times (e.g. see Apperley 1971; Lim 1968; Wakeman 1972), the emphasis throughout was on the development of instrumentation and no meaningful astronomical investigations were undertaken with the various radio telescopes that were built. This would seem to be a clear case of a missed opportunity. If the Engineering Department had thought to network with the Physics Department, then graduate students in physics surely could have used some of the radio telescopes developed by their engineering counterparts for viable post-graduate research projects.

Finally, for a decade, from 1972, Professor Paul Edwards and his colleagues from the Physics Department at the University of Otago (in Dunedin), used meridian transit instruments operating in the 110–160 MHz range to observe flare stars and radiation from the direction of the Galactic Centre. A number of publications resulted from this work (see Edwards 1975).

24.4 Concluding Remarks

In 1948 New Zealand played an important role in elucidating the true nature of 'radio stars', through investigations carried out at Pakiri Hill and Piha by two scientists from the Division of Radiophysics, C.S.I.R. (Sydney). Both of these scientists, John Bolton and Gordon Stanley, went on to achieve international prominence in radio astronomy, as did the third member of the team, Bruce Slee, who remained in Sydney conducting parallel observations at the time. Stanley was born in Cambridge, New Zealand, but this fact seems to have been of no interest to the newspaper reporters and others who visited the little mobile radio telescope while it was stationed at Pakiri Hill or Piha.

As we have seen in the previous chapter, the Australian 'radio star' project came near the end of an intensive period of solar radio astronomy in New Zealand, initiated by Dr Elizabeth Alexander (Fig. 23.1) and her independent detection and subsequent investigation of solar radio emission in 1945 (Alexander 1945a, b, 1946). This research was followed in 1947–1948 by the 'Canterbury Project' at Wakanui Beach and Ashburton, involving the DSIR's Dr Robert Unwin and Carter Observatory Director, Ivan Thomsen; by Thomsen's (1948) investigation in Wellington of the association between solar radio flux and sunspot activity, which culminated in the publication of a paper in *Nature*; and by Maxwell's (1948) investigation of 100 MHz solar radio emission at what is now Auckland University for what was then one of the first M.Sc. theses in radio astronomy ever written. But this promising start in solar radio astronomy in New Zealand was not destined to continue, as there was not a critical mass of scientists present to sustain these scattered isolated efforts, and the research enthusiasm fostered by the development of radar in New Zealand during WWII (Fraser 2005; Unwin 1992; World War II 1948) was not able to continue once the war ended (Galbreath 1998).[2] Consequently, all research in solar radio astronomy had ceased by the end of 1948 following the departure of Alexander and then Maxwell from New Zealand and Unwin's need to focus on radio meteorology *in lieu* of radio astronomy. Nor did the Australian 'radio stars' work inspire any immediate research efforts in non-solar radio astronomy in New Zealand.

---

[2]Peter Robertson (personal communication 2014) has suggested that

> … one factor in the 'NZ lost opportunity' is that, in contrast to the statutory body CSIRO, DSIR was part of the public service. There was a fair amount of freedom to carry out fundamental research in CSIRO, but less so in DSIR.

We can say, therefore, that from 1945 until the end of 1948 radio astronomy flourished in New Zealand, but it was like a supernovae: it made a sudden short-lived appearance, rose rapidly to prominence, attracted much attention, and then quietly subsided. Despite this promising start in the 1940s, radio astronomy would remain the 'poor cousin' of optical astronomy in New Zealand for decades, until the emergence of Professor Sergei Gulyeav and the Institute of Radio Astronomy and Space Research at the Auckland University of Technology, New Zealand's successful participation in VLBI experiments (see Gulyaev et al. 2005), and New Zealand's partnership with Australia in hosting a part of the Square Kilometre Array (see Head 2010).

Acknowledgements I am grateful to the following people for the assistance: J.B. Buchanan (*The New Zealand Herald*, Auckland, NZ), the late John Bolton (Queensland, Australia), Dr Jessica Chapman (CSIRO Astronomy and Space Sciences, Sydney), Dr Dan Lewis (Huntington Library, San Marino, USA), Professor Alan Maxwell (Harvard-Smithsonian Center for Astrophysics, USA), Dr Bruce Slee (CSIRO Astronomy and Space Sciences, Sydney), Professor Woody Sullivan (University of Washington, USA) and especially Peter Robertson (University of Melbourne). I also wish to thank Peter Robertson for reading and commenting on the first draft of this chapter, and CSIRO Astronomy and Space Sciences, the Huntington Library and Bruce Slee for kindly providing Figs. 24.2, 24.3, 24.4 and 24.8.

References

Alexander, F. E. S. (1945a). *Long Wave Solar Radiation*. DSIR, Radio Development Laboratory (Report).

Alexander, F. E. S. (1945b). *Report on the Investigation of the "Norfolk Island Effect"*. DSIR, Radio Development Laboratory Report dated 1 August (R.D. 1/518).

Alexander, F. E. S. (1946). The Sun's radio energy. *Radio & Electronics*, *1*(1), 16–17, 20.

Allen, C. W. (1947). Solar radio-noise at 200 Mc./s. and its relation to solar observations. *Monthly Notices of the Royal Astronomical Society, 107*, 386–396.

Apperley, M. D. (1971). *Digital Data Processing in Radio Astronomy*. Unpublished Ph.D. Thesis Department of Electrical Engineering, University of Auckland, New Zealand.

Baade, W., & Minkowski, R. (1954). Identification of the radio sources in Cassiopeia, Cygnus A and Puppis A. *Astrophysical Journal, 119*, 206–214.

Bolton, J. C. (1948). Discrete sources of galactic radio frequency noise. *Nature, 162*, 141–142.

Bolton, J. C., & Stanley, G. J. (1948a). Observations of the variable source of cosmic radio frequency radiation in the constellation of Cygnus. *Australian Journal of Scientific Research, A1*, 58–69.

Bolton, J. C., & Stanley, G. J. (1948b). Variable source of radio frequency radiation in the constellation of Cygnus. *Nature, 161*, 312–313.

Bolton, J. C., & Stanley, G. J. (1949). The position and probable identification of the source of the galactic radio-frequency radiation Taurus-A. *Australian Journal of Scientific Research, A2*, 139–148.

Bolton, J. C., Stanley, G. J., & Slee, O. B. (1949). Positions of three discrete sources of galactic radio-frequency radiation. *Nature, 164*, 101–102.

Bolton, J. G., & Slee, O. B. (1953). Galactic radiation at radio frequencies. V. The sea interferometer. *Australian Journal of Physics, 6*, 420–433.

Bolton, J. C. (1955). Australian work on radio stars. *Vistas in Astronomy, 1*, 568–573.

Bolton, J. G. (1982). Radio astronomy at Dover Heights. *Proceedings of the Astronomical Society of Australia, 4*, 349–358.
Christiansen, W. N. (1984). The first decade of solar radio astronomy in Australia. In Sullivan, 1984a, pp. 113–131.
Cosmic 'noise' from region of the Milky Way. *New Zealand Herald*, 26 June 1948.
Edwards, P. J. (1975). Astronomy at the University of Otago Physics Department. *Southern Stars, 26*, 90–92.
Fraser, G. J. (2005). The antecedents and subsequent development of scientific radar in New Zealand. *Journal of Atmospheric and Solar-Terrestrial Physics, 67*, 1411–1418.
Galbreath, R. (1998). *DSIR. Making Science Work for New Zealand. Themes from the History of the Department of Scientific and Industrial Research*. Wellington: Victoria University of Wellington Press, in association with the Historical Branch, Department of Internal Affairs.
Gledhill, K. R., & Liley, B. S. (1984). The Hamilton radio telescope. *Southern Stars, 31*, 41–52.
Gulyaev, S., Natusch, T., Addid, B., Tingay, S., & Deller, A. (2005). Development of radio astronomy and trans-Tasman long baseline array. *Southern Stars, 44*(4), 12–16.
Hanbury Brown, R., & Hazard, C. (1951). A radio survey of the Cygnus region. I. The localized source Cygnus (I). *Monthly Notices of the Royal Astronomical Society, 111*, 576–584.
Head, M. (2010). A brief history of radio astronomy in Aotearoa New Zealand. *Southern Stars, 49*, 7–19.
Hey, J. S., Parsons, S. J., & Phillips, J. W. (1946a). Fluctuations in cosmic radiation at radio-frequencies. *Nature, 158*, 234.
Hey, J. S., Phillips, J. W., & Parsons, S. J. (1946b). Cosmic radiations at 5 metres wave-length. *Nature, 157*, 296–297.
Kellermann, K., & Sheets, B. (Eds.). (1983). *Serendipitous Discoveries in Radio Astronomy*. Green Bank, National Radio Astronomy Observatory.
Kellermann, K. I., Orchiston, W., & Slee, B. (2005). Gordon James Stanley and the early development of radio astronomy in Australia and the United States. *Publications of the Astronomical Society of Australia, 22*, 13–23.
Lim, J. C. (1968). *Non-uniformly Spaced Arrays of Directional Elements*. Unpublished Ph.D. Thesis, Department of Electrical Engineering, University of Auckland.
Lovell, B. (1977). The effects of defence science on the advance of astronomy. *Journal for the History of Astronomy, 8*, 151–173.
Lovell, B. (1983). Impact of World War II on radio astronomy. In Kellermann and Sheets, pp. 89–104.
Mackrell, T. F. (1963). A survey of galactic radio noise at 144 Mc/s between declination 10° and 50° south. *New Zealand Journal of Geology and Geophysics, 6*, 728–734.
Mackrell, T. F. (1967). Amateur radio astronomy: some New Zealand observations. *Southern Stars, 22*, 58–71.
Maxwell, A. (1948). *Enhanced Solar Radiation at 3 Metre Wavelengths*. M.Sc. Thesis, Physics Department, Auckland University College.
Mills, B. Y. (1949). Letter to R. Minkowski, dated December 16. National Archives of Australia, 972099—C3830—A1/1/1 Part 4 Box 1.
Mills, B. Y., & Thomas, A. B. (1951). Observations of the sources of radio frequency radiation in the constellation of Cygnus. *Australian Journal of Scientific Research, A4*, 158–171.
Mills, B. Y. (1952). The positions of six discrete sources of cosmic radio radiation. *Australian Journal of Scientific Research, A5*, 456–463.
Mills, B. Y., Slee, O. B., & Hill, E. R. (1958). A catalogue of radio sources between declinations +10° and −20°. *Australian Journal of Physics, 11*, 360–387.
Mills, B. Y., Slee, O. B., & Hill, E. R. (1960). A catalogue of radio sources between declinations −20° and −50°. *Australian Journal of Physics, 13*, 676–699.
Mills, B. Y. (1984). Radio sources and the log N–log S controversy. In Sullivan, 1984a, pp. 146–165.
Mills, B. Y. (2006). An engineer becomes astronomer. *Annual Review of Astronomy and Astrophysics, 44*, 1–15.

Minkowski, R. (1949). Letter to B. Y. Mills, dated December 25. National Archives of Australia, 972099–C3830–A1/1/1 Part 4 Box 1.

New Zealand Lost Opportunity. (1948). Study of Radiations. Newspaper article dated August, 11 1948 (source unspecified, but probably the *New Zealand Herald*). In: Bolton Papers, Series 3, Folder 1, Newspaper cuttings 1948–93, National Library of Australia, Canberra.

Orchiston, W. (1993). New Zealand's role in the identification of the first 'radio stars'. *Southern Stars, 35*, 46–52.

Orchiston, W. (1994). John Bolton, discrete sources, and the New Zealand field-trip of 1948. *Australian Journal of Physics, 47*, 541–547.

Orchiston, W., & Slee, B. (2002). The flowering of Fleurs: an interesting interlude in Australian radio astronomy. *ATNF News, 47*, 12–15.

Orchiston, W. (2004a). From the solar corona to clusters of galaxies: the radio astronomy of Bruce Slee. *Publications of the Astronomical Society of Australia, 21*, 23–71.

Orchiston, W. (2004b). The rise and fall of the Chris Cross: a pioneering Australian radio telescope. In W. Orchiston, R. Stephenson, S. Débarbat, & I.-S. Nha (Eds.), *Astronomical Instruments and Archives from the Asia-Pacific Region* (pp. 157–162). Seoul: IAU Commission 41.

Orchiston, W. (2005a). Dr Elizabeth Alexander: first female radio astronomer. In W. Orchiston (Ed.), *The New Astronomy: Opening the Electromagnetic Window and Expanding our View of Planet Earth* (pp. 71–92). Dordrecht: Springer.

Orchiston, W. (2005b). Sixty years in radio astronomy: a tribute to Owen Bruce Slee. *Journal of Astronomical History and Heritage, 8*, 3–10.

Orchiston, W., & Slee, B. (2005). The Radiophysics field stations and the early development of radio astronomy. In W. Orchiston (Ed.), *The New Astronomy: Opening the Electromagnetic Window and Expanding our View of Planet Earth* (pp. 119–168). Dordrecht: Springer.

Orchiston, W., & Slee, B. (2006). Early Australian observations of historic supernova remnants at radio wavelengths. In K.-Y. Chen, W. Orchiston, B. Soonthornthum, & R. Strom (Eds.), *Proceedings of the Fifth International Conference on Oriental Astronomy* (pp. 43–56). Chiang Mai: Chiang Mai University.

Orchiston, W., & Kellermann, K. I. (2008). Bolton, John Gatenby. *Dictionary of Scientific Biography* (pp. 332–337). New York: Gale.

Orchiston, W., & Mathewson, D. (2009). Chris Christiansen and the Chris Cross. *Journal of Astronomical History and Heritage, 12*, 11–32.

Orchiston, W., George, M., Slee, B., & Wielebinski, R. (2016). Early low frequency radio astronomy in Australia. In Y.-L. Shi (Ed.), *Astronomical Heritages in Asia-Pacific Areas: Proceedings of the Eighth International Conference on Oriental Astronomy*. Hefei: University of Science and Technology of China.

Osterbrock, D. E. (1983). *Biographical Memoir. Rudolph Minkowski 1895–1976*. Washington: National Academy of Science.

Osterbrock, D. E. (2001). *Walter Baade—A Life in Astrophysics*. Princeton: Princeton University Press.

Reber, G. (1983). Radio astronomy between Jansky and Reber. In Kellermann and Sheets, pp. 71–78.

Reber, G. (1984). Early radio astronomy at Wheaton, Illinois. In Sullivan, 1984a, pp. 43–66.

Robertson, P. (1992). *Beyond Southern skies: Radio Astronomy and the Parkes Radio Telescope*. Cambridge: Cambridge University Press.

Robertson, P., Cozens, G., Orchiston, B., Slee, B., & Wendt, H. (2010). Early Australian optical and radio observations of Centaurus A. *Publications of the Astronomical Society of Australia, 27*, 402–430.

Robertson, P., Orchiston, W., & Slee, B. (2014). John Bolton and the discovery of discrete radio sources. *Journal of Astronomical History and Heritage, 17*, 283–306.

Ryle, M., Smith, F. G., & Elmore, B. (1950). A preliminary survey of the radio stars in the Northern Hemisphere. *Monthly Notices of the Royal Astronomical Society, 110*, 508–523.

Smith, F. G. (1951). An accurate determination of the positions of four radio stars. *Nature, 168*, 555.

Smith, F. G. (1984). Early work on radio stars at Cambridge. In Sullivan, 1984a, pp. 236–248.

Stanley, G. J., & Slee, O. B. (1950). Galactic radiation at radio frequencies. II. The discrete sources. *Australian Journal of Scientific Research, A3*, 234–250.

Stephenson, F. R., & Green, D. A. (2002). *Historical Supernovae and their Remnants*. Oxford: Oxford University Press.

Stephenson, F. R., & Green, D. A. (2003). Was the supernova of AD1054 recorded in European history? *Journal of Astronomical History and Heritage, 6*, 46–52.

Sullivan, W. T. (Ed.). (1982). *Classics in Radio Astronomy*. Dordrecht: Reidel.

Sullivan, W. T. (Ed.). (1984a). *The Early Years of Radio Astronomy*. Cambridge: Cambridge University Press.

Sullivan, W. T. (1984b). Karl Jansky and the discovery of extraterrestrial radio waves. In Sullivan 1984a, pp. 3–42.

Sullivan, W. (2009). *Cosmic Noise. A History of Early Radio Astronomy*. Cambridge: Cambridge University Press.

Thomsen, I. L. (1948). Solar radio emissions and sunspots. *Nature, 161*, 134–136.

Unwin, R. S. (1992). The development of radar in New Zealand in World War II. *The Radioscientist, 3*(1), 8–18.

Verschuur, G. L. (1987). *The Invisible Universe Revealed*. New York: Springer.

Wakeman, P. E. D. (1972). *A Dual Channel Solar Radiometer*. Unpublished M. Eng. Thesis, Department of Electrical Engineering, University of Auckland.

Wendt, H., Orchiston, W., & Slee, B. (2011). The contribution of the division of radiophysics Potts hill field station to international radio astronomy. In W. Orchiston, T. Nakamura, & R. Strom (Eds.), *Highlighting the History of Astronomy in the Asia-Pacific Region* (pp. 379–431). New York: Springer.

World War II Narrative No. 3. Radar. Wellington, DSIR. (1948). Original Manuscript in National Archives, Wellington (W3424, Box 16).

Index

W. Orchiston, *Exploring the History of New Zealand Astronomy*,
Astrophysics and Space Science Library 422,
DOI 10.1007/978-3-319-22566-1